2

SOURCE BOOK ON POWDER METALLURGY

SOURCE BOOK ON POWDER METALLURGY

A comprehensive collection of outstanding articles from the periodical and reference literature

Compiled by
Consulting Editor:

Samuel Bradbury
Hoeganaes Corporation
Subsidiary of Interlake, Inc.

 **American Society for Metals**
Metals Park, Ohio 44073

Library of Congress Cataloging in Publication Data

Main entry under title:
Source book on powder metallurgy.

Includes bibliographical references and index.
1. Powder metallurgy. I. Bradbury, Samuel.
TN695.S645   671.3′7   78-24466
ISBN 0-87170-030-1

PRINTED IN THE UNITED STATES OF AMERICA

Preface

In common with many of the other newer materials, powder metallurgy (P/M) alloys and components have received extensive coverage in the periodical technical literature. This may have contributed to the substantial and progressive increase in the production of powder metallurgy parts over the past decade. Nevertheless, it is generally contended that not enough designers, metallurgists and manufacturing engineers have taken the time to familiarize themselves with the fundamentals of powder metallurgy or the advantages and limitations of parts produced by P/M processes. It is intended that SOURCE BOOK ON POWDER METALLURGY will promote a broader interest and understanding of these subjects by providing all of the necessary background information in a single, convenient and authoritative source.

One of the myths that continue to disturb P/M producers is that all P/M parts are necessarily expensive and therefore noncompetitive with other fabricating methods. In *fact,* P/M parts are often highly cost-competitive and are used in many products that are subject to stringent manufacturing-cost controls, among them automobiles, home appliances, business machines, farm machinery, trucks and trailers. The cost factors in P/M have been accurately defined. Powder costs vary primarily as a function of basis metal and alloy content. Other costs generally increase directly with increases in strength and density requirements — which, in turn, dictate processing techniques. In other words, overdesigning in terms of strength and density increases costs. These are among the many enlightening subject areas covered in this book. The contents of the book, which comprise 45 articles arranged in 12 sections, are briefly summarized in the following paragraphs.

An Introduction to Powder Metallurgy. The introductory article by Kurt Miska reviews all of the essentials, including types, forms and cost of ferrous and nonferrous metal powders; the engineering properties (density, porosity, permeability, mechanical properties and corrosion resistance); processing and fabrication; and design details.

Ferrous Powders. This section consists of five authoritative articles on ferrous powders. Two key subjects covered by individual articles are the design of ferrous alloys for powder forging and a comparison of steel compacts prepared from prealloyed, premixed elemental powders with compacts prepared from mixtures of prealloyed and elemental hybrid powders. Other articles are devoted to the fatigue properties of sintered nickel steels; the effect of phosphorus additions on some mechanical properties of sintered steels based on sponge iron powder and high-purity, atomized powder; and iron-carbon behavior in sintering.

Nonferrous Powders. The six articles in this section cover the nonferrous alloys of principal interest — aluminum-base, copper-base, titanium-base, and superalloys. Much of the data provided on these alloys is indicative of the advanced technology currently available to the designer of nonferrous parts. Aluminum P/M alloys, for example, are available in tensile strengths ranging from 11 to 50 ksi (76 to 345 MPa), depending on composition, density and heat treatment. Similar technological breakthroughs have been achieved with the other nonferrous P/M alloys.

Consolidation. This, the second longest section in the book, contains eight outstanding articles on consolidation techniques, including hot rolling, extrusion, hot forming, and forging. Most of the articles emphasize advanced technology and suggest potentials that can be achieved in developing part configurations, better mechanical properties, higher densities and other significant improvements. The successful production of large net shapes, as described by N. P. Pinto, and the developments in fabricating complex gear configurations, as described by H. W. Antes, are indicative of the future of P/M design and processing.

Sintering – Homogenization. The articles in this section reflect both fundamental and highly practical aspects of sintering and homogenization. The introductory article considers the effects of heat treatment and deformation on the homogenization of compacts of blended powders and deals specifically with rates of homogenization in one-phase and multi-phase binary systems. In another article, the rapid burn-off system for practically eliminating lubricant residue problems and replacing the preheat furnace in a sintering line is described. Finally, an article by Cornelius Durdaller offers comprehensive coverage of the subject of furnace atmospheres employed in P/M processing.

Powder Mixing and Blending. "There are more problems involved in mixing of two or more powders than are usually recognized," according to Henry H. Hausner, co-author of the article in this section. Admittedly, powder mixing is a very complex process that involves a large number of variables. This article identifies the twelve major variables in the mixing process and then proceeds to analyze each in terms of information presently available and areas that require investigation.

Tooling for P/M. The tool materials selected for use in the production of P/M parts often account for the profit or loss in a given operation. Selection, in turn, depends on tool design, tool construction and subsequent processing, subjects that are discussed in an enlightening and authoritative article by Robert Kunkel of Ford Motor Co.

Impregnation. The greatest potential for impregnated P/M parts is for pressure-tight components such as valves, pumps, meters, compressors and hydraulic systems. Impregnation also permits P/M components to be successfully plated and painted and improves their machinability. Vacuum impregnation opens up major new application opportunities for P/M parts, as described in an article from *Precision Metal*.

Sizing – Coining. The stresses that occur during pressing and sizing operations, especially the frictional stresses, affect not only dimensions but also internal residual stress buildup and the possibility of microcracking. Microcracking is seldom encountered in small parts unless they are subjected to fatigue conditions; it is of greater importance in large parts. An article by M. Eudier considers three types of re-pressing — drawing, sizing and coining — and provides examples that show their influence on mechanical properties, especially fatigue limit.

Machining. This section consists of two articles, the first providing information on both machining processes and machinability, and the second offering some of the latest information on machining P/M parts. In combination, the articles serve to supplement each other in their coverage of the machining processes and the pertinent variables encountered.

Properties and Applications. This section, which contains 14 articles, is the longest in the book. Among the properties covered in separate articles are fracture toughness, fatigue life, electrical conductivity and magnetic behavior. The applications described cover a broad spectrum: gears and pinions, hot-formed parts, bearings, gun parts, heavy-duty parts of complicated shape, parts for the farm-equipment industry, filters, high-conductivity electrical contacts, and parts for magnetic applications.

Metallographic Preparation and Microstructure. Of special interest to metallurgists, metallographers, and others concerned with metallurgical structures, this concluding section fulfills the need for expert guidance in the preparation of metallographic specimens of P/M parts, and presents an atlas of typical microstructures of ferrous P/M alloys.

For his invaluable contribution in selecting and organizing the articles in this book, the American Society for Metals extends its grateful appreciation to Mr. Samuel Bradbury. His performance as consulting editor of the book reflects his long experience with, and intimate knowledge of, powder metallurgy. Most grateful acknowledgment is extended to the many authors whose work appears in this book and to their publishers.

Paul M. Unterweiser
Staff Editor
Manager, Publications Development
American Society for Metals

William H. Cubberly
Director of Reference Publications
American Society for Metals

Cover photo courtesy of Interlake, Inc.

Contributors to This Source Book*

JOHN S. ADAMS
Federal Mogul

M. M. ALLEN
Pratt & Whitney Aircraft

H. W. ANTES
Hoeganaes Corp.

R. L. ATHEY
Pratt & Whitney Aircraft

M. BALASUBRAMANIAM
Drexel University

JOHN K. BEDDOW
Polytechnic Institute of Brooklyn

G. T. BROWN
Guest Keen and Nettlefolds Group Technological Centre

T. L. BURKLAND
Deere and Co.

D. J. BURR
International Nickel Ltd.

R. G. BUTTERS
University of British Columbia

R. J. CAUSTON
Davy-Loewy Research and Development Centre

W. S. CEBULAK
Alcoa Technical Center

J. P. COOK
Hoeganaes Corp.

A. P. CREASE, Jr.
Drever Co.

GAIL F. DAVIES
Gould, Inc.

R. J. DEANGELIS
University of Kentucky

C. L. DOWNEY
Cincinnati Inc.

J. J. DUNKLEY
Davy-Loewy Research and Development Centre

CORNELIUS DURDALLER
Hoeganaes Corp.

P. C. ELOFF
Gleason Works

E. J. ESPER
Robert Bosch GmbH

M. EUDIER
Société Metafram

C. E. EVANS
AMAX Metal Powders

W. F. FOSSEN
Centre for Powder Metallurgy, Ontario Research Foundation

GERALD FRIEDMAN
Nuclear Metals Inc.

ERNST GEIJER
Hoeganaes Corp.

DOUGLAS GLOVER
Federal Mogul

G. GOLLER
AMAX Base Metals

P. ULF GUMMESON
Hoeganaes Corp.

H. P. HATCH
Army Materials and Mechanics Research Center

*Affiliations given were applicable at date of contribution.

CONTENTS

SECTION I:
An Introduction to Powder Metallurgy

Powder metal parts

Powder metal parts, commonly called P/M parts, are produced by blending metal powders, compacting the mixture in a die and then heating or sintering the compacted powder in a controlled atmosphere to bond the particles into a strong shape.

P/M parts are made from a wide range of materials and may weigh from less than an ounce to nearly 1,000 lb (450 kg); however, most P/M parts weigh less than 5 lb (2.3 kg). Shapes may be simple cylindrical bearings and washers or highly complex configurations impractical to make by other processes. P/M parts may be subsequently heat treated, repressed, forged, machined, impregnated and finished to improve mechanical or physical properties, increase dimensional accuracy or enhance decorative appeal. They also may be joined by brazing or mechanical means.

P/M parts are practically well suited for fast, high volume production of countless parts in the automotive, appliance, agricultural equipment, business machine, electrical and electronics, power tool, ordnance and machine tool industries. However, they also find uses in the aerospace, nuclear and other industries.

The most common standards and specifications for metal powders and P/M parts are those of the Metal Powder Industries Federation. In addition to those shown in Table 1, ASTM, SAE and the Government have specifications covering these materials.

Types and forms

Numerous ferrous and nonferrous elemental and prealloyed metal powders are available in different grades to meet a wide range of requirements.

Most metal powders are produced by atomization, reduction of oxides, electrolysis or chemical reduction. Available metals include iron, tin, lead, nickel, copper and aluminum, as well as the refractory and reactive metals. Metals can be physically mixed to produce alloys or they may be prealloyed, in which case each particle is in itself an alloy. It is also possible to combine metal and nonmetallic powders to provide a composite with the desirable properties of both in the finished part.

Ferrous metals

Iron powders are the most widely used P/M materials for structural parts. The powder may be used alone but frequently small additions of carbon, copper or nickel, singly or in combination, are included to improve mechanical properties of parts.

Plain carbon steel P/M compositions consist of mixtures of iron and graphite. When compacts are sintered, carburization takes place to produce a carbon steel structure with carbon contents up to approximately 0.75%. Low, medium and high density parts can be produced from these powders.

Adding copper to iron powder increases strength and tends to increase hardness. However, copper additions decrease ductility. Copper steels, covered by MPIF Standard 35, contain from 1.5 to 10.5% copper and up to 1.0% carbon. Low, medium and high density parts, and bearings are produced from copper steels. Recently, Hoeganaes announced a high compressibility iron powder containing 0.45% phosphorus and no more than 0.03% carbon for the production of parts with improved ductility, impact strength and magnetic properties.

Nickel steels contain from 2 to 8% nickel, with or without copper. A more complex nickel steel is INCO's IN-861, Fe-2Ni-1.0Mo-0.7Mn, which is a pre-mixed, medium carbon, air hardening alloy that provides good properties without heat treating. A developmental low carbon P/M nickel steel, Fe-2Ni-1Mo-1Mn, is expected to offer very flexible processing combined with very high tensile strengths. In general, the nickel steels are used to produce parts of exceptionally high strength, combined with good toughness and fatigue strength.

Steel powders for P/M forging include AISI 1025, 1080, 4620 and 4650. Modifications of 4600-type, Ni-Mo, prealloyed steel powders either contain 0.40 to 0.60% nickel or 1.75 to 1.90% nickel.

Austenitic and martensitic stainless steels, conforming to AISI compositional limits, are being specified more widely than ever before, especially in applications requiring good corrosion resistance. AISI grades 303, 316 and 410 now are covered by MPIF Standard 35. Other stainless steels for P/M parts are 304 and 434.

Porous ferrous parts can be infiltrated with lower melting materials, such as copper and some brasses. Copper content in such parts can range from about 8 to 25%, reducing residual porosity virtually to zero. Besides improving strength, infiltration also is used to obtain more uniform density in parts that are difficult to press to uniform density.

Nonferrous metals

Major nonferrous P/M metals and alloys include copper, aluminum and titanium.

Pure copper powder is used primarily where high electrical and/or thermal conductivity is required. Prealloyed brasses are available in a wide range of compositions with zinc contents ranging from approximately 10% to as high as 30%. Leaded brasses contain 1 to 2% lead and 77 to 80% copper. Machinability of leaded brasses is comparable to wrought and cast material of the same composition.

The P/M bronzes may be made from mixtures of elemental copper and tin powders or from prealloyed powders. Elemental P/M bronzes, with 18 to 30% controlled porosity, are used for oil-impregnated bearings. High strength prealloyed bronzes are usually specified for high strength structural parts.

Copper-nickel-zinc P/M compositions, like their wrought counterparts, also are referred to as nickel-silvers. These prealloyed powders contain between 16 and 19% nickel. Leaded nickel-silvers contain 1.8% lead. Properties of nickel-silver P/M parts are similar to those of P/M brasses but have improved corrosion resistance.

Several commercially available aluminum alloy powders are being produced from elemental powders. The most commonly specified grades contain copper, magnesium and silicon. Copper content ranges from 0.25 to 4.4%, magnesium is between 0.5 and 2.5% and silicon level is between 0.3 and 0.9%. Parts made from these compositions have tensile properties similar to those made from wrought alloys 2014, 6061 and 7075. Two promising developmental aluminum P/M compositions are Alcoa's Al-8.0Zn-2.5Mg-1.62Co-1.0Cu and Al-6.5Zn-2.42Mg-1.6Cu-0.37Co, with the first having no conventional wrought counterpart. Parts made from these compositions are expected to show a 20% increase in tensile properties while retaining excellent corrosion and fatigue resistance and toughness.

Titanium P/M is still relatively new and Gould is the producer of the only two available grades. These are commercially pure (CP) and Ti-6Al-4V alloy, with the latter being produced either from elemental or prealloyed powders.

Prices
Prices of metal powders commonly used in the production of P/M parts are given in Table 2.

Engineering properties

The engineering properties of P/M parts depend on the material, density, design configuration and whether or not the material can be heat treated. Tables 3 through 6 list mechanical properties of ferrous and nonferrous P/M parts.

Density
Most properties of P/M parts are closely related to final density, which is expressed in grams per cubic centimeter (g/cu cm). Normally, density of mechanical or structural parts is specified on a dry, unimpregnated basis, while the density of bearings is reported on a fully oil impregnated basis. MPIF Standard 42 outlines a way to compute density.

Density also is expressed as percent of theoretical density, which is defined as the ratio of a P/M part's density to that of its wrought counterpart. P/M parts having theoretical densities less than 75% are considered to be low density; those above 90% are high density; and those in between are classified as medium density. Generally, structural parts will have densities ranging from 80% to above 95%. Oil-impregnated bearings have densities of approximately 75% and filter parts as low as 50%.

1 Specifications for P/M materials

Material	MPIF Designation	ASTM Spec.	SAE Spec.	Military Spec.
Ferrous				
Copper steel	FC-0808-N	B426, Grade 3 Type I	866A	—
Iron copper	FC-1000-N	B222, B439, Grade 3	862	B-5687-C Type II, Comp B
Iron nickel	FN-0200-R	B484, Grade 1, Type I, Class A	—	—
Iron nickel	FN-0200-S	B484, Grade 1, Type II, Class A	—	—
Iron nickel	FN-0200-T	B484, Grade 1, Type III, Class A	—	—
Nickel steel	FN-0205-R	B484, Grade 1, Type I, Class B	—	—
Nickel steel	FN-0205-S	B484, Grade 1, Type II, Class B	—	—
Nickel steel	FN-0205-T	B484, Grade 1, Type III, Class B	—	—
Nickel steel	FN-0208-R	B484, Grade 1, Type I, Class C	—	—
Nickel steel	FN-0208-S	B484, Grade 1, Type II, Class C	—	—
Nickel steel	FN-0208-T	B484, Grade 1, Type III, Class C	—	—
Iron nickel	FN-0400-R	B484, Grade 2, Type I, Class A	—	—
Iron nickel	FN-0400-S	B484, Grade 2, Type II, Class A	—	—
Iron nickel	FN-0400-T	B484, Grade 2, Type III, Class A	—	—
Nickel steel	FN-0405-R	B484, Grade 2, Type I, Class B	—	—
Nickel steel	FN-0405-S	B484, Grade 2, Type II, Class B	—	—
Nickel steel	FN-0405-T	B484, Grade 2, Type III, Class B	—	—
Nickel steel	FN-0408-R	B484, Grade 2, Type I, Class C	—	—
Nickel steel	FN-0408-S	B484, Grade 2, Type II, Class C	—	—
Nickel steel	FN-0408-T	B484, Grade 2, Type III, Class C	—	—
Iron nickel	FN-0700-T	B484, Grade 3, Type I, Class A	—	—
Iron nickel	FN-0700-S	B484, Grade 3, Type II, Class A	—	—
Iron nickel	Fn-0700-T	B484, Grade 3, Type III, Class A	—	—
Nickel steel	FN-0705-R	B484, Grade 3, Type I, Class B	—	—
Nickel steel	FN-0705-S	B484, Grade 3, Type II, Class B	—	—
Nickel steel	FN-0705-T	B484, Grade 3, Type III, Class B	—	—
Nickel steel	FN-0708-R	B484, Grade 3, Type I, Class C	—	—
Nickel steel	FN-0708-S	B484, Grade 3, Type II, Class C	—	—
Nickel steel	FN-0708-T	B484, Grade 3, Type III, Class C	—	—
Infiltrated iron	FX-2000-T	B303, Class A	870	—
Nonferrous				
Aluminum	—	B595[a]	—	—
Bronze	CT-0010-N	B438, Grade 1, Type I	840	—
Bronze	CT-0010-R	B438, Grade 1, Type II	841	B-5687-C Type I, Comp A[b]
Bronze	CT-0010-S	B255, Type II	842	—
Nickel silver	CZN-1818-U	B458, Grade 1, Type I	—	—
Nickel silver	CZN-1818-W	B458, Grade 1, Type II	—	—
Nickel silver, leaded	CZNP-1618-U	B458, Grade 2, Type I	—	—
Nickel silver, leaded	CZNP-1618-W	B458, Grade 2, Type II	—	—

[a]Specifies 3 grades and 3 types. [b]Does not allow 1% max Fe.

Porosity and permeability

Porosity is the percentage of void volume in a P/M part. For example, a part of 85% theoretical density will have 15% porosity. Porosity can take the form of interconnected pores that extend to the surface like that of a sponge or porosity can take the form of closed holes within the part. Interconnected porosity is required for effective self-lubricated bearings.

Porosity is controllable and depends on the base material and how it is processed. Parts can be produced with uniform porosity or with variations in porosity (and density) from one section to another to provide a range of properties. For example, a gear can be very strong in one area and self-lubricating in another. MPIF Standard 35 outlines a method for calculating pore volume or oil content of self-lubricating P/M parts in terms of interconnected porosity.

Permeability, or the ability to pass fluids or gases, is a unique property of P/M parts, which stems from the ability to control porosity. Depending on forming and sintering techniques, a part can be made to have any permeability from zero to 60%. This property is important when designing filters. Filters can be designed with permeabilities that will separate materials selectively.

Tensile, fatigue, impact properties

Density, pore size, shape and distribution, and the extent of sintering strongly influence the final tensile properties of P/M parts. For this reason, mechanical property data are commonly given in graphs showing the relationship between the property and density. A typical graph, Fig 1, shows the increase of strength with increasing density for a typical part made from a P/M copper steel composition.

In general, yield strength of P/M parts is 65 to 85% of ultimate strength. Therefore, yield strengths of P/M parts are generally closer to ultimate strengths than for wrought metals. Also, the yield strength of many P/M parts, particularly stainless steels, may be higher than that of the wrought form.

The relationship of fatigue strength to density is shown in Fig 2. Fatigue strength is best at high densities. For similar P/M and wrought parts, the ultimate tensile strength to fatigue strength ratios are the same. However, fatigue strengths of P/M parts generally are more stable and uniform than for wrought parts. Parts containing nickel show

improved fatigue resistance compared to iron-carbon steels, and high density nickel steel parts can be case hardened to improve wear and fatigue properties.

The tensile strengths for non-heat treated carbon steel P/M parts range from 16,000 to 60,000 psi (110 to 414 MPa), and can be increased to over 90,000 psi (620 MPa) by heat treatment. Tensile strengths for copper steel parts range from 20,000 to over 80,000 psi (138 to 552 MPa) before heat treatment and to over 100,000 psi (689 MPa) after heat treatment. Strengths of 90,000 to 100,000 psi (621 to 690 MPa) can be achieved in nonheat treated nickel steel parts, and as high as 180,000 psi (1241 MPa) by heat treatment. Standard copper infiltrated iron

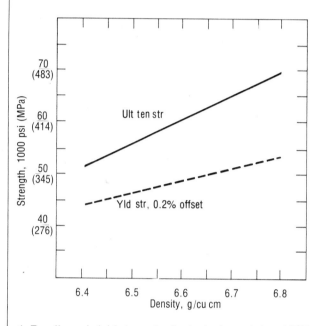

1 Tensile and yield strengths for typical, as-sintered P/M copper steel parts increase with density.

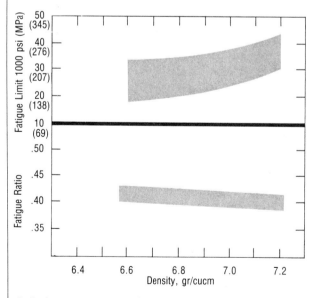

2 Fatigue strength of P/M parts is best at high densities. For similar P/M and wrought compositions, the ultimate tensile strength to fatigue strength ratios are the same.

2 Prices of metal powders[a]

Metal or alloy	$/lb ($/kg)
Aluminum, atomized,	
100 mesh	0.54-0.57 (1.19-1.25)
200 mesh	0.56-0.60 (1.23-1.32)
Brass, 80-20	0.923 (2.03)
Bronze, 90-10 prealloyed	1.476 (3.24)
Copper, spherical	1.035 (2.28)
Iron,	
sponge	0.175 (0.385)
atomized	0.180 (0.396)
electrolytic	0.720 (1.58)
Lead	0.075-0.12 (0.16-0.26)
Nickel silver	1.498 (3.29)
Stainless steel	0.82 (1.80)
Tin	0.17-0.21 (0.37-0.46)
Zinc	0.53 (1.17)

[a]Prices as of June 1976.

3 Mechanical properties of iron and steel P/M parts

Composition,[a] %	MPIF designation	Density g/cu cm	Cond.[b]	Ult ten str, 1000 psi (MPa)	Yld str, 1000 psi (MPa)	Elong,[c] %	Mod of elast 10⁶ psi (10⁴ MPa)	Apparent hardness, Rockwell	Impact str, smooth Charpy, ft·lb (J)
Iron									
99.7-100Fe	F-0000-N	5.6-6.0	A.S.	16 (110)	11 (76)	2.0	10.5 (7.23)	10R$_b$	13.0 (4.07)
99.7-100Fe	F-0000-P	6.0-6.4	A.S.	19 (131)	14 (96)	2.5	13.0 (8.96)	70R$_b$	4.5 (6.1)
99.7-100Fe	F-0000-R	6.4-6.8	A.S.	24 (165)	16 (110)	5.0	16.0 (11.0)	80R$_b$	9.5 (12.9)
99.7-100Fe	F-0000-S	6.8-7.2	A.S.	30 (207)	21.5 (148)	9.0	19.0 (13.1)	15R$_b$	15.0 (20.3)
99.7-100Fe	F-0000-T	7.2-7.6	A.S.	40 (276)	26 (179)	15.0	25.0 (17.2)	30R$_b$	23.0 (31.2)
Steel									
97.4/99.7Fe-0.3/0.6C	F-0005-N	5.6-6.0	A.S.	18 (124)	14.5 (100)	1.0	10.5 (7.23)	5R$_b$	2.5 (3.4)
	F-0005-P	6.0-6.4	A.S.	25 (172)	20 (138)	1.5	13.0 (8.96)	20R$_b$	3.5 (4.7)
	F-0005-R	6.4-6.8	A.S.	32 (221)	23 (159)	2.5	16.0 (11.0)	45R$_b$	5.0 (6.8)
	F-0005-R	6.4-6.8	H.T.	60 (414)	57 (393)	<0.5	—	100R$_b$	—
	F-0005-S	6.8-7.2	A.S.	43 (296)	28 (193)	3.5	19.0 (13.1)	60R$_b$	9.0 (12.2)
	F-0005-S	6.8-7.2	H.T.	80 (552)	75 (517)	<0.5	—	25R$_c$	—
97.0/99.1Fe-0.6/1.0C	F-0008-N	5.6-6.0	A.S.	29 (200)	25 (172)	0.5	10.5 (7.23)	35R$_b$	2.0 (2.7)
	F-0008-N	5.6-6.0	H.T.	42 (290)	—	<0.5	—	90R$_b$	—
97.0/99.4Fe-0.6/1.0C	F-0008-P	6.0-6.4	A.S.	35 (241)	30 (207)	1.0	13.0 (8.96)	50R$_b$	3.0 (4.07)
	F-0008-P	6.0-6.4	H.T.	58 (400)	—	<0.5	—	100R$_b$	—
	F-0008-R	6.4-6.8	A.S.	42 (290)	36 (248)	1.5	16.0 (11.0)	65R$_b$	3.5 (4.7)
	F-0008-R	6.4-6.8	H.T.	74 (510)	—	<0.5	—	25R$_c$	—
	F-0008-S	6.8-7.2	A.S.	57 (393)	40 (276)	2.5	19.0 (13.1)	75R$_b$	7.0 (9.5)
	F-0008-S	6.8-7.2	H.T.	94 (648)	91 (627)	<0.5	—	30R$_c$	—
Copper iron									
Fe-2.7Cu-0.15C	FC-0200-P	6.0-6.4	A.S.	23 (159)	17 (117)	2.5	13.0 (8.96)	80R$_h$	5.5 (7.5)
	FC-0200-R	6.4-6.8	A.S.	30 (207)	21 (145)	4.0	16.0 (11.0)	15R$_b$	7.0 (9.5)
	FC-0200-S	6.8-7.2	A.S.	37 (255)	23 (159)	7.0	19.0 (13.1)	30R$_b$	17 (23.0)
Copper steel									
Fe-2.7Cu-0.45C	FC-0205-P	6.0-6.4	A.S.	40 (276)	34 (234)	1.0	13.0 (8.96)	45R$_b$	3.5 (4.7)
	FC-0205-R	6.4-6.8	A.S.	50 (345)	38 (262)	1.5	16.0 (11.0)	70R$_b$	5.5 (7.5)
	FC-0205-R	6.4-6.8	H.T.	85 (586)	81 (558)	<0.5	—	30R$_c$	—
	FC-0205-S	6.8-7.2	A.S.	62 (427)	45 (310)	3.0	19.0 (13.1)	80R$_b$	9.5 (12.9)
	FC-0205-S	6.8-7.2	H.T.	100 (689)	95 (655)	<0.5	—	35R$_c$	—
Fe-2.7Cu-0.8C	FC-0208-N	5.6-6.0	A.S.	33 (227)	30 (207)	<0.5	10.5 (7.23)	45R$_b$	2.5 (3.4)
	FC-0208-N	5.6-6.0	H.T.	43 (296)	—	<0.5	—	95R$_b$	—
	FC-0208-P	6.0-6.4	A.S.	45 (310)	41 (283)	<0.5	13.0 (8.96)	60R$_b$	3.0 (4.07)
	FC-0208-P	6.0-6.4	H.T.	55 (379)	—	<0.5	—	25R$_c$	—
	FC-0208-R	6.4-6.8	A.S.	60 (414)	48 (331)	1.0	16.0 (11.0)	70R$_b$	5.0 (6.8)
	FC-0208-R	6.4-6.8	H.T.	80 (552)	—	<0.5	—	35R$_c$	—
	FC-0208-S	6.8-7.2	A.S.	80 (552)	57 (393)	1.5	19.0 (13.1)	80R$_b$	8.0 (10.8)
	FC-0208-S	6.8-7.2	H.T.	100 (689)	95 (655)	<0.5	—	40R$_c$	—
Fe-5.0Cu-0.45C	FC-0505-N	5.6-6.0	A.S.	35 (241)	30 (207)	0.5	10.5 (7.23)	50R$_b$	3.0 (4.07)
	FC-0505-N	5.6-6.0	H.T.	—	—	—	—	90R$_b$	—
	FC-0505-P	6.0-6.4	A.S.	50 (345)	42 (290)	1.0	13.0 (8.96)	60R$_b$	4.5 (6.1)
	FC-0505-P	6.0-6.4	H.T.	—	—	—	—	95R$_b$	—
	FC-0505-R	6.4-6.8	A.S.	66 (455)	55 (379)	1.5	16.0 (11.0)	75R$_b$	5.0 (6.8)
	FC-0505-R	6.4-6.8	H.T.	—	—	—	16.0 (11.0)	25R$_c$	—
Fe-5.0Cu-0.8C	FC-0508-N	5.6-6.0	A.S.	48 (331)	43 (296)	<0.5	10.5 (7.23)	60R$_b$	3.0 (4.07)
	FC-0508-N	5.6-6.0	H.T.	—	—	—	—	95R$_b$	—
	FC-0508-P	5.6-6.0	A.S.	62 (427)	57 (393)	<1.0	13.0 (8.96)	65R$_b$	3.5 (4.7)
	FC-0508-P	5.6-6.0	H.T.	—	—	—	—	30R$_c$	—
	FC-0508-R	6.4-6.8	A.S.	75 (517)	70 (483)	1.0	16.0 (11.0)	85R$_b$	4.5 (6.1)
	FC-0508-R	6.4-6.8	H.T.	—	—	—	—	35R$_c$	—

parts have as-sintered tensile strengths ranging from approximately 65,000 to 90,000 psi (448 to 621 MPa) and heat treatment will increase strengths to 130,000 psi (896 MPa).

Of the nonferrous P/M compositions, aluminum P/M parts have tensile strengths from about 15,000 to 50,000 psi (103 to 345 MPa). Sintered brass parts have tensile strengths up to 40,000 psi (276 MPa). Bronze bearings have tensile strengths of 8,000 to 18,000 psi (55 to 125 MPa) and high strength bronze structural parts, made from prealloyed powders, have tensile strengths up to 50,000 psi (350 MPa).

Ductility
Ductility of P/M parts tends to be relatively low because of the presence of pores. Measured in terms of percent elongation values for ductility are generally less than 10% for ferrous P/M materials; however, some P/M brasses reach 15 to 20% elongation. Ductility of most P/M materials can be increased substantially by hot or cold repressing followed by additional sintering.

Hardness
Because of the difference in structure, gross indentation hardness values of wrought and P/M parts cannot be compared directly. The hardness value for a P/M part is referred to as "apparent hardness," using a standard hardness tester. Apparent hardness is the combination of powder particle hardness and porosity, as outlined in MPIF Standard 43. Microhardness tests, such as Knoop or DPH, measure true particle hardness.

Because of possible density variations, it is necessary to specify the area in which hardness measurements are made. Precautions must also be taken

Composition,[a] %	MPIF designation	Density g/cu cm	Cond.[b]	Ult ten str, 1000 psi (MPa)	Yld str, 1000 psi (MPa)	Elong,[c] %	Mod of elast 10^6 psi (10^4 MPa)	Apparent hardness, Rockwell	Impact str, smooth Charpy, ft·lb (J)
Fe-8.5Cu-0.8C	FC-0808-N	5.6-6.0	A.S.	36 (248)	—	<0.5	—	55R$_b$	—
Fe-10Cu-0.15C	FC-1000-N	5.6-6.0	A.S.	30 (207)	—	0.5	—	70R$_f$	—
Iron nickel									
Fe-2.0Ni-1.25Cu-0.15C	FN-0200-R	6.4-6.8	A.S.	28 (193)	18 (124)	4.0	17.0 (11.7)	38R$_b$	14.0 (18.9)
	FN-0200-S	6.8-7.2	A.S.	38 (262)	25 (172)	7.0	21.0 (14.5)	42R$_b$	32.0 (43.4)
	FN-0200-T	7.2-7.6	A.S.	45 (310)	30 (207)	10.5	23.0 (15.8)	51R$_b$	50.0 (67.8)
Fe-4.25Ni-1.0Cu-0.15C	FN-0400-R	6.4-6.8	A.S.	36 (248)	22 (152)	5.0	17.0 (11.7)	40R$_b$	16.0 (21.7)
	CN-0400-S	6.8-7.2	A.S.	49 (338)	30 (207)	6.0	21.0 (14.5)	60R$_b$	35.0 (47.5)
	FN-0400-T	7.2-7.6	A.S.	58 (400)	36 (248)	6.5	23.0 (15.8)	67R$_b$	50.0 (67.8)
Fe-7.0Ni-1.0Cu-0.15C	FN-0700-R	6.4-6.8	A.S.	52 (358)	30 (207)	2.5	17.0 (11.7)	60R$_b$	12.0 (16.3)
	FN-0700-S	6.8-7.2	A.S.	71 (489)	40 (276)	4.0	21.0 (14.5)	72R$_b$	21.0 (28.5)
	FN-0700-T	7.2-7.6	A.S.	85 (586)	48 (331)	6.0	23.0 (15.8)	83R$_b$	26.0 (35.2)
Nickel steel									
Fe-2.0Ni-1.25Cu-0.45C	FN-0205-R	6.4-6.8	A.S.	37 (255)	23 (159)	3.0	17.0 (11.7)	47R$_b$	10.0 (13.6)
	FN-0205-R	6.4-6.8	H.T.	82 (565)	65 (448)	0.5	17.0 (11.7)	32R$_c$	6.0 (8.13)
	FN-0205-S	6.8-7.2	A.S.	50 (345)	31 (214)	3.5	21.0 (14.5)	74R$_b$	18.0 (24.4)
	FN-0205-S	6.8-7.2	H.T.	110 (758)	88 (607)	1.0	21.0 (14.5)	42R$_c$	16.0 (21.7)
	FN-0205-T	7.2-7.6	A.S.	61 (421)	37 (255)	4.5	23.0 (15.8)	85R$_b$	32.0 (43.4)
	FN-0205-T	7.2-7.6	H.T.	134 (924)	105 (724)	2.0	23.0 (15.8)	46R$_c$	28.0 (38.0)
Fe-2.0Ni-0.1Mn-1.0Mo-0.45C	IN-861[d]	6.8	A.S.	65 (448)	40 (276)	2.0	15.0 (10.3)	77R$_b$	7.5 (10.2)
	IN-861[d]	7.3	A.S.	97 (669)	50 (345)	—	22.0 (15.1)	90R$_b$	27.0 (36.6)
Fe-2Ni-1.25Cu-0.75C-	FN-0208-R	6.4-6.8	A.S.	48 (331)	30 (207)	2.0	17.0 (11.7)	62R$_b$	8.0 (10.8)
	FN-0208-R	6.4-6.8	H.T.	100 (689)	94 (648)	0.5	17.0 (11.7)	34R$_c$	6.0 (8.1)
	FN-0206-S	6.8-7.2	A.S.	65 (448)	41 (283)	3.0	21.0 (14.5)	79R$_b$	14.0 (19.0)
	FN-0208-S	6.8-7.2	H.T.	135 (931)	128 (882)	0.5	21.0 (14.5)	45R$_c$	12.0 (16.3)
	FN-0208-T	7.2-7.6	A.S.	79 (545)	50 (345)	3.5	23.0 (15.8)	87R$_b$	22.0 (29.8)
	FN-0208-T	7.2-7.6	H.T.	160 (1103)	155 (1069)	0.5	23.0 (15.8)	47R$_c$	18.0 (24.4)
Fe-4.25Ni-1.0Cu-0.45C	FN-0405-R	6.4-6.8	A.S.	45 (310)	26 (179)	3.0	17.0 (11.7)	63R$_b$	10.0 (13.6)
	FN-0405-R	6.4-6.8	H.T.	112 (772)	94 (648)	0.5	17.0 (11.7)	27R$_c$	6.0 (8.1)
	FN-0405-S	6.8-7.2	A.S.	62 (427)	35 (241)	4.5	21.0 (14.5)	72R$_b$	15.0 (20.3)
	FN-0405-S	6.8-7.2	H.T.	154 (1062)	128 (882)	1.0	21.0 (14.5)	39R$_c$	10.0 (13.6)
	FN-0405-T	7.2-7.6	A.S.	74 (570)	43 (296)	6.0	23.0 (15.8)	80R$_b$	30.0 (40.7)
	FN-0405-T	7.2-7.6	H.T.	180 (1241)	154 (1062)	1.5	23.0 (15.8)	44R$_c$	14.0 (19.0)
Fe-4.25Ni-1.0Cu-0.45C	FN-0408-R	6.4-6.8	A.S.	57 (393)	42 (290)	1.5	17.0 (11.7)	72R$_b$	6.0 (8.1)
	FN-0408-S	6.8-7.2	A.S.	77 (531)	57 (393)	3.0	21.0 (14.5)	88R$_b$	10.0 (13.6)
	FN-0408-T	7.2-7.6	A.S.	93 (641)	68 (469)	4.5	23.0 (15.8)	95R$_b$	16.0 (21.7)
Fe-7.0Ni-1.0Cu-0.45C	FN-0705-R	6.4-6.8	A.S.	54 (372)	35 (241)	2.0	17.0 (11.7)	69R$_b$	9.0 (12.2)
	FN-0705-R	6.4-6.8	H.T.	107 (703)	80 (552)	0.5	17.0 (11.7)	24R$_c$	8.0 (10.8)
	FN-0705-S	6.8-7.2	A.S.	76 (524)	48 (331)	3.5	21.0 (14.5)	83R$_b$	17.0 (23.0)
	FN-0705-S	6.8-7.2	H.T.	140 (965)	110 (758)	1.0	21.0 (14.5)	38R$_c$	15.0 (20.3)
	FN-0705-T	7.2-7.6	A.S.	90 (620)	57 (393)	5.0	23.0 (15.8)	90R$_b$	24.0 (32.5)
	FN-0705-T	7.2-7.6	H.T.	168 (1158)	130 (896)	1.5	23.0 (15.8)	40R$_c$	20.0 (27.1)
Fe-7.0Ni-1.0Cu-0.45C	FN-0708-R	6.4-6.8	A.S.	57 (393)	41 (283)	1.5	17.0 (11.7)	75R$_b$	6.0 (8.1)
	FN-0708-S	6.8-7.2	A.S.	80 (552)	55 (379)	2.5	21.0 (14.5)	88R$_b$	12.0 (16.3)
	FN-0708-T	7.2-7.6	A.S.	95 (655)	66 (455)	3.0	23.0 (15.8)	96R$_b$	16.0 (21.7)

[a]Nominal [b]A.S.—as sintered; H.T.—heat treated. [c]ASTM E 8. [d]International Nickel Co.

4 Mechanical properties of stainless steel P/M parts

Composition, grade	MPIF designation	Density g/cu cm	Ult ten str, 1000 psi (MPa)	Yld str, 1000 psi (MPa)	Elong,[a] %	Impact str, smooth Charpy ft·lb (J)	Apparent hardness, Rockwell
AISI 303	SS-303-P	6.0-6.4	35 (241)	32 (221)	1.0	—	—
AISI 303	SS-303-R	6.4-6.8	52 (358)	47 (324)	2.0	—	—
AISI 316	SS-316-P	6.0-6.4	38 (262)	32 (221)	2.0	—	—
AISI 316	SS-316-R	6.4-6.8	54 (372)	40 (276)	4.0	—	—
AISI 410	SS-410-N	5.6-6.0	42 (290)	41 (283)	0-1.0	—	—
AISI 410	SS-410-P	6.0-6.4	55 (397)	54 (376)	0-1.0	—	—
AISI 304L[b]	—	6.7-7.0	64-79.5 (441-548)	—	5-19	4-44 (5.4-59.6)	—
AISI 316L[b]	—	6.6-7.1	60-75.5 (414-520)	—	4-21	4-45 (5.4-61)	—
AISI 410[b]	—	6.7-7.3	67-71 (462-489)	—	2-7	1.5-3 (2.0-4.1)	—
AISI 316[c]	—	—	47 (324)	25 (172)	16[e]	—	67R$_f$
AISI 304[d]	—	—	38 (262)	21 (145)	9[e]	—	70R$_f$
303L[f]	—	6.2-7.0	45-70 (310-483)	35-48 (241-331)	4-10	—	41-75R$_b$
Fe-20Cr-30Ni-2.5Mo-3.5Cu-1.0Si-0.2Mn[b]	—	6.2-7.0	48-70 (331-482)	30-48 (207-331)	5-13	—	44-80R$_b$
434L[f]	—	6.2-7.0	50-91 (345-627)	42-66 (290-455)	2-9	—	70-85R$_b$

[a]Per ASTM E 8. [b]Remington Arms Co. [c]IPM SS-100. [d]IPM SS-101. [e]In 1 in. (25.4 mm). [f]SCM Glidden.

Source: *Materials Engineering*, Aug 1976

when interpreting surface hardness values on case hardened parts because the indentations sometimes penetrate the case.

Corrosion resistance

Corrosion resistance of P/M parts is affected by the presence of voids. Entrapment of corrosives can lead to internal corrosion but corrosion resistance can be improved by compacting to higher density.

Stainless steel and titanium P/M parts have relatively good corrosion resistance in the atmosphere and in weak acids. Nonferrous P/M compositions have good atmospheric corrosion resistance. Steam treating of ferrous P/M parts imparts a corrosion resistant, blue-black iron oxide surface. Finishes and coatings are used also to improve corrosion resistance.

5 Mechanical properties of infiltrated steel and iron P/M parts[a]

Composition,[b] %	MPIF designation	Cond[c]	Ult ten str 1000 psi (MPa)	Yld str 1000 psi (MPa)	Elong,[d] %	Mod of elast 10^6 psi (10^5 MPa)	Apparent hardness, Rockwell	Impact str, unnotched Charpy, ft·lb (J)
Fe-11.5Cu-0.45C	FX-1005-T	A.S.	83 (572)	64 (441)	4.0	20.0 (1.38)	75R_b	14.0 (19.0)
	FX-1005-T	H.T.	120 (827)	107 (738)	1.0	20.0 (1.38)	35R_c	7.0 (9.5)
Fe-11.5Cu-0.8C	FX-1008-T	A.S.	90 (620)	75 (517)	2.5	20.0 (1.38)	80R_b	12.0 (16.3)
	FX-1008-T	H.T.	130 (896)	105 (724)	<0.5	20.0 (1.38)	40R_c	7.0 (9.5)
Fe-20Cu-0.15C	FX-2000-T	A.S.	65 (448)	—	1.0	—	60R_b	15.0 (20.3)
Fe-20Cu-0.45C	FX-2005-T	A.S.	75 (517)	50 (345)	1.5	18.0 (1.24)	75R_b	9.5 (12.9)
	FX-2005-T	H.T.	115 (793)	95 (655)	<0.5	18.0 (1.24)	30R_c	6.0 (8.1)
Fe-20Cu-0.8C	FX-2008-T	A.S.	85 (586)	75 (517)	1.0	18.0 (1.24)	80R_b	10.0 (13.6)
	FX-2008-T	H.T.	125 (862)	107 (738)	<0.5	18.0 (1.24)	42R_c	5.0 (6.8)

[a]Density=7.2 to 7.6 g/cu cm. [b]Nominal [c]A.S.-as sintered; H.T.-heat treated. [d]Per ASTM E 8.

6 Mechanical properties of nonferrous P/M parts

Composition,[a] %	MPIF designation	Density, g/cu cm	Cond	Ult ten str 1000 psi (MPa)	Yld str 1000 psi (MPa)	Elong,[c] %	Mod of elast 10^6 psi (10^4 MPa)	Apparent hardness, Rockwell	Impact str, unnotched Charpy. ft·lb (J)
Brasses and bronzes									
Cu-10Sn-0.85C-0.5Fe	CT-0010-N	5.6-6.0	A.S.[b]	8.0 (55)	—	1.0	—	—	—
	CT-0010-R	6.4-6.8	A.S.	14.0 (96)	—	1.0	—	—	—
	CT-0010-S	6.8-7.2	A.S.	18.0 (124)	—	2.5	—	—	—
Cu-19.6Zn-1.5Pb-0.15Fe	CZP-0220-T	7.2-7.6	A.S.	24.0 (165)	11.0 (76)	13	—	55R_H	—
	CZP-0220-U	7.6-8.0	A.S.	28.0 (193)	13.0 (90)	19	—	68R_H	—
	CZP-0220-W	8.0-8.4	A.S.	32.0 (221)	15.0 (103)	23	—	75R_H	—
Cu-10.15Zn-0.15Fe	CZ-0010-T	7.2-7.6	A.S.	20.0 (138)	9.0 (62)	13	—	57R_H	—
	CZ-0010-U	7.6-8.0	A.S.	27.0 (186)	10.0 (69)	18	—	70R_H	—
Cu-29.6Zn-0.15Fe	CZ-0030-T	7.2-7.6	A.S.	31.0 (213)	13.0 (90)	20	—	76R_H	—
	CZ-0030-U	7.6-8.0	A.S.	37.0 (255)	15.0 (103)	26	—	85R_H	—
Cu-10.1Zn-1.5Pb-0.15Fe	CZP-0210-T	7.2-7.6	A.S.	18.0 (124)	7.0 (48)	14	—	46R_H	—
	CZP-0210-U	7.6-8.0	A.S.	25.5 (176)	8.0 (55)	20	—	60R_H	—
Cu-29.6Zn-1.5Pb-0.15Fe	CZP-0230-T	7.2-7.6	A.S.	28.0 (193)	11.0 (76)	22	—	65R_H	—
	CZP-0230-U	7.6-8.0	A.S.	34.0 (234)	13.0 (90)	27	—	76R_H	—
Nickel silver									
Cu-18Ni-18Zn	CZN-1818-U	7.6-8.0	A.S.	30.0 (207)	—	10	14.0 (9.6)	75R_H	10.0 (13.6)
	CZN-1818-W	8.0-8.4	A.S.	37.0 (255)	—	12	14.0 (9.6)	85R_H	13.0 (17.6)
Cu-18Ni-16.6Zn-1.4Pb	CZNP-1618-U	7.6-8.0	A.S.	30.0 (207)	—	10	13.0 (8.96)	75R_H	9.0 (12.2)
	CZNP-1618-W	8.0-8.4	A.S.	35 (241)	—	12	14.0 (9.6)	85R_H	12.0 (16.2)
Aluminum									
Al-0.8Mg-0.5Si-0.5Cu	ASTM B 595	2.3-2.45	T1[d]	12.0 (83)	9.0 (62)	4.0[e]	—	60-65R_H	—
		2.3-2.45	T4[d]	14.0 (97)	11.5 (79)	3.5	—	65-70R_H	—
		2.3-2.45	T6[d]	20.0 (138)	19.0 (131)	0.5	—	80-85R_H	—
		2.45-2.6	T1	18.5 (128)	10.0 (69)	6.0	—	80-85R_H	—
		2.45-2.6	T4	23.0 (159)	15.0 (103)	5.0	—	50-55R_E	—
		2.45-2.6	T6	30.0 (207)	28.0 (193)	2.0	—	65-70R_E	—
Al-2.0Cu-0.7Mg-0.5Si	ASTM B 595	2.3-2.45	T1	19.0 (131)	13.0 (90)	4.0	—	80-85R_H	—
		2.3-2.45	T4	23.0 (159)	17.5 (121)	5.0	—	50-55R_E	—
		2.3-2.45	T6	26.0 (179)	25.0 (172)	1.0	—	55-60R_E	—
		2.45-2.6	T1	20.0 (138)	15.5 (107)	4.0	—	80-85R_H	—
		2.45-2.6	T4	24.5 (169)	18.5 (127)	3.5	—	55-60R_E	—
		2.45-2.6	T6	28.0 (193)	27.0 (186)	2.0	—	60-65R_E	—
Al-4.25Cu-0.5Mg-1.2Si	ASTM B 595	2.3-2.45	T1	20.0 (138)	14.0 (97)	2.0	—	80-85R_H	—
		2.3-2.45	T4	24.0 (165)	21.0 (145)	2.0	—	55-60R_E	—
		2.3-2.45	T6	30.0 (207)	25.0 (172)	0.5	—	65-70R_E	—
		2.45-2.6	T1	22.0 (152)	17.0 (117)	3.0	—	85-90R_H	—
		2.45-2.6	T4	26.0 (179)	22.0 (152)	2.5	—	55-60R_E	—
		2.45-2.6	T6	35.0 (241)	33.0 (228)	1.0	—	70-75R_E	—
		2.60	T1	25.0 (172)	22.0 (152)	3.0	—	55-60R_E	—
		2.60	T4	32.0 (221)	26.0 (179)	2.5	—	70-75R_E	—
		2.60	T6	42.0 (290)	40.0 (276)	2.0	—	80-85R_E	—
Titanium									
Commercially pure	—	95.2	—	60.0 (414)	47.0 (324)	10	—	—	—
Ti-6Al-4V	—	97.5	—	127.0 (876)	114.0 (786)	8	—	—	—

[a]Nominal [b]As sintered. [c]ASTM E 8. [d]T1—as sintered; T4—sol'n heat treated at 940 to 970 F (777 to 794 K), cold water quench and aged min 4 days at RT; T6—sol'n heat treated at 940 to 970 F (777 to 794 K), cold water quench and aged 18 hr at 320 to 350 F (433 to 450 K). [e]All aluminum alloys in 1 in. (25.4 mm).

Processing and fabrication

Processing and fabrication of P/M parts falls into two major categories—primary operations, which include blending of powders, compacting and sintering, and secondary operations, which include repressing or forging, impregnation, heat treating, machining, injection molding, brazing and finishing.

Repressing and forging

Repressing, sometimes called coining or sizing, is used to increase density or provide greater dimensional accuracy. It can be used also to produce more complex shapes not possible from a single-press operation or to reshape or emboss the surface.

Sintered parts also may be subsequently forged or hot-formed. Like repressing, forging or hot forming also increase density and dimensional accuracy; however, it also enables forming configurations which cannot be made from a single pressing. There are two methods of producing P/M forgings:
- Forming, rather than forging, involves forming a preform in punch and die type tooling. This method already is in use.
- Forging involves the use of P/M preforms, rather than bar stock, in conventional closed die forging. This P/M process is still developmental.

Both methods differ from repressing in that a substantial amount of the final part shape is imparted during the forming operation which may involve at least a moderate amount of lateral flow in addition to vertical compaction.

Heat treatment

Ferrous P/M parts, containing 0.3% or higher combined carbon, can be quench-hardened for increased strength and wear resistance. The percentages of carbon and other alloying elements effectively combined in the material, density of the part, determine the degree of hardening possible for any given quench condition. Surface hardness of 500 to 600 Knoop (file hard) is possible with quench hardening.

Ferrous parts without carbon can be carburized by standard methods. Low density parts carburize throughout while high density parts develop a distinct carburized case. Very high density parts respond well to fused salt carbonitriding but density must be high enough to prevent absorption of salt into the pores. Low and medium density parts absorb brines and salts during salt bath carbonitriding which can lead to subsequent corrosion. Thus, oil quench hardening is recommended for low- and medium-density parts.

The properties of as-sintered aluminum P/M parts are improved by a series of thermal treatments. Aluminum P/M parts achieve higher strength by solution and precipitation of soluble alloying elements. As-sintered strength is affected by the rate of cooling from sintering temperature. Parts cooled very slowly (about 50 F, 27.7 K per hr) will develop the lower strengths of annealed tempers.

Impregnation and infiltration

The controlled porosity of P/M parts makes it possible to infiltrate them with another metal or impregnate them with oil or a resin either to improve mechanical properties or to provide other performance characteristics. Oil impregnated P/M bearings hold from 10 to 30% oil by volume. Properties resulting from infiltration with another metal depend upon the metals which constitute the structure of the infiltrated part, together with the way and proportions in which they are combined. Infiltration is used to improve mechanical properties, seal pores prior to electroplating, improve machinability and make parts gas or liquid tight.

Machining

Though normally produced to final shape, P/M parts sometimes are machined to produce special shapes or to achieve closer tolerances. Machining characteristics of P/M parts are similar to cast materials. Small amounts of lead, sulfur, copper or graphite are common additives which improve the machinability of ferrous P/M parts. Lead is also commonly used to increase the machinability of nonferrous parts.

Machining speeds and feeds for high density parts (above 92% theoretical density) are similar to those for wrought metals. Lower density parts require adjustment of feed and speed to obtain optimum results. In general, high speeds, light feeds and extremely sharp carbide tools are recommended. Lubricants and coolants should be used with caution, especially when machining porous parts, to avoid entrapping solutions which could cause corrosion. Grinding of P/M parts is similar to grinding wrought materials; however, where surface porosity is required, grinding tends to reduce porosity.

Finishing

Practically all common finishing methods are applicable to P/M parts. When tumbling, rust inhibitors should be added to the water. After tumbling parts should be spun dry and heated to speed the evaporation of the water from the pores. Tumbling should be done after machining to avoid abrasive pick-up in the pores that might result in excessive tool wear.

Burnishing can be used to improve part finish and dimensional accuracy or to work-harden surfaces. Compared to burnishing wrought parts, closer tolerances can be held on P/M parts because the surface porosity allows metal to be displaced more easily.

Ferrous P/M parts can be colored by several methods. For indoor corrosion resistance, parts are blackened by heating to the blueing temperature and then cooled. Oil dipping gives a deeper color and slightly more corrosion resistance. Ferrous P/M parts also can be blackened chemically, using one of several commercial liquid salt baths. On parts with a density lower than 7.3 g/cu cm care must be taken to avoid entrapping the salt bath. Nickel- and copper-bearing parts are adversely affected by blackening baths.

Steam treating increases corrosion resistance of ferrous P/M parts. The process involves heating parts to 750 to 1100 F (672 to 866 K) and exposing them to superheated steam under pressure. Parts may be oil-dipped after cooling to further enhance corrosion and wear resistance, as well as appearance. By filling some of the interconnecting porosity and much of the surface with the oxide coating, density is increased, which leads to higher compressive strength. The oxide coating also increases hardness and wear resistance but somewhat at the expense of increased brittleness and reduced machinability. Heat-treated parts are seldom steam-treated because the treatment anneals them.

Copper, nickel, chromium, cadmium and zinc plating may be applied to P/M parts. High density (7.2 g/cu cm) and infiltrated parts can be plated using the same methods as on wrought parts. Lower density parts should be sealed to avoid entrapment of plating solutions. Electroless nickel plating can be used as well as electroplating which is applicable to nonimpregnated ferrous parts in the 6.6 to 7.2 g/cu cm density range.

Injection molding

In a newly developed process, powdered metals are mixed with a proprietary binder and formed into larger-than-final-dimension parts using a standard plastic injection molding machine. The binder is removed by another proprietary process. Densities ranging from 94 to 98% of theoretical and tolerances of ±0.003 to 0.005 in. (±0.076 to 0.127 mm) are claimed. Components with intricate configurations, which currently require secondary operations, can be produced in one operation. Production speeds compare to those for plastic injection molding.

Brazing

P/M parts can be brazed with a Ni-Mn-Cu alloy. The process limits penetration to approximately 0.0625 to 0.093 in. (1.58 to 2.36 mm) for densities between 5.7 and 6.8 g/cu cm and less at higher densities. The braze alloy, developed by SKC P/M Engineering, develops bond strengths between 55,000 and 60,000 psi (379 and 414 MPa).

Design details

The P/M process has distinct design capabilities and there are specific guidelines that must be considered in producing sound, economical parts.

Part size

Although there is no known theoretical limit to the size to which P/M parts can be pressed, maximum practical size is set by available presses and powder characteristics. Most parts range in area from 0.125 to 25 sq in. (10 sq mm to 0.015 sq m) and are from 0.031 to 6 in. (1 to 150 mm) long. Maximum surface area depends on the powder, part density and press capacity.

Shapes

The most suitable shapes are those that have uniform dimensions in the direction of pressing. These include simple cylindrical, square and rectangular shapes as well as odd shapes in which the contour is a plane at right angles to the direction of pressing. For example, cams and gears with no changes in thickness are relatively simple to press.

Perfect spheres cannot be made by the P/M process. Therefore, spherical P/M parts are designed with straight or flat areas around the equator. Parts that must fit into ball sockets are repressed to produce a more spherical shape. Spherical depressions up to a hemisphere also are possible.

Multi-level shapes

Because metal powders hardly flow laterally during compaction, this imposes certain limitations on variations in thickness, numbers of steps, flanges, slots or grooves. Differences in thickness of up to 10 or 15% are feasible, if variations in density are acceptable.

When only limited nonuniformity in density is acceptable, or when the overall thickness variation is too great, multiple punch tooling, stepped dies or both must be used. In such cases the part is held captive in the tools and special press motions are required for proper ejection.

Flatness

Total measured flatness depends on surface area. It is affected by part thickness, with thin parts more subject to distortion than thick parts during sintering or heat treatment. Also, flatness is maintained more easily in simple shapes and cross sections. Flatness can be improved in soft metal parts by repressing and by grinding with hard materials.

Holes

Holes in the direction of pressing are readily incorporated. Round holes are the least expensive. Splines, keyways, keys, D-shapes, square and hexagonal holes can be produced at very little added cost. Blind holes or blind steps in holes, and tapered holes also are produced readily. Side holes, or holes not parallel to the direction of pressing, must be made by secondary machining.

Maximum diameter of holes depends on wall thickness and minimum diameter depends on hole length. Lightening holes are used frequently in large parts to reduce effective pressing area, with lower weight and reduced pressing forces being added benefits.

Wall thickness

Minimum wall thickness of P/M parts is governed by overall size and shape of the part. For parts of any appreciable length, walls should not be less than approximately 0.060 in. (1.6 mm) thick. A wall less than 0.030 in. (0.8 mm) thick generally is considered too thin in all but the smallest part where thicknesses are also very small. In cases where the ratio of length-to-wall thickness is as high as 8-to-1 or more, special precautions should be taken to get a uniform fill and even then, density variations are almost unavoidable.

Tapers and drafts

Drafts generally is not required or desired on straight-through P/M parts as is usually the case with die castings. Tapered side walls can be produced but production speed is slower in order to keep powder from being wedged between the taper of the die and other tool parts during filling. Tapered sections usually require a short, straight land to prevent the upper punch from running into the taper in the die wall or on the core rod.

Fillets, radii, chamfers and bevels

Generous fillet radii are desirable and economical. A true radius is not possible at the juncture of a punch face and the die wall, since this requires that the punch edge or skirt be feathered to zero. Therefore, a full radius would have to be approximated by hand or machine finishing.

Chamfers, rather than radii, are preferred on part edges to prevent burrs. A preferred chamfer is 30 deg. The cost of chamfering varies with part shape and tool design.

Flanges, bosses and hubs

A small flange, step or overhang can be produced by a step or shelf in the die. Beyond a certain point, when the overhang becomes too great to permit ejection without breaking the flange, special tooling must be considered.

Bosses generally can be located on top or bottom of a part. These must be small compared to overall part thickness. Round shapes are preferred and draft angles should be at least 45 deg.

Hubs on gears or sprockets are produced readily and a variety of shaped holes in the longitudinal direction are possible. Set screw holes at right angles to the drive axis must be produced by secondary machining. Generous radii between hub and flange are desirable. Hubs that serve as self-lubricating surfaces can be produced at a lower density than other portions of the part for high oil capacity.

Undercuts, threads

Undercuts on the horizontal plane cannot be produced because they prevent die ejection of the part. Threads also prevent part ejection and thus cannot be made by conventional P/M methods. Similarly, annular grooves around a part must be machined.

Tolerances

Tolerances of P/M parts compare very favorably with those of other conventional fabricating processes. Very often P/M parts can be held to closer tolerances than sand castings, stampings and conventional forgings. Tolerances on P/M parts depend on the metal powder itself, size and dimension of the part, run-out, heat treatment and the extent of coining and repressing.

Kurt H. Miska, Associate Editor

Acknowledgements

The author is grateful to the following companies for supplying information for this manual: Alcan, Alcoa, Bundy, Burgess Norton, Brockway Pressed Metals, Chrysler Amplex, Engineered Sinterings and Plastics, Federal Mogul, Gould, Greenback Industries, GKN, Glidden Durkee, INCO, IPM, Interlake (Hoeganaes), Metal Powder Industries Federation, Midwest Sintered Products, National Sintered Alloys, New Industrial Techniques, NJ Zinc, Remington Arms, Sintered Metals, SKC, P/M Engineering, Stanadyne, Stellite Div. of Cabot and A.O. Smith-Inland. All graphs and drawings courtesy of MPIF.

Development of alloy systems for powder forging

G. T. Brown

Precision forgings are now being produced from preforms pressed from steel powders. There are a greater number of factors influencing the choice of alloy composition than in traditional wrought technology. Since strength is frequently controlled by heat treatment, hardenability is of importance. The elements that are most useful in promoting hardenability at lowest cost are manganese and chromium. However, because of their affinity for oxygen they are not the most satisfactory elements for powder making. The author gives the reasoning behind the design of alloys for the powder-forging process. It is shown how mechanical strength and cost are considered to be controlling variables. MT/272A

©1976 The Metals Society. The author is with Guest Keen and Nettlefolds Group Technological Centre, Wolverhampton.

The powder-forging process is technically capable of producing a wide range of engineering components. A number of manufacturing facilities are now in existence, and components are being made in both pilot and production size batches. In Fig.1 is shown a number of parts that are either currently in production or undergoing pilot testing by customers. From this it will be seen that relatively complicated shapes can be made in non-axially symmetric forms as well as the more obvious round parts. All the parts illustrated are as-forged and require considerably less machining to finish than if they had been made by the traditional processes of forging, machining, casting, etc., thus presenting the user with the possibility of a cost saving. Although powder is a relatively expensive material, this is offset by a higher efficiency of utilization than is possible with other processes.

In adopting powder-fabricating methods, certain constraints have been found to exist in the nature of the alloys that may be produced, bearing in mind the essential cost factors and that the mechanical performance of the final components must be able to meet a specification appropriate to the service environment. In the case of a connecting rod, for instance, the fatigue performance is expected to be at least as good as that known to be obtained with the standard drop-forged article. Gear components are expected to have core strength and case-bearing properties equal to those currently produced by machining from solid bar or forged blanks. In many cases the service requirements are assessed only by reference to existing material

Table 1 Compositions of wrought steels typically used in component engineering

Specification	Composition, wt-%				
	C	Mn	Ni	Cr	Mo
080A37*	0·38–0·40	0·7–0·9	..	..	..
530A36†	0·34–0·39	0·6–0·8	..	0·9–1·2	..
606M36†	0·32–0·4	1·3–1·7	..	..	0·22–0·32
637M17†	0·14–0·2	0·6–0·9	0·85–1·25	0·6–1·0	..
SAE8620†	0·17–0·23	0·6–0·95	0·35–0·75	0·35–0·65	0·15–0·25

* BS 970 :1970 lists 21 specifications based on carbon-content range
† BS 970 :1970 lists 4 specifications based on carbon content

standard specifications. Since powder-forged materials now have some important advantages that may not be apparent from the first examination of testpiece results, it is sometimes necessary to challenge the relevance of traditional materials specifications.[1]

The selection of steels for the traditional forging and machining routes is based primarily on strength and cost. In Table 1 is shown a number of steels in common engineering use for through-hardening and carburizing applications. The dividing line for the two classes of steel is usually determined by the carbon content. Steels for carburizing applications, however, are frequently of higher alloy content, partly to make up for the loss of hardenability owing to the lower carbon content in the core. The steels produced by powder forging have a non-metallic inclusion structure different from that encountered in wrought steel. The nature and number of inclusions present are a function of the manufacturing route and exert an influence on properties. The alloys used are therefore designed within the limits imposed by this influence. Four of the compositions in current use, for example, are listed in Table 2; comparison with Table 1 shows some important differences. It is the author's purpose to examine the reasons for these differences and to relate them in particular to the commercial development, with cost as an overriding influence.

Outline of powder forging

BASIC PROCESS ROUTE

The powder- (or sinter-) forge basic process route is illustrated diagrammatically in Fig.2 and has the following steps:

(i) mix powders with graphite to adjust carbon content
(ii) meter powder into die cavity and apply pressure to produce a green preform
(iii) heat preform in controlled-atmosphere furnace

1 Selection of parts made by powder forging illustrating range of shapes possible (by courtesy of GKN Forgings Ltd)

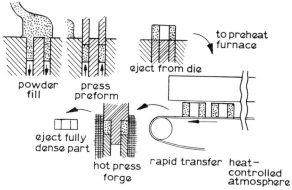

2 Schematic diagram of powder-forging process

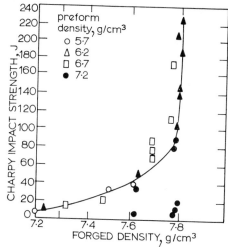

3 Impact strength as function of density in powder-forged samples (after Moyer[2])

Table 2 Low-alloy steel compositions currently being used in powder forgings

Alloy code	Composition, wt-%			
	Mn	Cr	Ni	Mo
W4	0·3–0·4	0·1–0·25	0·2–0·3	0·25–0·35
W32	0·4–0·5	0·25–0·35	0·2–0·3	0·25–0·35
W135	0·25–0·4	..	0·4–0·5	0·55–0·65
W78	0·25–0·35	..	1·8–2·2	0·45–0·55

(iv) remove preform directly from hot zone and transfer to press/forge

(v) complete densification by single stroke in heated, totally enclosed die set.

It is important that the transfer time between furnace and press is at a minimum, since no atmosphere protection is used at this point. Too long an exposure would cause penetration of oxide along the interconnecting pores of the preform, which is still relatively porous at this stage.

The design of the dies for the final hot compaction is such that the throwing of a flash, as in traditional drop forging, is eliminated. The design of the preform and the powder-forging conditions, i.e. temperature and pressure of the final forming stage, are such as to ensure complete densification throughout the component.

MECHANICAL PROPERTIES: RELATIONSHIP TO OTHER PROCESSES

It is important to appreciate the basic differences between powder-forged and traditionally sintered material on the one hand and rolled and forged steel ingot on the other.

The distinction between powder-forged and 'sintered-only' steel is simply one of density. Reference to impact performance serves as a good comparison for the two types of material, and Fig.3 shows the remarkable improvement demonstrated at high density compared with the less dense structure of a sintered part (i.e. at a density of >7·8 compared with 7·3 g/cm³). Table 3 compares the tensile and ductile properties of plain iron, forged to full density, and

the same material in the as-sintered condition. Clearly, the powder-forged material can be used in a much more highly stressed environment than is possible without the addition of the final hot-compacting stage of the process.

The distinction between powder-metallurgical products and wrought steel is in homogeneity and directionality of properties. The non-metallic inclusion structure of powder-forged steel is random and has virtually no directionality except in cases where the preform is deliberately made to flow during forging (as, for example, in the case of gear teeth forming from plain preforms, Fig.4). Wrought steel, by comparison, exhibits very marked directionality. This is illustrated in Table 4, which compares, at the same strength level, wrought steel tested transversely and longitudinally to the direction of rolling with a powder-forged steel. It will be noted from this that the ductility and impact properties of the powder-based material are intermediate

Table 3 Comparison of properties of powder-forged and 'sintered-only' plain-iron powder

	UTS, N/mm²	Elongation, %	RA, %	Impact, J
Powder A	388	40	70	116
Powder B	365	34	70	110
Sintered only	120–200	3–6	..	3–5 *

* Unnotched

4 Base of gear tooth, powder forged from a plain preform showing evidence of flow around the tooth form

×30

Source: *Metals Technology*, May-June 1976

Table 4 Comparison of properties of wrought and powder-forged steel

	UTS, N/mm²	Elonga-tion, %	RA, %	Impact, J
Mn–Mo billet steel				
Longitudinal	920–980	17–19	60–62	100
Transverse	910–950	5–12	8–24	10
Ni–Mo powder forged	900–930	13–15	40–50	27

between those of the wrought and the cast materials. The explanation again lies with the non-metallic inclusions, which now present the same length of path to a propagating crack, whatever the direction of testing. In wrought steel, crack propagation is assisted in the one direction by the elongated inclusions and diverted in the other when the crack is propagating perpendicular to them.

In the case of fatigue, the same relationship of powder-forged to wrought steel has been demonstrated.[3]

The materials philosophy

POWDER MAKING
In order to discuss alloy design in relation to powder forging, it is appropriate first to examine briefly the principal processes by which powder may be made. For ferrous materials there are two basic processes: (a) the direct-reduction or sponge method and (b) the atomizing of liquid iron or steel. In the former method, a high-purity powdered iron ore is reduced by CO directly to a sponge or sintered cake of iron. This sponge is then crushed into a powder, which is refined by a final reduction anneal giving soft, easily compressed particles with a high green strength. By far the largest tonnage of ferrous powders currently in use for powder-metallurgical purposes is produced by this method. A limitation is that it is very difficult to separate the last traces of silica and alumina present in the original ore; the non-metallic inclusion content of such powders is therefore high, e.g. over 1% by volume. In atomizing processes the starting point is a liquid melt, which is converted into powder directly by an atomizing medium under high pressure. When water is the medium the powder is subsequently annealed in a hydrogen-bearing atmosphere, both to soften the particles and to reduce the thin oxide layer that forms owing to the oxidizing nature of the atomizing water in direct contact with the molten particles.

In order to produce compactable powder with good green strength, irregularly shaped particles are essential. In the case of atomizing processes, the operating conditions both at the atomizing stage and during subsequent processing by annealing, grinding, etc., must be carefully controlled so that a majority of the particles produced are irregular. Figure 5 shows the nature of powder particles produced by sponge iron and water-atomizing processes and illustrates their similarities.

Other atomizing processes utilize air or an inert gas to break up the metal stream. In the case of inert gas, the formation of a surface oxide is avoided, but the particles produced are spherical in character and so have no cohesion in cold pressing. This difficulty is overcome when air is used to atomize, because a thick oxide film is produced, and during subsequent annealing and processing it is possible to produce a sponge that will result in a pressable powder.

In atomizing from the liquid state, it is possible to carry out alloying during melting. The resultant powder particles thus have the alloying elements in solution. With sponge powders, alloying is possible only by solid-state diffusion, and since economics control the heating cycle the final element distribution is generally poor.

IMPORTANT PROPERTIES REQUIRED
While the aim is to show how the composition of the alloys chosen is influenced by the restrictions of the processing route, it must be borne in mind that both of these factors are significantly influenced by the requirements of the final component in terms of its properties. Over and above the technical choices, the question of cost imposes a moderating influence.

The two attributes of most concern in producing acceptable and economic powder forgings are machinability and 'strength'. The strength parameters of most concern are yield and fatigue, although most specifications traditionally use ultimate strength as their main parameter. Even though one is dealing with a precision-forging method, some finish machining is still necessary for most components, e.g. cross holes, screw threads, parting off, broaching, etc. Most of the parts require the 'strength' to be developed by heat treatment, either by quenching and tempering or by case carburizing. Only rarely is high resistance to impact damage critical, although naturally the aim is to produce structures with at least a modicum of impact strength as demonstrated in the standard Charpy test.

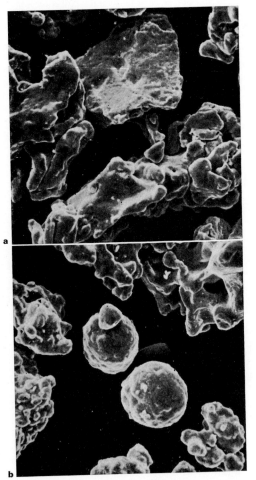

a sponge iron; *b* water atomized
5 Stereoscan photographs of iron powder particles
×375

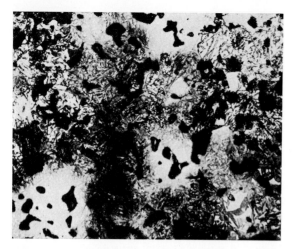

6 Alloy steel sintered powder, showing residual porosity and microstructural heterogeneity ranging from ferrite/carbide through martensite and retained austenite etched, ×450

First consider the choices governing the selection of the type of powder to be used. In traditional powder metallurgy the highest density is achieved by pressing the softest possible powders; thus sponge iron with no carbon in solution may be alloyed with elemental nickel, copper, etc. From the point of view of storage of raw material, this arrangement has many advantages, since it reduces the number of separate specifications in stock (neglecting for the moment that it is not unusual to use powder from different sources with different shrinkage characteristics on sintering, etc.). However, as stated previously, the sintering times in industrial use allow relatively little diffusion, with the result that the final microstructure is extremely inhomogeneous (Fig.6). Thus sponge-iron powders are not considered to be the prime raw material for powder-forging purposes for the following reasons:

(i) inhomogeneity of alloying element gives rise to inefficient hardenability control

(ii) mechanical properties have been shown to be inferior to more homogeneous structures[4]

(iii) hard spots within the structure are very difficult to machine

(iv) carburized cases would contain excessive retained austenite and soft spots.

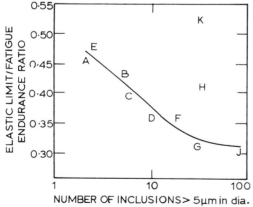

7 Relationship of inclusion structure to fatigue performance of powder-forged through-hardened steels (after Brown and Steed[3])

The disadvantages outlined above are eliminated when water-atomized materials are used, and this is therefore considered to be the most useful process for producing powder for powder forging.

As stated above, the two principal requirements for powder forging materials are an ability to develop an appropriate hardenability to guarantee strength and the control of fatigue performance by reference to microstructural features and in particular the non-metallic inclusion content. A relationship between fatigue performance and the non-metallic inclusion content of powder-forged steels has been established tentatively and is reproduced in Fig.7. The origins and control of non-metallic inclusions are thus important.

In terms of designing an alloy for fabrication by the powder-forging route, the conditions controlling optimum hardenability tend to work in opposition to those promoting high fatigue and ductility performance. To describe the conflicting requirements, it is necessary to consider the origins of the non-metallic inclusions and their history at each stage of processing.

The normal steelmaking inclusions resulting from slag entrapment, refractory breakdown, and deoxidation products are atomized along with the steel and present an angular appearance in the final microstructure. In addition, there is a series of complex oxides formed by virtue of the impingement on the molten metal of the atomizing medium. During atomizing with high-pressure water jets, the environment of the first-formed droplets is essentially oxidizing. In the current discussion, the alloys considered are dilute, and since iron is the base material, the first-formed oxides would naturally be expected to contain a preponderance of this element.

During water atomizing, some spherical particles are produced as well as the more angular ones so necessary for the generation of adequate green strength. The selection of the atomizing conditions, i.e. pouring rate, degree of superheat, composition of the water, composition of the steel, water pressure, angle of impingement, etc., are all variables that control the shape of the product. Examination of the as-atomized material reveals that the rounded particles often exhibit a thicker shell of oxide than the angular particles. It is also possible for the oxide to become detached completely and to form separate discrete particles.

After atomizing, the powder is separated from the water and dried in hot air. By virtue of a small amount of corrosion that inevitably takes place, the oxygen content is increased at this stage. After drying, the powder is annealed in dissociated ammonia and the thin oxide films are reduced. Reference to the elementary thermodynamic data (Fig.8) shows that oxides of iron, molybdenum, and nickel are easily reduced but that the reduction of oxides of manganese and chromium is more difficult. If a high temperature is used in order to aid the reduction, the powder particles sinter together strongly and need an unacceptable degree of grinding to restore flow properties. If too low a temperature is selected, an impossibly 'dry' atmosphere would be needed with adverse influence on the economics.

In producing plain-iron powders, the thin oxide film is reduced completely back to iron, whereas with alloy steels a progressive change in composition takes place. As annealing continues, the proportion of iron oxide decreases until a state is reached in which the remaining inclusion particles contain a relatively high proportion of manganese, chromium, etc., compared with the original oxide layer. Figure 9 shows the partial reduction of a film surrounding a spherical particle.

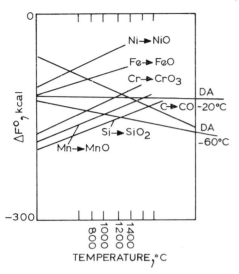

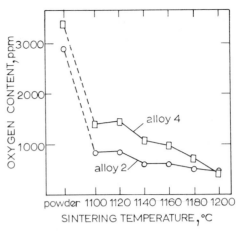

8 Thermodynamic data for a number of oxides; dissociated ammonia (DA) lines for two dewpoints are included

10 Oxygen content as function of sintering temperature in hydrogen atmosphere: alloy 4 contains 1·22%Mn, 0·36%Cr; alloy 2 contains 0·47%Mn, 0·23%Cr (after Lindskog and Grek[3])

The reduction process continues during heating for forging by virtue of CO generated from the graphite that is added to adjust the carbon content and is aided by the atmosphere of the sinter–forge reheat furnace, which is of course arranged to be in balance with the carbon content of the preforms. Much has now been written concerning the virtues of using high-temperature heating in either pure-hydrogen or dissociated-ammonia atmospheres. It has been shown that the Cr and Mn oxide-bearing inclusions are reduced at 1120° and 1160°C, respectively (Fig.10). Although some improvement in impact performance has been demonstrated (*see* Fig.11), it is inappropriate to overstate the benefits without considering some other implications such as heating time. Time is a factor not only in relation to the reduction equilibrium but also in relation to the diffusion of the elements released. The situation is analogous to that of mixed elemental powders and the possibility of hard spot formation owing to inhomogeneous alloy distribution. Under commercial conditions in which cost is a major controlling influence, the use of long heating times, or of temperatures over 1125°C, tends to represent a barrier; this is also the case with the use of very dry reducing gases, which are expensive in comparison with endo gas. Induction heating offers the possibility of raising the temperature further, but unfortunately other difficulties, such as the design of coil for non-spherical parts and problems in handling work through the furnace present barriers.

Even if all the chromium and manganese oxides were reduced, there would still be the problem of the silicon and aluminium, which tend to occur in the form of complex alloy silicates and alumina particles which do not reduce. For example, published work[6] on a 0·5%Mn steel sintered at 1250°C, still shows a significant number of 30 μm particles, presumably composed of silicates, etc., which will exert a significant influence on fatigue performance.

To summarize the position on powder-forging temperature selection, it seems that it may be possible theoretically to ignore oxide inclusions and to rely on an appropriate heating cycle to modify or remove them, but commercially this is not expedient, and more work is needed to determine the effects of the inclusions *in toto* and their relative importance.

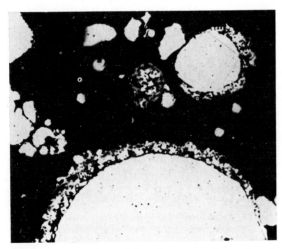

9 Water-atomized particle after annealing, showing incomplete reduction of oxide 'shell'
unetched, ×300

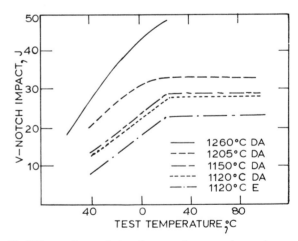

11 Effect of presinter temperature on impact performance in dissociated ammonia (DA) and endothermic (E) atmospheres during heating (after Cook[6])

Development of the alloy strategy

ALLOY DESIGN

In considering the alloying of powder-formed steels, it is to be expected that the effect of elements can be predicted by reference to wrought-steel technology. However, since powder particles are so very rapidly quenched, some differences will exist. 'Ingotism' is now on a submicroscopic scale, and grain size is generally very fine (e.g. ASTM 10–16). As yet, very little real, fundamental knowledge exists regarding the nature of the previous powder/particle interfaces and their influence in the fully dense structure.

It has been indicated that the control of non-metallic inclusions is critical in the achievement of a high standard of mechanical-property integrity. It is therefore imperative to choose alloy systems that produce the minimum number of non-metallic inclusions or, alternatively, to balance the alloy selection against the properties acceptable within the service environment of the component. Clearly, by this argument, it would be desirable to use nickel and molybdenum as the principal alloying elements, since their oxides are easily reduced, and to eliminate manganese and chromium because of the refractory oxide problem. However, in basic cost terms, nickel and molybdenum are exceedingly expensive alloy additions compared with manganese and chromium. From a hardenability point of view, manganese, chromium, and molybdenum are very efficient promoters, whereas nickel is not. The strategy, therefore, is to restrict nickel and molybdenum because of cost and manganese and chromium because of the oxide-forming tendencies. It is fortunate that the trend over recent years has been toward much leaner compositions of steels.

In designing on a hardenability basis, the objective is to produce as narrow a band as possible, in order to achieve consistent heat-treated properties from batch to batch. Unfortunately, a further variable to be considered is the variation that exists in the processes of different powder manufacturers. A specification that one manufacturer accepts readily may be a cause of some concern to another. However, to be meaningful, a specification should be capable of interpretation by whoever uses it. Therefore, in determining the appropriate alloy compositions, a degree of caution is advisable. It will be necessary in due course to reappraise the limits set. However, the limits that currently appear suitable will now be reviewed with an explanation of the underlying considerations.

Sulphur

Traditionally, sulphur gives rise to 'shortness' during hot working. During powder forging, cracking of the preform can take place but is mainly a function of its porosity. It is not clear whether sulphur exerts an additional influence. However, since some manganese will usually be present, inclusions of the iron/manganese sulphide type are formed, and, in fact, these are often associated with silicates or other non-metallic inclusions. With relatively high sulphur contents (free-cutting grades) discrete globules appear (Fig.12). To guard against the possibility of accidentally high levels, the sulphur limit is therefore usually set at 0·03%, which ensures that the general level of inclusions is not increased.

Nickel

The top limit for nickel is set as much by its cost as by other factors. It is usually accepted that up to 2% is a

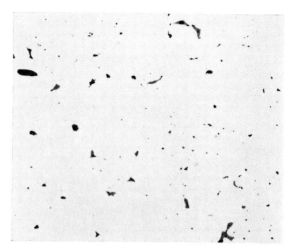

12 Sulphide inclusions in free-cutting plain-iron powder-forged steel   **unetched, ×300**

reasonable compromise between the attainment of high hardenability and low cost. It has been suggested[7] that 1% gives an optimum hardenability; however, for deeper hardening the extra 1% is undoubtedly beneficial. The problem of retained austenite is also a factor that inhibits the use of more than 2% nickel. At the lower end of the scale, a 0·25% addition gives a degree of hardenability enhancement, particularly in relation to the so-called 'synergistic' effect.[8,9]

Molybdenum

A minimum quantity of 0·25% is necessary to affect hardenability. Up to 0·7% is currently used, but beyond this level molybdenum becomes too expensive, except for special-purpose alloys. Also, at high levels there is a tendency to form a massive molybdenum carbide, which forms on relatively slow cooling.[9] This carbide poses a problem in machining, and on subsequent hardening, when a long soaking time is necessary in order to dissolve it completely. Apart from other considerations, a small percentage of molybdenum would be expected to be beneficial from the point of view of temper embrittlement.

Manganese

Although a high manganese content is extremely desirable from the point of view of hardenability and cost its affinity for oxygen limits its use. Its effect in promoting a high inclusion count is illustrated in Fig.13. To restrict oxide formation, an upper limit of about 0·55% is appropriate; in order to obtain any significant effect on hardening, not less than 0·25% is necessary. It has been suggested[9] that such a small percentage would be oxidized almost entirely. However, probe analysis results suggest that most of the manganese content is retained within the individual powder particle and that only the surface suffers depletion. The results for a 1%Mn steel are shown in Fig.14. The hardenability curves in Fig.15 illustrate the improvement in hardenability performance gained by the addition of a small amount of manganese, which tends to invalidate the 'total loss' theory.

Chromium

Since chromium has such a strong influence on hardenability, it would be desirable to include as much as possible, compatible with the conditions of oxide formation and

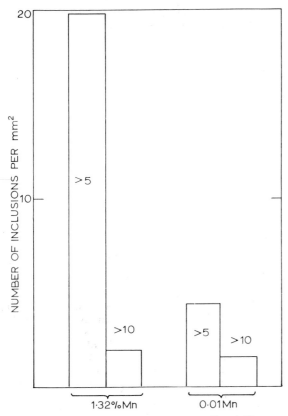

13 Number of inclusions greater than 5 and 10 μm as function of manganese content

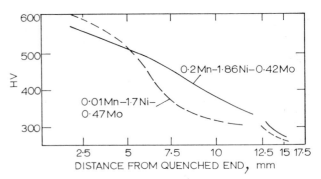

15 Jominy end-quench curves illustrating change in hardenability with the addition of 0·2% manganese (after Brown and Steed[3])

control. Although chromium is theoretically easier to reduce than manganese, it is known to form a coherent and tenacious oxide film. Since both elements appear to be additive when considering oxidation effects, it might be thought preferable to add chromium rather than manganese, in view of its lower reduction temperature. However, manganese is still the first choice, owing to its effect on sulphur and, hopefully, for its effect on impact transition, although as yet this latter attribute is obscured by the inclusion influence. Studies are currently being conducted to examine in more detail the relative merits of Cr *v.* Mn.[10] A minimum of 0·15% is used in order to contribute to the synergistic effect; the upper limit is not yet clearly defined,

but at the 0·5% level there is certainly a marked increase in inclusion content.

Copper

Copper is of lower cost than nickel but has only a small effect on hardenability. Up to 0·4% has been shown[7] to aid hardenability and would doubtless promote some corrosion resistance. Currently, very few of the compositions in use contain copper additions.

NON-METALLIC INCLUSION CONTENT

The inclusion structure in powder-forged steel consists of a few relatively large particles and rather more fine ones, which consititute a background effect (Fig.16). Fatigue performance responds primarily to the large inclusions and impact more to the background, although there is a response to an unusually high number of large particles.

The quantitative assessment of inclusions can be carried out using the QTM. An alternative is to determine the oxygen content; however, to be meaningful, a total oxygen determination must be supplemented by an inclusion count, because the former figure gives no indication of the presence of a few very large inclusions.

Steels in current use

As mentioned above, it is a relatively simple matter to alter the carbon content of powder-forging steels by admixture of graphite. This gives the possibility of changing the carbon content critically and therefore of altering the final performance of the material considerably. The knowledge of the service requirements of particular components analysed over a period of time has shown that it is possible to alter the hardenability and hence performance by quite small changes in carbon content, which must be matched

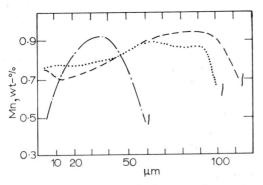

14 Electron microprobe analysis trace of manganese distribution within as-atomized particles: chemically determined average manganese level is 0·98%

Table 5 Mechanical properties of some powder-forged steels

Type	UTS, N/mm²	Elongation, %	RA, %	Impact, J
W4 *	700	18	45	28
W32†	690	19	47	32·5
W135	740	18	58	46·5
W78†	800	15	47	33

* Hardened and tempered
† Heat treated as 'core' of case-hardened part

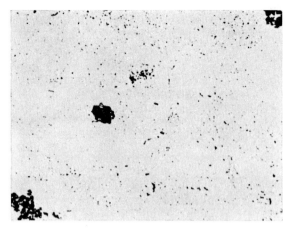

16 Inclusion structure of powder-forged steel containing 0·5%Mn and 0·6%Cr, showing large isolated inclusions and smaller 'background' particles
unetched, ×300

by appropriate control throughout processing. The powder forger has in effect a dimension of control not available to the wrought-steel metallurgist.

The steels in current use for powder-forged applications are listed in Table 2. The W4 material is intended primarily as a hardened and tempered material (e.g. for connecting rods), whereas the other three are intended to serve for both carburizing and through-hardening applications, depending upon the particular requirements.

In designing steels specifically for carburizing applications, a study of the components submitted for assessment revealed that the SAE 8600 hardenability band was the one most likely to be required. The approximate hardenability bands of the three materials used principally for carburizing applications are illustrated in Fig.17, from which it will be seen that steel W78 approximates to SAE 8600. The mechanical properties obtained from powders made to the chemical specifications of Table 2 are shown in Table 5.

Summary

In designing steels for powder forging, the overriding factor is that of cost. In heat-treated components 'strength'

is a function of hardenability but the elements that promote deep hardening performance at lowest cost also promote oxide inclusions in the original powder. Modification of the processing route in terms of high-temperature preheating of the preforms offers the possibility of an added dimension in oxide control, but this upsets the basic economics. The alloy design is thus a delicate balance of the lower-cost elements to promote hardenability, but limited by their oxide-forming tendencies, and additions of elements with less oxygen affinity, but restricted by cost and other technical factors.

The fact that powder-forged structures are isotropic in mechanical properties is a very strong point in their favour, but this is not always appreciated by designers who are accustomed to using the longitudinal properties of wrought steel as a measure of materials performance. This is a false premise, and it has already been shown[11] that the fatigue performance of powder-forged components when tested as components may be superior to the equivalent articles produced by other means. An extension of this conclusion is the proposition that if the service requirements of a component could be accurately defined the powder forging of steels could be designed to meet these parameters. It is probable that because wrought steel has such anisotropic properties the 'factors of safety' commonly used are much too high; a more homogeneous powder-based steel could well overcome these disadvantages.

Acknowledgment

The author wishes to thank Dr T. L. Johnston, Technical Director of GKN Group Technological Centre, for permission to present this paper and also for much helpful discussion during its writing.

References

1. G. T. BROWN and T. B. SMITH: 'Modern developments in powder metallurgy', Vol.7, (ed. H. H. Hausner and W. E. Smith), 9; 1974, Princeton, NJ, MPIF and APMI.
2. K. H. MOYER: 'The effect of preform density on the impact properties of atomized iron P/M forgings', Proc. 1971 Fall Power Metallurgy Conference, APMI and MPIF.
3. G. T. BROWN and J. A. STEED: *Powder Met.*, 1974, **17**, (33), 157.
4. R. T. CUNDILL *et al.*: *ibid.*, 1970, **13**, (26), 165.
5. P. LINDSKOG and S. E. GREK: 'Modern developments in powder metallurgy', Vol.7, (ed. H. H. Hausner and W. E. Smith), 285; 1974, Princeton, NJ, MPIF and APMI.
6. J. P. COOK: *Progress Powder Metallurgy*, 1974, **30**, 173.
7. Y. E. SMITH and R. PATHAK: 'New hardenability data for application in low alloy ferrous powder forging', Proc. 1971 Fall Powder Metallurgy Conference, APMI and MPIF.
8. G. T. BROWN: *Powder Met.*, 1974, **17**, (33), 116.
9. S. MOCARSKI: 'Modern developments in powder metallurgy', Vol.7, (ed. H. H. Hausner and W. E. Smith), 303; 1974, Princeton, NJ, MPIF and APMI.
10. Research in progress at GKN.
11. G. T. BROWN and J. A. STEED: *Powder Met.*, 1973, **16**, (32), 405.

General references
J. W. WISKER and P. K. JONES: 'Modern developments in powder metallurgy', Vol.7, (ed. H. H. Hausner and W. E. Smith), 33; 1974, Princeton, NJ, MPIF and APMI.
P. K. JONES: *Powder Met.*, 1970, **13**, (26), 114.
G. T. BROWN: 'The powder forging process: application to high and low cost material', Proc. 2nd National Scientific–Technical Conference, 1972, Bulgaria.

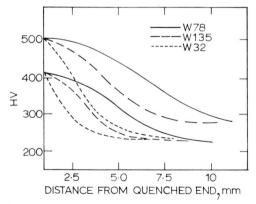

17 Approximate hardenability bands of three carburizing grades of powder-forged steel; carbon range is 0·17–0·23%

Prealloyed, Premixed, and Hybrid Sintered Steels

D. J. BURR*

ABSTRACT

A comparison is made between steel compacts prepared from prealloyed, premixed elemental and mixtures of prealloyed plus elemental Hybrid powders. In order to obtain optimal properties, a selection is to be made concerning the three types of powders and the processing steps. Results of original work are compared with the data presented in the literature.

Introduction

Comparisons[1,2] of the properties of Type 4600 sintered steels made from prealloyed powder and mixed elemental powders respectively differ significantly.

Heck[1] reported that with a density in the range 6.5-7.2 g/cm^3, P/M steels in the as-sintered condition had superior tensile properties if made from prealloyed powder, but that steels given a post sintering heat treatment had better properties if made from mixed elemental powders. As would be expected, the steels made from prealloyed powder possessed relatively uniform structure, but those made from elemental powders were rather heterogeneous.

Using different starting powders, Holcomb and Lovenduski[2] found that with a density in

the range 7.0 to 7.4 g/cm^3, materials produced using prealloyed powders exhibited higher strengths than mixed elemental materials of the same nominal composition; not only in as as-sintered condition but also after heat treatment. They also examined a Type 4600 P/M steel composition by adding nickel powder to a low-alloy prealloyed powder. This gave properties similar to or better than the fully prealloyed powder. By replotting their results as a function of compacting pressure, Holcomb and Lovenduski concluded that prealloyed powder and mixed elemental powders would give similar properties in steels produced by a fixed processing route.

It was suggested by Holcomb and Lovenduski that the differences between their results for heat treated steels and those of Heck were influenced by the latter's particular prealloyed powder, which contained more carbon (0.15%) and less molybdenum (0.25%) than typical commercial powders, and also tended to laminate at high compacting pressures.

Some further information relevant to the above mentioned questions is available from a series of tests on steels made from Type 4600 and other compositions with a different selection of prealloyed and elemental powders. It involved high quality commercial prealloyed powders from a source which differed from those utilized by the referenced investigations. The Type 4600 powder did not

*International Nickel Ltd., Birmingham, England.

suffer from the drawbacks cited above; it had a molybdenum content of 0.5% and less than 0.01% carbon; no lamination could be found in the compacts. The work also differed from previous investigations in that a constant pressing procedure was employed. This of course resulted in a higher density for the mixed elemental materials. But it was considered that a comparison of steels made by a fixed processing route would provide the most useful information to manufacturers. It is significant that the steels tested included samples made with a low alloy prealloyed powder mixed with additional elemental nickel, i.e. similar to the steels mentioned by Holcomb and Lovenduski. Such materials will be referred to as "hybrid" steels. They were included because it was recognized that the prealloyed powder and mixed elemental powders represented two extremes, and that the optimum steel structure might be achieved with an intermediate selection of powders.

An additional feature of the work to be reported below was the inclusion of impact tests on the steels, which included low nickel compositions as well as Type 4600.

Experimental Procedure

The raw materials used are listed in Table 1.

Seven steels were made, in three compositional groups. The first group had a nominal nickel content of 1.9% and was composed of a prealloyed steel made with Type 4600 powder, a hybrid steel made with low-alloy prealloyed powder plus 1.5% elemental nickel, and a mixed elemental steel. As shown in Table 2, both had almost the same composition but the hybrid steel had a high manganese content. A further comparison of hybrid and mixed elemental materials with constant manganese was therefore provided by the second group (nominal nickel 0.9%) consisting of a steel made with low alloy powder plus 0.5% nickel,

TABLE 1 Raw Materials Used

Atomized Prealloyed Powders

Type 4600	Fe - 1.9% Ni - 0.5% Mo - 0.1% Mn - <0.01%C 97% / −100 mesh.
Low alloy	Fe - 0.4% Ni - 0.5% Mo - 0.4% Mn - <0.01%C 96% / −100 mesh

Elemental powders

Iron (Atomized)	>99% Fe −100 mesh
Nickel (Type 123)	>99.7% Ni 4-7 μm
Graphite	99.7%C −350 mesh
Ferromolybdenum	Fe - 68% Mo −100 mesh
Ferromanganese	Fe - 70% Mn −100 mesh

a mixed elemental equivalent. A third group (nominal nickel 0.4%) was made to give a futher comparison of low alloy prealloyed steel with a mixed elemental equivalent. The three groups thus gave the additional advantage of comparisons materials with three nickel levels.

Carbon was added to all steels as elemental graphite, the level of addition being sufficient to give 0.5% in the finished bars. The prealloyed or elemental metallic powders were blended with the graphite and a lubricant (0.75 ^{w}/o zinc stearate) for one hour in a double-cone mixer. It will be noted from Table 1 that in the mixtures of elemental powders ferromolybdenum and ferromanganese were used rather than elemental molybdenum and manganese. Previous work[3] has shown that the use of ferromolybdenum causes no change of properties in sintered elemental steels, and ferromanganese is already widely used as an addition.

The pressing and sintering routes were as follows:

Pressing	45 tsi (620 N/mm^2)
Dewaxing	1/2h/500 C
Presintering	1/2h/800 C
Repressing	45 tsi (620 N/mm^2)
Sintering	1/2h/1120 C

TABLE 2 Composition of Steels Used

Steel	Description	Composition wt%			
		Ni	Mo	Mn	C
1	Prealloyed (Type 4600)	1.86	0.53	0.1	0.50
2	Hybrid (low alloy pre-alloyed + 0.5% Ni)	1.93	0.56	0.4	0.50
3	Mixed elemental	1.93	0.53	0.1	0.48
4	Hybrid (low alloy pre-alloyed + 1.5% Ni)	0.93	0.56	0.4	0.45
5	Mixed elemental	0.93	0.56	0.4	0.48
6	Prealloyed (low alloy)	0.43	0.56	0.4	0.51
7	Mixed elemental	0.43	0.56	0.4	0.46

Dewaxing and sintering occurred in cracked ammonia (dewpoint −40 C) in a laboratory tube furnace.

The steels were tested in the as-sintered condition and after heat treatments consisting of austenitising 1/2h at 900 C, oil-quenching, and tempering for 1/2h at 150, 350 and 550 C. Duplicate tensile tests were performed using "dog-bone" (BS 2590) specimens. An extensometer recorded the 0.2% proof stress but elongation was determined by the crude but most widely used method of fitting the broken specimens together and measuring across the fracture. Unnotched Charpy impact values were determined on specimens machined from rectangular bars. Metallographic examination was carried out on cross sections of selected tensile test samples, and was supported by electron probe microanalysis.

Results

Table 3 shows densities and tensile properties determined on all specimens.

The order or merit for as-sintered tensile strength in the first group (1.9% nickel steels) was (1) hybrid steel, (2) prealloyed steel, and (3) mixed elemental steel. This order was confirmed by the tests on the 0.9% and 0.4% nickel steels in the other groups.

In the heat treated condition the order was completely reversed, i.e., in the first group the elemental specimens gave best properties, followed by prealloyed steel and finally by hybrid steel. The results from the other groups were in accord with this as illustraed in Fig. 1. Many of the samples showed no measurable tensile elongation, but the elemental steels consistently had more ductility than the other types.

By comparing the results for the different compositional groups it can be seen that increasing nickel content (in either elemental or pre-alloyed form) brought an increase in UTS. This was true for as-sintered as well as for heat treated steels. Either an increase in nickel content or the use of elemental powders increased the relative value of tempering at 150 C and this, rather than 550 C, became the most effective temperature for achieving high strength.

The impact values shown in Table 4 indicate that, with both as-sintered and also with heat treated samples, the elemental steels

TABLE 3 Densities and tensile properties of the steels

| | | | | Tensile properties As sintered | | | Tensile properties - heat treated ½h/900°C, oil quenched & tempered | | | | | | | | |
| | | | | | | | Temper ½h/150°C | | | Temper ½h/350°C | | | Temper ½h/550°C | | |
Alloy	Ni %	Type	Density g/cm³	0.2% PS psi x 10³	UTS	Elong %	0.2% PS	UTS psi x 10	Elong %	0.2% PS psi x 10³	UTS	Elong %	0.2% PS psi x 10³	UTS	Elong %
1	1.9	Prealloyed	7.25	58	73	2	141	143	<0.5	*	130	<0.5	106	112	<0.5
2	1.9	Hybrid	7.25	62	81	2	*	138	<0.5	*	113	<0.5	*	116	<0.5
3	1.9	Elemental	7.45	38	71	8	141	178	1	128	141	1.5	91	99	3.5
4	0.9	Hybrid	7.25	55	71	2	*	90	<0.5	*	86	<0.5	101	106	<0.5
5	0.9	Elemental	7.45	32	59	5	*	129	0.5	*	113	<0.5	90	93	1
6	0.4	Prealloyed	7.25	52	59	1.5	*	78	<0.5	*	83	<0.5	*	104	<0.5
7	0.4	Elemental	7.45	35	55	4.5	91	107	0.5	*	106	<0.5	81	83	1

*Elongation too low for measurement of 0.2% proof stress

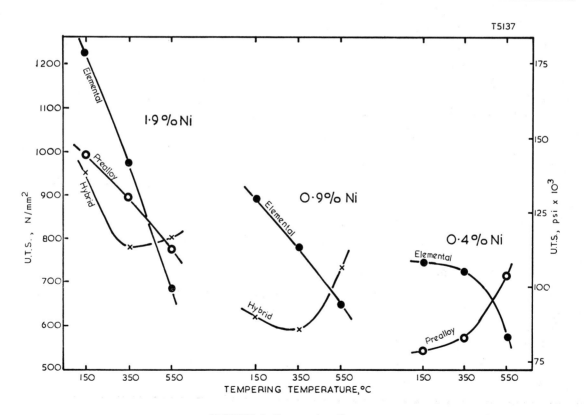

FIGURE 1 Tempering Curves

TABLE 4 Charpy unnotched impact values (Joules)

Alloy	Type	Nickel %	As-sint.	H.T.* (temp 150°C)
1	Prealloyed	1.9	17	15
2	Hybrid	1.9	16	16
3	Elemental	1.9	27	27
4	Hybrid	0.9	19	13
5	Elemental	0.9	24	22
6	Prealloyed	0.4	17	17
7	Elemental	0.4	33	25

*Heat treated ½h/900°C/oil quench/½h/150°C

were markedly superior to the pre-alloyed and hybrid versions, which did not differ significantly from each other.

The microstructures of the prealloyed and elemental 1.9% Ni steels in the as-sintered condition and after tempering at 150 C were essentially the same as those described by Fischer and Heck[4] for equivalent single pressed and sintered materials. The present alloys however had a slightly larger grain size. The as-sintered structures consist of pearlite and ferrite, finely and fairly uniformly distributed in the pre-alloyed sample but segregated into irregular regions of 10-20 μm diameter in the elemental sample. After quenching and tempering the prealloyed material was uniformly bainitic, while the elemental steel was very heterogenous with martensite, bainite/pearlite and ferrite regions. These structures have been well illustrated by Fischer and Heck. Microprobe analyses confirmed the heterogeneous nickel distribution which they reported for elemental steel, and the only new structural observation in this material was of isolated small, hard, white-etching areas. These were shown to consist predominantly of molybdenum and nitrogen, and were probably due to segregated

ferromolybdenum forming nitrides during sintering.

The microstructure of the hybrid 1.9% Ni steel was almost identical to that of the elemental steel, in both as-sintered and heat treated conditions. Examination of the 0.9 and 0.4% Ni steels also indicated that the microstructure of the hybrid steel was related closely to that of the elemental materials. Microprobe analysis indicated that the heterogeneity in nickel distribution in the 1.9% nickel hybrid steel was almost as great as for elemental steel.

Discussion

By comparing prealloyed and mixed elemental powders specimens, it is apparent from data in Table 3 and Fig. 1 that the relationship between the properties of the different steels made by a constant processing route are similar to those obtained by Heck[1] using a constant density. With a constant processing route, the greater compressibility of the mixed elemental powders accentuages their superiority in heat treated steels, but this density advantage does not overcome the lower as-sintered strength exhibited by these materials compared to their prealloyed counterpart. The present results are also consistent with those of Holcomb and Lovenduski[2] for as-sintered steels but not for heat treated materials, in which these authors found prealloyed powders to give superior properties at a constant density, indicating that approximate equality would occur if a constant processing route was adopted. This disagreement between the present results plus Heck's results on the one hand, and Holcomb and Lovenduski's on the other, does not appear to be due to the differences in steel composition, since both the present results and those of Holcomb and Lovenduski showed self-consistent trends over a range of compositions; it is clearly not due to the particular characteristics of Heck's prealloyed powder cited earlier. It is possible however that other characteristics of the prealloyed powders are important, or that the properties are affected by differences in the heat treatment schedules or even processing routes. Work on other steels has indicated that varying degrees of heterogeneity, determined by mixing procedure and sintering time, can produce significantly different properties in specimens with mixed elemental materials[5].

The reversal in relative behavior from as-sintered to heat-treated conditions observed here and in Heck's tests can not easily be explained. Experience with higher alloyed compositions indicates a rather complex relationship between properties and hetero-geneous microstructure[5]. However, considering that at nearly full density there is little difference in behavior between heat treated prealloyed and elemental steels[1] a possible explanation deals with the soft regions in heterogeneous elemental sintered steels which are critical factors. While reducing the strength of as-sintered samples, they might improve stronger but less ductile heat treated

materials by helping to accommodate the stress-raising effect of pores. When porosity is decreased, e.g. by hot forging, this benefit disappears.

The properties of the hybrid steel, made with a mixture of elemental and prealloyed powder, are most interesting; in the as-sintered condition the hybrid steel has better strength than the prealloyed or mixed elemental materials. However, its heat treated properties were lower than those of the alternative materials. Holcomb and Loven-duski[2] did not discuss their results for comparable steels in detail, but their fixed density data appear to be consistent with the trends observed here with a fixed processing route. It would be interesting to see a direct comparison of the properties of the hybrid steel with those of a steel of the same composition made with diffusion bonded powder, since this is partially prealloyed and represents another form of hybrid material. Lindskog and Skoglund have demonstrated that this too gives superior properties to full prealloyed powder in as-sintered steels[6].

An observation at the present work, common to both as-sintered and heat treated conditions indicated that the properties of the hybrid steels were closer to those of fully prealloyed steel than to those of the mixed elemental type. In constrast to this, the microstructure of the hybrid steel was very similar to that of mixed elemental steel. A possible explanation of the observed behavior might be found in terms of the different physical properties of the elemental iron and pre-alloyed Fe-Ni-Mo-Mn powders used as a basis for the different steels. Property advantages shared by the two pre-alloyed powders might be more important than microstructural effects. On the other hand both iron and prealloyed powders were produced by a similar route, and it does not appear reasonable to suggest that the pre-alloyed constituent controls the behavior of hybrid steels in view of the clear effect of the elemental nickel addition on strength. (Compare steels 2 and 4 - Tables 2 and 3).

It is not quite clear whether the superiority of elemental steel over hybride steel in the heat-treated condition is entirely due to its density advantage, bearing in mind that Heck (but not Holcomb and Lovenduski) found a similar superiority of elemental over prealloyed steel even at equal densities. Another possibility is that it is advantageous to add elemental molybdenum (or, rather ferrmolybdenum) instead of having it fully prealloyed; the observed molybdenum nitride however was not distributed in a manner that would contribute to strength. Other work was shown that no such advantage follows from separate manganese or ferromanganese additions.

Density is known to be a more significant factor for impact properties than for tensile properties. The higher impact values recorded in Table 4 for elemental steel compared with the two other types, in both as-sintered and heat treated conditions, could thus simply be a function of density. This emphasizes the advantage of compressibility offered by the mixed elemental powders.

Conclusions

With the constant double pressing and sintering route employed in this work, the best sintered steel properties are obtained with the following raw material types

Condition	Property	Best Raw Material
As-sintered	Tensile	Mixed pre-alloyed and elemental
As-sintered	Impact	Mixed elemental
Heat Treated	Tensile	Mixed elemental
Heat Treated	Impact	Mixed elemental

For the range of sintered steels investigated, (0.4 to 1.9% nickel) it is thus advantageous to add all or part of the alloying additions as elemental powder. However, it must be noted that this work did not include a comparison with partially prealloyed diffusion bonded powders, and also that variations in processing may modify the steel behavior.

References

1. F. W. Heck, Modern Development in P/M, *5*, 453-469, (1971).
2. R. T. Holcomb and J. Lovenduski, Modern Development in P/M, *8*, 85-107, (1974).
3. D. J. Burr & G. M. Krishnamoorthy, Powder Met. *16*, p. 33 (1973).
4. J. J. Fischer and F. W. Heck, Modern Development in P/M, *5*, 471-480, (1971).
5. R. G. Faulkner and D. J. Burr, To be published.
6. P. Lindskog and G. Skoglund, Powder Met. Euro Symp. Supplement, 375-396, (1971).

Fatigue Properties of Sintered Nickel Steels*

A. F. KRAVIC** and **D. L. PASQUINE****

ABSTRACT

Few data on the fatigue behavior of sintered metals, that is, P/M parts, have been published. This paper summarizes the results of a comprehensive study on the fatigue properties of sintered nickel steels made from mixed elemental powders. Sintered nickel steels possess distinct fatigue limits, occurring between 10^6 and 10^8 cycles. The fatigue ratio of sintered nickel steel has an average value of 0.4 up to 150,000 psi tensile strength. The fatigue limit of sintered nickel steel increases with higher sintered density, alloy content, and any heat treatment that increases tensile strength.

INTRODUCTION

Sintered nickel steels are gaining in popularity for many industrial products, notably autos, appliances and business machines. Design engineers, when selecting a sintered steel for use in the production of a particular part, must consider all of the physical and mechanical properties for the various sintered steels before an intelligent decision can be made. At present there is a lack of comprehensive data on the relative fatigue properties of sintered steels.(1) As a result, the International Nickel Company initiated a research program to supply this much needed information for sintered nickel steels.

This program was divided into two basic phases covering some of the more common factors which may influence the fatigue performance of sintered metals: (1) the effect of composition and heat treatment at varying sintered densities and (2) the effect of external notches. The alloy contents selected for study were those presently being standardized by the Powder Metallurgy Parts Association, i.e. 2, 4 and 7 per cent nickel.

Fatigue studies were carried out with small laboratory fatigue specimens tests in rotary bending, supplemented by tension, hardness, and room-temperature impact tests.

PROCESS CONDITIONS

Electrolytic iron powder was mixed with carbonyl nickel powder in varying percentages by weight (2, 4 and 7%). Table I gives the chemical and physical properties of these powders. Zinc stearate (1% by weight of the total mix) was added as a compacting lubricant in all mixes, which were then blended for 30 minutes.

All compacts were pressed in double-acting hardened steel dies to 6.6 ($\pm$0.1) g/cc green density. Sintering was performed in a three zone, mesh belt, gas fired sintering furnace using an endothermic atmosphere (dew point +28°F). All specimens were pre-heated 24 minutes at 1500F and sintered 30 minutes at 2050F. The cooling rate after sintering was 40-50F per minute to room temperature. Those samples requiring densities over 6.6 g/cc were presintered at 1650F for 30 minutes, re-pressed to the requisite density and then sintered.

Natural graphite was used for carbon additions. Some loss of carbon occurs during sintering, due to oxygen content of the powders. Therefore, a

---

* Preprinted as Technical Paper EM 68-140 by The American Society of Tool and Manufacturing Engineers

** The International Nickel Company, Inc.

International Journal of Powder Metallurgy 5 (1) 1969

TABLE I.   Raw Powders Used

Element	Iron	Nickel
Type	Electrolytic	Carbonyl
Apparent Density, g/cc	2.74	1.61
Flow, sec.	30.2	no flow
Screen Analysis, %		
— 80 + 100	3.6	—
— 100 + 150	11.2	trace
— 150 + 200	18.3	0.4
— 200 + 250	15.6	0.9
— 250 + 325	17.5	2.8
— 325	33.6	95.9
Chemical Analysis, %		
Fe or Ni	99.0	99.6
Carbon	0.02	0.14
Oxygen	0.32	0.10

set of trial bars was made to determine the exact loss. The addition of 0.55% graphite resulted in a final carbon content of 0.48% while 1.0% graphite resulted in 0.80% final carbon.

Heat treatment consisted of quenching in oil from 1600F followed by tempering for 30 minutes at temperatures ranging from 400 to 1200F. All specimens were air cooled from the tempering temperature.

All test specimens were accurately machined from sintered rectangular bars. Tensile test bars and Charpy impact test bars were made in accordance with American Society for Testing and Materials Standards E8-61T and E23-64. The impact bars were unnotched. All test specimens were less than 0.5 inch section size.

Fatigue tests were conducted on conventional R. R. Moore rotating-beam machines at speeds of 3,600 or 10,000 rpm. The smooth, i.e. unnotched, and notched (2.2 Kt) specimens employed are shown in Fig. 1. All unnotched specimens were carefully ground and lapped longitudinally to remove all traces of circumferential scratches before testing The notched specimens were formed with a fine abrasive grinding wheel (Norton No. 38A80-K8V), freshly dressed to produce the proper contour while great care was taken in the final grinding steps to minimize residual surface stress. Eight fatigue specimens were tested to establish each S-N (applied stress versus number of cycles to failure) curve. The fatigue limit was taken as the maximum stress borne for 100 million reversals without failure. The curves were derived as *lines of best fit* with the exception that the fatigue limit was based on test results which showed no failure after 100 million cycles and was always less than the individual test results of specimens which had failed.

AS-SINTERED MATERIALS   Although sintered nickel steels are more generally used in the quenched and tempered condition to secure the maximum benefit from the alloy content,

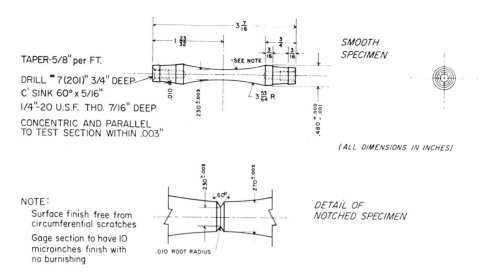

FIGURE 1.   R. R. Moore fatigue specimen dimensions

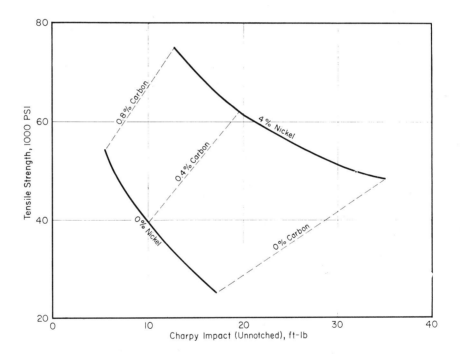

FIGURE 2.   Toughness of as-sintered steels with and without nickel at 7.0 g/cc density

there still remains a wide field of usefulness for as-sintered nickel steels. (3) Higher strengths can, of course, be obtained by increasing the carbon content of plain carbon steel sinterings, but the corresponding losses in ductility and toughness limit the practical use of carbon as a strengthener. It is, therefore, often better to keep the carbon content relatively low and to achieve the desired improvement in strength and toughness through appropriate nickel additions. Fig. 2 illustrates this point.

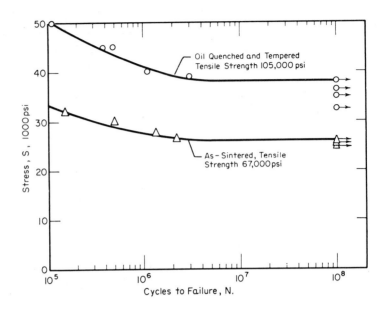

FIGURE 3. S-N Diagrams representing fatigue behavior of 7.0 g/cc density, 4 Ni–0.48 C steels

TABLE II. Effect of Sintered Density on the Mechanical Properties of As-Sintered 4% Nickel 0.48% Carbon Steel

Sintered Density (g/cc)	Brinell Hardness Number	Tensile Strength (ksi)	Yield Strength 0.2% Offset (ksi)	Elong. (%)	Fatigue Limit (ksi)	Fatigue Ratio	Charpy Unnotched Impact (ft-lb)
6.6	95	48.8	27.3	3.0	19.0	0.39	10
7.0	115	66.8	33.2	6.5	26.0	0.39	18
7.2	130	74.4	33.4	10.0	30.5	0.41	44

One of the characteristics of the fatigue behavior of wrought steels is that the S-N curve usually shows a distinct fatigue limit. This is most marked in wrought plain carbon steels and usually occurs between 10^5 and 10^7 cycles. A typical S-N curve for an as-sintered nickel steel is shown in Fig. 3. As-sintered nickel steels possess distinct fatigue limits occurring between 10^6 and 10^8 cycles.

When dealing with wrought materials, designers are well aware of the significance and the limitations of nominal mechanical properties. When dealing with powder metal (P/M) parts, however, another and highly significant variable, density or porosity, must be taken into account. The variation in mechanical properties due to changes in sintered density are shown in Table II. All mechanical properties, including the fatigue limit, increase with increased sintered density.

TABLE III. Effect of Final Carbon Content on the Mechanical Properties of As-Sintered 4% Nickel Steel at 6.6 g/cc Density

Carbon Content (%)	Brinell Hardness Number	Tensile Strength (ksi)	Yield Strength 0.2% Offset (ksi)	Elong. (%) (ksi)	Fatigue Limit	Fatigue Ratio	Charpy Unnotched Impact (ft-lb)
0	75	35.3	19.4	6.0	17.0	0.48	15
0.48	95	48.8	27.3	3.0	19.0	0.39	10
0.80	110	56.4	36.5	1.0	25.6	0.45	8

TABLE IV. Effect of Nickel Content on the Mechanical Properties of As-Sintered 0.48% Carbon Steels

Sintered Density (g/cc)	Nickel Content (%)	Brinell Hardness Number	Tensile Strength (ksi)	Yield Strength 0.2% Offset (ksi)	Elong. (%)	Fatigue Limit (ksi)	Fatigue Ratio	Charpy Unnotched Impact (ft-lb)
6.6	2	80	40.0	23.0	4.0	16.5	0.41	11
6.6	4	95	48.8	27.3	3.0	19.0	0.39	10
6.6	7	120	63.3	33.2	2.0	25.0	0.40	9
7.2	2	120	62.1	30.5	13.0	27.0	0.43	48
7.2	4	130	74.4	33.4	10.0	30.5	0.41	44
7.2	7	160	94.1	40.4	8.0	34.0	0.36	30

Considerable efforts have been made to correlate the results of fatigue tests with other mechanical properties. Since it is generally accepted that the fatigue strength of a metal is related to its tensile strength, it follows that any factor which increases tensile strength, such as the addition of alloying elements, will result in an increase in fatigue strength. To illustrate the effect of chemical composition on the properties of as-sintered nickel steels, Tables III and IV are provided. Here the fatigue limit increases with the nickel and carbon contents, as do strength and hardness; however, ductility and toughness decrease at higher alloy contents.

The relation between the fatigue limit of sintered nickel steels and their tensile strength, that is, the *fatigue ratio*, is shown in Fig. 4. The smooth fatigue ratio of as-sintered nickel steels has an average value of 0.4 up to a tensile strength of 95,000 psi. This ratio is essentially independent of alloy content and density level.

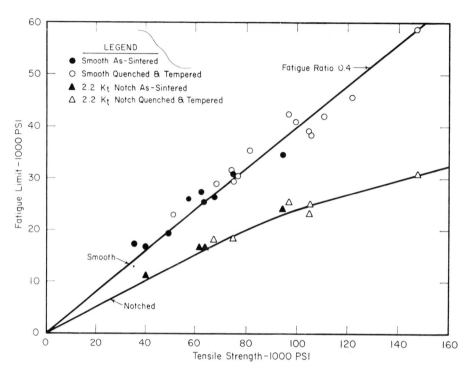

FIGURE 4.   Relationship between fatigue limit and tensile strength (fatigue ratio) of sintered nickel steels

HEAT TREATED
MATERIALS

Sintered nickel steels respond especially well to the ordinary process of heat treatment and provide a wide range of mechanical properties to meet varied design requirements.(3) The hardening characteristics imparted by nickel are reflected in excellent mechanical properties after mild quenching and tempering treatments. The nickel steels provide, at a given strength, greatly improved impact resistance over compositions strengthened solely with carbon. This benefit is illustrated in Fig. 5.

A typical S-N curve for one of the quenched and tempered sintered nickel steels studied, selected at random, is also shown in Fig. 3. Like as-sintered products, heat treated sintered nickel steels also possess distinct fatigue limits occurring between 10^6 and 10^8 cycles.

Tables V and VI demonstrate the effects of density and nickel content on the properties of quenched and tempered sinterings. Again, all mechanical properties increase with higher sintered density. Higher nickel contents raise hardness and strength including fatigue but decrease ductility and toughness. Increasing the tempering temperature, as shown in Table VII improves toughness and elongation to fracture but hardness, strength and the fatigue limit decrease.

A plot of the fatigue ratio, Fig. 4, indicates an average smooth value of 0.4 up to 150,000 psi tensile strength. Thus the average fatigue ratio for sintered nickel steel is 0.4 which is apparently independent of density level,

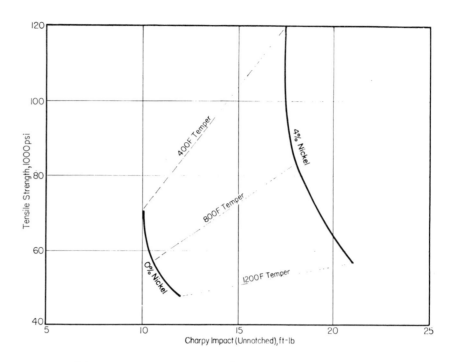

FIGURE 5. Toughness of heat treated 0.40% carbon sintered steels with and without nickel at 7.0 g/cc density

TABLE V. Effect of Sintered Density on the Mechanical Properties of Oil Quenched and Tempered 4% Nickel 0.48% Carbon Steel

Sintered Density (g/cc)	Brinell Hardness Number	Tensile Strength (ksi)	Yield Strength 0.2% Offset (ksi)	Elong. (%)	Fatigue Limit (ksi)	Fatigue Ratio	Charpy Unnotched Impact (ft-lb)
6.6	140	81.2	58.7	2.0	35.0	0.43	10
7.0	190	105.0	67.7	3.0	38.0	0.36	17
7.2	210	121.2	71.5	3.5	45.0	0.37	27

Quenched from 1600F, tempered 30 minutes at 400F

alloy content, and state of heat treatment and therefore, can be used to predict the fatigue behavior of other sintered nickel steels.

Tables V and VI demonstrate the effects of density and nickel content on the properties of quenched and tempered sinterings. Again, all mechanical properties increase with higher sintered density. Higher nickel contents raise hardness and strength including fatigue but decrease ductility and toughness. Increasing the tempering temperature, as shown in Table VII

TABLE VI. Effect of Nickel Content on the Mechanical Properties of
Oil Quenched and Tempered 0.48% Carbon Steels

Sintered Density (g/cc)	Nickel Content (%)	Brinell Hardness Number	Tensile Strength (ksi)	Yield Strength 0.2% Offset (ksi)	Elong. (%)	Fatigue Limit (ksi)	Fatigue Ratio	Charpy Unnotched Impact (ft-lb)
6.6	2	130	74.8	55.7	3.0	29.0	0.39	11
6.6	4	140	81.2	58.7	2.0	35.0	0.43	10
6.6	7	160	96.2	60.9	1.0	42.0	0.44	6
7.2	2	175	104.5	65.1	6.0	38.5	0.37	35
7.2	4	210	121.2	71.5	3.5	45.0	0.37	27
7.2	7	265	147.0	82.7	3.0	58.0	0.40	20

Note: Quenched from 1600F, tempered to 30 minutes at 400F

TABLE VII. Effect of Tempering Temperature on the Mechanical Properties of
Oil Quenched 4% Nickel 0.48% Carbon Steels

Sintered Density (g/cc)	Tempering Temp. (°F)	Brinell Hardness Number	Tensile Strength (ksi)	Yield Strength 0.2% Offset (ksi)	Elong. (%)	Fatigue Limit (ksi)	Fatigue Ratio	Charpy Unnotched Impact (ft-lb)
6.6	400	140	81.2	58.7	2.0	35.0	0.43	10
6.6	600	135	74.5	55.9	1.5	31.0	0.42	11
6.6	800	125	68.0	53.2	2.0	28.5	0.41	12
6.6	1200	90	51.4	36.6	3.5	22.5	0.44	13
7.2	400	210	121.2	71.5	3.5	45.0	0.37	27
7.2	600	190	110.0	67.3	5.0	41.5	0.38	28
7.2	800	165	99.3	66.8	6.0	40.5	0.41	29
7.2	1200	130	76.8	46.7	8.0	30.0	0.39	31

Note: Quenched from 1600F, tempered 30 minutes

improves toughness and elongation to fracture but hardness, strength and the fatigue limit decrease.

A plot of the fatigue ratio, Fig. 4, indicates an average smooth value of 0.4 up to 150,000 psi tensile strength. Thus the average smooth fatigue ratio for sintered nickel steel is 0.4 which is apparently independent of density level, alloy content, and state of heat treatment and therefore, can be used to predict the faitgue behavior of other sintered nickel steels.

EFFECT OF AN
EXTERNAL NOTCH
The detrimental effect of stress raisers on fatigue strength and the importance of avoiding or reducing stress concentrations in parts subjected to

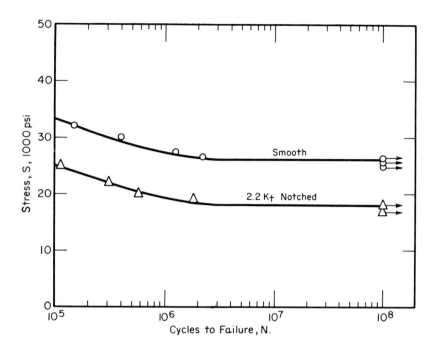

FIGURE 6. Effect of an external notch on the fatigue properties of as-sintered 44 Ni–0.48 C steel at 7.0 g/cc density

fluctuating stress are well known. In this phase of the investigation, notched fatigue curves were established for some of the materials tested in the unnotched condition. Typical notched fatigue curves for as-sintered and quenched and tempered material (Fig. 6 and 7) exhibited the same general shape, that is, a continuously decreasing stress with increasing number of cycles to failure with the notched fatigue limit occurring between 10^6 and 10^8 cycles.

Table VIII illustrates the effect of an external notch on as-sintered products while Table IX shows the effect of an external notch on heat-treated products. In all cases the fatigue limit is lowered by the addition of a stress concentration. The fatigue ratio also decreases from the average smooth value of 0.4 to 0.26 — 0.28 for as-sintered material and 0.21 — 0.26 for heat-treated material, indicating that heat treatment increases notch sensitivity. Similar effects have been reported for wrought and cast steels.

Sintered nickel steels can also be compared by their notch sensitivity (q), used to express the difference between the theoretical stress concentration factor (K_t) and the actual factor (K_f). By definition:

$$q = \frac{K_f - 1}{K_t - 1} \qquad (4)$$

Thus when (q) is progressively lowered in numerical value, a material is said to be less and less notch sensitive. A knowledge of the value of (q) should permit better consideration of the calculated stress concentration

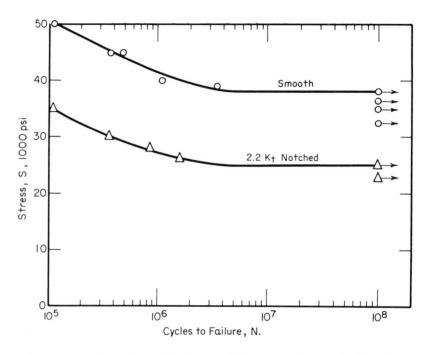

FIGURE 7. Effect of an external notch on the fatigue behavior of quenched and tempered 4 Ni–0.48 C sintered steel at 7.0 g/cc density

TABLE VIII. Effect of an External Notch on As-Sintered Sintered Nickel Steels

Sintered Density (g/cc)	Nickel Content (%)	Carbon Content (%)	Tensile Strength (ksi)	Smooth Fatigue Limit (ksi)	Smooth Fatigue Ratio	Notched Fatigue Limit (ksi)	Notched Fatigue Ratio	Fatigue Notch Factor (Kf)	Fatigue Notch Sensitivity (q)
6.6	2	0.48	40.0	16.5	0.41	11.0	0.28	1.50	0.42
6.6	7	0.48	63.3	25.0	0.40	16.5	0.26	1.51	0.43
7.0	4	0.48	66.8	26.0	0.39	18.0	0.27	1.44	0.36
7.2	2	0.48	62.1	27.0	0.43	16.5	0.27	1.64	0.53
7.2	7	0.48	94.1	34.0	0.36	24.0	0.26	1.42	0.35

Note: Stress concentration factor = 2.2

factors in a design; however, it is not yet possible to predict the value of (q) for a given situation. Tables VIII and IX also list values for the K_f, K_t and q factors. The value of q, for a given alloy and density, increases when that material is heat-treated, indicating increased notch sensitivity. However, these tables do not show that notch sensitivity increases with higher tensile

TABLE IX.   Effect of an External Notch on Quenched and Tempered Sintered Nickel Steels

Sintered Density (g/cc)	Nickel Content (%)	Carbon Content (%)	Tensile Strength (ksi)	Smooth Fatigue Limit (ksi)	Smooth Fatigue Ratio	Notched Fatigue Limit (ksi)	Notched Fatigue Ratio	Fatigue Notch Factor (Kf)	Fatigue Notch Sensitivity (q)
6.6	2	0.48	74.8	29.0	0.39	18.0	0.24	1.61	0.51
6.6	7	0.48	96.2	42.0	0.44	25.0	0.26	1.68	0.57
7.0	4	0.48	105.0	38.0	0.36	25.0	0.24	1.52	0.43
7.2	2	0.48	104.5	38.5	0.37	23.0	0.21	1.68	0.57
7.2	7	0.48	147.0	58.0	0.40	30.5	0.21	1.90	0.75

Note (1): Oil quenched from 1600F, tempered 30 minutes at 400F
Note (2): Stress concentration factor = 2.2

strength, which one would expect, illustrating the variability of K_f, K_t, and q factors.

CONCLUSIONS

1. Sintered nickel steels possess distinct fatigue limits, occurring between 10^6 and 10^8 cycles.
2. The smooth fatigue ratio has an average value of 0.4 up to 150,000 psi tensile strength and is independent of sintered density, alloy content and state of heat-treatment.
3. The fatigue limit increases with higher sintered densities and alloy content within the ranges studied.
4. Any heat treatment that increases tensile strength also increases the fatigue limit of sintered nickel steels within the ranges studied.
5. Sintered nickel steels are affected by external notches which lower the fatigue limit and therefore, the fatigue ratio. The effect was more pronounced on heat treated material.

REFERENCES

1. Kravic, A. F., "The Fatigue Properties of Sintered Iron and Steel," Intl. J. of Powder Metallurgy, 3, 7, April (1967).
2. "Manual on Fatigue Testing," ASTM Spec. Tech. Publ. No. 91, 1949.
3. Anon., "The Tensile and Impact Properties of Nickel Alloy Steels Sintered at 2050F," The International Nickel Company, Inc., August 1966.
4. "Guide for Fatigue Testing and Statistical Analysis," ASTM Spec. Tech. Publ. No. 91-A, 1963.

ACKNOWLEDGMENT   The authors wish to thank their colleagues, D. L. Townsley, for preparing the many test specimens, and C. L. Verona for testing the specimens.

THE EFFECT OF PHOSPHORUS ADDITIONS ON THE TENSILE, FATIGUE, AND IMPACT STRENGTH OF SINTERED STEELS BASED ON SPONGE IRON POWDER AND HIGH-PURITY ATOMIZED IRON POWDER*

Per Lindskog†

The mechanisms operating during the sintering of iron–phosphorus PM alloys are discussed, as well as the factors contributing to the unique combination of strength, ductility, and toughness that is characteristic of these materials. Alloying methods are reviewed with special reference to powder compressibility, tool wear during compaction, and homogenization during sintering. The preferred production method is to add phosphorus in the form of a fine Fe_3P powder to iron powder. The mechanical properties of a number of sintered steels made with and without Fe_3P additions to sponge iron or to high-purity atomized iron powders are reported. Use of atomized powder makes it possible to reach extremely high density by single pressing and the resulting phosphorus-containing sintered steels have very high ductility and impact strength. The fatigue strength is related linearly to the tensile strength, with a correlation coefficient of 0·91. It is concluded that structural factors other than those that control ductility and toughness are responsible for the fatigue-resistance of sintered steels.

PHOSPHORUS, which is considered as a harmful 'tramp' element in conventional steelmaking, was found to be a useful additive in ferrous powder metallurgy by Dr Fritz Lenel[1] as early as 1939. He discovered that the strength of sintered iron was increased $\sim 70\%$ by the addition of phosphorus. Even more remarkable was his discovery that the improvement in strength was not accompanied by a loss of ductility, as with most other alloying elements. The seeming contradiction between the effects of phosphorus in conventional and sintered steels can be resolved by considering the mechanisms that operate when the alloy is formed. The liquidus and solidus curves bounding the two-phase region $\alpha +$ liquid in the iron–phosphorus system (Fig. 1) are widely separated and the conditions are thus such as to produce considerable segregation of phosphorus during the solidification of a steel ingot. The resulting

* Manuscript received 25 June 1973. Contribution to a Symposium on 'PM Alloys and Properties', to be held in Eastbourne on 19–21 November 1973.

† Höganäs AB, Höganäs, Sweden.

POWDER METALLURGY, 1973, Vol. 16, No. 32

regions of high phosphorus content are widely spaced and the diffusivity of the phosphorus is not sufficiently great to allow homogenization on subsequent heat-treatment. There is therefore a risk of forming brittle iron phosphide at the grain boundaries of the high-phosphorus regions, with consequent embrittlement of the steel.

Sintering Mechanisms

Although the pressed powder compact is heterogeneous the diffusion distances are much smaller than in a segregated steel ingot and considerable homogenization may therefore be achieved during sintering. If the phosphorus is added to the iron powder as elemental phosphorus or an iron–phosphorus alloy with > 2·0–2·8% P a liquid phase is formed during sintering at temperatures above 1050°C. The liquid phase wets the iron well and is therefore distributed by capillary action throughout the pore system of the compact, thus providing a favourable starting condition for the diffusion of phosphorus into the iron particles.

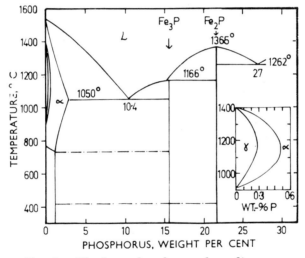

FIG. 1. The iron–phosphorus phase diagram.

As long as the liquid phase is present it also forms an easy path for material transport by solution and reprecipitation, which markedly increases the rate of sintering.[2] Although this is of benefit for the mechanical properties of the material it has an unavoidable detrimental consequence, in that the shrinkage increases and dimensional control becomes more difficult. It is, however, possible to establish such conditions that good dimensional control is combined with improved mechanical properties.[3]

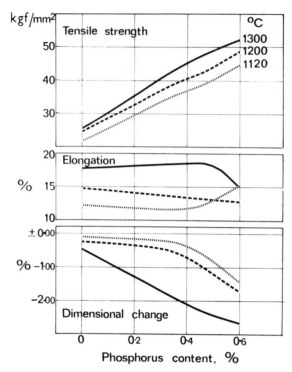

FIG. 2. Tensile properties and shrinkage as functions of the phosphorus content for different sintering temperatures. Sponge iron powder grade NC100.24. Compacting pressure 589 N/mm². Sintering time 1 h.

Fig. 2 shows how mechanical properties and shrinkage are related to both phosphorus content and sintering temperature. It is seen that at a sintering temperature of 1120°C the phosphorus content can be increased to 0·45% without causing the shrinkage to exceed 0·5%.

Returning now to the sintering mechanism, the diffusion of phosphorus from the liquid phase into the iron powder particles results in a gradual reduction of melt volume and a simultaneous precipitation of solid α-iron. The time required for complete solidification of the melt is a function of both the phosphorus content and the sintering temperature. The photomicrographs in Fig. 3 indicate that solidification of a 1·5% phosphorus alloy sintered at 1200°C is complete in ∼15 s. (Differences in phosphorus content have been made visible by etching with Oberhoffer reagent.)

When all the liquid phase has disappeared the solid-state diffusion of phosphorus continues from ferrite regions (with up to 2·3% P at 1120°C) towards austenite regions poor in phosphorus. Complete homogeneity is never achieved in materials with phosphorus contents of 0·25–0·55%

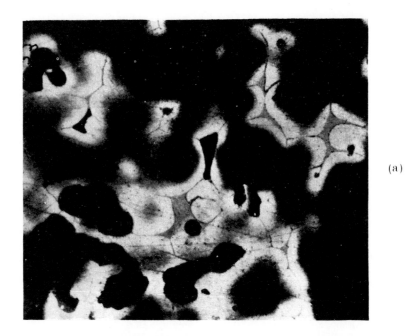

(a)

(b)

FIG. 3. Micrographs of 1·5% phosphorus–iron alloys sintered at 1200°C. Etched with Oberhoffer reagent. Light-etching areas are rich in phosphorus. Note the presence of eutectic penetrating into the grain boundaries in specimen sintered for 5 s. The eutectic has disappeared after 15 s. sintering. Sintered for (a) 5 s.; (b) 15 s. ×400.

FIG. 4. Micrograph of specimen with 0·45%P sintered at 1120°C for 1 h. Etched with Ober-hoffer reagent. ×320.

because these compositions fall within the two-phase region, $\alpha + \gamma$, in the phase diagram (Fig. 1).

A typical well-diffused structure of material with 0·45% P sintered at 1120°C for 60 min is shown in Fig. 4. Here light grains with 0·55% P alternate with darker grains containing 0·25% P or less. During the solid-state phase of the sintering process the rate of material transport is greater in compacts containing phosphorus than in pure iron. This is due to the presence in the phosphorus steel of body-centred cubic ferrite at the sintering temperature.[4] Pure iron is austenitic at this temperature and the rate of self-diffusion is about two orders of magnitude lower in austenite than in ferrite.

It is of course important to avoid the formation of large pools of liquid phase in compacts during sintering. Figs. 5–7 are scanning electron micrographs of the fracture surface of a tensile test-bar, made from sponge iron powder and ferro-phosphorus, which failed at a lower stress than normal. The (low magnification) micrograph in Fig. 5 shows a crater of ~ 0.5 mm dia. surrounded by a 0·2 mm-wide zone with brittle intergranular fracture. Fig. 6 shows intergranular cleavage facets most clearly to the left near the crater edge and signs of ductile behaviour to the lower right. The walls of the crater were partly covered by corrosion products, but under these was a rod-type Fe–Fe$_3$P eutectic structure (Fig. 7). Evidently at this location there had been

Fig. 5. Scanning electron micrograph of fracture surface of a specimen that failed prematurely. × 63.

too large a particle of ferro–phosphorus for the surrounding material to absorb during sintering.

This example shows that it is essential to prevent the formation of agglomerates or the presence of large primary particles in the green powder compacts. When phosphorus is added as ferro-phosphorus, it is relatively easy to avoid large primary particles because the ferro-phosphorus is extremely brittle and can be comminuted to a very small particle size. One must, however, be careful not to cause agglomeration during mixing. The choice of pressing lubricant is important in this respect. Tests with different lubricants showed that zinc stearate gave better sintered properties than stearic acid and, although with a smaller margin, than synthetic wax. The sticky nature of the latter lubricants

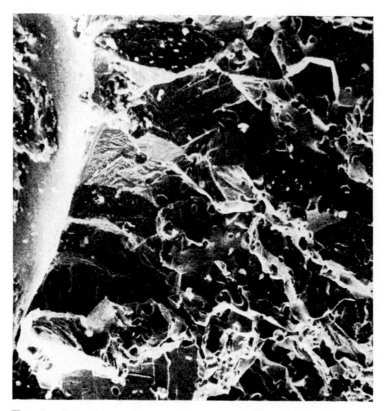

Fig. 6. Detail from the brittle zone in Fig. 5. The crater edge is to the left. × 265.

probably tends to make the small ferro-phosphorus particles adhere to larger ones, forming agglomerates by a sort of 'snowballing effect'.

Alloying Methods

The subject of alloying technique has received a great deal of attention in the literature. Phosphorus has been added in forms ranging from completely prealloyed powder having the desired final phosphorus content to elemental red phosphorus. Good accounts of the different methods employed have been given by Eisenkolb[5] and Rebsch.[6] Red phosphorus can be dismissed because when mixed with iron powder it may ignite. The other extreme, a fully prealloyed powder, has an inherent advantage in that there is no need for interdiffusion. Moreover 'rat-nests', surrounded by material with too high a phosphorus content, as shown in Figs. 5–7, cannot be formed during sintering. However, the solid-solution-strengthening effect[7] of the dissolved phosphorus reduces the compressibility to an unacceptably low level. By increasing the phosphorus content of the prealloyed powder to 3%,

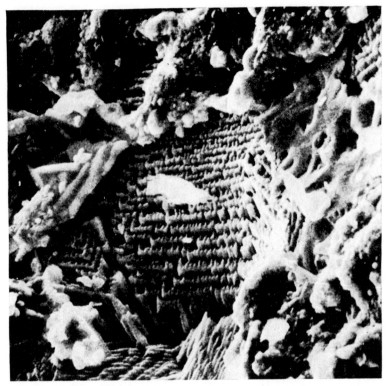

Fig. 7. Detail from the inner wall of the crater in Fig. 5. Rod-type eutectic solidification structure. ×1440.

as suggested by Naeser et al.[2], some improvement in compressibility is achieved, but the mixtures are difficult to compact on account of the large percentage of hard powder that has to be used. Also, such a master-alloy powder consists essentially of ferrite and is thus not easily comminuted to the fine particle size desirable for rapid interdiffusion. Very little or no liquid phase is formed during sintering and, although this brings the advantage of smaller dimensional change, it makes diffusion paths longer.

The method originally chosen by Lenel was to use fine ferro-phosphorus powder mixed with iron powder. Ferro-phosphorus normally contains 20–25% phosphorus and is a very brittle material consisting primarily of Fe_2P and, depending on the total phosphorus content, some Fe_3P or a higher phosphide of unknown composition as a minor constituent. Its brittleness makes it easy to grind it to powder of the desired fine particle size. Since the ferro-phosphorus is relatively rich in phosphorus only small amounts need to be added and hence the compressibility of the iron powder is little affected.

During practical production of sintered parts from this type of powder

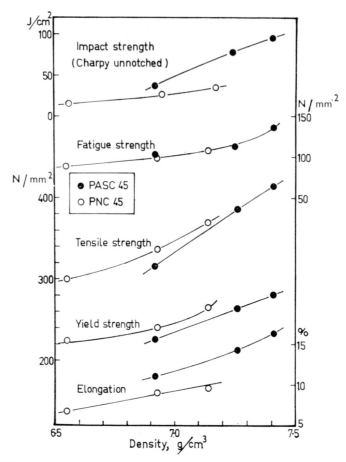

Fig. 8. Mechanical properties of PNC45 and PASC45 as functions of sintered density. Test-specimens were sintered at 1120°C for 60 min in dissociated ammonia.

an increase in the rate of tool wear during compaction was observed. It was evident that the high wear was connected with the hardness of the ferro-phosphorus particles. A series of specimens with phosphorus contents ranging from 15 to 20% was made in the author's laboratory by remelting 25% ferro-phosphorus with additions of pure iron. The specimens were sectioned, polished, and etched. The microhardness (100 g load) was then measured in different grains. The predominant constituent of the specimen with 20% P had a hardness of 1150, whereas most of the grains in the specimen with 15% P showed values of only 700–800. It seems reasonable to conclude that Fe_2P (with 21·7% P) is the phase responsible for the high rate of tool wear and that the tool wear should diminish if a pure Fe_3P powder (with 15·6% P) were used instead.

Fe_3P proved to be as brittle and easy to comminute as the other ferro-

TABLE 1. Mechanical Properties of Sintered Steels

Material*	Compacting Pressure,‡ N/mm²	Sintered Density,‡ g/cm³	Tensile Strength,‡ N/mm²	Yield Strength,‡ (0·2% offset), N/mm²	Elongation,‡ %	Impact Energy,‡ J	Fatigue Limit N/mm²	Dimensional Change during Sintering,‡ %
NC100.24 (Sponge Fe)	589	6·89	207(5)†	108(2)†	12·3(1·0)†	26·6(1·8)†	85	−0·17(0·01)
ASC100.29 (Atomized Fe)	589	7·21	234(5)	130(2)	15·9(1·4)	53·2(5·0)	90	−0·10(0·02)
PNC 30	589	6·86	301(5)	204(4)	10·1(0·7)	35·1(2·4)	95	−0·26(0·01)
PNC 45	392	6·55	300(7)	224(5)	6·7(0·6)	14·6(1·9)	90	−0·80(0·04)
"	589	6·93	336(3)	240(1)	9·0(0·4)	26·0(3·2)	100	−0·75(0·03)
"	785	7·14	370(3)	265(3)	9·6(1·0)	35·0(3·7)	110	−0·69(0·02)
PASC 45	392	6·92	316(7)	225(4)	11·0(0·9)	36·7(2·7)	105	−0·60(0·02)
"	589	7·26	385(5)	264(4)	14·3(1·5)	78·8(5·9)	115	−0·54(0·02)
"	785	7·41	415(2)	280(3)	16·3(1·6)	94·3(9·0)	140	−0·53(0·02)
NC100.24+2% Cu	589	6·82	269(4)	204(4)	6·6(0·4)	22·4(2·6)	110	+0·15(0·02)
PNC 45+2% Cu	589	6·84	388(4)	329(5)	4·7(0·6)	15·4(2·3)	110	−0·21(0·01)
NC100.24+2% Cu+ 0·7 C	589	6·82	517(9)	397(9)	2·2(0·2)	14·4(1·5)	150	+0·17(0·02)
PNC 45+0·7% C	589	6·76	478(14)	366(8)	3·3(0·3)	14·8(1·7)	140	+0·09(0·03)

* Materials were sintered for 1 h at 1120° C either in dissociated ammonia or (if containing graphite) in closed gettered boxes.
† Standard deviation values in brackets.
‡ Values are the average of measurements from 7 tests.

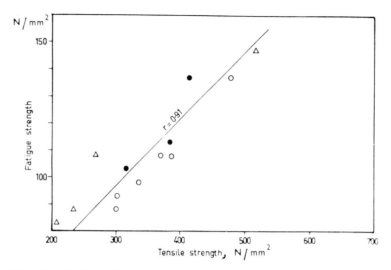

FIG. 9. Fatigue limit of different sintered steels as a function of tensile strength. Triangles are values for materials without phosphorus; open circles correspond to PNC materials; closed circles to PASC materials.

phosphorus compound. To study the effect of Fe_3P addition on tool wear a comparative wear test as described by Bockstiegel[8] was carried out, using sponge iron powder (Grade NC100.24) mixed with either Fe_3P or 25% P ferro-phosphorus powder. In both batches the phosphorus content of the mix was 0·45% and 0·8% zinc stearate was added. Pure sponge iron powder with 0·8% zinc stearate was used as a reference standard. Cylindrical dies (8 mm bore) made from high-speed steel with a hardness of R_c64 were used for the tests and compacts 8 mm high were pressed to a density of 6·75 g/cm³. The rate of die wear with the powder containing 25% P ferro-phosphorus was 2·2 times that obtained with pure iron powder. The corresponding factor for the powder with Fe_3P was 1·4.

Fe_3P powder[9] thus offers a number of advantages as a means of adding phosphorus to sintered steel. In combination with sponge iron powder, grade NC100.24, it is used for medium-density applications. A new high-purity atomized iron powder, ASC100.29, makes it possible to reach 7·2–7·3 g/cm³ density with a single pressing/single sintering operation. In spite of its higher cost it has found wide use for high-density applications where it is possible to eliminate the costly double pressing/double sintering practice. When Fe_3P is added to this powder the beneficial effects of high density and phosphorus on the ductility and toughness of the sintered material are fully utilized.

Mixes of Fe_3P and sponge iron powder grade NC100.24 are termed

PNC, followed by two digits denoting the phosphorus content in hundredths of 1%. Correspondingly, mixes based on high-purity atomized iron powder grade ASC100.29 are denoted by the prefix PASC in the following discussion.

Mechanical Properties

The mechanical properties of thirteen different materials with and without additions of Fe_3P, some based on sponge iron powder and others on atomized powder were determined. Tensile test-bars and impact specimens were pressed in carbide dies and sintered in a belt furnace under dissociated ammonia atmosphere. The sintering temperature was 1120°C and the time at temperature was 1 h. Tensile testing was performed in an Instron testing machine at a rate of deformation of 5·0 mm/min. The strain was measured with a strain-gauge extensometer over a 26 mm gauge-length. The Charpy impact fracture energy was determined with unnotched specimens. The fatigue limit was determined after 10^7 cycles of reverse bending in Schenck machines. The specimens used were ordinary flat MPIF tensile test-bars from which any flash was removed by grinding.

The complete test results are listed in Table I. The properties of PNC45 and PASC45 are given as functions of the sintered density in Fig. 8, and in Fig. 9 the fatigue limits of all the different materials are plotted vs. tensile strength.

The data clearly illustrate the pronounced effect of phosphorus on tensile and yield strength. Of all elements that form substitutional solid solution with iron, phosphorus has the greatest influence on strength. At tensile strengths of 300–400 N/mm² the materials with phosphorus possess unique ductility. The impact strength is increased by the addition of phosphorus, especially for the pure atomized iron powder whereas additions of carbon, and to a somewhat lesser extent of copper, cause a sharp drop of both impact strength and ductility. This difference in behaviour is attributed to the high sintering rate obtained with phosphorus (primarily due to the existence of ferrite at the sintering temperature) which brings about a more rounded pore shape than with carbon or copper additions. It seems reasonable to assume as a first approximation that the elimination of sharp corners would contribute to an increased fatigue strength. The fatigue data obtained in this investigation do not, however, correlate at all with either ductility or impact strength. There is good correlation between fatigue strength and tensile strength (coefficient of correlation, $r = 0·91$, see Fig. 9), indicating that overall porosity and the strength of the metallic phase have greater importance than the pore shape on the fatigue strength of sintered steels. The ratio of fatigue limit to tensile strength is ~ 0.3 for all materials except the softest ones, i.e. the pure irons and the iron–2% copper alloy, which show a ratio of ~ 0.4.

Acknowledgements

The author gratefully acknowledges the assistance given by Jan Mårtenson, Lars-Erik Svensson, and Otto Struglics, who all contributed to the experimental parts of this paper. Thanks are also due to Höganäs AB for permission to publish.

References

1. F. Lenel, USA Patent No. **2226520**, 1940.
2. F. Naeser *et al.*, UK Patent No. **1155918**, 1969.
3. P. Lindskog and A. Carlsson, *Powder Met. Int.*, 1972, **4**, 1.
4. N. Dautzenberg, 'Second European Symposium on Powder Metallurgy' 1968, Stuttgart, paper No. 6–18.
5. F. Eisenkolb, *Arch. Eisenhüttenwesen*, 1953, **24**, 257.
6. H. Rebsch, *Neue Hütte*, 1965, **10**, 335.
7. W. A. Spitzig, *Met. Trans.*, 1972, **3**, 1183.
8. G. Bockstiegel, *Powder Met.*, 1969, **12**, 316.
9. Patents pending.

IRON-CARBON BEHAVIOR DURING SINTERING

By P. Ulf Gummeson and Athan Stosuy

INTRODUCTION

Sintered steel has been commercially produced for many years, but relatively little data has been published about its behavior during processing. This report discusses important variables which must be controlled and how they affect several of the more important properties of P/M parts. The combined carbons noted in this paper are metallographic estimates based on the relative amount of pearlite formed.

REACTIVITY OF IRON

Reactivity refers to the ability of iron powder to react with graphite additions to form combined carbon. Surface impurities or admixed impurities can have a detrimental effect on carbon pick up (reactivity) but the most important factor is the oxide content of the powder.

During sintering, the major portion of oxides in an iron particle have to be reduced by the graphite and the sintering atmospheres before combination of iron and graphite takes place effectively. Although it is not necessary that the powder have a low hydrogen loss, it is important to know how much, since the amount of graphite addition must vary accordingly. It is usually desirable that the sintered part have a combined carbon content corresponding to eutectoid composition of around .8-.9% and generally the graphite addition will have to be higher than this figure. Figure 1 shows the combined carbon content of a compact made of an iron having a hydrogen loss of .32%. The transverse rupture strength (TRS) is 71,000 psi. Figure 2 shows the combined carbon of a compact made from an iron having a hydrogen loss of .73%. The TRS of this sample was 47,500 psi. In both cases 1.25% graphite was mixed with the iron powder. Obviously, a slightly higher graphite addition should have been used for the second iron.

REACTIVITY OF GRAPHITE

There is a wide variation in the ability of various graphites to react with iron. Generally speaking, graphite should be of small particle size and be free of silicon carbide.

Fig. 1 H$_2$ loss Effect 800x
H$_2$ loss .32% Combined carbon .75%

Fig. 2 H$_2$ Loss Effect 800x
H$_2$ loss .73% Combined carbon .50%

Source: Technical Bulletin D164, Hoeganaes Corp., Dec 1972

Table I lists a number of different grades of graphites and one iron sintered simultaneously at two different densities and at two different graphite additions. For the graphites, the origin, particle size and ash content and for the resulting bars, the dimensions, hardness and strength are reported. Sintering was done at 2070°F (1132°C) for 40 minutes in purified exothermic atmosphere in a production furnace. The actual values of dimensions, hardness and strength are immaterial in this case. Only the comparison between the various graphites is of interest at this point. We find that synthetic graphites can be ruled out as unsuitable in spite of the fact that they are all of a particle size comparable to the natural graphites and with very low ash contents. The reason for this is not clear.

While natural and synthetic graphites have the same crystal structure, one important difference is that the ashes in synthetic graphites are mainly silicon carbide while in natural graphite they might be mica, clay, etc. Therefore, one-half of one percent of ashes in synthetic graphites can be worse from powder metallurgy standpoint than 5% of ashes in natural graphites.

Carbon black or lamp black are not very reactive in spite of a particle size ranging from .5 - .005 microns. They are also completely detrimental to compressibility. One explanation is that they have a highly disordered lattice and an extremely fine crystal structure. Another reason for their non-reactivity might be their very high affinity for oxygen. Oxygen is strongly absorbed on the crystal's surface with a bond that gets stronger at increased temperatures.

When selecting a graphite, consideration must be given to high reactivity for fast reduction of oxides and fast carburization and an easily distributed graphite that gives good flow and widest range of mixed densities as needed. Therefore, the choice is narrowed to natural graphites of low ash content (less than 5%) and fine particle size. Natural graphites originating in Ceylon, Madagascar, Bavaria and Texas produce good results in hardness and strength of P/M parts.

EFFECT OF COMBINED CARBON

The effect of combined carbon on the tensile strength of wrought steel is well known and is shown in Figure 3. As carbon content is increased, strength increases quite rapidly up to 1.00%, then gradually reaches a maximum at the highest carbon content of 1.6%.

A similar plot was made for sintered steel. Combined carbon was varied by adding increasing amounts of graphite. Test bars were pressed to a density of 6.3 g/cm^3, then sintered 30 minutes at 2050°F in dissociated ammonia. Figure 4 shows transverse rupture strength as a function of combined carbon. As with solid steel, there is a rapid initial increase in strength with increasing combined carbon content, but a maximum is reached at about the eutectoid composition of .85% carbon, then between .90 and 1.00% the strength drops catastrophically.

To explain this sudden decrease in strength, a series of specimens of varying combined carbon contents was prepared for metallographic examination. (The technique is described in the appendix) Micrographs (Figures 5 to 16) taken at 600 x illustrate the increase in combined carbon. Bar graphs illustrating the strength relative to the maximum strength attained in the series is also shown. As the amount of pearlite which contains the combined carbon increases, strength and hardness correspondingly

Sintered Properties of Iron-Graphite Compositions

Influence of Graphite Quality

M.P.I.F. transverse rupture test bars 1.250" x .500" were sintered at 2070° F (1132° C) for 40 minutes in purified exothermic atmosphere in a production furnace

Column groups are headed by **Green Density g/cm³** and **Graphite Addition %**. The four sub-columns of each property are labelled below as *Green Density (g/cm³) / Graphite Addition (%)*.

Mfg.	No.	Source	Graphite Addition	Ash %	Dim. Change, % 6.1/1.1	6.1/1.4	6.4/1.1	6.4/1.4	Hardness R_B 6.1/1.1	6.1/1.4	6.4/1.1	6.4/1.4	Transv. Rupture P.S.I. 6.1/1.1	6.1/1.4	6.4/1.1	6.4/1.4
A	(1)	Ceylon	5 Micron Average	2.8 Ash	.06	.10	.18	.19	27	22	53	52	51,400	53,300	72,200	83,000
A	(2)	Ceylon	10 Micron Average	2.2 Ash	.10	.06	.18	.21	24	4	51	47	46,600	44,100	70,100	80,600
A	(3)	Ceylon	97% <325 Mesh	5.2 Ash	.06	.06	.10	.15	17	16	41	49	39,300	43,900	63,700	65,300
A	(4)	Ceylon	97% <325 Mesh	3.5 Ash	.03	.18	.14	.22	-5	20	34	42	33,600	34,800	49,500	55,700
B	(1)	Ceylon	5 Micron Average	1.7 Ash	.07	.14	.14	.23	22	28	50	55	49,600	56,700	74,100	81,900
B	(2)	Ceylon	20 Micron Maximum	1.6 Ash	.06	.14	.14	.26	17	24	45	55	42,100	51,800	72,500	80,800
B	(3)	Ceylon	20 Micron Maximum	2.9 Ash	.02	.18	.11	.30	17	36	46	62	47,100	53,800	70,400	81,200
B	(4)	Ceylon	98% <325 Mesh	1.5 Ash	.06	.21	.26	.14	8	37	44	57	36,500	53,700	72,000	80,000
B	(5)	Ceylon	100 Mesh, 60-75% <325 Mesh	2.1 Ash	.10	.22	.24	.31	27	33	50	62	47,700	53,000	75,100	80,900
B	(6)	Madagascar	25 Micron Maximum	2.7 Ash	-.02	.13	.09	.28	14	37	46	61	48,100	53,100	70,000	83,300
B	(7)	Bavaria	20 Micron Maximum	3.8 Ash	-.02	.09	.14	.17	22	33	53	57	48,800	54,500	72,900	88,900
B	(8)	Bavaria	20 Micron Maximum	4.2 Ash	.06	.10	.15	.23	21	35	51	61	50,300	58,900	72,700	89,300
C	(1)	Texas	98% <325 Mesh	3.6 Ash	.16	.10	.10	.24	-8	15	22	56	34,100	39,000	56,100	73,000
C	(2)	Texas	95% <8 Micron	2.9 Ash	.07	.14	.09	.32	16	38	43	63	49,300	58,100	72,600	89,500
C	(3)	Texas	85% <2 Micron	3.5 Ash	.05	.22	.14	.22	23	37	52	62	46,900	46,900	73,500	91,800
D	(1)	?	60% <8 Micron	1.8 Ash	.06	.18	.14	.26	24	35	52	63	48,800	54,700	74,700	85,400
D	(2)	?	99% <325 Mesh	4.8 Ash	.02	.03	.13	.14	20	17	51	48	47,800	51,400	72,000	79,600
D	(3)	?	98% <325 Mesh	2.3 Ash	-.05	.10	.10	.18	6	28	38	54	41,000	48,800	61,400	69,800
D	(4)	Mexican	70% <8 Micron	17.6 Ash	-.02	-.01	.10	.17	-24	-24	14	10	30,300	26,600	39,000	32,100
D	(5)	Mexican	99% <325 Mesh	18.6 Ash	.10	.18	.23	.39	-34	-32	6	6	21,700	19,400	32,800	25,400
D	(6)	Mexican	98% <200 Mesh	18.2 Ash	.23	.34	.42	.58	-49	-53	-6	-15	15,800	12,800	22,800	17,300
D	(7)	Mexican	98% <200 Mesh	18.9 Ash	.31	.47	.47	.58	-58	-64	-11	-19	14,000	11,100	20,000	15,600
Synthetic	(1)	Electro	Colloidal	.9 Ash	-.02	-.02	-.10	-.06	-34	-24	6	12	20,600	10,500	31,300	-
Synthetic	(2)	Electro	25 Micron Maximum	1.0 Ash	.81	1.18	.62	.86	-22	-17	28	35	11,700	10,500	23,400	25,000
Synthetic	(3)	Electro	98% <325 Mesh	1.7 Ash	.81	1.11	.58	.82	-14	-4	35	42	13,000	12,500	27,200	27,700
Synthetic	(4)	Electro	98% <200 Mesh	1.6 Ash	1.00	1.43	.78	1.07	-24	-27	24	32	9,100	9,300	22,000	22,400

TABLE I

Source: Technical Bulletin D164, Hoeganaes Corp., Dec 1972

increase until at .80% carbon (Figure 11) the entire structure is pearlitic. The pearlite spacing becomes finer as the carbon is increased, until it is too fine to be resolved at 600 x. Above .80% carbon a carbide network begins to form along the grain boundaries and at 1.10% carbon (Figure 14) the network is entirely continuous. Unfortunately, many grain boundaries are located between many of the pores. These boundaries were initially formed during sintering of adjacent particles and under the sintering conditions used were not given enough energy to be substantially moved from their original position.

Figure 17 shows the start of the formation of a carbide network originating at a pore which is the most likely location of the admixed graphite. Figure 18 shows a greater amount of grain boundary carbide which is almost continuous from pore to pore, while Figure 19 shows a separation or crack through a massive carbide grain boundary. While not often seen, it illustrates the brittle nature of the material that has formed the bonds between almost all the particles in the specimen. Figure 20 shows another extreme condition of a carbide layer lining the pores of the specimen. It should be mentioned that at the highest carbon contents an appreciable amount of free ferrite is formed adjacent to the carbide network. This "abnormality" further explains the poor strength of these materials.

It can be concluded that the drop in strength is caused by formation of a brittle carbide network which is closely linked with the stress inducing porosity and which forms a majority of the particle-to-particle bonds responsible for the strength of any sintered compact.

Although the importance of combined carbon to the strength of sintered steel has been illustrated, there are several factors which determine the amount of combined carbon besides the type and amount of graphite added. The furnace atmosphere must be controlled at a high enough carbon potential to prevent excessive loss of carbon to the atmosphere since decarburization results in correspondingly poor properties.

EFFECT OF ATMOSPHERE COMPOSITION

The tendency of parts to oxidize or reduce and to carburize or decarburize, and the rates at which these take place during sintering, depend on the sintering temperature and the proportion of various gases in the sintering atmosphere.

Carbon Potential

The sintering atmosphere has a strong influence on the amount of combined carbon formed during sintering. If the carbon potential is not controlled, then carbon may be lost to the atmosphere or excessive carburization may take place.

The carbon potential is determined by the ratio of H_2O to H_2, CO_2 to CO and CH_4 to H_2. The amount of graphite in the parts also contributes to the carbon potential in the furnace atmosphere. Figure 21 illustrates equilibrium ratios at various temperatures for oxidation-reduction and decarburization-carburization reactions. Note that the CO_2 to CO ratio can be fairly high at all temperatures without causing oxidation. However, it is necessary to keep the CO_2 content low to prevent decarburization.

COMBINED CARBON VS TENSILE STRENGTH SOLID STEEL

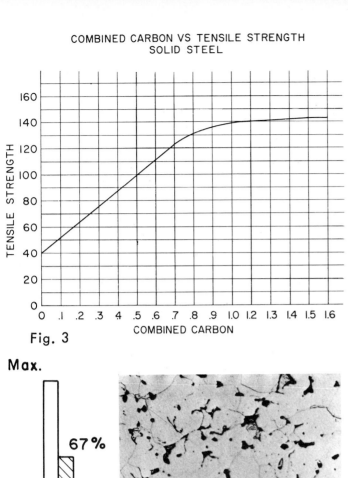

Fig. 3

COMBINED CARBON VS. TRANSVERSE RUPTURE STRENGTH SINTERED STEEL

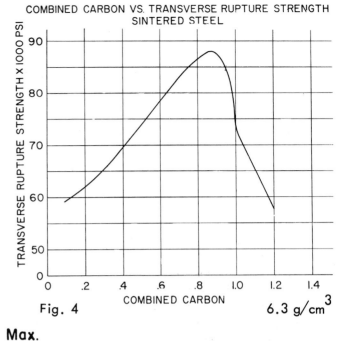

Fig. 4   6.3 g/cm^3

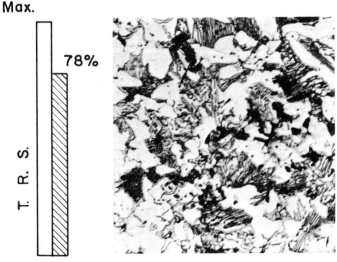

Fig. 5 Carbon Content Effect   600x
6.3 g/cm^3 Combined carbon – trace

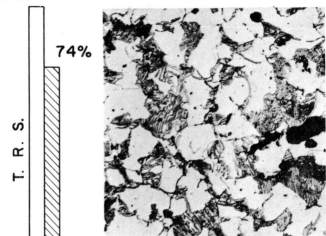

Fig. 6 Carbon Content Effect   600x
6.3 g/cm^3 Combined carbon   .30%

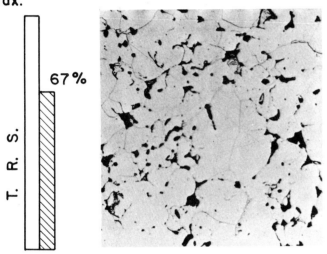

Fig. 7 Carbon Content Effect   600x
6.3 a/cm^3 Combined carbon   40%

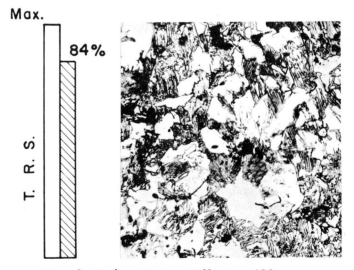

Fig. 8 Carbon Content Effect   600x
6.3 a/cm^3 Combined carbon   .50%

Source: Technical Bulletin D164, Hoeganaes Corp., Dec 1972

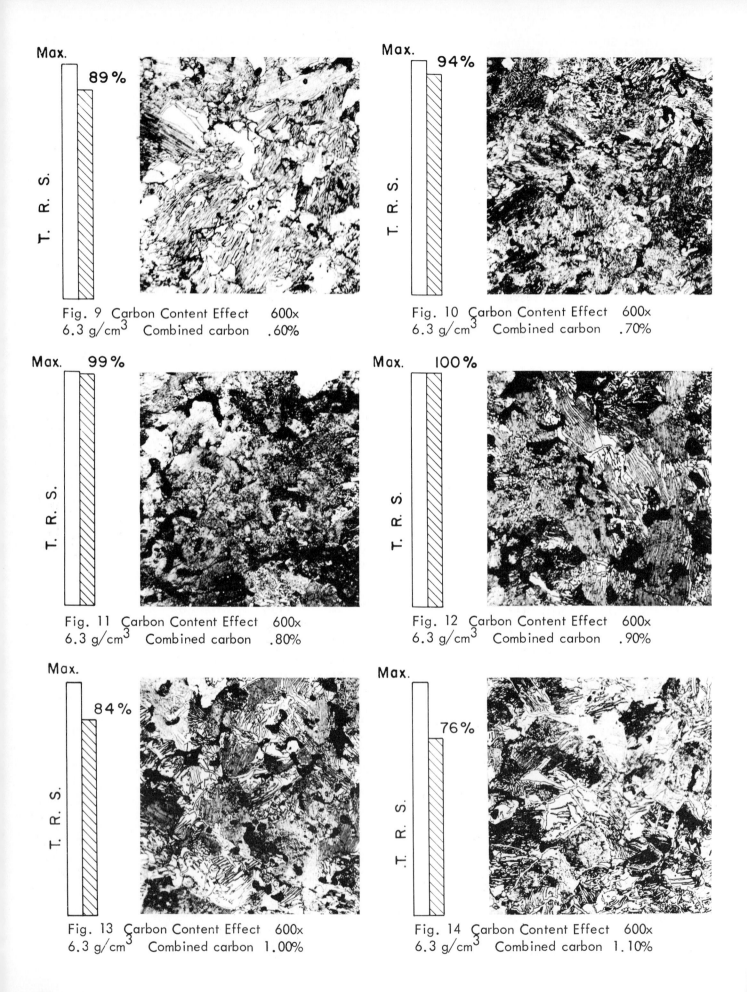

Max. 89%

T. R. S.

Fig. 9 Carbon Content Effect 600x
6.3 g/cm³ Combined carbon .60%

Max. 94%

T. R. S.

Fig. 10 Carbon Content Effect 600x
6.3 g/cm³ Combined carbon .70%

Max. 99%

T. R. S.

Fig. 11 Carbon Content Effect 600x
6.3 g/cm³ Combined carbon .80%

Max. 100%

T. R. S.

Fig. 12 Carbon Content Effect 600x
6.3 g/cm³ Combined carbon .90%

Max.

84%

T. R. S.

Fig. 13 Carbon Content Effect 600x
6.3 g/cm³ Combined carbon 1.00%

Max.

76%

T. R. S.

Fig. 14 Carbon Content Effect 600x
6.3 g/cm³ Combined carbon 1.10%

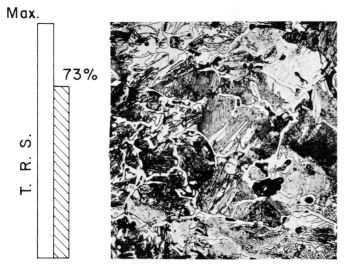

73%

T. R. S.

Fig. 15 Carbon Content Effect   600x
6.3 g/cm^3 Combined carbon 1.20%

Max.

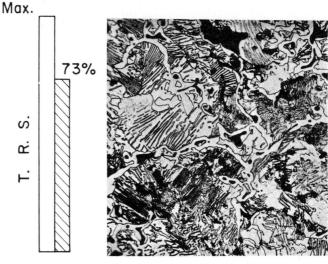

73%

T. R. S.

Fig. 16 Carbon Content Effect 600x
6.3 g/cm^3 Combined Carbon 1.20%

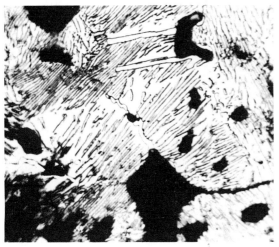

Fig. 17 Carbide Network   2000x
6.3 g/cm^3 Combined carbon 1.00%

Fig. 18 Carbide Network   2000x
6.3 g/cm^3 Combined carbon 1.10%

Fig. 19 Carbide Network   2000x
6.3 g/cm^3 Combined Carbon 1.20%

Fig. 20 Carbide Network   2000x
6.3 g/cm^3 Combined Carbon 1.20%

Source: Technical Bulletin D164, Hoeganaes Corp., Dec 1972

The CO_2 content is kept low by maintaining a low enough dew point (H_2O content) to balance the otherwise continuously reversible "water-gas" reaction $CO_2 + H_2 \rightleftharpoons H_2O + CO$. A dew point of 25° F to 30°F at the generator is usually low enough to maintain a .70% to .90% carbon potential during continuous sintering of iron-graphite mixes.

If lower combined carbon is desired, then a smaller graphite addition is made or a higher dew point is used to raise the CO_2 content.

The ratio of CH_4 to H_2 also determines the tendency of parts to gain or lose carbon. At normal sintering temperatures, even small amounts of CH_4 in the furnace atmosphere will increase the carburizing tendency. Increased amounts of CH_4 will further increase this tendency.

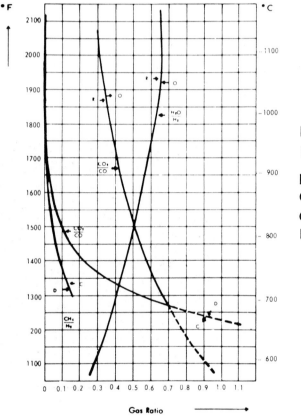

Fig. 21

Equilibrium
Reactions

R= reducing
C= carburizing
O= oxidizing
D= decarburizing

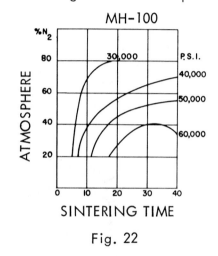

Transverse Rupture Strength
vs Sintering Time and Atmosphere

MH-100

Fig. 22

Speed of Reaction

Atmospheres having better than equilibrium ratios are required unless long sintering times or high sintering temperatures can be permitted. Also, the nitrogen content must be kept as low as is practical.

For example, purified exothermic gas (high nitrogen content) can be used but often larger graphite additions are required to attain the desired combined carbon. Endothermic gas (low nitrogen content) on the other hand promotes rapid carbon pick up and less graphite is needed. Figures 22 and 23 show the influence of the inert nitrogen on strength and hardness.

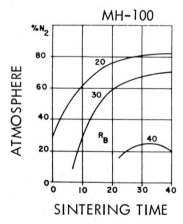

Hardness vs Sintering
Time and Atmosphere

MH-100

Fig. 23

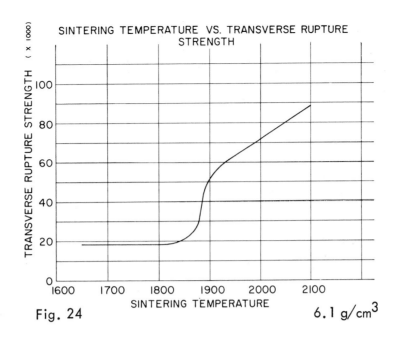

SINTERING TEMPERATURE VS. TRANSVERSE RUPTURE STRENGTH

Fig. 24   6.1 g/cm^3

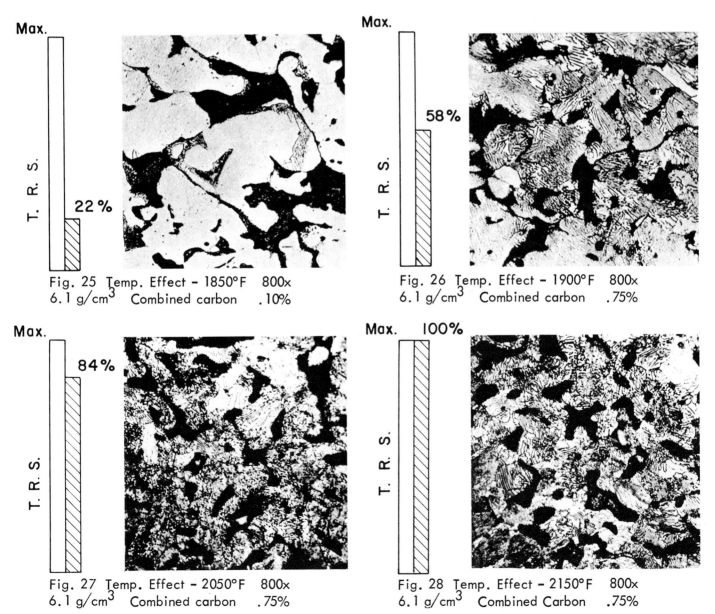

Max.   22%   T. R. S.

Fig. 25 Temp. Effect – 1850°F   800x
6.1 g/cm^3 Combined carbon   .10%

Max.   58%   T. R. S.

Fig. 26 Temp. Effect – 1900°F   800x
6.1 g/cm^3 Combined carbon   .75%

Max.   84%   T. R. S.

Fig. 27 Temp. Effect – 2050°F   800x
6.1 g/cm^3 Combined carbon   .75%

Max.   100%   T. R. S.

Fig. 28 Temp. Effect – 2150°F   800x
6.1 g/cm^3 Combined Carbon   .75%

Source: Technical Bulletin D164, Hoeganaes Corp., Dec 1972

EFFECT OF TEMPERATURE

Although sintering temperature has a profound effect on the amount of combined carbon formed in a given time, graphite combines quite readily under sintering temperatures and times normally encountered. This is shown by a series of bars made of iron plus 1.25% graphite, pressed to 6.1 g/cm^3 and sintered for 30 minutes at temperature. Figure 24 illustrates the effect of the sintering temperature on the transverse rupture strength. No strengthening occurs as the sintering temperature is raised from 1650°F to 1850°F, but above this temperature a substantial increase is noted. Figures 25 to 28 illustrate the changes in microstructure. The accompanying bar graphs show the strength relative to the maximum attained in the series. At low sintering temperatures no carbon is combined as evidenced by complete absence of pearlite. At 1850°F (Figure 25) some carbide has just begun to form while at 1900°F (Figure 26) the maximum combined carbon content is achieved. Additional strengthening at higher temperatures is caused by an increased degree of sintering as evidenced by elimination of grain boundaries and spheroidization of pores shown by Figure 27 at 2050°F and Figure 28 at 2150°F.

EFFECT OF TIME

Time of sintering also affects the amount of combined carbon formed, but at the sintering times normally used, the maximum amount of combined carbon is usually formed. A very noticable effect on strength and dimensional change during sintering was noticed when a series of similar bars was sintered at several temperatures for times varying from 5 to 120 minutes. Figure 29 shows a plot of sintering time vs. transverse rupture strength. There is a fairly rapid increase in strength up to 30 minutes followed by continuing increases at longer times. Figure 30 shows a plot of sintering time vs. dimensional change for several temperatures. These are typical size change curves showing a maximum growth at relatively short times followed by a gradual reduction at longer times. Figures 31 to 33 show the microstructure as influenced by sintering time at 2050°F. Figure 31, a microstructure of an unsintered bar shows the individual particles, grains within particles, and graphite located mostly at the pores formed between particles. Figure 32 shows almost complete formation of pearlite in 5 minutes, but numerous grain boundaries are visible and the porosity is quite angular. Figure 33 shows some disappearance of grain boundaries and slight spheroidization of pores after 30 minutes, while Figure 34 shows almost complete absence of grain boundaries and substantial spheroidization of pores after 120 minutes. It might be mentioned that sintering for 10 minutes at 2200°F results in strength equal to that attained after 120 minutes at 2050°F.

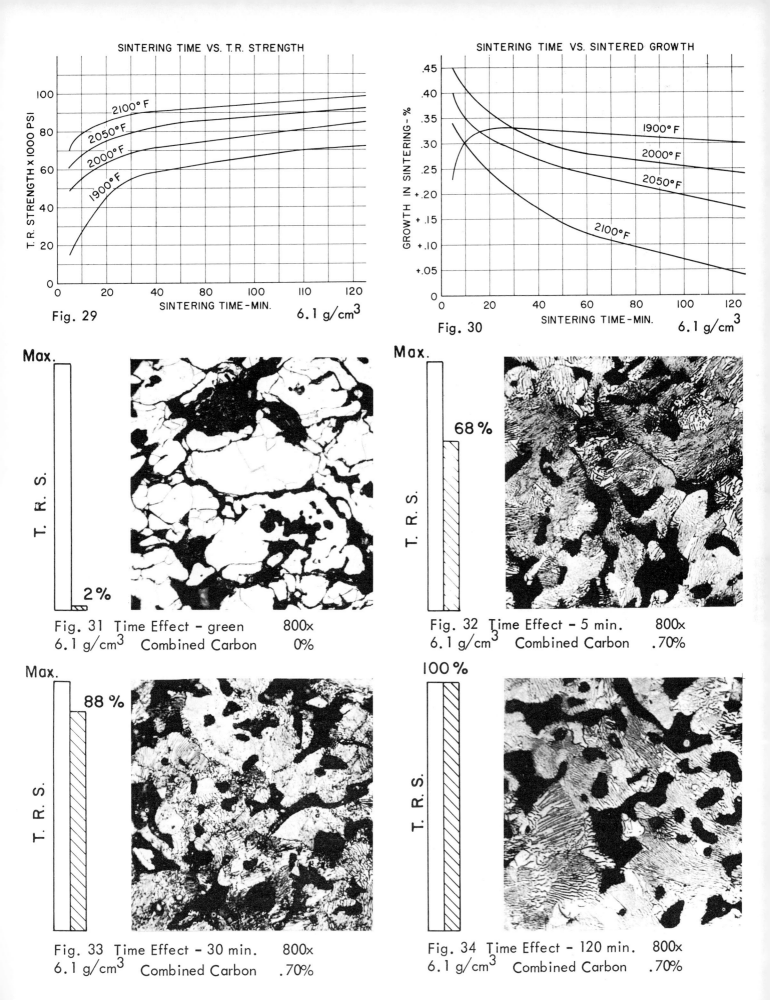

SINTERING TIME VS. T.R. STRENGTH

2100° F
2050° F
2000° F
1900° F

T. R. STRENGTH x 1000 PSI

SINTERING TIME-MIN.

Fig. 29   6.1 g/cm³

SINTERING TIME VS. SINTERED GROWTH

1900° F
2000° F
2050° F
2100° F

GROWTH IN SINTERING - %

SINTERING TIME-MIN.

Fig. 30   6.1 g/cm³

Max.

2%

T. R. S.

Fig. 31 Time Effect – green   800x
6.1 g/cm³   Combined Carbon   0%

Max.

68%

T. R. S.

Fig. 32 Time Effect – 5 min.   800x
6.1 g/cm³   Combined Carbon   .70%

Max.

88%

T. R. S.

Fig. 33 Time Effect – 30 min.   800x
6.1 g/cm³   Combined Carbon   .70%

100%

T. R. S.

Fig. 34 Time Effect – 120 min.   800x
6.1 g/cm³   Combined Carbon   .70%

Source: Technical Bulletin D164, Hoeganaes Corp., Dec 1972

EFFECT OF RATE OF COOLING

The influence of cooling rate on strength in wrought steel is well known. Samples similar to those used in the study of effect of time and temperature were sintered for 30 minutes at 2050°F followed by cooling at different rates through the range of 1350°F to 1000°F. The three rates used and the resulting mechanical properties are shown in Table II. As expected, faster cooling rates cause increased strength and hardness. All three sets of bars show substantially 100% pearlite, but the pearlite spacing is much finer for the faster cooling rates. This is illustrated by Figures 35, 36 and 37 showing the differences in fineness of pearlite. Figure 35 shows a very coarse pearlite structure which was formed by allowing the parts to cool in the hot zone of the furnace after the power was turned off. Figure 36 shows a finer structure caused by a faster cooling rate controlled by the speed of withdrawal from the hot zone. Figure 37 shows the structure formed when the parts are shoved immediately into the cooling chamber. Here the pearlite structure is extremely fine. In fact, the structure could not be clearly resolved by the microscope.

EFFECT OF RATE OF COOLING

	COOLING RATE	TR STRENGTH	HARDNESS	PEARLITE SPACING
1.	3.5° F/MIN.	67000 PSI	RB 37	VERY COARSE
2.	115	82000	50	MEDIUM
3.	225	87000	57	VERY FINE

TABLE II

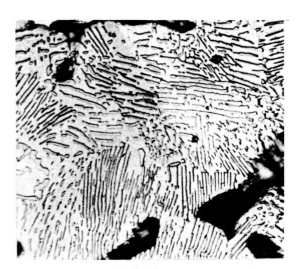

Fig. 35 Cooling Rate Effect   2000x
6.1 g/cm^3 3.5°F per minute

Fig. 36 Cooling Rate Effect   2000x
6.1 g/cm^3 115°F per minute

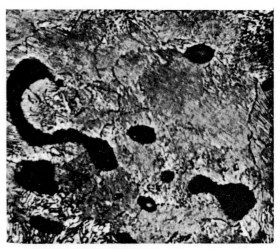

Fig. 37 Cooling Rate Effect   2000x
6.1 g/cm^3 225°F per minute

CONCLUSIONS

It has been shown that conscientious control of sintering conditions i.e. time, temperature, carbon potential and cooling rate and the use of the proper graphite will result in high quality sintered steel P/M parts. The suggestions for selection of graphite and control of the sintering process together with the photomicrographs showing examples of both good and poor sintering should help solve many powder metallurgical problems as they arise.

This paper is the first of several planned on the metallography of sintered materials.

APPENDIX – METALLOGRAPHIC TECHNIQUE

In preparing these porous sintered samples, they were first impregnated by soaking them in molten Acrawax C at 350°F for 2 to 4 hours to preserve the true pore structure and to avoid staining due to absorbed moisture. After cooling and removing the surface wax, the samples were ground and mounted in Bakelite. Transparent mounting material is avoided since the etches normally used react with the resin.

Grinding was done on a belt sander followed by hand grinding on successively finer metallographic paper. Rough polishing was accomplished on a 250 rpm wheel using a hard cloth (Dur) and 15 micron diamond paste. Final polishing was done on a slightly softer cloth (Mol) using .1 micron alumina (Linde).

After being cleaned with soap and water and blown dry, the specimen is ready for etching with 4% picral to which .5% nitric acid is added. For best results, repolishing and re-etching are recommended.

Metallographic examination should be done at 400 x or greater since the structure normally encountered is too fine to be analyzed at lower magnification. This reduces the visible field to the extent that observations must be based on several areas of the samples and mentally averaged. With practice this is not difficult to do.

SECTION III:
Nonferrous Powders

High Strength Aluminum P/M Mill Products

W. S. CEBULAK,* E. W. JOHNSON,* AND H. MARKUS**

ABSTRACT

Wrought aluminum powder metallurgy (P/M) mill products in Al-Zn-Mg-Cu-Co alloys develop combinations of properties that offer unique advantages over existing wrought ingot metallurgy (I/M) aluminum alloys.

P/M alloy products have stress corrosion cracking resistance superior to commercial I/M alloy products. P/M alloy MA87 develops superior fracture toughness to those of I/M Alloys. 7049, 7050 and 7175 at comparable yield strengths. The higher Co-containing P/M alloy MA67 has higher strength and SCC resistance than P/M MA87 and the I/M Alloys 7050 and 7075, with lower fracture toughness than P/M alloy MA87.

The property advantages of the P/M wrought products are due to the fine grain structure resulting from rapid solidification achieved in powder atomizing.

Properties demonstrated are attractive to warrant a production-size scale up of this process. This scale up will lead to fabrication of hot-pressed 3300-lb. (1,500 kg) billets and production fabrication of plates, extrusions and die forgings in production facilities for detailed evaluation.

Introduction

The development of wrought powder metallurgy (P/M)* products has spanned many years and explored $Al-Al_2O_3$, Al-Fe and other alloy systems (1, 2). Development work by Roberts (3, 4) explored these and other compositions and confirmed unpublished Alcoa studies showing that only Al-Zn-Mg-Cu P/M alloys with high Zn and Mg were capable of higher room temperature strength than commercial alloys. This work, however, was troubled by low quality extrusions.

Continued development by Towner (5) led to a process that gave high quality, blister-free products. Haarr (6) used this P/M process to develop alloys with combinations of high strength and stress corrosion cracking (SCC) resistance superior to commercial ingot metallurgy (I/M)* alloys. Atmospheric tests of transverse tensile bars from Haarr's extrusions after nearly eight years (2857 days) exposure show that an Al-8Zn-2.5Mg-1Cu-3.2Fe-5Ni-0.1Cr alloy has immunity to SCC with 78 ksi (538 MPa) transverse yield strength. However, P/M products fabricated by the Towner process with inert-gas preheat had substantially lower transverse fracture toughness than I/M products at the same strength.

Cebulak (8) refined the process by optimizing the fabricating process and alloy compositions to achieve marked improvements in fracture toughness, scaling up the process to make 170-lb. (77 kg) billets.

The P/M processes are shown in Table 1, comparing the substitution of vacuum pre-

---

* Alcoa Technical Center, Alcoa Center, Pa.
** Frankford Arsenal, Philadelphia, Pa.
* The abbreviation P/M will be used in this paper to indicate wrought products fabricated from powder and I/M for those from cast ingots.

TABLE 1

Comparison of Vacuum Preheat and
Inert Gas Preheat

Vacuum Preheat Process	Inert Preheat Process
Melt and Alloy	Melt and Alloy
Atomize	Atomize
Cold Compact	Cold Compact
Encapsulate Compact	
Vacuum Preheat	Flowing Argon Preheat
Vacuum Hot Press	Air-Argon Hot Press[1]
Scalp	Scalp
Hot Work	Hot Work
Heat Treat, Age	Heat Treat, Age

NOTE: 1. Immediately after argon preheat, compact
is hot pressed in air in a heated die.

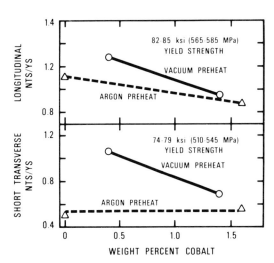

FIGURE 1 Effect of preheat method on fracture toughness (NTS/YS) of P/M Al-8Zn-2.5 Mg-1Cu-variable Co die forgings. All aged 24 hours at 250°F (394 K) + 6 hours at 325°F (435 K).

heating for inert gas preheat and vacuum hot press for inert and air atmosphere hot press. The vacuum process resulted in improved fracture toughness, especially for low cobalt alloys (Fig. 1). Note that the improvement is most striking in the short transverse direction.

Vacuum-preheated material was at least of the same quality as the argon-preheated material in SCC resistance, based on limited directly comparable data shown in Fig. 2. Both in time-to-failure and in percent surviving, die forgings from vacuum preheated MA67 alloy (Al-8Zn-2.5Mg-1Cu-1.5Co) were at least equal or possibly superior to argon preheated MA67 forgings. Therefore, atmospheric SCC experience generated for wrought products from inert-preheat processed material will be used to demonstrate the level of SCC performance of these P/M materials, with preheat process distinction being made only in reference to fracture toughness.

This paper will report the property performance of aluminum P/M mill products and compare properties to existing I/M alloys, where nearly comparable data exist. The P/M alloys to be discussed (listed in Table 2) are those that offer potentially useful combinations of high strength, SCC resistance and fracture toughness. The status of the continuing scale up of the process and potential applications will be reviewed.

P/M Mill Product Properties

Strength and Stress Corrosion*

The long standing advantage of P/M wrought products has been corrosion resistance, as demonstrated for extrusions in accelerated (5, 7, 9, 10) and atmospheric (7, 9) tests and discussed above. In 4.1 year atmospheric tests, MA49 and MA67 alloy extrusions offer 8 to 10 ksi (55 to 69 MPa) higher transverse yield strength than I/M 7075 extrusions with equal SCC resistance

---

*SCC tests used 0.125 in. (3.2 mm) diam. specimens.

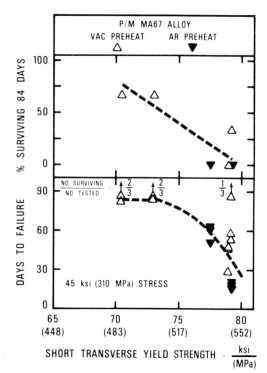

FIGURE 2 Effect of preheat method on time to fail and survival rate of tensile-bar stress-corrosion specimens exposed 84 days in 3.5% NaCl alternate immersion test. Yield strength of Ma67 alloy die forgings varied by aging at 325°F (435 K) for varying times.

TABLE 2

Compositions of Powder Metallurgy
Wrought Products

Alloy	Weight Per Cent						
	Zn	Mg	Cu	Co	Fe	Ni	Si
MA87	6.5	2.5	1.5	0.4	0.05		0.06
MA66	8.0	2.5	1.0		0.04		0.05
MA67	.8.0	2.5	1.0	1.5	0.07		0.06
MA49	8.0	2.5	1.0		0.8	0.8	0.13

(Fig. 3). Cobalt-bearing alloys are being explored in the scale-up phase of on-going P/M mill product development.

The above P/M MA67 to I/M 7075 comparison for extrusions is the longest atmospheric exposure (4.1 years) test available for MA67. In more recent tests of

extrusions that include other P/M and I/M alloy comparisons (650 days exposure to date). P/M MA66 alloy (Al-8Zn-2.5Mg-1Cu), with no Co or Fe + Ni additions, offered better SCC performance than I/M 77178 (Fig. 4). The cobalt in MA67 does impart an 8 ksi (55 MPa) strength advantage over MA66 with equal SCC performance and a 12 ksi (83 MPa) Y.S. advantage over I/M 7178 extrusions.

While transverse SCC specimens from round extrusions can provide meaningful information on expected product performance, a true short transverse structure in a SCC susceptible alloy-temper develops SCC failure earlier at lower stresses than the transverse specimens from round extrusions, as shown for I/M 7075 products in Fig. 5. This means that longer time exposure of transverse specimens from round extrusions is necessary to establish SCC differences between alloys as compared to

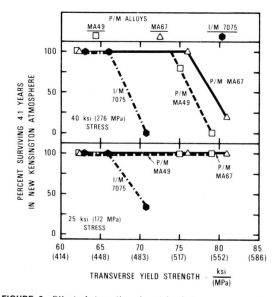

FIGURE 3 Effect of strength and sustained stress on percent surviving 4.1 years in New Kensington atmosphere. Transverse tensile bar specimens from 2″ (51 mm) dia extruded rod (10:1 extrusion ratio).

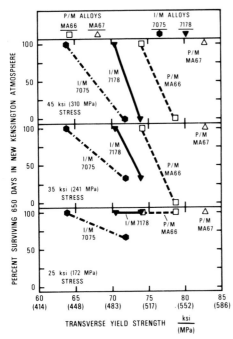

FIGURE 4 Effect of strength and sustained stress on percent surviving 650 days in New Kensington atmosphere. Transverse tensile bar specimens from 1.58″ (39.6 mm) octagonal extruded bar (17:1 extrusion ratio).

short transverse specimens from plate or forgings. These observations indicate that SCC tests of the die forgings and plate, discussed below, are more rigorous than the above described tests on extrusions.

Atmospheric SCC test results of P/M MA66 and MA67 web-flange die forgings shown in Fig. 6 are compared to I/M 7049 and 7050 alloy forgings (11) in Fig. 7. The tests to date show P/M MA67 offering a 7 to 12 ksi (48 to 83 MPa) short transverse yield strength advantage over I/M 7050 and a greater strength advantage over I/M 7049 with equal SCC performance. It should be noted that this type of web-flange die forging represents a very rigorous SCC test for I/M alloys because of the very distinct grain elongation developed in the flange in the flash plane (See Ref. 9, Fig. 11 and 12). The relative performance differ-

ence between MA67 and the I/M alloys is similar after 500 days in atmosphere or after 30 days of the 3.5% NaCl solution alternate immersion (A.I.) test (Federal Test Method 823), shown in Fig. 8.

For P/M 1.5-in. (38 mm) thick plates described in Ref. 12, 650 days of atmospheric exposure have been completed, with only a limited number of tempers of the P/M alloys in test (Fig. 9). P/M MA67 plate showed a 7 and 12 ksi (48 or 83 MPa) short transverse yield strength advantage

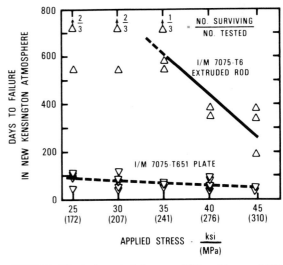

FIGURE 5 Effect of product form on SCC resistance of I/M 7075. Short transverse specimens from plate compared to transverse specimens from 1.56″ (39.6 mm) octagonal rod.

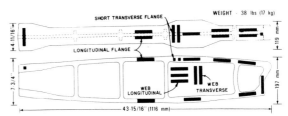

FIGURE 6 Rib-web forging Alcoa die 9078 with schematic illustration of specimen locations.

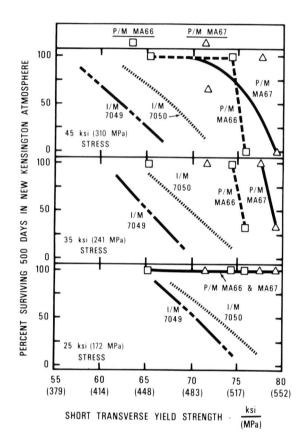

FIGURE 7 Effect of strength and sustained stress on percent surviving 500 days in New Kensington atmosphere. Tensile bar specimens from rib of rib-web type die.

over 7050 and 7075, respectively, with equal SCC resistance. These results are similar to the performance shown for 30 day exposure to the A.I. test in Fig. 10.

In summary, P/M MA 67 wrought products with over one year atmospheric stressed exposure offer a 7 to 12+ ksi (48 to 83 + MPa) strength advantage over 7050 and 7049 alloy products with equal SCC resistance. The level of SCC resistance of the P/M products increases with decreasing yield strength as in the commercial I/M alloys.

All of the SCC performance discussed above was for P/M products from inert gas preheat process material. Turning to the

stress corrosion performance of vacuum preheated P/M alloys MA87 and MA67, extrusions and die forgings are in atmospheric test but have not been exposed for a

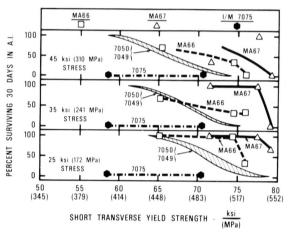

FIGURE 8 Effect of strength and applied stress on percent surviving 30 days in the alternate immersion stress corrosion test. Tensile bar specimens from die forgings.

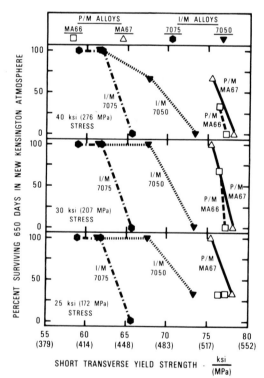

FIGURE 9 Effect of strength and sustained stress on percent surviving 650 days in New Kensington atmosphere short transverse tensile bar specimens from 1.5 to 2.5" (38–64 mm) thick plate.

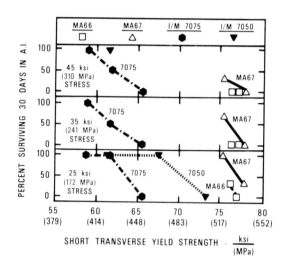

FIGURE 10 Effect of applied stress and yield strength on percent surviving 30 days in alternate immersion SCC test. Short transverse tensile bars from 1.5–2.5″ (38–64 mm) thick P/M and I/M 7050 and I/M 7075 plate.

sufficient time to give meaningful results. Test results are, however, available from the accelerated A.I. test for 84 day exposure (Federal Test Method 823) and are shown in Fig. 11 and 12 for extrusions and die forgings, respectively.

P/M alloys MA87 and MA67 developed combinations of yield strength and SCC resistance that are better than the control I/M 7050 alloy. The relative performance of I/M 7075 forgings is indicated by results shown in Ref. 12. These P/M alloys are capable of 5 to 9 ksi (34 to 62 MPa) higher transverse strength than I/M 7050 with equal SCC resistance, and 8 to 14 ksi (55 to 97 MPa) higher strength than I/M 7075 with equal SCC resistance.

In the more rigorous SCC test of short transverse specimens from die forgings, P/M alloys MA87 and MA67 developed strength and SCC resistance better than I/M 7050 and I/M 7075 at all test stresses in the 84 day A.I. test (Fig. 12). These P/M alloys offer 7 to 12 ksi (48 to 83 MPa) higher short transverse yield strength than I/M

7050 control die forgings with equal SCC resistance. This is nearly the same strength advantage as after 30 days A.I. exposure.

These accelerated SCC test results on vacuum preheated products are expected to be verified in atmospheric exposure, based on atmospheric test results of argon-preheat products.

Fracture Toughness

The improvement in fracture toughness of aluminum P/M wrought products shown in Fig. 1 with vacuum preheat has recently been verified for extrusions and die forgings (13, 14). Because this development is relatively recent, comparable fracture toughness data on the new, high toughness I/M alloys is as yet very limited. One significant comparison for extrusions was previously reported (Ref. 8, Table 10)

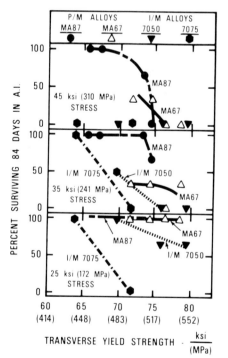

FIGURE 11 Effect of strength, applied stress and alloy on percent surviving 84 days in the alternate immersion stress corrosion test. Transverse tensile bars from 2″ (51 mm) diameter extruded rod.

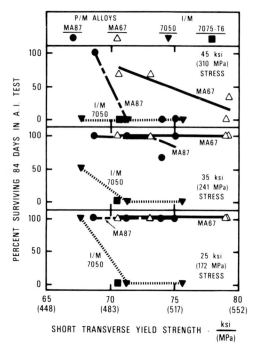

FIGURE 12 Effect of strength and applied stress on resistance to stress corrosion cracking of P/M alloy die forgings. Compared to control I/M 7075-T6 and 7050-T7X die forgings.

even at these lower levels, MA67 can match the median fracture toughness at equal strength of I/M 7049 forgings while offering improved SCC resistance.

Properties from one limited experiment with MA87 plate 1.5-in. (38 mm) thick are summarized in Table 3. The ductility in all test directions and the short transverse fracture toughness of the MA87 plate are all respectably high. Compared to I/M high strength plate (Table 4), P/M MA87 can match the fracture toughness of 7475 and 7050 and exceed the toughness of 7075 and 2124.

With the vacuum preheat process, P/M mill products have equal or superior fracture toughness compared to commercial I/M alloys, while offering higher useable yield strengths than those of existing alloys because of concurrent improved SCC resistance.

showing P/M extrusions matching the toughness of 7050 alloy extrusions.

In die forgings, MA87 offers equal or superior fracture toughness at equal or higher strengths compared to I/M 7050, 7049 and 7175 forgings (11) in both longitudinal and short transverse directions. Fig. 13 shows this relationship between yield strength and fracture toughness for a number of alloy forgings, with the candidate plane strain stress intensity factor K_Q being the fracture toughness measure. This level of fracture toughness at yield strengths above those proposed for 7050, 7049 and 7075 represents a marked improvement in capability for high strength aluminum alloy mill products.

MA67 alloy, with substantially higher Co and Zn than P/M MA87, shows somewhat lower fracture toughness. However,

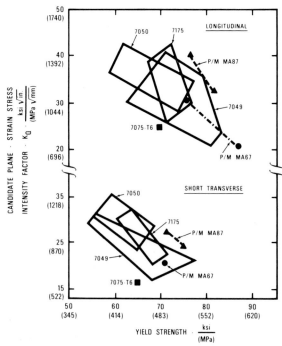

FIGURE 13 Comparison of fracture toughness of P/M MA67 and MA87 die forgings to commercial I/M alloy forgings.

TABLE 3

Properties of P/M MA87 - 1.5" (38mm) Thick Plate

Direction	Sample A[1] Y.S. (KSI)	% El. in 4D	K_{I_C} KSI√in.	Sample B[2] Y.S. (KSI)	% El. in 4D	K_{I_C} KSI√in.
Longitudinal	73.1	15.6	N.D.[3]	73.6	14.9	N.D.
Long Transverse	72.5	15.6	N.D.	73.0	15.6	N.D.
Short Transverse	71.4	9.4	32.9[4]	72.6	5.5	25.6

Direction	Y.S. MPa	% El. in 4D	K_{I_C} MPa√mm	Y.S. MPa	% El. in 4D	K_{I_C} MPa√mm
Longitudinal	504	15.6	N.D.	507	14.9	N.D.
Long Transverse	500	15.6	N.D.	503	15.6	N.D.
Short Transverse	492	9.4	1145[4]	500	5.5	891

Notes: 1. Hot Pressed Dwell = 10 minutes
2. Hot Press Dwell = 1 minute
3. N.D. = Not Determined
4. K_q - Specimen too thin per ASTM E399

TABLE 4

Plane-Strain Fracture Toughness of P/M MA87 Alloy
Plate, Compared to Plate of Various I/M Alloys[1]
Short Transverse Direction Only

Alloy	K_{I_C} KSI√in.	Y.S.-KSI	K_{I_C} MPa√mm	Y.S.-MPa
P/M MA87-A	33[2]	71	1150[2]	490
P/M MA87-B	26	73	905	503
I/M 7475-T651	27	65	940	448
I/M 7475-T7351	33	54	1150	372
I/M 7050-T73651	26	63	905	432
I/M 7075-T651	18	65	625	448
I/M 7075-T7351	20	54	695	372
I/M 2124-T851	22	61	765	420

NOTES: 1. From Ref. 15.
2. K_Q Specimen too thin per ASTM E399.

Fatigue

Only limited fatigue testing of P/M mill products has been completed to date and only on extrusions from argon preheat process products. Rotating beam smooth and notched specimen and axial stress smooth specimen tests were reviewed in Ref. 7. It was noted in those tests that some P/M alloys could exceed the performance of I/M 7075 or match the upper level of the 7075 S-N curve fatigue band. Additional axial stress notched specimen fatigue tests on MA66 and MA67 extrusions are discussed here to update the fatigue performance of these materials. Both MA66 and MA67 show a substantial notched axial-stress fatigue endurance-limit (@ 10^7–10^8 cycles) advantage over the only commercially fabricated I/M alloy extrusion fatigue data from the same fatigue test conditions.

These P/M alloys appear to offer a 40% improvement in notched fatigue strength over commercial I/M 7075-T6510. Compared to the smaller, laboratory fabricated I/M 7075-T6 extrusion, the P/M extrusions show a 20% notched fatigue advantage. MA67, with 1.5% cobalt (equivalent to 3.6 volume % Co_2Al_9) was superior to I/M 7075-T6510 over the entire S-N range tested, while MA16 showed clear superiority only over 10^5 cycles at stresses under 25 ksi. The cobalt appears to enhance low cycle ($< 10^7$ cycles) fatigue resistance of P/M products in these tests.

The unusually high fatigue results in these notched specimen tests illustrate potentially superior performance. The fatigue testing to date suggests that the P/M wrought products at least match the fatigue performance of I/M 7075. Confirmation of an overall fatigue performance improvement will require substantially more fatigue testing of P/M wrought products.

Potential Applications for Wrought P/M Mill Products

Extruded rod and web-flange die forgings fabricated from vacuum-processed

MA87 and MA67 alloys are the only products that represent the full property potential of the aluminum P/M mill products. The properties of these materials in extruded 2-in. (51 mm) diameter rod (Table 5) and in the Alcoa Die 9078 forgings (Table 6) are summarized here to show the broad property potential of these materials in different product forms and in different tempers. The application potential of these materials discussed below is based on these properties.

Ordnance

A significant factor contributing to U. S. Army interest in the development of these materials was the potential for improved hardware performance through (1) improved corrosion resistance at required strength level, (2) weight reduction permitted by higher strength or (3) maximizing the combination of strength and fracture toughness.

One program to investigate this P/M product potential used the M16 rifle receiver forgings as a test vehicle to determine whether improved resistance to exfoliation corrosion could be obtained without sacrificing any of the strength of the standard production I/M 7075-T6 forgings. Among all the materials tested, P/M MA49 developed the highest strength with high exfoliation resistance, (17) clearly demonstrating the potential of P/M forgings in a temper that offers low residual stresses (to reduce machining distortion) and a yield strength that is 20+ ksi (138 MPa) higher than the minimum specified for this application. The P/M wrought alloys certainly have potential for use in future firearm applications.

The higher strength of P/M alloy prod-

TABLE 5

COMPARATIVE PROPERTIES OF EXTRUSIONS IN P/M ALLOYS
MA67 AND MA87 WITH I/M 7075 (TEST RESULTS)[5]

Alloy/ Temper		MA67 [1a]	MA67 [1b]	MA87 [1c]	MA87 [1d]	7075-T6	7075-T73
Tensile Str.	L	97	90	89	82	99	80
(KSI)	T	86	82	80	75	80	72
Yield Str.	L	93	86	85	76	87	73
(KSI)	T	79	75	73	67	72	64
% Elong.	L	11	10	12	16	10	12
	T	6	9	11	12	8	8
NTS/TYS[2]	L	1.14	1.20	1.34	1.39	1.31	1.35
	T	0.76	0.93	1.13	1.28	1.05	1.15
K_{IC}	L	29	38	42[3]	38[3]	35[3]	-
(KSI√in.)	T	-	19	22[3]	-	22	-
SCC --	45	31,38,61	32,61,P84	84,2P84	2P84	2,3,3	61,73,82
Days to	42	-	-	-	3P84	-	-
Fail in	35	39,79,P84	83,84,P84	3P84	2P84	3,3,6	3P84
AI at	25	3P84	3P84	3P84	2P84	43,57,60	3P84
Stress (KSI)							

Notes:
1. P/M alloys solution heat treated 2 hours at 910-920°F, cold water quenched, and first step aged 24 hours at 250°F. Second step aged at 325°F for indicated time:
   a. 2 hours   c. 6 hours
   b. 6 hours   d. 14 hours
2. NTS/TYS: Notch Tensile Strength/Tensile Yield Strength Ratio.
3. K_Q: Invalid per ASTM E399. Specimen too thin for K_Q and Y.S.
4. Federal Test Method 823: XP84 indicates number surviving 84 days in test.
5. See Table 1, Appendix for S.I. units.

TABLE 6

COMPARATIVE PROPERTIES OF DIE FORGINGS IN P/M ALLOYS
MA67 AND MA87 WITH I/M 7075 (TEST RESULTS)[6]

Alloy/Temper		MA67[1a]	MA67[1b]	MA87[1c]	MA87[1d]	7075-T6	7075-T73[3]
Tensile Str.	L	93	85	88	82	93	73
(KSI)	ST	86	80	83	77	80	-
Yield Str.	L	87	78	81	74	83	63
(KSI)	ST	79	73	75	69	71	-
% Elong.	L	10	12	12	14	12	13
	ST	4	8	11	9	9	-
NTS/TYS[2]	L	0.90	1.14	1.25	1.34	1.34	1.45
	ST	0.67	0.96	1.09	1.25	0.94	-
K_{I_C}	L	21.1	-	32.8[5]	-	-	-
	ST	-	-	24.5	-	-	-
SCC[4]- Days	45	29,46,47	84,2P84	71,78,84	2P84	1,2,2	-
To Fail in	42	-	-	-	3P84	-	No SCC
Al at	35	3P84	3P84	3P84	2P84	2,3,3	Failures
Stress	25	3P84	3P84	3P84	2P84	2,3,3	Anticipated
KSI							

Notes:

1. P/M alloys solution heat treated 2 hours at 910-920°F, cold water quenched, and first step aged 24 hours at 205°F. Second step aged at 325°F for indicated time:

   a. 1 hour   c. 4 hours
   b. 12 hours   d. 18 hours

2. NTS/TYS: Notch Tensile Strength/Tensile Yield Strength Ratio.
3. No 7075-T73 forgings evaluated. Value shown are considered average or typical.
4. Federal Test Method 823: XP84 indicates number surviving 84 days in test.
5. K_Q - Specimen too thin per ASTM E399.
6. See Table 2, Appendix for S.I. units.

ucts at satisfactory levels of toughness and corrosion resistance, offers the reduction of parasitic weight (bases, bodies, ogives) in artillery shells and rocket warheads.

Cartridge cases require a maximum combination of high fracture toughness and high yield strength. I/M Alloy 7475 is widely accepted as the best commercially available aluminum material for this application. Because P/M MA87 plate appears capable of matching the short transverse toughness of 7475 at somewhat higher strength (Table 4), P/M alloys may be the aluminum cartridge case materials of the future.

Aerospace

The very high strength P/M products of Alloy MA67 are of particular interest in members subjected to predominantly compressive stresses in use, where strength and corrosion resistance are the most significant design considerations. Upper wing skins, from plate or extrusions, stiffeners, longerons and landing gear forgings are only a few of the types of aerospace applications where the highest strength alloys could be useful.

A lower strength alloy like P/M MA87 offers broad potential, ranging from structurals that experience predominantly compressive loading for the high strength tempers to applications where tensile loading in any direction are well accommodated. The high fracture toughness and SCC resistance of the lower strength tempers of MA87, at yield strengths equal to 7075-T6, clearly make these materials useful for lower wing components, helicopter rotors and other major airframe primary load-carrying structural members.

Status of Aluminum P/M Mill Products

Alcoa Laboratories, on a U. S. Army

Table 1, Appendix
Comparative Properties of Extrusions in P/M Alloys
MA67 and MA87 with I/M 7075 (Test Results)

Alloy- Temper		MA67[1a]	MA67[1b]	MA87[1b]	MA87[1c]	7075-T6	7075-T73
Tensile Str.	L	669	620	614	565	682	552
(MPa)	T	593	565	552	517	552	496
Yield Str.	L	641	593	586	524	600	503
(MPa)	T	545	517	503	462	496	441
% El. in 4D	L	11	10	12	16	10	12
	T	6	9	11	12	8	8
NTS/TYS[2]	L	1.14	1.20	1.34	1.39	1.31	1.35
	T	0.76	0.93	1.13	1.28	1.05	1.15
KIC	L	1009	1322	1462[3]	1322[3]	1218[3]	-
(MPA√mm.)	T	-	661	766[3]	-	766	-
SCC[4] -	310	31,38,61	32,61,P84	84,2P84	2P84	2,3,3	61,73,82
Days to Fail	290	-	-	-	3P84	-	-
in Al at	241	39,79,P84	83,84,P84	3P84	2P84	3,3,6	3P84
Stress(MPa)	172	3P84	3P84	3P84	2P84	43,57,60	3P84

Notes:
1. P/M alloys solution heat treated 2 hrs. at 760-766 K, cold water quenched, and aged 24 hrs. at 394 K + second step age at 436 K for indicated time:
   a. 2 hours   b. 6 hours   c. 14 hours
2. NTS/TYS: Notch Tensile Strength/Tensile Yield Strength Ratio.
3. K_Q: Invalid per ASTM E399. Specimens too thin for K_Q and Y.S.
4. Federal test method 823: XP84 indicates number surviving 84 days in test.

Table 2, Appendix
Comparative Properties of Die Forgings in P/M Alloys
MA67 and MA87 with I/M 7075 (Test Results)

Alloy- Temper		MA67[1a]	MA67[1b]	MA87[1c]	MA87[1d]	7075-T6	7075-T73[3]
Tensile Str.	L	641	586	607	565	641	503
(MPa)	ST	593	552	572	531	552	-
Yield Str.	L	600	538	558	510	572	434
(MPa)	ST	545	503	517	476	490	-
% El. in 4D	L	10	12	12	14	12	13
	ST	4	8	11	9	9	-
NTS/TYS[2]	L	0.90	1.14	1.25	1.34	1.34	1.45
	ST	0.67	0.96	1.09	1.25	0.94	-
KIC	L	734	-	1141	-	-	-
(MPA√mm)	ST	-	-	853	-	-	-
SCC - Days to	310	29,46,47	84,2P84	71,78,84	2P84	1,2,3	-
Fail in A.I.	290	-	-	-	3P84	-	No Scc
At Stress	241	3P84	3P84	3P84	2P84	2,3,3	Failures
(MPa)	172	3P84	3P84	3P84	2P84	2,3,3	Anticipated

Notes:
1. P/M alloys solution heat treated 2 hrs. at 760-766 K, cold water quenched and aged 24 hrs. at 394 K + second step aged at 436 K for indicated time:
   a. 1 hour   b. 12 hours   c. 4 hours   d. 18 hours
2. NTS/TYS = Notch Tensile Strength/Tensile Yield Strength Ratio.
3. No 7075-T73 forgings evaluated. Values shown are considered average or typical.
4. Federal test method 823: XP84 indicates number surviving 84 days in test.
5. K_Q- specimen too thin per ASTM E399.

contract, is currently scaling up the aluminum P/M mill product fabrication to 3300-lb. (1497 kg) hot pressed compact, which will net 2900-lb. (1315 kg) billet for mill product fabrication.

Products being fabricated for this evaluation include hot rolled plate, a die forging and an extruded shape. All phases of the fabrication except powder making will be performed in production equipment, while all testing will be performed at Alcoa Laboratories.

Conclusions

1. P/M Alloy MA67 die forgings offer a 7 to 12 ksi (48 to 83 MPa) yield strength advantage over I/M 7050 with equal atmospheric SCC resistance and an even greater strength advantage over I/M 7049 and 7075 forgings.

2. P/M MA67 plate offers a 7 or 12 ksi (48 or 83 MPa) yield strength advantage over I/M 7050 or 7075 plate, respectively, with equal SCC resistance.

3. P/M MA87 die forgings are capable of equal or higher fracture toughness at equal or higher strengths compared to I/M Alloys 7050, 7049 and 7175 forgings in longitudinal and short transverse directions.

4. P/M MA67 die forgings are capable of comparable fracture toughness to I/M 7049 forgings at equal strength.

5. P/M MA87 plate is capable of equal fracture toughness compared to I/M 7475 and 7050 plate and superior toughness compared to I/M 7075 and 2124.

6. In notched specimen axial stress (R = 0.0) fatigue tests, P/M MA67 alloy is capable of 40% higher maximum stress to failure compared to I/M 7075-T6510 extrusions at 10^7–10^8 cycles.

Acknowledgments

This work was supported by the Aluminum Company of America and by a Manufacturing Methods and Technology contract from the U. S. Army, Frankford Arsenal. Constructive technical discussions and encouragement from J. P. Lyle, Alcoa Laboratories, are gratefully acknowledged.

References

1. J. P. Lyle, Jr., Metal Progress December 1952, Page 109.
2. R. J. Towner, Metals Progress, May 1958, Page 70.
3. S. G. Roberts, WADC Report No. 56-481, April, 1957.
4. S. G. Roberts, Summary Project Report No. MSPR61-69, Contract No. DA-04-200-507-ORD-886, November 15, 1961.
5. R. J. Towner, Annual Progress Report for September 29, 1961 to September 30, 1962, Project No. 593-32-004, Contract No. DA-36-034-ORD-3559RD, October 16, 1962.
6. A. P. Haarr, Final Report—Section III, Contract No. DA-36-04-ORD-3559RD, May 31, 1966.
7. J. P. Lyle, Jr. and W. S. Cebulak, "Properties of High Strength Aluminum P/M Products," *Metals Engineering Quarterly*. Feb., 1974.
8. J. P. Lyle, W. S. Cebulak and K. E. Buchovecky, Progress in Powder Metallurgy, 1972 28, Metal Powder Industries Federation, N.Y., 1972, Page 93.
9. J. P. Lyle and W. S. Cebulak, "Powder Metallurgy Approach for Control of Microstructure and Properties," presented at TMS-AIME Spring Meeting, June 1, 1973.
10. J. P. Lyle and W. S. Cebulak, "Fabrication of High-Strength Aluminum Products from Powder; *Powder Metallurgy for High Performance Applications*, Syracuse University Press, Syracuse, N. Y., 1972, P. 231–254.
11. J. T. Staley, Technical Report AFML-TR-73-34, May, 1973, Page 70.
12. W. S. Cebulak and D. J. Truax, Phase III Final Report, Contract DAAA25-70-C0358, September 29, 1972.
13. W. S. Cebulak, Second Quarterly Report, Contract No. DAAA25-72-C0593, February 28, 1973.
14. W. S. Cebulak, Third Quarterly Report, Contract No. DAAA25-72-C0593, June 11, 1973.
15. L. W. Mayer, "Alcoa Green Letter: Alcoa Alloy 7050," GL220, April 1973. Application Engineering Division, Aluminum Company of America.
16. J. G. Kaufman, et al, "Fracture Toughness, Fatigue and Corrosion Characteristics of X7080-T7E41 and 7178-T651 Plate and 7075-T73510, 7080-T7E42 and 7178-T6510 Extruded Shapes, Technical Report AFML-TR-69-255, November, 1969.
17. J. V. Rinnovatore, K. F. Lukens and J. D. Corrie, "Aluminum Alloys with Improved Resistance to Exfoliation," *Metals Engineering Quarterly*, August, 1973, p. 49–52.

Aluminum P/M parts are strong, economical and they save weight

Conventionally pressed and sintered aluminum powder metal parts have been commercially available for many years. They offer excellent strength, light weight and are economical. Also, properties can be improved through any one of several optional manufacturing steps. Here is an update.

by **Kurt H. Miska,** Associate Editor

In addition to the obvious advantages of light weight, excellent corrosion resistance and good finishing properties, aluminum P/M parts also offer:

■ Strength nominally equal to and often greater than most medium-density iron parts, and they compress to high densities with lower compacting pressures. This results in good green strength and permits easy handling.

■ More economy than copper-infiltrated iron parts.

■ Less energy is used for production because sintering is faster and at lower temperatures.

In addition, uncertain mechanical properties of finished parts is no longer a problem. American Powdered Metals Co. now guarantees minimum tensile and yield strength and elongation by alloy, density and heat treatment. Further guidance is offered by ASTM Standard B 595, "Standard Specification for Sintered Aluminum Structural Parts."

Heat treating, coining, sizing, cold forming and hot forging are optional manufacturing steps in the aluminum P/M process and the steps in the complete process are detailed in Fig 1.

Selecting powder compositions
Commercially available aluminum powder alloy compositions (Table 1) consist of blends of elemental powders with or without a lubricant and these cost from 62 to 66¢/lb ($1.36 to $1.45/kg). At this time there are no prealloyed aluminum P/M compositions commercially available.

Alloys 601AB and MD 69 are similar to wrought 6061. They offer strength, good ductility, corrosion resistance and can be specified for anodized parts. Grade 601AC aluminum powder is the same as 601AB but without an admixed lubricant and is used for isostatic and die wall lubricated compaction.

Alloys 201AB, AD 24 and Sinteral 1 are similar to wrought 2014. Al-coa's 201AB contains no manganese but Alcan's AD 24 contains 0.4%. Both compositions develop high strengths, especially when thermally treated. Alloy 201AC is the same as 201AB but without the lubricant.

When good conductivity is required, 602AB should be specified. Conductivity of this material ranges from 42.0 to 48.5% IACS (72.3 x 10^{-8} to 83.51 x 10^{-8} ohm·m) depending on heat treatment. 202AB is a relatively new composition developed for parts requiring high strength and high ductility and this grade is expected to be used for P/M forging preforms as well as for conventionally pressed and sintered parts. Alcan's MD 76 is a fairly complex P/M composition, which develops properties similar to wrought 7075.

For porous aluminum P/M parts, Chrysler Amplex and Alcan offer grades for oil-filled bearings and filters. Deletion of the lubricant provides advantages in forging, especially for larger parts. Material costs are reduced and sintering rates are speeded because lubricant burnoff is eliminated. Sintered strength and ductility are increased, especially for parts having cross-section thickness of greater than 1 in. (25.4 mm). However, lubrication of the die in compaction is necessary. This is not difficult with simple shapes but may be more difficult with complex and intricate parts.

Three ways to sinter
Aluminum P/M parts can be sintered in a controlled, inert atmosphere, in a vacuum or in air. Air sintering remains relatively controversial but there are indications that the approach is as valid as sintering in a protective atmosphere.

1 Composition of aluminum P/M blends

Grade	Cu	Mg	Si	Mn	Cr	Zn	Lub[b]
601AB (Alcoa)	0.25	1.0	0.6	—	—	—	1.5
201AB (Alcoa)	4.4	0.5	0.8	—	—	—	1.5
602AB (Alcoa)	—	0.6	0.4	—	—	—	1.5
601AC (Alcoa)	0.25	1.0	0.6	—	—	—	—
201AC (Alcoa)	4.4	0.5	0.8	—	—	—	—
202AB (Alcoa)	4.0	—	—	—	—	—	1.5
22 (Alcan)	2.0	1.0	0.3	—	—	—	1.5
24 (Alcan)[c]	4.4	0.5	0.9	0.4	—	—	1.5
69 (Alcan)[d]	0.25	1.0	0.6	—	0.10	—	1.5
76 (Alcan)[e]	1.6	2.5	—	—	0.20	5.6	1.5
Sinteral[f]	4.0	0.6	—	—	—	—	1.5[f]
Bearing[g]	3.5	0.65	0.10	—	—	—	—
MD4160[h]	1.0	—	1.0	—	—	—	—
MD4090[h]	—	—	2.0	—	—	—	—

[a] Balance is aluminum. [b] Lubricant. [c] Equivalent to wrought 2014. [d] Equivalent to wrought 6061. [e] Equivalent to wrought 7075. [f] Powder Metallurgy Inc.; proprietary lubricant. [g] Special Chrysler Amplex blend for oil-filled bearings; also contains 2.5-2.8% Sn and 1.75-2.25% Pb. [h] Premixes for porous parts; Alcan compositions.

According to Alcoa and Alcan, parts are sintered between 1100 and 1150 F (866 and 894 K) for 30 minutes, about one-half that of iron parts and two-thirds that of copper P/M parts. Time and temperature influence the final mechanical properties considerably. The recommended protective atmosphere is nitrogen with a dew point of −45 F (230 K). Dissociated ammonia can also be used but nitrogen is preferred since it results in the highest sintered properties and costs very little in large quantities.

Alcan's sintering recommendations roughly parallel those made by Alcoa. The temperature range is from 1095 to 1155 F (863 to 898 K) for periods up to 30 minutes. The sintering temperature must be controlled to within ±9 F (±5 K) with the nitrogen having a dew point of −40 F (233 K) or drier. Alcan grade MD-22 develops good properties with a wider sintering temperature range of ±14 F (±7.5 K).

Sintering aluminum P/M parts in air results in notable cost reduction, according to Samuel Storchheim, a leading exponent of the process. He notes that surface oxidation that occurs in air sintering stiffens the shape and keeps it from sagging. The result is that tolerances can be kept within 0.0005 in. (0.0127 mm) without coining.

Storchheim reduces the problem of internal oxidation by compacting parts to 95% density and eliminating the lubricant, which, when it burns off, results in porosity. He also speeds furnace heating to reduce sintering time. Again, however, this process requires die wall lubrication systems in pressing.

Wide range of properties
Aluminum P/M parts sintered in nitrogen develop ultimate tensile strengths ranging from 11 to 50 ksi (76 to 345 MPa), depending on the composition, density and thermal treatment. For example, a 601AB part compressed to 95% density and heat treated to the T4 condition has the following properties: tensile strength 22 ksi (152 MPa), yield strength 15.0 ksi (103 MPa) and 5% elongation in 1 in. (25.4 mm). The same part made of alloy 201AB, at the same density and thermal treatment, has a tensile strength of 38.0 ksi (262 MPa), yield strength of 31.0 ksi (214 MPa) and 5% elongation in 1 in. (25.4 mm).

Since these two alloys achieve properties by solution and precipitation of soluble alloying elements, the as-sintered strengths of both grades are affected by the cooling rates from the sintering temperature. Parts cooled very slowly, i.e., 50 F/hr (28 K/hr) will develop the low strengths typical of the soft, annealed temper. Rapid cooling, as by water quenching, will precipitation harden at room temperature to T4 temper properties. Actual cooling rates are somewhere between these extremes.

In the T6 temper, high density 601AB parts develop tensile strengths of approximately 35.0 ksi (241

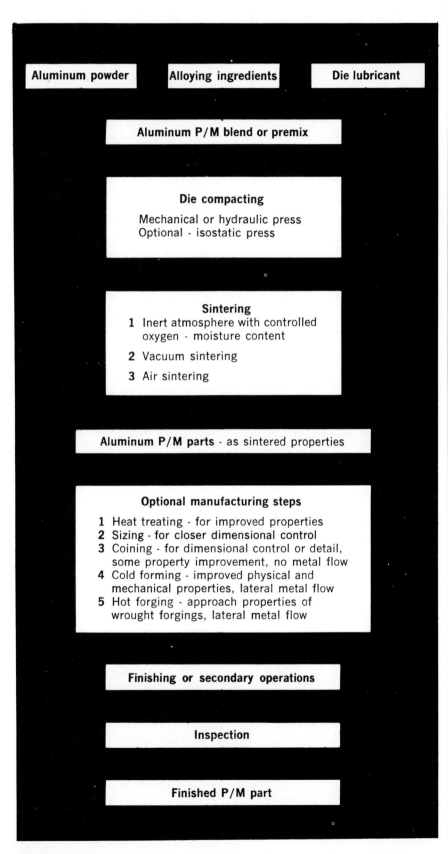

| Aluminum powder | Alloying ingredients | Die lubricant |

Aluminum P/M blend or premix

Die compacting
Mechanical or hydraulic press
Optional - isostatic press

Sintering
1 Inert atmosphere with controlled oxygen - moisture content
2 Vacuum sintering
3 Air sintering

Aluminum P/M parts - as sintered properties

Optional manufacturing steps
1 Heat treating - for improved properties
2 Sizing - for closer dimensional control
3 Coining - for dimensional control or detail, some property improvement, no metal flow
4 Cold forming - improved physical and mechanical properties, lateral metal flow
5 Hot forging - approach properties of wrought forgings, lateral metal flow

Finishing or secondary operations

Inspection

Finished P/M part

1 Process flow chart for aluminum powder metal parts production.

Source: *Materials Engineering*, Apr 1975

MPa), with slightly higher strengths achieved if the parts are repressed prior to heat treatment. Similar 201AB parts develop 48.0 ksi (331 MPa) tensile strength in the T6 condition and this increases to 52.0 ksi (358 MPa) with repressing.

Air sintered aluminum P/M parts made from Sinteral 1 alloy develop strengths comparable to nitrogen sintered parts. Tensile strengths range from 13.2 to 36.8 ksi (91 to 254 MPa), the low value being for parts with a green density of 85% and heat treated to the T1 condition. The high value is for parts compacted to 97% density and heat treated to T4 condition. Elongation in 1 in. (25.4 mm) reaches 3.0% for 97% dense parts heat treated to T1 condition. The least elongation is 0.7% for a 95% dense part heat treated to the T4 condition.

Additional property information on sintered aluminum P/M parts is given in Table 2.

Hot forging moves metal
Although compacting and sintering produces practical aluminum P/M parts, powder forging combines the advantages of both P/M and forging. The fabrication sequence is simple. Powder is compacted into a preform shape and sintered in a controlled atmosphere. The preform is then hot forged within a confined die in one step to the final part shape. Powder forging provides cost savings over conventional forging through reductions in labor and tooling requirements, greater materials utilization, elimination of trimming and fewer secondary operations.

Alcoa compositions 601AB, 602-AB, 201AB and 202AB are designed for forgings. The last of these is said to be especially suited for cold forging and all of the aluminum powder alloys respond to strain hardening and precipitation hardening, providing a wide range of properties.

For example, hot forging at 800 F (700 K) followed by heat treatment gives tensile strengths of 32.0 to 38.0 ksi (221 to 262 MPa) ultimate and 20.0 ksi (138 MPa) yield along with 6 to 16% elongation (in 1 in.; 25.4 mm) with alloy 601AB-T4. Heat treated to T6 condition, 601AB has ultimate tensile strengths from 44.0 to 50.0 ksi (303 to 345 MPa), yield is 44.0 to 46.0 ksi (303 to 317 MPa) and up to 8% elongation. Forming pressure and percentage of

2 Range of sizes of practical aluminum P/M parts encompasses 0.75-in. long pinion gear to complex plate approximately 4.125 in. across and 0.75 in. thick.

2 Sintered properties of aluminum P/M compositions

Comp	Green density, %	gm/cu cm	Thermal cond	Ten str ksi (MPa)	Yld str ksi (MPa)	Elong in in., %	Rockwell hardness
601AB[a] (Alcoa)	85	2.29	T1	16.0 (110)	7.0 (48)	6.0	55-60R$_H$
			T4	20.5 (141)	14.0 (96)	5.0	80-85R$_H$
			T6	26.5 (183)	—	1.0	70-75R$_H$
	90	2.42	T1	17.5 (121)	8.0 (55)	7.0	60-65R$_H$
			T4	—	—	—	80-85R$_H$
			T6	32.5 (224)	31.0 (214)	2.0	75-80R$_H$
	95	2.55	T1	18.0 (124)	8.5 (59)	8.0	65-70R$_H$
				17.0 (117)[f]	9.0 (62)[f]	3.0[f]	—
			T4	22.0 (152)	15.0 (103)	5.0	85-90R$_H$
				21.0 (145)[f]	14.0 (96)[f]	4.0[f]	—
			T6	36.5 (252)	35.0 (241)	2.0	80-85R$_H$
				28.0 (193)[f]	—	—	—
201AB[b] (Alcoa)	85	2.36	T1	24.5 (169)	21.0 (145)	2.0	60-65R$_E$
			T4	30.5 (210)	26.0 (179)	3.0	70-75R$_E$
			T6	36.0 (248)	—	—	75-80R$_E$
	90	2.50	T1	29.2 (201)	24.6 (170)	3.0	70-75R$_E$
				20.0 (138)[f]	18.0 (124)[f]	1.5[f]	—
			T4	35.6 (245)	29.8 (205)	3.5	75-80R$_E$
				24.0 (165)[f]	19.0 (131)[f]	2.0[f]	—
			T6	46.8 (323)	—	—	85-90R$_E$
				32.0 (221)[f]	—	—	—
	95	2.64	T1	30.3 (209)	26.2 (181)	3.0	70-75R$_E$
			T4	38.0 (262)	31.0 (214)	5.0	80-85R$_E$
			T6	48.1 (332)	47.5 (327)	2.0	85-90R$_E$
202AB[c] (Alcoa)	90[d]	2.5	T1	23.2 (160)	10.9 (75)	10.9	—
			T4	28.2 (194)	17.2 (119)	8.0	—
			T6	33.0 (227)	21.3 (149)	7.3	—
	90[e]	2.5	T2	33.9 (234)	31.4 (216)	2.3	80R$_E$
			T4	34.3 (236)	21.5 (148)	8.0	70R$_E$
			T6	39.8 (274)	25.1 (173)	8.7	85R$_E$
			T8	40.6 (280)	36.2 (250)	3.0	87R$_E$
MD-22 (Alcan)	90	2.45	T1	24.0 (165)	16.0 (110)	6.0	83R$_E$
			T6	38.0 (262)	29.0 (200)	3.0	74R$_E$
MD-24 (Alcan)	90	2.50	T1	24.0 (165)	14.0 (96)	5.0	80R$_E$
			T6	35.0 (241)	28.0 (193)	3.0	72R$_E$
MD-69 (Alcan)	90	2.42	T1	18.5 (128)	10.0 (69)	10.0	66R$_H$
			T6	30.0 (207)	28.0 (193)	2.0	71R$_E$
MD-76 (Alcan)	90	2.52	T1	30.0 (207)	22.0 (152)	3.0	90R$_H$
			T6	45.0 (310)	40.0 (276)	2.0	80R$_E$

[a] Sinter 30 min at 1150 F (894 K) in N$_2$; avg dewpoint −45 F (230 K). [b] Sinter 30 min at 1100 F (866 K) in N$_2$; avg dewpoint −45 F (230 K). [c] Presinter 3 hr at 700 to 800 F (644 to 700 F) in N$_2$, sinter 15 to 30 min at 1165 F (902 K) in N$_2$ with avg dewpoint −45 F (230 K). [d] Compacts - preforms. [e] Cold-formed parts; 19% strain. [f] Guaranteed minimum; American Powdered Metals Co.

3 Forging preform for watchcase (left) and forged watchcase (right) are experimental parts.

4 Typical of very small aluminum P/M parts. Pinions are for small toothed belts in office equipment. The small pinion measures 0.75 in. long and is 0.56 in. in diameter.

reduction during forging influence final properties.

Properties of 52.0 to 58.0 ksi (358 to 400 MPa) ultimate, 37.0 to 38.0 ksi (255 to 262 MPa) and 8 to 18% elongation are possible with 201AB heat treated to the T4 condition. When heat treated to T6, the tensile strength of 201AB increases 57.0 to 63.0 ksi (393 to 434 MPa). Yield for this condition is 56.0 to 60.0 ksi (386 to 414 MPa) and elongation ranges from 0.5 to 8%.

Cold forming betters properties

Properties of cold formed aluminum P/M alloys are increased by a combination of strain hardening densification and improved interparticle bonding. Alcoa has evaluated cold-formed alloys 601AB, 602AB, 201AB and 202AB.

Alloy 601AB achieves 37.3 ksi (257 MPa) tensile strength and 34.9 ksi (241 MPa) yield strength after forming to 28% upset. Properties for the T4 and T6 condition do not change notably between 3 and 28% upset. Alloy 602AB has moderate properties with good elongation. Strain hardening (28% upset) results in 32.0 ksi (221 MPa) tensile and 29.4 ksi (203 MPa) yield. The

5 Typical of large complex aluminum P/M parts. At left are two machine parts measuring 4.5 in. by 3.125 in. with a thickness of 1 in. The gear in the foreground is 3.75 in. in diameter and 0.44 in. thick. It is intended for a copier. Part at right is 4.125 in. across and 0.75 in. thick.

surface finishes and outstanding tool life. Small broken chips result when machining 90% dense parts from 601AB and 201AB alloy.

Some current applications

Even though the total amount of aluminum powder going into P/M parts is very small, literally millions of pressed and sintered parts are being produced.

One example is for an automotive suspension application and is being produced by the hundreds of thousands. Formerly this part was made from iron P/M and now the part is being made by American Powdered Metals Co. from alloy 201AB heat treated to the T4 condition. When the customer made his decision to use aluminum instead of the iron powder, the part had to pass exactly the same tests as the iron part. Available iron tooling is used to produce the part. Cost savings using aluminum in this application are expected to be at least 10%.

A relatively complicated part for a truck antiskid braking system was made of 6061 aluminum, solid extruded followed by machining. Now, alloy 601AB is being used and this provides savings from 50 to 100%.

A fuze part in the final stages of development was formerly an aluminum impact extrusion but now this part is scheduled for production at a rate of nearly 1.5 million units per month from 601AB heat treated to the T4 condition. The aluminum powder part measures 1.56 in. (39.7 mm) in diameter, 0.625 in. (15.8 mm) high and with a wall thickness of 0.125 in. (3.17 mm).

Figure 2 shows a representative range of sizes possible with aluminum P/M press and sinter techniques. The small pinion is for a piece of office equipment and measures only 0.75 in. (19 mm) long. The other small part is for a sewing machine. The large plate is under development and measures 4.125 in. (104.7 mm) across, is 0.75 in. (19 mm) thick and weighs 0.75 lb (0.34 kg).

Figure 3 shows two watchcases. At left is a P/M forging preform compacted from 602AB and at right is the forged watchcase ready for further fabricating. This case is not yet in production because of some special finishing requirements. Figures 4 to 6 also show what can be done with aluminum P/M. □

T6 parts temper achieves 37.0 ksi (255 MPa) tensile and 33 ksi (227 MPa) yield.

Highest cold formed properties are achieved by 201AB. In the as-formed condition, yield strength increases from 30.3 ksi (209 MPa) for 92.5% density to 40.7 ksi (281 MPa) for 96.8% density.

Alloy 202AB is best suited for cold-forming. The T2 temper, or as cold-formed, increases the yield strength significantly. In the T8 temper, 202AB develops 40.6 ksi (280 MPa) tensile and 36.2 ksi (250 MPa) yield with 3% elongation at the 19% upset level.

Design details simplified

Basic design details for aluminum P/M parts are the same as those for iron, copper and other powder metal compositions.

Aluminum can be substituted directly for iron but it is recommended that the part be designed to take specific advantage of the properties of aluminum. For example, American Powdered Metals Co. frequently tries their new tooling for iron parts by directly substituting aluminum.

The lower compacting pressures required for aluminum permit wider use of existing presses, and depending on the press, a larger part can often be made by taking advantage of maximum press force. An example illustrates this. A part with a 20-sq-in. (12,904-sq-mm) surface area and 2 in. (50.8 mm) deep is formed readily on a 500-ton press but the

6 Cold formed connecting rod at left is made from P/M preform at right. Preform is 90% dense and cold forming increases density to 99+%. Other preforms and cold formed parts are in the background. (Photo, Aluminum Company of America)

same part in iron would require a 600-ton press. The point is that the same part made out of aluminum, as might be made out of iron, requires a lower compacting pressure. The same holds for intricate shapes. Since aluminum responds better to compacting and it moves more readily in the die, more complex shapes can be produced. For this reason it is also possible to put more precise and smaller detail into aluminum P/M parts.

To hold tolerances, just as on iron parts, more sizing may be required. Parts with walls of 0.100 in. (2.54 mm) or less parallel to the vertical stroke of the press can prove to be a problem but the same wall thickness at right angles to the vertical stroke is no problem at all in aluminum.

Sintered aluminum parts normally do not have the smooth surfaces of most other P/M parts but repressing is one way to overcome this problem. Appearance is improved by mechanical and chemical treatments, etching to achieve texture, plating, anodizing or coloring.

Aluminum P/M parts offer many of the important machining advantages of wrought aluminum, including high cutting speeds, smooth

Acknowledgements: The author appreciates assistance provided by Alcan, Alcoa, American Powdered Metals Co., Metal Powder Industries Federation and Powder Metallurgy, Inc.

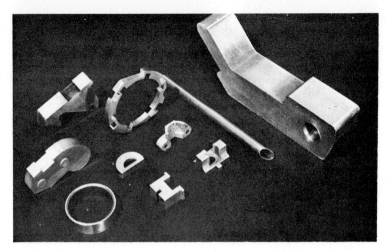

Fig. 1—These copper P/M parts demonstrate the design and manufacturing versatility of the process.

Production of P/M parts from copper powders

Copper parts for electrical and thermal application such as wires, contacts, tubing, etc., customarily have been produced from wrought copper; only a few have been made by P/M. Today, thin sheet and refrigeration tubing can be produced by P/M methods to give products with properties comparable to those made by conventional methods. As a matter of fact, a finer grain size is usually obtained in foil or tubing using P/M technology.

By P. W. TAUBENBLAT* W. E. SMITH** C. E. EVANS***

Copper has a combination of properties not found in any other metal. It has exceptionally high electrical and thermal conductivity. It has excellent resistance to corrosion, is highly ductile, exhibits good strength, and is totally nonmagnetic. In addition, it can be welded, brazed, and soldered without difficulty. It can also be plated. Its pleasing color finds many applications in the decorative field. All these factors make copper a most desirable P/M material.

An increasing number of P/M manufacturers are producing pure cop-

*Mr. Taubenblat is Manager, New Product Development AMAX Base Metals R&D, Inc.

**Mr. Smith is Research Group Leader AMAX Base Metals R&D, Inc.

***Mr. Evans is Technical Sales Representative AMAX Metal Powders

per parts. Intricate electrical parts and thermal components are being made by combining the advantages of copper with the advantages of P/M technology. The great potentials of this growth area are being developed.

Selection of copper powder. A variety of copper powders is now available in the commercial market. Their properties are strongly related to the methods by which they are produced. These are: atomization, chemical precipitation, and electrolytic deposition. Atomized powders are generally characterized by a high apparent density good flow rates, and moderate green strength. Compared to atomized powders, chemically precipitated powders usually have a lower apparent density, slower flow rate, and higher green strength.

The chemistry of atomized and precipitated powders and the amount of residual impurities present depend on the raw material used for melting (in the case of atomized powders), and on the purification techniques of mother solutions (in the case of chemically precipitated copper powders). Residual impurities such as phosphorus, iron, silicon, and arsenic must be extremely low since any of these, when present even in very small amounts in solid solution, will appreciably reduce both electrical and thermal conductivity. For example, 0.005 percent (50 ppm) of iron in solid solution reduces the electrical conductivity of pure copper by about 4 percent IACS, as may be seen by the wrought copper curve in figure 2.

The presence in solid solution of any one of the four elements shown in Table I in amounts greater than 0.005 percent would adversely affect both the electrical and thermal conductivity. However, not all the impurities found in a P/M product are necessarily in solid solution. For example, iron may be present either in solid solution or in the form of discrete particles or-

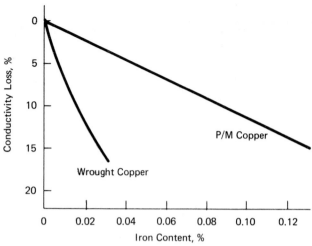

Fig. 2 — Effect of Iron on Conductivity

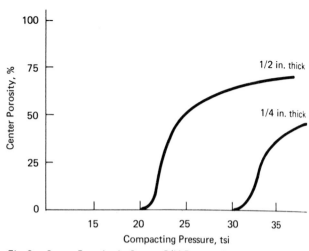

Fig. 3 — Center Porosity in Copper P/M Parts vs Compacting Pressure

iginating either from the copper production process or from mechanical contamination during powder blending or sintering. If it is present in solid solution, conductivity will follow the curve for wrought copper figure 2. If it is present as a mechanical contaminant, conductivity will follow the curve for P/M copper, or lie somewhere close to it.

Phosphorus has an even greater impact than iron on conductivity, and should be specially avoided.

Electrolytic copper powder can be produced to a variety of apparent densities and particle sizes by adjusting the conditions in the electrodeposition tanks. Since pure electrolytic copper in the form of cathodes is used as anodes for powder production, the composition of the final product is uniform from lot to lot. Typical composition and properties of an electrolytic copper powder are shown in Table I.

High purity electrolytic copper powder is also characterized by the high green strength, which is a major consideration for the production of complicated shapes, especially those with thin-walled sections. The green strength of about 2300 psi, obtainable at 20-tsi compacting pressure, is highly advantageous for most intricate applications.

Selection of processing conditions. Type LO electrolytic copper powder (1) was used to determine the optimum processing conditions for production of P/M parts. Electrical conductivity was used to check degree of sintering and part soundness.

Compacting pressure. The compacting pressure was found to be a function of the thickness of the parts made. A compacting pressure above 20 tsi should be avoided for parts more than ½-inch thick.

Figure 3 shows the effect of compacting pressure on uniformity of porosity distribution throughout the part. At a compacting pressure of 15 to 20 tsi no difference in the amount of porosity was detected between the center and the rest of the part.

Figure 4 illustrates a poorly sintered center structure, the result of using too high a compacting pressure. Figure 5 shows a sound center structure, required for strong, uniform, high conductivity sintered parts.

Gas atmosphere. To obtain optimum conductivity and strength, dissociated ammonia or hydrogen are best suited for the sintering atmosphere. Endo gas and exo gas can be used, but will produce parts with slightly less conductivity and strength. Table II illustrates the effect of various atmosphere gases on conductivity.

Sintered density. Once a high purity copper powder has been selected, sintered density is the next most important factor affecting properties. To achieve optimum strength and conductivity, sintered density must be uniform throughout the part. When this is accomplished (by using a proper compacting pressure) conductivity will follow approximately the curve shown in figure 6, while strength and elongation will follow those of figure 7. In a single-press, single-sintering cycle tensile strength of 30,000 psi and elongation of 33 percent can

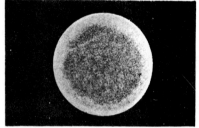

Fig. 4—Excessive compacting pressure causes poorly sintered center structure.

Fig. 5—Proper compacting pressure produces sound center structure.

be obtained. By using a double-pressing, double-sintering cycle, conductivity of 95 percent, strength of 35,000 psi, and elongation of 35 percent in 1 inch can be achieved at a density of 8.45 g/cc.

Table III suggests optimum conditions for obtaining sound, high-strength, high-conductivity parts from electrolytic copper powder. Figure 1 shows several parts presently produced by various manufacturers from this powder, and points out the versatility of design and manufacturing techniques possible with P/M. [pm]

(1)Type LO copper powder is produced by United States Metal Refining Co., an AMAX subsidiary.

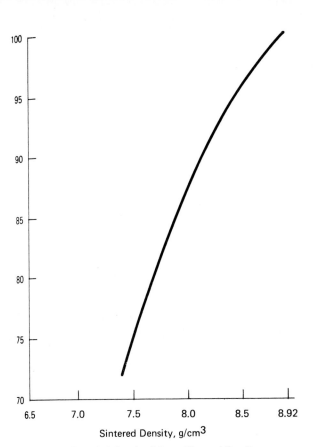

Fig. 6 – Electrical Conductivity vs Sintered Density

TABLE I

TYPICAL PROPERTIES OF ELECTROLYTIC COPPER POWDER FOR HIGH CONDUCTIVITY PARTS *	
Chemical Analysis	
Copper	99.8 percent
Phosphorus	< 0.001
Iron	< 0.005
Silicon	< 0.002
Arsenic	< 0.0005
Screen Analysis	
+100 mesh (percent)	0.1
−325 mesh (percent)	50
Apparent Density	2.6 g/cc
Flow	29 sec/50 g
Green Strength at 20 tsi	2300 psi

* AMAX type "LO" Copper Powder

TABLE II

EFFECT OF SINTERING ATMOSPHERE ON CONDUCTIVITY	
Atmosphere	**Conductivity percent IACS**
Hydrogen	87
Dissoc. Ammonia	86
Endothermic	85
Exothermic	83

Conditions: Compacted at 20 tsi, sintered at 1000° C (1830° F) for 30 min.

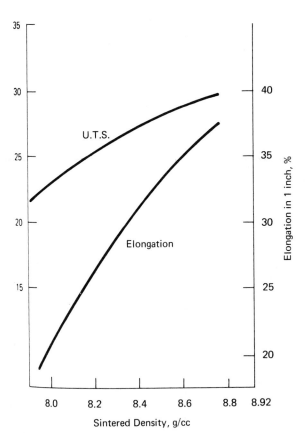

Fig. 7 – Strength and Elongation of Copper P/M Parts as a Function of Sintered Density

TABLE III

OPTIMUM PROCESSING CONDITIONS *

1. **Raw Material** — High Purity Copper Powders, avoid especially presence of P, Fe, Si and As

2. **Compacting Pressure** — Between 15 and 30 tsi depending on:
   a. The amount of volatiles present (including lubricant)
   b. Thickness of the finished part

3. **Sintering Conditions** — 1830° F (1000° C) for 30 min at temperature, preferably, in dissociated NH_3 or H_2

NOTE: Steps 4 and 5 are required only if conductivities above 85-90 IACS are desired

4. **Coin** at least 10 tsi higher than the original compacting pressure

5. **Re-sinter** at 1830° F (1000° C) for 15 min at temperature, preferably in dissociated NH_3 or H_2

* These conditions will produce high strength, high conductivity P/M parts.

The Powder Metallurgy of High-Strength Ti Alloys

By GERALD FRIEDMAN

For critical aerospace applications, conventional powders and press-sinter techniques are not suitable. Prealloyed powders made by the Rotating Electrode Process are preferred over other types.

WHEN it is realized that normal hot-work practices require forging billets weighing 10 to 20 times as much as the final-machined parts, it is easy to understand the rationale for a powder metallurgy approach to cost reduction in aerospace applications.

In addition, as designers press for higher and higher-strength alloys and the physical metallurgist responds with more highly alloyed compositions, problems of chemical segregation and workability increase to the point where properties obtained in lab samples or sub-scale production cannot be realized in the required large sizes. Use of prealloyed powder eliminates these problems because each powder particle contains the full complement of alloying elements in the correct proportion. This leads to greater workability and uniformity in mechanical properties.

Mr. Friedman is with the Technical & Engineering Dept., Nuclear Metals Inc., Concord, Mass.

Two types of prealloyed powder have been examined for their suitability in aerospace applications. One is made by the hydrid-dehyride method. The powder is of reasonable purity and has sufficient green strength to be cold compacted in either a die or isostatic press. In comparison with powder made by the Rotating Electrode Process (REP), it is less pure with respect to both interstitial elements and metal species originating in the crushing apparatus, has a low bulk density, and is more expensive.

Background — There is a tendency to think of P/M as a press-and-sinter operation, with the manufacturing sequence beginning with a fine, high-surface-area powder and ending with a precisely formed but slightly porous final object.

For critical aerospace applications, however, the structure and properties produced via press-sinter techniques are inadequate; and it is not surprising that titanium alloy powders differ considerably from the usual ferrous and nonferrous metal powders. The processing of titanium powder must be carefully controlled to eliminate oxidation or reaction with lubricants. Unlike the case with iron powder, it is not possible to reduce titanium oxide in a sintering furnace. For this reason, high-strength alloys are consolidated without lubricants, and are never exposed to even a partial pressure of air at elevated temperatures.

The residual porosity inherent in pressed-and-sintered parts confers a loss in ductility and fracture toughness, while the local chemical heterogeneities resulting from the use of mixed elemental powders cause local physical and chemical anomolies that are unacceptable in highly stressed parts.

In addition, the use of elemental powder, either alone or mixed with other metals, presents problems relating to the residual sodium or magnesium chloride left from the Kroll reduction of titanium tetrachloride. The chloride is normally removed in the arc melting stage, but since the crushed sponge normally used for press-sinter titanium P/M has not been subjected to this pyrometallurgical refining, the chloride remains, causing weld porosity and a lowering of corrosion resistance. For all of these reasons, the technology for structural parts is based entirely on the attainment of full density from high-purity prealloyed powder.

Process — The Rotating Electrode Process was developed in answer to a need for pure uranium and zirconium alloy powders. Because of their chemical reactivity, these metal powders must be produced

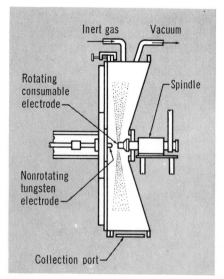

Fig. 1 — Apparatus for the Rotating Electrode Process.

Inert gas — Vacuum

Rotating consumable electrode —

— Spindle

Nonrotating tungsten electrode —

Collection port —

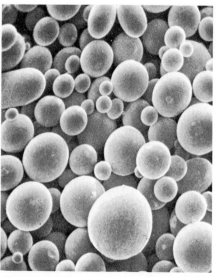
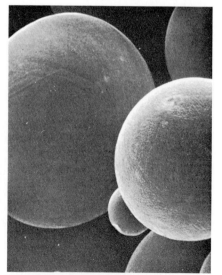

Fig. 2 — Two scanning electron photomicros of Ti-6Al-2Sn-4Zr-6Mo REP powder. Left is at 50×; right is at 200×.

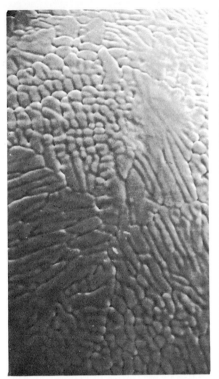

Fig. 3 — Structure of a Ti-6Al-2Sn-4Zr-6Mo REP particle at 1000×.

in an inert atmosphere. The process has proved to be equally well suited for producing powders from titanium and other reactive and refractory metals, as well as the more common engineering alloys. Powders produced by this process include most metals on the periodic chart.

REP powder is produced by the action of an electric arc impinging on the face of a rapidly rotating electrode, a titanium bar. The rotating electrode and a nonrotating, nonconsumable electrode are contained within a large circular chamber that has been evacuated and then filled with helium (Fig. 1).

As the titanium electrode rotates, centrifugal force causes the molten metal produced by the arc to fly off in the form of fine droplets that freeze in flight into spherical microcastings as they fall to the floor of the chamber. Particle size is controlled by the electrode diameter and the speed at which it is rotated; a typical titanium electrode is 63.5 mm (2½ in.) in diameter. Since the titanium electrode is melted in the absence of any crucible, and the melting takes place in a high-purity static gas atmosphere, the powder is of high purity, with a chemical analysis virtually identical to that of the starting material.

Powder particles are predominantly spherical, with only a few ellipsoidal in shape. They are very smooth, with a specific surface of approximately 0.009 m^2/g, and seldom consist of twins or particle clusters (Fig. 2).

Particle size distribution falls within a normal Gaussian distribution, with an average size of 200 to 250 microns. Virtually no particles are finer than 325 mesh (44 microns).

The bulk (loose) density of this powder is typically 60 to 62% of theoretical. Individual particles only infrequently contain shrinkage cavities and typically display the full density of the parent alloy. The powder mass can be vibrated to a tap density of 64 to 66% of theoretical. This high density is extremely advantageous in consolidation by the hot isostatic process.

Microstructure — Because these powder particles are formed "from the melt," all prior processing history is erased, and the microstructure is that of a chill casting showing a fine dendritic structure (Fig. 3), with typical dendrite arm spacings in the order of 3 to 6 microns. A natural corollary of the rapid freezing experienced by these particles is the fact that, for the case of two-phase alloys, the particles are in the solution-treated condition.

Partsmaking — Based on the powder attributes noted above — individual spheres, in the quenched state — it is not difficult to understand that REP titanium powder cannot be cold compacted into a cohesive mass. In fact, all consolidation is performed at elevated temperatures.

Hot consolidation techniques currently employed include vacuum hot pressing, hot isostatic pressing (HIP), extrusion, and forging. The first three techniques are frequently used as preliminary operations for forging. Except for hot pressing in a vacuum apparatus, in all of the above operations the powder is enclosed in a disposable evacuated canister, usually made of low-carbon steel.

Hot pressing has been successful in an extrusion press with the die replaced by a blank plate. Consolidation parameters included temperatures of 900 to 1010 C (1650 to 1850 F) at pressures of 1034 and 1379 MPa (150 000 and 200 000 psi).

Hot pressing is more usually carried out by the HIP process, which confers the ability to simultaneously control time, temperature, and pressure and where forms other than right cylinders may be fabricated. HIP is the most versatile of the consolidation processes because it is independent of rigid tooling. REP powder is especially well suited to HIP consolidation since its high and uniform bulk density minimizes the size of canister required and offers good predictability of the consolidated size and shape.

The size and shape of HIP'ed bodies are described by the relationship governing the change in volume accompanying consolidation of a porous body. The change in linear dimensions for a porous body is approximately equal to one-third of the change in porosity. In the case of powder with a 64% tap density, tap porosity is 36%. Isostatic consolidation of such a powder mass should theoretically reduce all dimensions by 12%.

Actual measurements of HIP structures show a change in diameter and thickness of 13 to 15%, based on the minimum dimensions in each direction. The use of a "shrinkage" factor of 16 to 18% therefore makes possible the design of a HIP canister that will produce a consolidated body requiring only minimum cleanup machining.

Applications — HIP consolidation of REP powder has multiple end uses: cylindrical forge bars, forge preforms, and the production of shapes requiring no further hot forming.

Figure 4 illustrates REP Ti-6Al-2Sn-4Zr-6Mo forge preforms. Typical HIP consolidation parameters for REP titanium alloy powders include time in the autoclave from 15 min to 4 h, temperatures of 850 to 1000 C (1560 to 1830 F) at pressures of 30 to 100 MPa (5000 to 15 000 psi).

Hot extrusion has been used to consolidate loose REP powder to round forging stock, as well as near-net structural shapes. In forming a structural shape directly from loose powder, extrusion reduction ratios (R = Billet Area/Product

Area) as low as 6 to 9:1 at temperatures similar to those used for HIP consolidation have proven feasible. Extrusion under such conditions through conical-approach dies simultaneously consolidates the spherical powder to full density and confers a wrought microstructure on the body thus formed. Unlike bodies consolidated by HIP, extrusions are anisotropic, with higher mechanical property values in the longitudinal than in the transverse direction.

Because of its high packing density, REP alloy powder is ideally suited to the forming of long complex shapes by the filled billet extrusion technique. Filled billet extrusion entails the construction of a cylindrical billet containing one or more cavities of simple or complex shapes, with or without cores. The cavities are filled with spherical powder, following which they are evacuated, and the billet is extruded through a round die. Subsequently, the structural shape is exposed by dissolving away the surrounding filler material in a suitable acid.

The extrusion of spherical powder into round forge bar is only a simplification of the procedures described above for complex shapes — powder encased in a steel can is extruded through a round die. Be-cause additional working will take place in the forging operation, the extrusion deformation and densification of forge bar need not be quite as great as for extruded-to-shape parts, and lower reduction ratios may be employed. Lower reduction ratios, in turn, result in lower extrusion forces and/or smaller billet and equipment sizes.

Evaluating Processes — A comparison of consolidation processes — with respect to time, temperature, pressure, and deformation mode — shows the following differences:

Extrusion exerts the highest pressures, over very short periods of time. The shear forces acting on the consolidated mass as it passes through the die are extremely effective in promoting particle welding; within a few seconds, the loose powder is transformed into a fully dense cohesive mass.

HIP consolidation utilizes pressures an order of magnitude lower than in extrusion for periods of time one to two orders of magnitude greater. HIP is an isothermal process. Rather close monitoring of actual processing temperatures can be exercised. The longer HIP cycles make good use of titanium oxide's solubility in titanium. Any high-oxygen surface layer resulting from an adsorbed gas film or a small leak in the canister can be easily diffused inward at the same time that bonding occurs at the particle surface. In short cycles where the hot billet is under pressure for only a few minutes, a post-consolidation heat treatment may be required to achieve this diffusion-removal of the high oxygen alpha case.

As noted above, as-produced two-phase powders are in a metastable quenched condition. Under usual processing conditions, consolidation is done high in the alpha-plus-beta or in the beta field, and heat treatment for full strengthening would follow consolidation and forming.

Nuclear Metals has consolidated some two-phase alloys (by rolling and by extrusion) in the lower alpha-beta regions at the aging temperature. Such processing, described as Micro-Quenched Age-Forming (MQAF) represents an attempt to further strengthen these alloys above the levels achievable by heat treatment alone.

Multipass rolling of pancake billets was used to produce Ti-7Al-2Cb-1Ta sheet from loose powder, using 30% reductions per pass. Extrusion consolidation proved more difficult because of the powder's great resistance to deformation at aging temperatures. One solution employed two or three low reduction (R = 4:1) sequential extrusions. A contemporary approach to this technique would utilize a low-temperature, high-pressure HIP cycle to deform and partially bond the particles, followed by a single extrusion to shear the particle surfaces.

REP titanium can be processed to the mechanical property level defined by specification for the alloy in question; the mechanical properties of current processed REP alloys equal or exceed those of their conventionally processed counterparts. Depending upon the interstitial level chosen, the powder can be manufactured and consolidated to produce bodies with high ductility and toughness (extra low interstitial grade) or high strength at lower levels of toughness (conventional oxygen content).

Fig. 4 — REP Ti-6Al-2Sn-4Zr-6Mo forge preforms after HIP.

Powder metallurgy: Thrifty shortcut to titanium parts

Advances in P/M techniques offer close control of part shape and dimension with minimal scrap loss, dramatically reduced machining cost, fine grained equiaxed structure. Non-aerospace applications seen accounting for up to 35% of titanium consumption by 1980.

TITANIUM powder metallurgy is an established art that makes available the unique properties of the metal in useful products at reasonable cost.

Why titanium? What will it buy that isn't available, for example, in a conventional ironbase P/M part? Titanium is relatively strong and lightweight, leading to a favorable strength-to-weight ratio. It is essentially chemically inert under a wide range of corrosive conditions.

Good performance at high temp

Alloys of the metal have good performance capabilities at sustained temperatures up to 1000F. Flame resistance is excellent, although it can contaminate with oxygen and become embrittled under prolonged exposure. Titanium is nonmagnetic, has low thermal conductivity, and a reasonable modulus of elasticity (15 to 17 million psi). It is essentially stainless and has long-lasting appeal.

Titanium powder now enjoys sustained availability and its base price is relatively stable. This is in sharp contrast to spiralling prices of most other metals.

Based on a paper presented at Conference on Designing for Powder Metallurgy at New York University by Edward L. Thellmann, Manager of Powder Metallurgy, and Gail F. Davies, Manager of Composite Development, Gould, Inc., Cleveland, OH 44108.

Why P/M process? The usual alternatives for procuring titanium parts are: 1) machining from prepared mill stock, 2) forging from mill stock and finish machining, and 3) employing P/M parts made from unalloyed, prealloyed, or alloys of elemental blended powders.

The powder metal method is often superior since P/M offers close control of shape and dimension with minimal scrap loss, greatly reduced or zero cost for secondary machining, and a fine-grained,

equiaxed structure with uniform strength in all directions.

So great are the versatility, range of products, performance capabilities, and economies of titanium powder metallurgy that it is difficult to sum up in one presentation.

Many applications for titanium involve exposure of high-surface electrodes, sparges, and filters to corrosive chemicals. Where permeability is required, methods for processing include gravity molding, slip casting, and pressing in dies

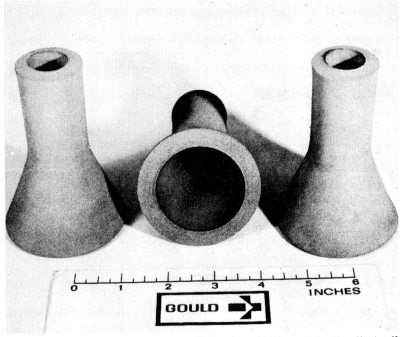

Fig. 1. Typical isostatically pressed titanium P/M part in the "green" as-pressed condition prior to sintering. Part weighs about ¾ lb.

Fig. 2. Gould has drummed up a brisk business in titanium P/M nuts, in a wide choice of sizes, for plating and anodizing applications.

that have limiting closure. Permeability and pore size are controlled by careful preselection of powder particle size and shape, and the parts are sintered in a vacuum under closely controlled conditions.

Properties of adequately worked P/M titanium compare favorably with those of arc melted wrought products. Typical properties of as-sintered unalloyed titanium (95% of theoretical density) are: 60,000 psi tensile strength, 45,000 psi yield, 15% elongation in 1 in., and elastic modulus of 15 x 10⁶. Typical properties of 6A1-4V alloy in as-sintered state (95.5% of theoretical density) are: 125,000 psi tensile strength, 105,000 min. yield, 5% min. elongation in 1 in., and elastic modulus of 17 x 10⁶.

Competitive with cast, wrought

The incorporation of forging or other wrought processes in conjunction with as-produced P/M forms yields a product competitive with its cast and wrought counterpart.

As early as 1954 hot-pressed and warm-coined titanium structures were evaluated in fatigue. Powder metal billets have been successfully extruded or swaged to optimum properties. More recently, preforms have been "gatorized" to provide acceptable compressor wheels and blades for aircraft engines. This approach, coupled with the high-volume capability of powder metallurgy to produce preforms, opens exciting possibilities for the future.

Probably the first product to gain industry recognition was the titanium filter. Titanium porous bodies ranging from 25% to 75% of theoretical density can be readily manufactured by powder metallurgy. The art of controlling permeability, pore size, and surface area is well developed.

Permeable, tubular titanium bodies have been produced in lengths up to 48 in. and diameters up to 7½ in. for handling films in liquid media. Other permeable titanium P/M bodies include sparges and a potentially high volume porous sheet component for electrical energy storage.

Titanium components having densities ranging from 85% to 95-98% of theoretical are usually produced either by hydrostatic or isostatic pressing or by high-volume mechanical pressing. The former include such shapes as preforms for subsequent forging and hardware approaching the dimensional tolerances of castings.

Proven marine use value

An example of a typical isostatically pressed titanium part is the funnel-shaped component shown in Fig. 1. This is a "green" (as-pressed) part prior to sintering. These parts weigh approximately ¾ lb.

Titanium has repeatedly proven its value in marine applications. As early as 1950, a predecessor company of Gould Inc. produced titanium powder metal valve seats for corrosion test at Kure Beach. After seven years' exposure to sea water, the seats showed the original machining marks. Additional tests were performed using many combinations of Teflon, stainless steel, and monel components with titanium. In every case, the titanium seats in bronze-bodied valves, when

Fig. 3. Approved aircraft fastener is pressed into donut shaped preform then coined into 12-point shape. Deformed blank (center) shows ductility.

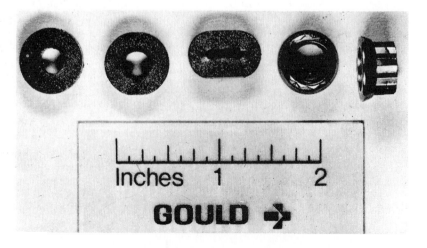

Inches 1 2

GOULD

Fig. 4. Holding fixture for soldering remains unwetted by molten metals. Machining of P/M part is minor compared with hogging part out of bar stock.

isolated from the bronze by a pitch coating, performed exceptionally well.

Quantity production of titanium P/M parts can be achieved by variations of the same approach used for powdered iron or bronze. Where quantity production is justified, the cost of dies can be readily amortized. Generally, hard tooling is desirable and requires experience with relief and titanium materials flow. Since titanium powders have relatively poor flow characteristics and are highly reactive, the choice of lubricant is important.

At Gould the production of titanium P/M nuts has become a substantial business. These are made from commercially pure titanium in a variety of sizes (Fig. 2) and are used in plating and anodizing operations, as are titanium spacers.

One of the more impressive applications of titanium P/M is shown in Fig. 3. In making this approved aircraft fastener, 6A1-4V

titanium alloy powder is pressed into the donut preform, then mechanically coined to the 12-point nut shape. The deformed blank (center) illustrates the ductility and formability of the fine-grained preform.

Hard to machine from bar

Titanium's surface stability is of interest to people employing liquid metals. The titanium P/M holding fixtures in Fig. 4 remain unwetted by most molten solders. Some minor secondary machining is necessary on this part before it is operational. However, its complexity illustrates the obvious difficulty of machining it completely from solid bar stock.

There is considerable interest by the Department of Defense in obtaining cost reductions through titanium and superalloy powder metallurgy techniques. The Air Force is sponsoring three major programs involving prototype structures in airframes and engines. These programs are exploring various parameters of 6A1-4V titanium alloy P/M parts production.

A contract with General Electric Co. has as its objective the establishment of manufacturing methods for hot isostatically pressed (HIP) 6A1-4V forging preforms for the J-79 engine compressor disc. This part is approx. 12 in. dia by 2.4 in. thick.

A contract with Crucible Materials Research Center concerns production of near net engine and airframe components. The engine component is a GEF 101 engine compressor stub shaft approx. 16 in. dia by 12 in. long, made from Ti 17. The airframe component is a 6A1-4V keel-splice fitting for the F-15 aircraft. It is a complex forged shape with webs and flanges,

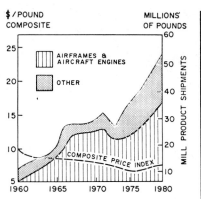

Fig. 5. Market projection shows non-aerospace applications taking up to 35% of titanium by 1980.

and is approx. 4- by 3- by 5-in. in size.

Non-aerospace use to 35%

A third contract with Pratt & Whitney Aircraft aims to establish manufacturing methods for production of near-net blended elemental titanium 6A1-4V components for engine applications. The part is the rear compressor synchronizer arm for the F-100 engine. It is approx. 14 in. long, 2 in. wide, and ½ in. thick.

A projection of where we believe titanium production and usage will be in the foreseeable future is shown in Fig. 5. It suggests that applications other than aerospace will account for perhaps 35% of total tonnage of titanium used by 1980. This will include chemical, structural, and possibly automotive applications. The projection was made at a time when the economy had slowed down and is considered conservative.

It is felt that current price trends for titanium and the economic and technical advantages of the powder metal process will help more users achieve objectives of better performance and greater durability at lower cost through titanium P/M. □

Application of Powder Metallurgy to Superalloy Forgings

M. M. Allen, R. L. Athey and J. B. Moore

IN THE development of certain advanced jet engines by Pratt & Whitney Aircraft it became apparent that Waspaloy and René 41, the most advanced nickel-base alloys then used for forgings, did not possess the high temperature strength required for turbine discs. Preliminary work indicated that a new wrought alloy known as Astroloy did exhibit the required strength and elevated temperature operating capability. The subsequent evaluation of several Astroloy forgings revealed that, while the desired potential was present, these forgings exhibited excessive scatter in mechanical properties due to gross structural segregation and lack of homogeneity. It was apparent that a considerable development effort would be required to produce Astroloy forgings with the quality and reliability necessary for advanced engines. A development program was therefore established by Pratt & Whitney Aircraft which dealt with all phases of disc production, including modification of the excessively broad Astroloy chemistry range, a comprehensive heat treatment investigation, full-scale ingot evaluation, and forging development.

The segregation noted in early Astroloy forgings could be traced back to the large columnar grain structure of the ingots being produced at that time. Metallurgical evidence in the form of macro- and microstructures of Astroloy, Waspaloy, and Inconel 718 from all major melting sources was compiled showing that segregation existing in conventional ingots was not eliminated by subsequent forging and thermal treatments. The problems directly attributable to the segregation of these columnar ingots included cracking during forging and heat treating, nonuniform strength and ductility, and a wide scatter of mechanical properties from disc to disc as well as within a given disc. The influence of ingot structure on final disc microstructure and mechanical properties was more evident in Astroloy than in weaker alloys.

Based on these findings, a full-scale Astroloy ingot evaluation program was initiated and funded by Pratt & Whitney Aircraft. The results of the evaluation showed that casting parameters yielding a uniformly fine, equiaxed grain structure minimized macrosegregation in the ingot. As a result of development aimed in this direction several melting vendors are now producing ingots with structure approaching that requested by Pratt & Whitney Aircraft. The development effort involving forging and heat treat practices in conjunction with the improved ingot has led to current production discs which have substantially more uniform properties and microstructure. While it is felt that further reduction in ingot segregation would yield additional improvements in uniformity and level of mechanical properties, it is doubtful that less segregated ingots than the improved product just discussed can be realized through conventional melting techniques.

The powder metallurgy approach to the production of large engine discs represents an extension of the ingot evaluation program to produce material with a minimum of macro segregation. Pratt & Whitney Aircraft, because of its need for higher strength wrought disc alloys than Astroloy, has performed development work on more complex alloy systems, the results of which further demonstrate the problem of ingot segregation. Through the use of ultrafine grain ingots, Pratt & Whitney Aircraft has demonstrated that the high temperature blade casting superalloys can be produced in wrought form. Small ingots having very fine equiaxed grain structure throughout have been produced from both IN 100 and Mar M-200 chemistries, Fig. 1.

These ingots were successfully forged into small discs. Macro- and microstructural examination revealed that while these discs had been produced from what could be considered an optimum fine grain ingot, an unacceptable degree of macro- and microsegregation was still present. Realizing that it was highly improbable that the structure of the small fine grain ingots could be reproduced in large ingots, it became apparent that more complex alloys would require a method of processing other than the large ingot approach.

The degree of macrosegregation in a small powder particle would be expected to be less by several orders of magnitude than that present in the structure of a large ingot conventionally melted and cast under optimum conditions. The amount of segregation in a large powder billet should be no more than that found in the individual powder particles making up the billet; therefore, in theory, there should be no limitation on powder billet size as regards the promotion of segregation. With a conventionally melted and cast ingot, however, we know that macro-segregation increases sharply with ingot diameter and eventually becomes prohibitive with

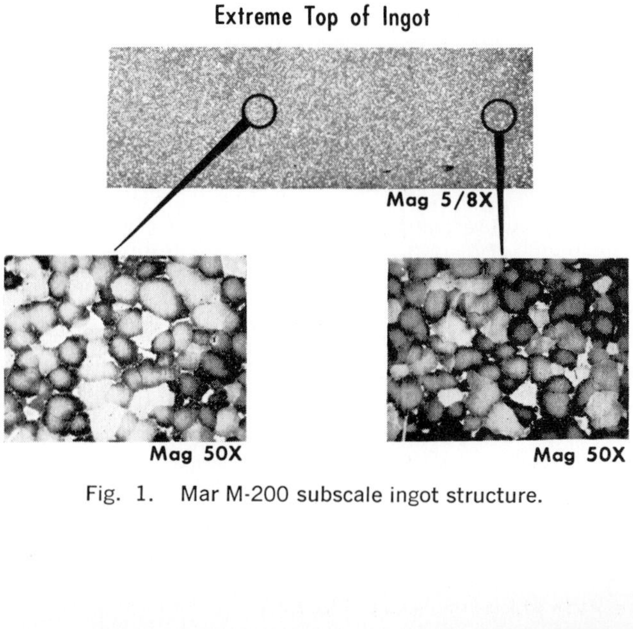

The authors are associated with Pratt & Whitney Aircraft, Florida Research & Development Center, Box 2691, West Palm Beach, Fla. 33402. This paper was presented at the 1968 Golden Gate Metals Conference, 25-27 September, San Francisco, Calif.

Extreme Top of Ingot

Fig. 1.   Mar M-200 subscale ingot structure.

superalloy compositions now in production. For these reasons, the powder metallurgy approach is a most promising method for producing advanced wrought products from the more highly segregation-prone alloys such as IN100 and Mar M-200.

Initial Development Program

It is our belief that the development of a production process for nickel-base superalloy powder products has been hampered by the oxidation of individual powder particles during processing. The formation of thin oxide films on powder particles of this class of alloys can be expected to take place even at room temperature. Unsatisfactory bonding would be expected during densification due to these films on individual powder particles. Past experience has shown that mechanical properties of hardware produced from contaminated powder exhibit a lack of reproducibility. Based on the reasoning just discussed, we felt that in order to obtain acceptable hardware a method of powder production, collection, and densification would have to be developed which would eliminate the problem of particle oxidation. Powder production methods in use at the time were, in general, similar to that shown schematically in Fig. 2.

To avoid oxidation of individual powder particles, Pratt & Whitney Aircraft metallurgists proposed an all-inert method of powder production, collection, and billet densification wherein the metal powders would not be exposed to the atmosphere prior to densification, Fig. 3. The investigation of this approach would involve the production and evaluation of powder, small densified billets, and forgings produced from these billets. Astroloy, the most advanced production nickel-base superalloy disc material, was selected for the investigation because of the excellent base line that Pratt & Whitney Aircraft had on conventional processed Astroloy discs and because the greatest gains using powder should be realized with such a highly alloyed material.

Powder Production and Densification. Pratt & Whitney Aircraft's initial concept of an "all inert" powder was a powder produced and hot pressed to a billet under vacuum, but equipment to implement such an approach was not known to exist in the industry. Pratt & Whitney Aircraft, therefore, decided that with certain modifications, the InFab

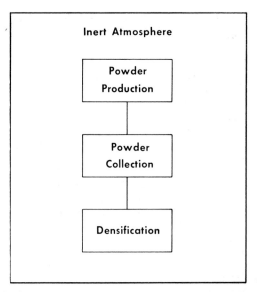

Fig. 3.   All-inert process of producing powder metal billets.

room, a government facility, operated by Universal Cyclops, could be used. This would allow production of the powder, its collection, and densification into small billets to be accomplished in a high purity argon atmosphere. Arrangements were made with Universal Cyclops and Federal Mogul and the work was initiated. Powder was produced in InFab, and, in addition, powder was obtained from Nuclear Metals Corp. This latter source, while not having the capability of handling the entire process in argon, was able to produce, collect, and package the powder without exposure to air. The basic material used by both sources for powder production was supplied by Pratt & Whitney Aircraft and conformed to PWA 1013 (Astroloy) chemistry requirements. Methods of powder production used by both sources yielded spherical particles. Microstructural evaluation revealed that each powder particle exhibited a fine dendritic structure. Samples of powder from both sources were analyzed for nitrogen, oxygen, and hydrogen content.

The results, while showing one vendor's powder to have substantially higher gas content (primarily oxygen), were considered to be within acceptable limits (less than 100 ppm). Prior to densification, the powder was rated as to screen size and percentage of each size present. Approximately 90% of the powder particles from both sources were within −40 to +250. Approximately 5% of the powder was finer than 325. The exact analysis for both sources is presented in Table 1. The range of powder particle sizes was not considered detrimental; on the contrary, it was

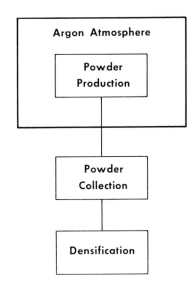

Fig. 2.   Standard method of producing powder metal billets.

Table 1.   Powder Characteristics

	Vendor 1	Vendor 2
Particle Shape:	Spherical	Spherical
Microstructure:	Cast	Cast
Gas Content, ppm		
N_2	60	50
O_2	95	30
H_2	—	—
Mesh, −40 + 100	25%	33%
−100 + 250	65%	55%
−250 + 325	5%	9%
−325	5%	3%

believed that full theoretical density could more readily be approached with varying particle sizes than with one uniform size.

Densification parameters, temperature and pressure, were deliberately selected to encompass a wide range of conditions, Table 2. Densification temperatures ranged from 2150 F to as high as 2350 F. Densification pressures ranged from 18,000 psi to as high as 40,000 psi. Densification was accomplished in a TZM die heated to the desired compacting temperatures by an external induction coil. A schematic representation of the densification procedure is shown in Fig. 4. The compacting force on the enclosed powder was achieved using a small hydraulic press. Fifteen billets, 2.5 in. in diam by 3.5 in. high were subsequently produced by this method. Each of the 15 billets had a porous surface appearance indicating incomplete densification. However, but for one exception, this obvious porosity was determined to be only a surface phenomenon and was eliminated by machining. This machining step reduced the billet dimensions to an average size of 2.2 in. in diam by 3.0 in. in length. The exception was the billet formed at 2350 F. When substantial machining failed to eliminate the porous appearance, the billet was sectioned and microscopically examined. The evaluation revealed various degrees of melting which apparently hindered densification and resulted in an unusable billet. The density of the remaining 14 billets was determined by weight-volume measurements. With one exception, the resulting densities exceeded 98%. A comparison of the density to that of conventional Astroloy ingot revealed no significant difference within the limitations of the calculations.

Two billets, one representing each powder source, were longitudinally sectioned and microstructurally examined. The examination revealed that there was no appreciable difference in the compacted billets between powder sources. The identity of the individual powder particles had not been eliminated. Although only very limited deformation of the powder particles had occurred, the dendritic microstructure

Table 2. Powder Consolidation Parameters and Resulting Billet Density

Billet Identity	Powder Source	Densification Temperature, F	Densification Pressure, ksi	Calculated Density, %
1	Vendor 1	2250	25.0	99.4
2	—	– –	Unknown	91.6
3	—	– –	30.0	96.1
4	—	– –	18.0	99.9
5	—	– –	30.0	99.9
6	—	– –	25.0	99.9
7	—	– –	40.0	98.0
1	Vendor 2	2250	40.0	99.9
2	—	2150	40.0	99.9
3	—	2250	30.0	99.9
4	—	2250	30.0	99.9
5	—	2350*	40.0	99.9
6	—	2250	40.0	99.9
7	—	2250	30.0	99.9
8	—	2150	40.0	99.5

* Incipient melting occurred.

and associated chemical segregation in each particle was altered to a more homogeneous condition.

The only exception to this occurred in the extreme top of each billet. This was attributed to a reduction in powder temperature caused by contact with the unheated ram during densification. (Similar to a "die lock" condition in conventional forgings.) Macro- and micro-structure of a typical densified billet are presented in Fig. 5. Further microstructural examination showed that wrought grains were present within the boundaries of the original powder particles, but in no instance did these grains grow across the prior particle boundaries. The interface between powder particles contained small discrete precipitates. Exposure of samples for prolonged periods at temperatures as high as 2150 F produced no solutioning or significant alteration in this precipitate. Isolated by digestion of the gamma and gamma-

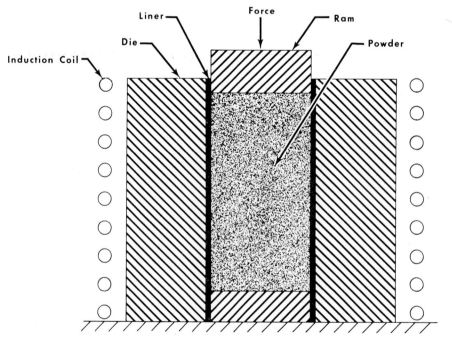

Fig. 4.   Powder densification technique.

Table 3. Temperatures and Reductions Used for Initial Forging Study

Billet Code	Billet Sub Code	Forging Temperature, F	Total Reduction, %
4	TB	1900	83
1	TA	2000	83
1*	BB	2050	83
1	BA	2050	83
4	TA	2100	83
4	BA	2200	83
1	TB	2250	83
4	BB	2100	50
		+2000	83

* Subsequently destroyed by attempted removal of the jacket by machining.

prime, the precipitate was determined to be titanium carbide (TiC) by x-ray diffraction analysis. The only obvious effect of the elevated temperature exposure was the solutioning of the gamma-prime precipitates. The powder particles retained their identity, and the recrystallized grains contained within the powder particles showed no growth.

Initial Forging Studies. Two billets were selected for the initial forging studies. Four pieces 1 in. in diam by 1½ in. in height were machined from each billet. These slugs were enclosed in ⅜-in. stainless steel packets. All eight pieces were upset to a total reduction of 83% as determined from starting height. On seven of the eight pieces this reduction was accomplished in one upset operation.

Seven upset temperatures ranging from a low of 1900 F to a high of 2250 F were used. The eighth piece, while being upset to a total reduction of 83%, was initially reduced 50% at 2100 F and finally upset at 2000 F. The parameters used for this initial forging study are presented in Table 3. The jacketed pancakes representing all the processing conditions were subjected to x-ray examination for soundness (bonding of the billet and jacket occurred during the upsetting and prevented the removal of the jackets). Results showed that absolutely no rupturing had occurred in the Astroloy during the upsetting operation.

The pancakes were cross-sectioned for a microstructural examination of the as-forged condition. The pancakes forged at 1900, 2000 and 2050 F exhibited a cold worked structure wherein the prior powder particles had maintained

their identity. The pancakes forged at 2100, 2200 and 2250 F exhibited a recrystallized structure having a grain size of approximately ASTM 6-7. In these pancakes, the powder particle identity had been eliminated. The pancake processed in two steps, while exhibiting a cold worked structure, was devoid of prior particle identity. Apparently the initial forging operation at 2100 F had resulted in the elimination of powder particle identity by recrystallization across the prior particle boundaries. The final operation at 2000 F cold worked the previously recrystallized structure. The as-forged microstructures representing the extremes of the forging parameters used are presented in Fig. 6. With the exception of the pancake forged at 1900 F where small voids formed in the peripheral area during upsetting, all pancakes exhibited excellent soundness. In the recrystallized structures, the only evidence that indicated a powder product had been employed was the presence of the elongated networks of TiC precipitate that had originally delineated the powder particles.

The pancakes were sectioned into pie-shaped segments for solution heat treatment studies. The solution temperatures investigated ranged from 2020 to 2225 F; each segment was held at the given temperature for 2 hr.

All heat treated segments were subjected to a thorough microstructural examination. The cold worked structures produced by forging at 1900, 2000 and 2050 F were completely recrystallized by solution temperatures of 2050 F and above. The powder particle identity still present in the as-forged structure was completely eliminated through the formation of the recrystallized grains. The same range of solution temperatures applied to the pancakes forged at 2100, 2200 and 2250 F produced moderate coarsening of the already recrystallized structure. The two-step upset (hot-cold work structure) also exhibited a uniform, fine recrystallized grain size.

The study revealed that a solution temperature of 2050 F or above produced a uniform, fully recrystallized structure irrespective of starting structure, having a grain size of approximately ASTM 4-5. Figure 7 illustrates the three conditions just described using a solution temperature of 2075 F. Solution temperatures in excess of 2075 F produced minor grain growth, with only an occasional grain larger than ASTM 4-5. Temperatures as high as 2225 F produced no incipient melting at the grain boundaries. The elongated

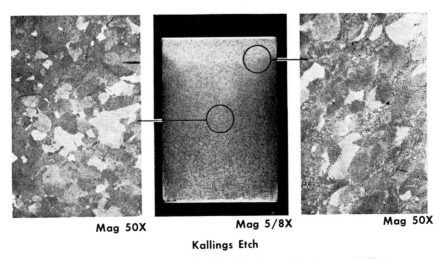

Mag 50X   Mag 5/8X   Mag 50X

Kallings Etch

Fig. 5. Cross section and microstructure of hot-pressed billet.

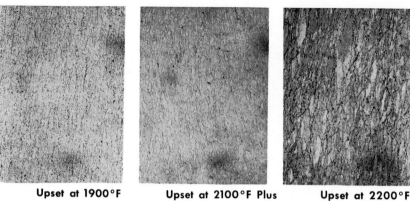

Upset at 1900°F   Upset at 2100°F Plus   Upset at 2200°F
Additional Upset at 2000°F

Kallings Etch (Mag 50X)

Fig. 6. Typical as-forged microstructure of three pancakes.

TiC networks remained unaltered to any significant extent over the entire temperature range investigated. Although all pancakes exhibited excellent structural characteristics, the two-step upset processing was selected for use on the full-size billets. This processing more nearly represents the sequence used for conventional Astroloy forgings.

Production of 5-In. Pancake Forgings. Four full-size billets approximately 2.2 in. in diam. and 3.0 in. in length representing both powder sources were selected for the scaled-up forgings. The only change in the processing was the wrapping of the billets in a $\frac{3}{16}$-in. blanket of thermofax to prevent bonding of the stainless jacket to the billet during upsetting, thereby affording easy removal of the jacket. The upset parameters were identical to those used previously, namely a 50% reduction at 2100 F followed by a further reduction at 2000 F for a total reduction in height of 83%. The upsetting was accomplished as before, between heated flat dies.

The thermofax insulation served the desired purpose, since jacket removal after pancaking was facilitated by the lack of bonding. There was some minor peripheral cracking, Fig. 8. Aside from the peripheral cracking, x-ray examination and sonic inspection showed all four pancakes to

be sound and intact. Subsequent evaluation of the four pancakes showed that the as-forged microstructure, uniformity, and grain growth characteristics were identical to those exhibited by the small piece forged using the same parameters. Material from all four pancakes was subjected to quantitative spectrographic analysis. The resulting chemical compositions were within current Astroloy (PWA 1013) chemical composition requirement.

Samples from all four upsets, along with samples from conventional Astroloy forgings, were analyzed for gas content. The oxygen as determined on the upset forgings was approximately twice that reported on the powder. Neglecting the unexplained difference in analysis from powder to upset forging it is noted that the two pancakes having the highest oxygen content were produced from the powder with the highest oxygen content. The two pancakes having oxygen contents of 60 and 90 ppm were found to be consistent with conventional forged Astroloy while the oxygen content of the remaining two pancakes was 110 and 140 ppm. The 4 pancake forgings were solutioned using a 2075 F(2) solution treatment and given the balance of the standard PWA 1013 heat treatment 1600 F(8)AC + 1800 F(4)AC + 1200 F(24)AC + 1400 F(8)AC. The

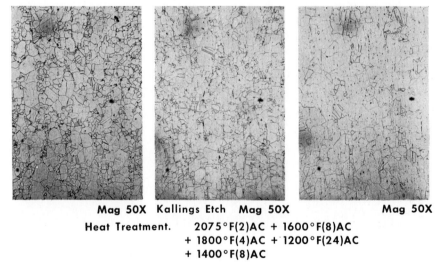

Mag 50X   Kallings Etch   Mag 50X   Mag 50X

Heat Treatment.   2075°F(2)AC + 1600°F(8)AC
+ 1800°F(4)AC + 1200°F(24)AC
+ 1400°F(8)AC

Fig. 7. Typical fully heat treated microstructure of three pancakes.

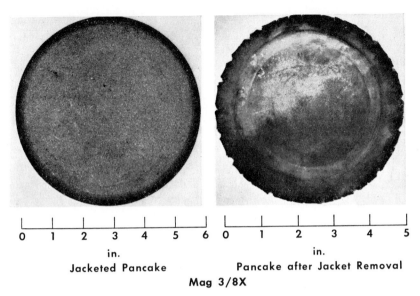

Jacketed Pancake | Pancake after Jacket Removal

Mag 3/8X

Fig. 8. Scaled-up powder Astroloy pancake.

typical "as forged" and fully heat treated microstructures are presented in Fig. 9.

Two tensile specimens, one combination smooth and notched stress-rupture specimen, and one creep specimen were machined from each of the four heat treated pancakes. The mechanical property evaluation was to our current PWA 1013 specification, namely a room temperature and 1400 F tensile test, a stress-rupture test at 1400 F —85,000 psi, and a creep test at 1300 F—74,000 psi to 0.1% extension. In general, the test results were good; however, it must be noted that the pancakes having the lowest oxygen content (less than 100 ppm) exhibited the highest level and most consistent mechanical properties, and with the exception of minor variations, the results conformed to PWA 1013 specification requirements and are representative of conventional forgings.

The two pancakes having oxygen contents in excess of 100 ppm exhibited substantially lower elevated temperature tensile ductility and stress-rupture life, and had one V-notch

failure in stress-rupture. The results of all the mechanical testing on the four pancakes are presented in Tables 4 and 5. The general high property level and microstructural uniformity obtained from these initial attempts to produce Astroloy forging by powder metallurgy techniques was felt to warrant additional work with larger scaled-up forgings.

Development of Scaled-Up Forgings

The initial program demonstrated that small Astroloy forgings made via the powder metallurgy technique could be satisfactorily processed over a wide temperature range and the subsequent solution heat treatment temperature range was not as restrictive as it is with conventional forgings for grain size control. As a result of the success with the "all inert" process, a vendor employing the same principles of powder production and collection as used in the Pratt & Whitney Aircraft program offered to produce and submit to Pratt & Whitney Aircraft small Astroloy pancake

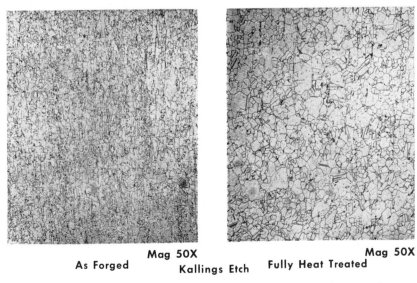

As Forged **Mag 50X** Kallings Etch Fully Heat Treated **Mag 50X**

Fig. 9. Typical microstructure of scaled-up powder Astroloy pancake.

Source: *Metals Engineering Quarterly*, Feb 1970

Table 4. Mechanical Properties of Scaled-Up Pancakes (RT-Tensile, 1400 F—Tensile)
(2075F (2) OQ + balance PWA 1013)

Billet Code	O₂ Content	Grain Size	0.2% Yield Strength, ksi	Ultimate Strength, ksi	Elonga-tion, %	Reduction Area, %
Room Temperature—Tensile						
4–1	140	6–7	145	212	23	32
7–1	110	5	137	210	22	23
4–2	90	6–7	145	213	25	36
7–2	70	5	141	211	23	31
PWA 1013G		4, Occ 3	140	195	16	18
1400 F—Tensile						
4–1	140	6–7	132	160	17	30
7–1	110	5	136	159	12	24
4–2	90	6–7	131	161	23	31
7–2	70	6–7	134	157	19	28
PWA 1013G		4, Occ 3	125	150	20	30

forgings for structural and mechanical property evaluation.

Production of 5-In. Pancake Forgings. Powder production, screening, and collection was performed in a high purity argon atmosphere. The collected powder was placed in stainless steel cans under a positive pressure of argon. These cans were subsequently evacuated and sealed, thus achieving a sequence of manufacture wherein the Astroloy powder was never exposed to a contaminating atmosphere. Densities of the canned powder ranged from 65 to 70%, the difference probably being due to the variation in powder particle size distribution from one can to another. Five forged pancakes were produced from the canned powder and were submitted to Pratt & Whitney Aircraft for evaluation.

Spectrographic analysis confirmed that the material conformed to Astroloy (PWA 1013) specification requirements. Results of gas analysis revealed that the oxygen contents were substantially below 100 ppm on all pancakes, ranging from 37 to 59 ppm. A radial etch slice was removed from each forging for metallographic examination and solution heat treatment studies. The as-forged structure, with

Table 5. Mechanical Properties of Scaled-Up Pancakes (Stress-Rupture, 0.1% Creep)
(2075F (2) OQ + balance PWA 1013)

Billet Code	O₂ Content	Grain Size	Temper-ature, F	Stress, ksi	Life, hr	Elonga-tion, %	Reduction Area, %
Stress Rupture							
4–1	140	6–7	1400	85	21	V-N	Failure
7–1	110	5	1400	85	22	31	9
4–2	90	6	1400	85	28	18	18
7–2	70	5	1400	85	35	21	19
PWA 1013G		4, Occ 3	1400	85	30	17	—
0.1% Creep							
4–1	140	6–7	1300	74	161		
7–1	110	5	1300	74	179		
4–2	90	6–7	1300	74	231		
7–2	70	6	1300	74	172		
PWA 1013G		4, Occ 3	1300	74	150 avg- 110 min		

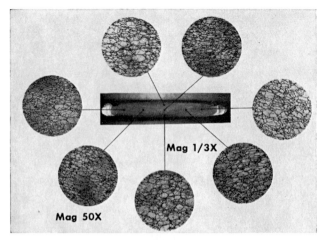

Fig. 10. Typical microstructure of a 5-in. diam. pancake produced from canned Astroloy powder.

the exception of the extreme periphery, was completely dense and uniformly fine grained. While some recrystallization had occurred during upsetting, it was confined within the boundaries of the prior powder particles.

The prior particle size was clearly evident in all etch slices. Subsequent heat treatment to PWA 1013 using a 2075 F solution temperature, while promoting additional recrystallization, did not eliminate the prior particle identity. Figure 10 represents a typical cross section and microstructure of a fully heat treated powder Astroloy pancake. The remainder of each fully heat treated forging was machined into four tensile and two combination stress-rupture specimens. Tensile testing was conducted at room temperature and 1400 F; stress-rupture tests were performed at 1400 F—85,000 psi per the current PWA 1013 specification. The tensile strength at both test temperatures was above that of conventional Astroloy forgings. Tensile elon-

Table 6. Typical Tensile Properties of 5-In. Diam Direct Upset Pancakes

Pancake No.	Temp., F	0.2% Yield Strength, ksi	Ultimate Strength, ksi	Elonga-tion, %	Reduction Area, %
1	70	160.6	218.0	15.8	17.2
		161.4	219.0	17.9	20.0
2	70	161.2	219.3	16.7	16.5
		161.4	215.8	12.7	12.8
3	70	156.2	208.4	17.5	20.3
		160.6	210.0	13.2	15.3
4	70	162.8	209.0	12.0	18.5
		161.4	215.5	12.8	14.3
5	70	161.6	212.5	12.2	15.2
		162.0	208.0	14.1	15.2
PWA 1013E	70	140.0	195.0	16.0	18.0
1	1400	144.9	171.0	18.9	29.2
		147.7	172.7	22.8	28.2
2	1400	141.3	165.4	10.5	14.2
3	1400	140.5	163.3	16.4	21.2
		142.3	170.0	20.5	34.2
4	1400	147.2	176.0	14.6	22.5
		146.5	175.1	14.8	25.4
5	1400	141.7	169.0	21.3	28.2
		138.3	163.4	18.7	29.2
PWA 1013E	1400	125.0	150.0	20.0	30.0

Table 7. Typical Stress-Rupture Properties of 5-In. Diam. Direct Upset Pancakes

Pancake No.	Temperature, F	Stress, ksi	Life, hr	Elongation, %
1	1400	85.0	38.6	10.5
			23.0	10.8
2	1400	85.0	36.4	18.5
			47.6	12.5
3	1400	85.0	25.5	– –
			39.2	– –
4	1400	85.0	40.7	12.0
			42.2	7.5
5	1400	85.0	46.2	8.5
			51.3	18.5
PWA 1013E	1400	85.0	30.0	17.0

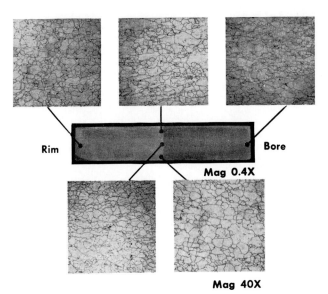

Fig. 12. Radial cross section of a fully heat treated 50-lb pancake produced from Astroloy powder.

gation, however, was generally low, ranging from 12 to 18% at room temperature and from 11 to 23% at 1400 F. Stress-rupture life exceeded the specification requirement of 30 hr, but again, similar to the tensile results, the stress-rupture elongation was low, ranging from 8 to 18%, Tables 6 and 7.

50-Lb Disc Forging. Based upon the initial results, a scaled-up forging weighing approximately 50 lb was produced. Powder production and all subsequent processing parameters were identical to those used to produce the initial five pancakes. This disc, after mechanical removal of the stainless steel can, measured 10 in. in diam by 1 in. thick. Machined to a square-cut outline the disc met x-ray and sonic inspection requirements. This forging provided the material required for a more comprehensive mechanical property evaluation, including 0.1% creep, low cycle fatigue (LCF) and high cycle fatigue (HCF). One-half of the forging was sectioned for this evaluation, Fig. 11. Prior to heat treating and machining of test specimens, the as-forged structure was metallographically evaluated and samples of material were submitted for spectrographic and gas analysis. The as-forged microstructure, grain size, and structural uniformity were similar to that of the previous five pancakes.

Chemistry conformed to PWA 1013 (Astroloy) requirements. The oxygen content, while being slightly higher than the previous pancakes, was still below 100 ppm. Previous studies had shown that a fine grain size could be maintained over a wide range of solution temperatures, and in an attempt to eliminate the prior powder particle identity, a solution treatment of 2200 F followed by the balance of the PWA 1013 heat treatment was used. This thermal treatment produced a uniformly fine grained, fully recrystallized structure having a grain size of ASTM 4-5. However, prior particle identity was not completely eliminated, Fig. 12. Initial property testing was conducted to the PWA 1013

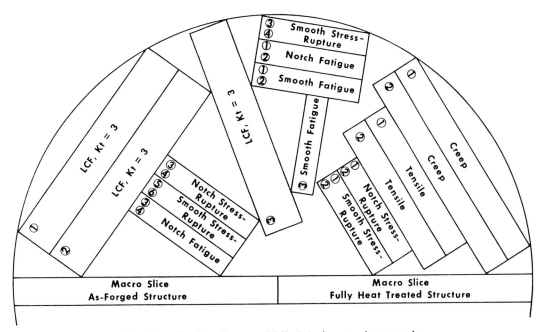

Fig. 11. Section diagram 50-lb Astroloy powder pancake.

Table 8. Tensile and Stress-Rupture Properties of a 50-Lb Astroloy Powder Disc

Process	Temperature, F	0.2% Yield Strength, ksi	Ultimate Strength, ksi	Elongation, %	Reduction Area, %
		Tensile			
Powder Forging	RT	148.6	196.5	11.5	10.1
PWA 1013E	RT	140.0	195.0	16.0	18.0
Powder Forging	1400	132.5	161.3	13.0	10.4
PWA 1013E	1400	125.0	150.0	20.0	30.0

Process	Temperature, F	Stress, ksi	Life, hr	Elongation, %
		Stress-Rupture		
Powder Forging	1400	85	58.9	10.8
			46.7	11.9
			156.3	(a)
			130.5	(a)
PWA 1013E	1400	85	30	17

(a) Notched specimen, test discontinued.

specification requirements. The room temperature and 1400 F tensile strength and the 1400 F—85,000 psi stress rupture life met the specification requirements with the corresponding elongations being low. Separate notched rupture specimens were tested at 1400 F—85,000 psi and discontinued at 130 and 156 hr, demonstrating that the low rupture ductility was not an indication of 1400 F notch-rupture sensitivity, Table 8. To be of value as a high temperature disc material, the powder product had to exhibit acceptable 0.1% creep capability.

Creep testing was conducted at 1300 F—74,000 psi, the specification requirement, and at 1400 F and 1500 F using stress levels at which comparative data from conventional Astroloy forgings were available. Specimens from the powder product exhibited a 25 to 50 F advantage over the temperature range investigated as shown in the Larson-Miller parameter plot, Fig. 13.

The question of fatigue capability, both LCF and HCF,

of the powder product remained to be answered. LCF testing was confined to notched ($K_t = 3$) specimens, as this was deemed the most critical condition. Testing was performed using standard methods at 1300 F and a combination of steady and vibratory stresses for which comparative data were available. Cycles to first indication (crack) were comparable to conventional material. Crack propagation as judged by the number of additional cycles from first indication to failure was slower than conventional material, Fig. 14.

Smooth and notched ($K_t = 3$) HCF testing was performed at 1300 F. Vibratory stress levels were again selected to facilitate a direct comparison. Test results were equivalent to conventional Astroloy, Fig. 15.

The capabilities exhibited by the scaled-up forging were very encouraging. Pratt & Whitney Aircraft however, still contended that complete elimination of the prior particle identity through additional thermomechanical work and/or subsequent heat treatment would increase the level of tensile and rupture ductility and further improve the strength. The remaining half of the disc forging was machined into forging blanks $3\frac{3}{8}$ in. in diam for additional forging studies. These blanks were upset approximately 50% (1 in. high to $\frac{1}{2}$ in. high) between flat dies. As-forged microstructures, while being uniformly fine grained ASTM 7-8, were only partially recrystallized and continued to exhibit evidence of the prior powder particle. Subsequent heat treatment using a solution temperature of either 2050 F or 2100 F produced a uniform, completely recrystallized fine-grained structure (ASTM 4-5) devoid of all prior particle identity, Fig. 16.

The resulting structure met all PWA 1013 specification grain size and microstructural requirements. Further microstructural evaluation using the electron microscope was unable to locate any unusual phases or oxide particles. The matrix exhibited a uniform, fine aging γ' with occasional areas of the larger, blocky γ'. All grain and twin boundaries exhibited globular, discrete $M_{23}C_6$ type carbides. A comparison of this structure with that of conventional Astroloy is shown in Fig. 17. Two tensile, two stress-rupture, and one creep specimen were machined from each of the two fully heat treated forgings. The room temperature and

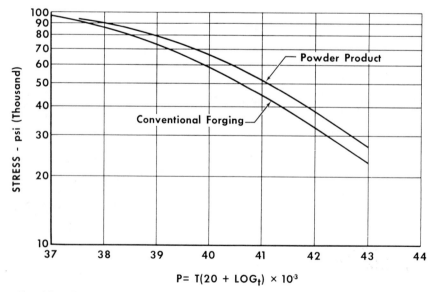

Fig. 13. Typical 0.1% creep strength of conventional and powder Astroloy.

$$P = T(20 + LOG_t) \times 10^{-3}$$

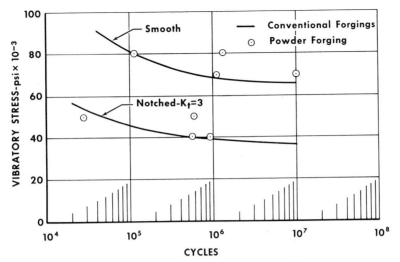

Fig. 14. Astroloy low-cycle fatigue $K_t = 3$ at 1300 F.

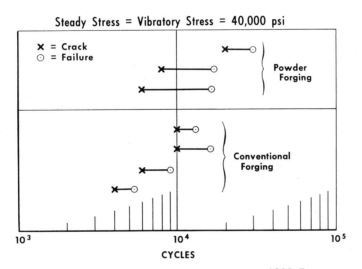

Fig. 15. Astroloy high-cycle fatigue at 1300 F.

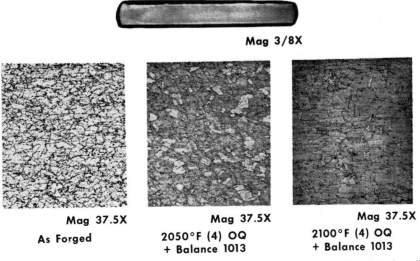

Mag 3/8X

| Mag 37.5X | Mag 37.5X | Mag 37.5X |
| As Forged | 2050°F (4) OQ + Balance 1013 | 2100°F (4) OQ + Balance 1013 |

Fig. 16. Small pancake forged from blank machined from 50-lb powder Astroloy disc.

Standard Cast Billet

Powder Billet

Standard Astroloy Forging
and Heat-Treating Sequence

Two-Step Subscale Forging
Forging Sequence
    Press Upset 72% at
    2150°F plus Hammer
    Forged 47% at 2050°F
Heat Treatment
    2200°F (4) OQ + Balance 1013

Mag 5,000X

Fig. 17. Comparison of electron microstructures.

Table 9. Mechanical Properties of Small Pancake Forged From Blank Machined From 50-Lb Powder Astroloy Disc

Process	Temperature, F	0.2% Yield Strength, ksi	Ultimate Strength, ksi	Elongation, %	Reduction Area, %
Tensile					
Powder Forging	RT	156.4	214.5	20.0	24.3
PWA 1013E	RT	140	195	16.0	18.0
Powder Forging	1400	140.4	161.7	18.7	38.1
PWA 1013E	1400	125	150	20.0	30.0

Process	Temperature, F	Stress, ksi	Life, hr	Elongation, %
Stress-Rupture				
Powder Forging	1400	85	66.1	21.1
PWA 1013E	1400	85	30.0	17.0
0.1% Creep				
Powder Forging	1300	74	200	–
PWA 1013E	1300	74	150 avg-110 min	–

1400 F tensile strengths, the 1400 F—85,000 psi stress-rupture lives, and the times to 0.1% creep at 1300 F—74,000 psi exceeded those seen in the best conventional forgings.

The most significant fact is that both the tensile and stress rupture ductility were significantly improved and, with minor deviations, met the requirements of PWA 1013, Table 9.

Summary

It has been demonstrated by Pratt & Whitney Aircraft that small Astroloy forgings possessing mechanical property capabilities and structural uniformity equalling or exceeding those of conventional forgings can be produced using "all inert" powder metallurgy techniques. Notwithstanding the most encouraging results that have been obtained thus far, Pratt & Whitney Aircraft metallurgists believe that the greatest gains that can be realized from this approach both from the standpoint of optimum mechanical properties and economics will come with the introduction of processes for the production of powder, its collection, and densification in vacuum. This will represent the next step forward.

SECTION IV:
Consolidation

Hot Rolling Behaviour of Iron Powder Preforms

* C. H. WEAVER, R. G. BUTTERS, AND J. A. LUND.

ABSTRACT

Slab-shaped preforms have been lightly sintered from loose and compacted iron powder and have been hot rolled between plane rolls, at temperatures of 850 and 1040 C, to obtain single-pass reductions of up to 72 per cent in thickness. Final densities of the hot rolled preforms ranged up to 99.8 per cent of solid. Mechanical properties, microstructure, and resistance to internal oxidation of the rolled preforms were studied as a function of final density.

INTRODUCTION

Cold powder rolling, as practiced commercially for nickel,[1] involves continuous roll compaction of powder to obtain a "green strip" which is subsequently sintered and hot rolled to more or less full density. At least to the extent that the green strip is a hot-working preform, the overall process has features in common with the more recently-developed hot forging of powder preforms.[2] In both processes it is found that sintering of the preform can be identified with heating for forming, and that the temperatures employed can be low relative to those used for conventional sintering. It also appears likely that the actual hot-working temperatures required are appreciably lower than is the case for the conventional hot working of cast or wrought preforms.

In both the rolling and the forging processes it may be technically impracticable, and quite unnecessary, to obtain a 100 per cent dense product in a single hot working step. Inadequate data is available relating properties to final density in the critical range of 95 to 100 per cent of solid, or theoretical, density. Most published information pertains to as-sintered or hot-pressed compacts, which are structurally different from hot worked preforms. For the case of the hot rolling of powder compacts, there is also a dearth of available information about the relationships between such processing variables as green density, sinter-preheat temperature, hot rolling temperature, and the reductions and pressures required in hot forming to attain a given final density.

It was primarily to provide some new data of this type that the present work was undertaken. Single-pass hot rolling has been applied to small, slab-shaped, sintered iron powder preforms, and several processing parameters have been varied. The work was extended to include preforms of very low initial density, obtained by sintering loose powder.

* Department of Metallurgy, University of British Columbia, Vancouver, Canada. C. H. Weaver is now a Metallurgical Engineer with the Aluminum Company of Canada, Arvida, Quebec, Canada.

Starting material for these experiments was an atomised iron powder, Easton RZ 365. Nominal screen and chemical analysis are as follows:

Tyler mesh	Wt. per cent		
On 100	5	Carbon	0.06
−100 +200	40	Manganese	0.22
−200 +325	30	Sulphur	0.02
−325	25	Phosphorus	0.01
		Silicon	0.04
		Hydrogen loss	0.5
		Acid insoluble	0.1

Three types of slab-shaped preforms were made from the powder, described as Series L, M, and H in Table I; they were 3 inches long, 1.25 inches wide, and from 0.12 to 0.15 inches thick. All were presintered very lightly; i.e. for 5 minutes at 900 C in cracked ammonia, and cooled to ambient temperature prior to being used in rolling experiments. Series L preforms were made from loose powder, sintered in a stainless steel tray. The other two series were statically compacted, according to the conditions in Table 1.

Preforms were hot rolled at 850 C (essentially ferritic) or at 1040 C (austenitic). Heating to the hot rolling temperature was effected in a muffle furnace using a cracked ammonia atmosphere. The furnace was located so that an extension of the muffle fed directly into the roll gap. A heated preform was pulled into the rolls from the hot zone of the furnace, by means of an attached wire. Another cracked ammonia-purged muffle received the hot-rolled specimen, thus ensuring that cooling also occurred in a reducing atmosphere. Optical metallography on sections of hot-rolled specimens established that the above technique was consistently effective in avoiding internal oxidation.

Total time in the furnace by the preforms prior to rolling was 5 minutes in all cases. Thickness reduction in a single pass was varied from 12 to 77 per cent by suitable adjustment of the roll gap. The two-high rolling mill had 4-inch diameter plane (unprofiled) rolls, internally heated to give a surface temperature of 200 C. The roll speed was 68.5 RPM. Calibrated strain-gauge force washers were mounted under the lower roll bearing blocks to permit the measurement of roll separating force.

TABLE I. Preforms

Series	Compacting Pressure (psi)	Thickness (inches)	Density, (% of solid Fe)
L (Low density)	NIL (loose powder)	0.140	39.0 ± 0.8
M (Med. density)	17000	0.147	62.0 ± 0.8
H (High density)	46000	0.120	77.5 ± 0.5

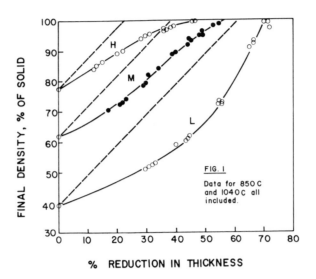

FIG. I

Data for 850 C and 1040 C all included.

FIGURE 1.   Final Density of Hot Rolled Preforms vs. Amount of Rolling Reduction in a Single Pass. (Broken lines indicate maximum density attainable).

Density values for preforms and hot-rolled specimens were based directly on their weights and measured dimensions. In the case of hot rolled strips, cracked edges were removed prior to making density determinations. All densities in this work are expressed as a "per cent of solid". This reference value for solid iron (ρ_s) was taken as 7.80 gms per cc., this being the highest density achieved after multiple rolling and heat treatment operations on an RZ365 iron powder compact.

Tensile specimens with a one inch long by 0.2 inch wide reduced gage section were spark machined from the hot-rolled strips. In most cases, the specimens were annealed for 30 minutes at 700 C in cracked ammonia prior to testing. However, some strips which had been rolled to high density from L series preforms at each of 850 and 1030–1040 C were tested in the as-rolled condition.

Specimens of low density were impregnated with an epoxy resin prior to polishing and grinding for metallographic preparation.

RESULTS AND DISCUSSION
Hot Rolling Behavior

Preforms of low and medium starting density (Series L and M) developed transverse cracks when given light reductions by hot rolling. For reductions greater than 20 to 30 per cent, only edge-cracking was encountered in the as-rolled strips. In the case of specimens rolled to >90 per cent of solid density, the extent of edge cracks was consistently 0.12 to 0.13 inches, essentially independent of original preform density or rolling temperature.

The density of hot rolled preforms is plotted in Fig 1 as a function of the amount of reduction in thickness imparted in one rolling pass. Results for both 850 and 1040 C rolling temperatures fitted a common line for each series. The broken lines of Fig 1 represent the density—per cent reduction

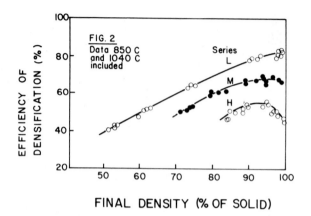

FIGURE 2.   Efficiency of Densification of Preforms vs. Final Density for single Pass Rolling.

relationship which would be expected if all the reduction had gone into closing the pores in the original preforms without any net increase in preform length. It is thus possible to define the "efficiency of densification" E as

$$E = \frac{\rho_f - \rho_0}{\rho_s R}(100)$$

Where ρ_0 is the original preform density, ρ_f is the density after hot rolling, ρ_s is the maximum density attainable ("solid" density), and R is the per cent reduction in thickness.

Efficiency of densification has been plotted against rolled density ρ_f in Fig 2 for the present results. Densification was most efficient for a single large reduction of a low initial density preform. For the preforms of higher initial density (Series M and H), the maximum efficiencies observed were much lower than for the loose powder preforms (Series L). As 100 per cent of solid density was approached, the efficiency decreased from its maximum value, this effect being most obvious in the case of preforms with high initial density.

In attempting to explain the above observations, we can first consider the processes which occur when a preform is rolled to high density; i.e., with relatively large reduction in thickness. In the early part of the roll gap after entry, the normal roll pressure is small because the resistance of the porous preform to compression is low. Slipping friction likely prevails at the roll-preform interface. Compaction of the powder particles occurs in this region largely by rearrangement as the original small sintered necks between particles are broken. Compaction is probably efficient under these conditions because there is high density material further ahead in the roll gap which acts to prevent longitudinal motion of the powder particles. As the preform

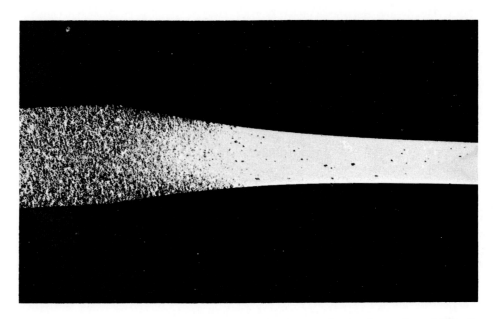

FIGURE 3. Polished Section of Partially-Rolled M-Series Preform. Rolled at 1030 C (Mag. 6.5×).

becomes more dense in the roll gap its compressive strength increases and the specific roll pressure increases. Sticking friction probably starts to prevail. Plastic deformation then becomes the dominant densification mechanism, but some of the flow goes into elongation without further closing pores. Thus the efficiency of densification in the latter part of the roll gap becomes low.

In rolling to high density, the efficiency of densification is higher if the starting density of the preform is lower. This is because a greater proportion of the densification is due to compaction by rearrangement, which is an efficient process under the specified conditions.

By contrast, if small rolling reductions are attempted, the compaction process will at best be inefficient due to a lack of dense material ahead of the rearrangement-compaction region of the roll gap.

Fig 3 shows a section through a partially hot-rolled Series M preform. The total reduction was 58 per cent in thickness, and the final density was 98 per cent of solid. Near the entry to the rolls, where slipping friction would be expected to prevail, cracks opened up at the surface of the preform. These subsequently healed further ahead in the roll gap where appreciable plastic deformation occurred. It is also evident from Fig 3 that early compaction was more efficient at the centre of the preform where the temperature was higher, and where there would likely be no effect of slipping between rolls and preform.

In Fig. 4 and 5, measured roll force values have been plotted against final density and efficiency of densification, respectively. Results for 850 and 1040 C fitted the same plots. It is interesting to note that the roll force re-

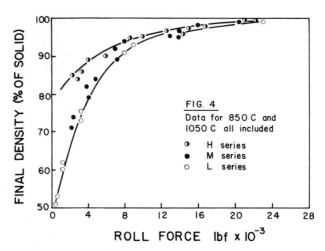

FIGURE 4. Final Density of Hot Rolled Preforms vs. Roll Force Required.

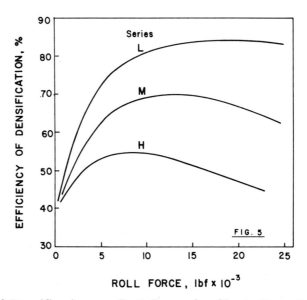

FIGURE 5. Efficiency of Densification vs. Roll Force for Single Pass Rolling.

quired to attain a density of 99 per cent of solid was little higher for preforms of low initial density than for preforms of high starting density. This observation is consistent with the earlier argument that compaction occurs largely by particle rearrangement at low roll pressures. With lower initial density preforms, larger reductions are necessary to attain high density, and the arc of contact between rolls and preform is greater. However, compaction by rearrangement is occurring over a larger fraction of the arc of contact. Most of the roll force is associated with the plastic deformation after compaction is well advanced. The results indicate that the length of

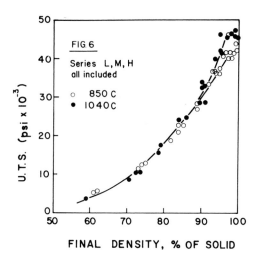

FIGURE 6.   Ultimate Tensile Strength of Hot Rolled Preforms vs. Final Density.

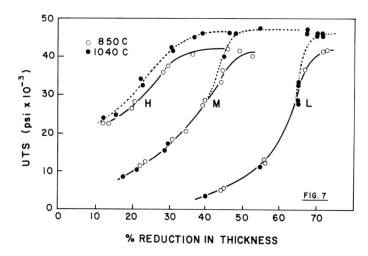

FIGURE 7.   Ultimate Tensile Strength of Hot Rolled Preforms vs. Reduction in a Single Pass.

that part of the arc of contact where roll pressures are high is essentially independent of total contact length.

Mechanical Properties of Hot Rolled Preforms
Ultimate tensile strength data for hot rolled and annealed specimens are plotted in Fig 6 against final density, and in Fig 7 against reduction in thickness by rolling. The data is summarised in Table II. The strength was independent of initial preform density, but not independent of rolling temperature. At densities greater than about 80 per cent of solid, specimens rolled at 1040 C were stronger after annealing than those rolled at 850 C. There was no significant increase in strength with increasing density beyond a level of about 95 per cent. The last entry in Table II is for fully dense material obtained by hot rolling and twice re-rolling at 1030 C an M-Series

TABLE II. Summary of U.T.S. and Elongation Data for Hot Rolled and Annealed Preforms

Final Density, % of solid	Rolling Temp., °C	Range of Tensile Data	
		U.T.S. (psi)	Elong. (%)
70–81	850	11,300–13,000	1–2
	1040	8,400–17,400	0–4
81.1–91	850	18,800–28,100	2–4
	1040	24,000–34,000	4–10
91.1–96	850	33,300–37,600	11–16
	1040	40,000–46,200	15–31
96.1–98	850	40,000–41,600	16–31
	1040	45,000–46,200	35–36
98.1–99.8	850	40,000–44,000	32–37
	1040	45,200–47,500	24–35
100*	1030	46,400–47,000	32–36

* This material was hot rolled three times, with final annealing. All other entries in the Table are for single-pass rolled specimens.

preform to give >65 per cent reduction in thickness. After annealing, the strength and ductility were closely comparable to those of M-Series preforms rolled with only 45–48 per cent reduction at 1040 C, and with final densities of 94–95 per cent. It is interesting to note from Table II that relatively high tensile strengths and elongations were exhibited by all specimens rolled to >91 per cent, this being particularly true for a rolling temperature of 1040 C.

Tensile elongation data is summarised in Table II. For specimens of >95 per cent of solid density, scatter was observed in the elongation values. It was difficult to obtain good axial alignment of the small strip tensile specimens in the grips, with the result that many of them fractured in the test by tearing from one edge, and very little necking occurred in some cases. It is believed that this behaviour explains the lower of the elongation values for specimens of high final density. The distribution of residual porosity, and of inclusions, probably played a minor rôle.

Some L-series specimens were tested in the as-rolled condition, i.e. without an annealing treatment. The tensile results are compared in Table III with those for annealed specimens of similar final density, and are discussed in relation to microstructure in the next section of the paper.

Metallographic Observations

Fig. 8 reveals the microstructure of an L-Series specimen as hot-rolled to high density at 850 C. Partial recrystallisation of the ferrite is evident in the photomicrograph. Near the surface, where the chilling effect of the rolls was greatest, the structure was essentially cold worked. The effect of annealing a similar specimen for 30 min. at 700 C is shown in Fig 9. Complete recrystallisation has occurred, with appreciably finer grain size near the surface.

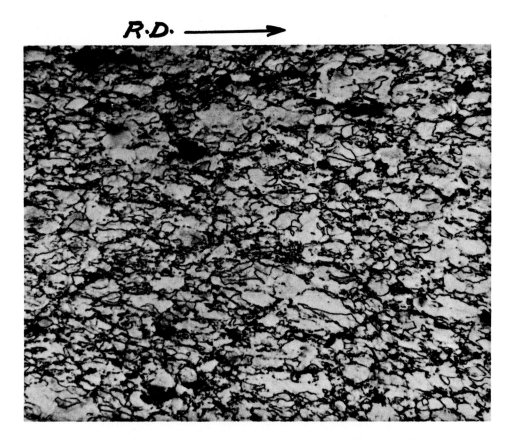

R·D· ⟶

FIGURE 8. Microstructure of Series L Preform, After Hot Rolling at 850 C to 99.5 Per Cent of Solid Density. As-Rolled (Not Annealed) (Mag. 114×).

Specimens rolled at 850 C were largely ferritic during deformation. By contrast, preforms rolled at 1030 to 1040 C were austenitic during deformation, except possibly near the roll-chilled surfaces. The microstructure of a specimen as-rolled at 1030 C is shown in Fig 10. There appears to have been complete recrystallisation, but the ferrite grain size is very fine as might be expected to be the result of deformation of initially fine-grained austenite followed by fairly rapid cooling through the transformation temperature. Annealing a similar specimen at 700 C produced the microstructure shown in Fig 11. The ferrite grain size has increased, but is still finer than that of the specimen rolled at 850 C and annealed (Fig. 9).

The high strength of the material as-rolled from 1030 C (Table III) must be attributed to its fine recrystallised grain size. In the case of the specimen as-rolled from 850 C, however, some of the high strength is clearly due to residual cold work. The low tensile elongation of the latter material is not unexpected. A higher elongation might have been anticipated for the specimen as-rolled from 1030 C in view of its microstructure. In fact the tensile specimen was observed to fail by tearing soon after the ultimate stress was

FIGURE 9.   Microstructures of Series L Preform, After Hot Rolling at 850 C to 98.3 Per Cent of Solid Density, Followed by Annealing at 700 C (Mag. 114×).

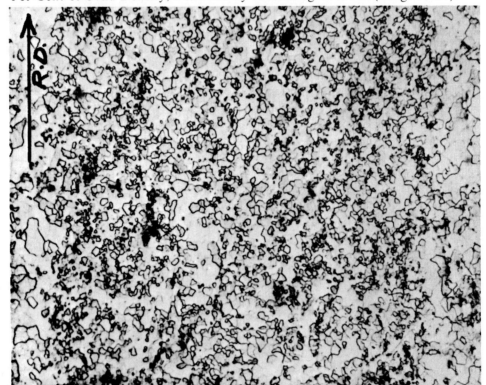

FIGURE 10. Microstructure, of Series L Preform, After Hot Rolling at 1030 C to 99.8 Per Cent of Solid Density. As-Rolled (Mag. 114×).

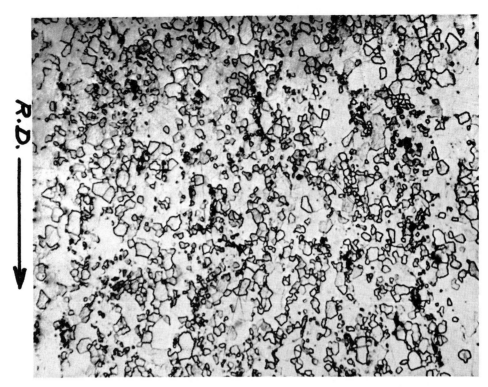

FIGURE 11. Microstructure of Series L Preform, After Hot Rolling at 1040 C to 99.5 Per Cent of Solid Density, Followed by Annealing at 700 C (Mag. 114×.)

TABLE III.   U.T.S. and Elongation Data for As-Rolled Preforms and Rolled-and-Annealed Preforms of High Density (L Series).

Final Density, % of solid	Rolling Temp. °C	Condition Tested	U.T.S. (psi)	Elong., %
98.3	850	Annealed	44,000	32
98.5	1040	Annealed	45,800	27
99.5	850	As-Rolled	56,700	10
99.8	1030	As-Rolled	54,000	11

attained. At the high stress level, it is possible that fracture was initiated prematurely by residual porosity or inclusions in the structure.

Specimens of high density. ≥95 per cent of solid, were heated in air to 900 C to test their resistance to internal oxidation. Metallography revealed no oxide from this source; the only inclusions present could be traced to the original iron powder, and were found in equal concentrations in non-oxidised specimens. Fig 8 to 11 reveal typical inclusions. Some of the apparent porosity in these microstructures is probably associated with the extraction of inclusions during metallographic preparation. Electron microprobe analysis has indicated that the larger inclusions were calcium-aluminum silicates.

CONCLUSIONS   1. Lightly sintered iron powder preforms of slab shape, whether made from loose or from compacted powder, can sustain large reductions by hot rolling at 850 or 1040 C between plane rolls.

2. The preforms can be hot rolled to a density at least as high as 99.8 per cent of solid in a single pass, and with modest roll force.

3. For a given final high density and thickness of rolled product, the roll force required is essentially independent of the starting density of the preform. The efficiency of densification in hot rolling is greater for low density preforms.

4. High density products of the single-pass hot rolling of preforms have high tensile strength and elongation, comparable to those of conventionally wrought material of the same composition and form. This high strength is relised in material which has been hot rolled to only $\sim$96 per cent of solid density.

5. Preforms hot rolled to $\geq$95 per cent of solid density do not internally oxidize when heated in air.

ACKNOWLEDGMENTS   This work was supported by the Defense Research Board of Canada, Grant No. 7510-69.

BIBLIOGRAPHY   1. M. H. D. Blore, V. Silins, S. Romanchuk, T. W. Benz, and V. N. Mackiw, "Pure Nickel Strip by Powder Rolling", Metals Eng. Quart., 6, (2) 1966, 54–60.
2. G. T. Brown and P. K. Jones, "Experimental and Practical Aspects of the Powder Forging Process", Int. J. of Powder Met., 6 (4), 1970, 29–42.

The P/M Extrusion of Tool Steel Bar

J. J. DUNKLEY* AND R. J. CAUSTON**

ABSTRACT

A new process for the bulk production of P/M high speed steel has been developed using hot extrusion to simultaneously consolidate the powder and work it to the finished section. This results in high yields and low costs since canning is not used. The properties of the extruded material are reviewed.

Introduction

For many years metallurgists working in the specialty steel industry have recognized the benefits of what P/M processing of high speed steels should bring. The basic problem in conventional processing is segregation which leads to ununiform properties, to poor hot working yields, distortion, poor grindability and erratic performance. Although progress has been made to sinter (ref. 1) or sinter forge (ref. 2) high speed steel powder compacts to full density, the technique is not widely used.

More successful has been the gas atomized HIP-route (ref. 3 & 4). This has allowed the production of highly alloyed grades and has been used for several years. However its more widespread application is hindered by economic considerations. Bearing this in mind a process has been developed aimed at competing economically with conventional production routes.

Powder Production

Water atomization is used for the production of the high speed steel powder (ref. 5). The powder analysis is 6.5%W, 4.7%Mo, 4.2%Cr, 1.9%V, 0.85%C, S & P < 0.03%, Si & Mn < 0.4%, 0:0.1 – 0.2%, N:0.01%. This is a rather inexpensive process capable of handling large volumes of material. The atomizer is 7 ft high and can process 230 kg heats of steel at up to 100 kg/min (6 tonnes/hr) using a 135 HP pump set. The powder is dried after decanting the atomizing water and has an oxygen content between 1,000 and 2,000 ppm. The as-atomized powder is too hard to compact readily and is thus annealed to reduce its hardness from over 500 HV to 250–300 HV. A typical high speed steel annealing cycle of 2 hrs at 850–900 C followed by cooling at less than 50 C/hr is used. The oxygen content of the powder is unaffected if a dry hydrogen, argon or nitrogen atmosphere is used. This allows billets of adequate green strength to be pressed without the need to resort to very high pressures which often lead to reliability problems under production conditions. The annealed powder is readily crushed and sieved through a 35 mesh (500 microns) sieve to remove or break-up agglomerates. Yields of dried powder from the furnace charge are typi-

* *Manager of P/M Development.*
** *Research Metallurgist, Davy-Loewy R. & D. Centre, Bedford, England.*

cally 95% and sieving losses 1–3% with a median powder particle size of 100 microns.

Billet Compaction

Since commercial hot extrusion presses process billets from 20 kg to over one ton and with a length to diameter ratio of typically 2.5–4.5, the production of extrusion billets by die pressing is not practical. Isostatic compaction is therefore used which allows the production of billets having adequate tolerances and the necessary nose profile. It is also possible to compact hollow billets for the production of tubes (ref. 6). The compacts have a density of 5.2–5.5 g/cc (62–67% TD) when pressed at approximately 30–40,000 psi (2100–2800 kg/cm²) and they have adequate strength.

Although there is a tendency to dismiss isostatic compacting as having a low productivity rate this is due mainly to the specialized fields in which they are used. Manufacturers are ready to quote presses capable of cycling 20–30 times per hour and this allows outputs of several tons/hr to be achieved.

Hot extrusion

The billets are heated to the extrusion temperature (1,000–1,200 C) in an inert or reducing atmosphere furnace. Investigations have been carried out into the induction heating of large compacts but problems still remain due to the variation of resistivity with temperature. The billets are coated with glass lubricant and hot extruded. With transfer times from furnace to press to 10–40 seconds no special precautions against oxidation are taken.

In the extrusion press two things occur; the billet is first compacted from 60% to about 98% TD and is then extruded through the die where it undergoes very

heavy shearing which eliminates all traces of porosity and shatters any residual surface oxide films that derive from the original powder. The pressure-extrusion ratio relationship is markedly different from that previously reported for the extrusion of solid materials namely,

$$P = K \log R \qquad (1)$$

where P is the extrusion pressure, K a constant and R the extrusion ratio (area of billet/area of extrude). However, a few comparative trials of solid and P/M HSS billets at ratio 13:1 showed no difference in pressure. Our observations suggest that

$$P = A + B \log R \qquad (2)$$

where A and B are constants, is a better approximation, certainly for $10 < R < 80$ (Fig. 1). When R = 1 equation (2) is inapplicable and it appears that the P vs R line does pass through the origin (fig. 1). In the region of transition from equation (2) to equation (1) i.e. as the extrusion ratio falls from 10 to 5 the quality of the extrusion degenerates. Transverse tearing of the surface begins and it is likely that inadequate densification has occurred before the metal passes

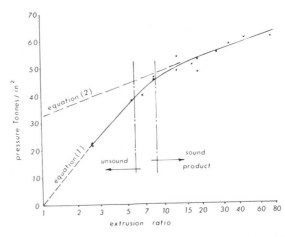

Figure 1 Extrusion Pressure/Ration Relation for M2 HSS Powder Billets.

through the die. It has proved possible to obtain sound extrusions from billets which have been previously hot compacted in the press to 99% density, at much lower ratios than green compacts of lower density.

The extrusion of a P/M HSS billet is more readily carried out than a conventional billet because the broader working temperature range is due to the absence of the low melting segregates which render the satisfactory extrusion of conventional ingots very difficult. Other advantages can be gained from the use of P/M billets. For instance very high extrusion ratios can be achieved by pressing a soft iron powder layer onto the nose of the billet. This reduces the peak extrusion pressure, permitting higher effective pressures to be used. Another possibility is the extrusion of composite bars having for example an M2 HSS layer on a 3% Mo steel base.

Product Quality

The main reason for the use of gas atomized powder in previous work on high speed steels has been the commonly held belief that for high speed steels, as for nickel based superalloys, a very low oxygen content is essential. However, it is very hard to locate significant amounts of published evidence on this subject (ref. 4). Comprehensive testing of the material under service conditions has been carried out with a view to establishing whether a high oxygen P/M HSS was inferior to a conventional HSS. The results of the test programme show that high oxygen (1,500–2,000 ppm) P/M M2 HSS grade is statistically speaking the same in cutting performance as conventional M2. In fact most results show a small superiority over the conventional product. Nevertheless by vacuum sintering before extrusion oxygen

contents of the order of 100 ppm can readily be achieved if required.

Microstructure

This structure is fairly typical of a P/M HSS with very fine grain size (S.G. No. 30–90) and carbide size (1–2 microns). The pronounced segregation apparent in conventional bars is completely absent. (Fig. 2 and 3)

Hardness

Hardness is generally slightly higher than usual (830–900 Hv for M2) especially at lower hardening temperatures, probably

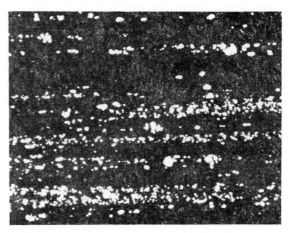

Figure 2 Annealed 1″ Conventional M2 Bar × 250.

Figure 3 Annealed 1″ PM Extruded M2 Bar × 250.

due to the more rapid solution of the fine carbides.

Toughness

Izod impact testing gave severe scatter on both P/M and conventional M2, and the anvil of the test machine was damaged several times. It was not possible to differentiate conventional and P/M products.

Grindability

As would be expected from previous work (Ref. 7, 4, 3) the grindability of P/M HSS is better than of the conventional material. Tests on both cylindrical and centreless grinding with cuts from 0.1–20 mils showed that the P/M M2 was as grindable as normal T1 and about twice as grindable as conventional M2. Grinding ratios and surface finishes were better, and higher metal removal rates could be achieved without burning the surface.

Turning Performance

Turning tests are the most economical and reproducible cutting tests. Several hundred separate tool tests have been carried out, the vast majority under interrupted cutting conditions. Tests with accelerating cutting speed (face plate type) showed an average advantage of 3.9%, over 20 separate extrusions each tested at least three times. Other tests showed similar results, namely a wide scatter of results giving an average performance for P/M M2 HSS slightly superior to the conventional product.

Drilling

Drilling tests have been carried out both by drill makers and independent research organizations; PERA.† No significant differences were detected.

End Milling

Tests carried out by PERA showed a life of 60.7" for a stock T1 cutter, 67.7" for an M2 cutter made and heat treated alongside the P/M M2 cutter which had a life of 98". This is not statistically significant in the test context.

These tests show clearly that the presence of oxygen in the form of finely dispersed oxides at levels from 1,500–2,000 ppm has no detrimental effect on the performance of high speed steel.

Economics

The conventional process route is compared to P/M extrusion in Fig 4. The reduction in the number of process steps and the amount of scrap makes the process considerably more economical than the conventional route. A detailed study of a 2,200 ton/year (single shift) plant showed a breakeven point of only 30% capacity and a return on capital at full output of 40–45%. This assumes that only M2, one of the cheaper grades, is processed. Grades T1, M2, M2S, M7, M15 and M42 have all been processed without any difficulties due to resulphurization or higher alloy content and these would yield greater economic returns.

Future Development

The process is technically capable of producing almost any alloy which can be hot worked. Thus stainless steels, die steels, cobalt, valve steels, nickel alloys and copper have been extruded as well as

---

† *Pera—Production Engineering Research Association, Melton Mowbray, Leicestershire.*

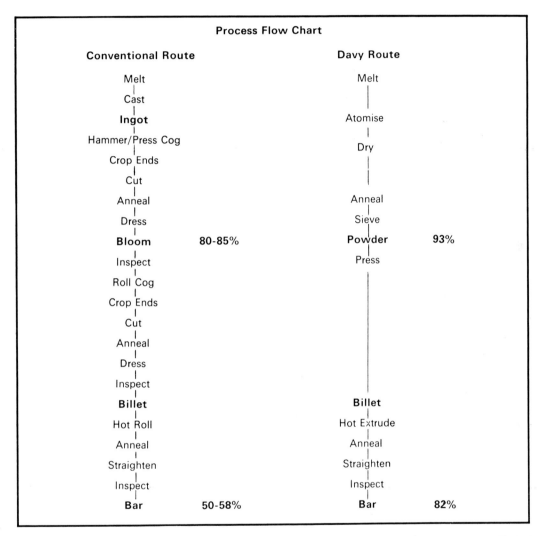

Process Flow Chart

Conventional Route		Davy Route	
Melt		Melt	
Cast			
Ingot		Atomise	
Hammer/Press Cog		Dry	
Crop Ends			
Cut			
Anneal		Anneal	
Dress		Sieve	
Bloom	80–85%	**Powder**	93%
Inspect		Press	
Roll Cog			
Crop Ends			
Cut			
Anneal			
Dress			
Inspect			
Billet		**Billet**	
Hot Roll		Hot Extrude	
Anneal		Anneal	
Straighten		Straighten	
Inspect		Inspect	
Bar	50–58%	**Bar**	82%

Figure 4 Process Flow Charts for the Conventional and PM Production of High Speed Steel bar with % Yields.

iron and low-alloy steels. Extrusion plants can have outputs of one million ton/year and thus bulk applications are feasible. It is likely that economic considerations will lead to high speed steel being produced first, followed by less costly alloys, but in the longer term quite large scale (10–50,000 ton/year) operations can be attractive, exploiting such things as composite bars and complex sections currently extruded from solid billets.

Conclusion

The P/M extrusion of high speed steels is an attractive alternative to conventional processing on economic grounds, as well as providing the technical advantages obtainable in P/M high speed steel. The basic process is so flexible that it could be extended to a wide range of alloys. It should also provide a boost for the hot extrusion press which has not made the progress once predicted as a hot working tool, mainly due to the high cost of billet preparation.

References

1. F. L. Jagger and W. J. C. Price—Powder Metallurgy 1971 p 407–424
2. T. Yamaguchi et al—Powder and Powder Metallurgy (Japan) *18*, 1971, No 2, p 56 –63 (Brutcher Translation No. 8672)
3. P. Hellman—Modern Development in Powder Metallurgy, vol 4, p 573–582, 1971
4. E. J. Dulis and T. A. Neumeyer—Materials for Metal Cutting (Iron and Steel Institute London 1970) p 112
5. J. J. Dunkley—Metal Powder Report, vol 30, No 1, p 2–7, Jan 1975
6. J. M. Siergiej—Proc. 4th Int. P.M. Conf. Brno Oct 1974, vol 1, p 359–372
7. Iron Age April 20 1972, p 90–91

Large Net Shapes By Powder Metallurgy

N. P. PINTO*

ABSTRACT

New methods are described for the manufacture of unconventional P/M parts, such as hollow cones or hemispheres. Discussion on the fabrication of porous parts. Hot isopressing is described.

Introduction

Powder metallurgy has expanded rapidly over the past few years particularly due to the development of new powder making methods and of new consolidation techniques. Superalloy, titanium alloy and other powders have better properties and therefore new applications when produced by the rotating electrode, hydride-dehydride and gas-atomization methods; improved grades of beryllium are made by impact-attritioning. In addition to consolidation by sintering, commercial parts are being produced by cold- and hot-isopressing,[3, 4] by plasma spraying and by hot pressing.

Because new types of powders are generally expensive and because powder metallurgy offers the availability of complex shapes, greater emphasis is placed on producing net shapes, that is, parts which have the same or about the same shape as the final piece and which have dimensions only slightly larger. Then, the final opera-

tion consists of a light forging or coining and/or machining to final tolerances.

The beryllium industry uses expensive powders to make complex parts and has pioneered in the use of innovative consolidation methods. A review of recent advances in the powder metallurgy of beryllium would include several examples of routine production parts made by processes which did not exist several years ago and other parts which have been made by advanced techniques but which, for other reasons, are not yet in production.

Re-Entry Vehicles

An example of a hollow shape is the experimental reentry vehicle made in 1972. It was a tubular body of about 8" diameter with a flared end opening to about 17" diameter, about 41" in overall length; the other end was closed with a spherically shaped end. The RV varied in wall thickness from about ¼" to 1", and it had a heavy internal boss for attachments; dimensions were held within .005" of the design.

Hollow parts are generally made by pressing the powder around a mandrel to assure better control of roundness and straightness; tubular shapes which are open on both ends are much simpler to produce than those with closed ends since the fixturing which supports the plastic enclosures which contain the powder can

---

* Kawecki Berylco Industries, Inc., Reading, Pa.

be attached to both ends of the mandrel rather than being cantilevered off one end.

The RV was made by hot-isopressing a previously cold-isopressed preform. Direct hot-isopressing of powder in a steel can did not appear practical since the spherical end varied in wall thickness from 1" to ¼" and there was concern about uneven collapsing of the steel can and loss of dimensional control. Special tooling supported the plastic outer container, with the mandrel positioned inside it. After beryllium powder was loaded into the annular space between mandrel and plastic bag, the powder was evacuated, and the assembly was cold-isopressed at 60,000 psi at the KBI plant in Hazleton, Pa. The outer bag was removed, and the preform was machined in the green state. Then, at Battelle Columbus Laboratories, it was encased in a fabricated steel sheetmetal jacket, evacuated, outgassed at 650 C (1200 F) to remove water vapor and $Be(OH)_2$;[7] both of these, at elevated temperatures, form hydrogen which is insoluble in beryllium and may add to swelling during post-isopressing heat treatment. Next, the preform was sealed and hot-isopressed at 915 C (1675 F). The stainless steel mandrel contracted more than the beryllium and was easily removed after cooling. The rough shape was clamped to the headstock of a lathe and the outer surfaces was first turned; then the inside was turned.

The outstanding accomplishment in making the 23-pound part consisted in the use of only a little over 60 pounds of powder. Machining from a cylinder would have required 670 pounds of powder; a rough, hollow blank might have been designed for a 510-pound charge.

It is important to note that hot-isopressing was required for this part since the high purity beryllium powder will not fully densify by conventional hot-pressing; the usual sintering aid (silicon),[6] an impurity in regular purity beryllium, is too low in concentration in the purer grade powder for satisfactory shrinkage. The mechanical properties were outstandingly good: ultimate tensile strength of 62,000 psi and yield strength of 38,000 psi; elongations measured in radial, circumferential and longitudinal directions were 3.6 to 4.2%, compared to the 2–2.5% which was typical of standard grades at that time. This part was the first in a series of parts which developed into a new high performance grade of beryllium, CIP/HIP-1.

Hollow Cones

A hollow frustrum or cone is a production item which is machined from blanks which are in the form of smooth surfaced cones about 8" in minor diameter, 17" major diameter and 36" long with 1.5" wall. Early in the production campaign, these parts were hot pressed as solid cylinders, and the blanks were machined chevron-style to extract more than one blank from a cylinder.

A process consisting of cold-isopressing followed by hot-pressing was modified to reduce the cost of the blank. A preform was cold-isopressed over a conical mandrel and machined lightly; then, the preform was positioned over a mandrel in the cylindrical cavity of a graphite hot pressing die, and pressure from the upper plunger of the die was transmitted to the outer surfaces of the preform via a medium such as graphite powder. Thus, the action of a hot-isopress was simulated in the hot-press. Whereas machining two cones from a solid cylinder might require a ratio of powder/machined part of about 6/1, the isopress/hot press method resulted in a ratio of about 3/1; machining time also was much less, of course.

Although relatively thin walled, the savings in powder weight and machining times were substantial. However, larger diameter, thicker walled cones have also been made; usually CIP/HIP (cold- followed by hot-isopressing) or direct hot-isopressing of powders has appeared most economical for short runs. Diameters up to about 10" and wall thicknesses up to about 2" have been isopressed using powder/part ratios between 5/1 and 3/1.

Although there is occasionally an advantage to isopressing a hollow cone as a free standing shape (without using a mandrel) for reasons of powder flow, can distortion, etc., hollow parts are usually pressed on a mandrel, and this provides excellent control of inside contour and dimensions. The taper provides easy removal of the mandrel which may be reused repeatedly; a straight sided (tubular) part would be more difficult and would require a mandrel material which contracts faster than the part during cooling from the isopressing temperature. Such hollow cylinders are more amenable to other processes and isopressing offers less attraction; of course, if the hollow cylinder has a closed end, isopressing is the preferred method. Mandrels need not have a straight contour, and there is latitude in designing an inner surface with depressions or bosses which vary the wall thickness, as long as reentrant angles are avoided which might complicate mandrel withdrawal.

The closed end tube or cone is ideally suited to isopressing, and very few processes can compete economically, especially if only prototypes or short runs are required. The tooling used in powder loading and the isopressing containers are more complex, and the accuracy with which wall thickness may be controlled is not as precise, but accuracy is quite good.

The design of the hollow part and the quantity dictate the process to be used, be it direct HIP, CIP/HIP, hot-pressing or some combination.

Hemispheres

Hemispheres and short cones are relatively simple to hot press or isopress and have been made in semiproduction quantities. But the shorter aspect ratio of length/diameter means that metal forming, e.g., spinning or deep drawing, may be less costly. P/M parts have been made by applying the pressure from the inside, forcing the powder against a hemispherical female die (Fig. 1), which eliminates problems of removing the mandrel from the part. By applying pressure in the opposite direction (against the outer surface), the part may be pressed on a mandrel. One modification is to use graphite powder as a pressure transmitting medium, or hot-isopressing is feasible. The hemisphere has also been made on a full production scale by hot pressing using a mandrel and a shaped female die; by stacking, as many as six are made in a single stroke. Hot pressed hemispheres generally have heavy walls, and thin walls are probably not practical.

Porous Parts

A controlled degree of porosity is often desirable for special applications. The porous bearing is a classic example. Another is the transpirationally cooled nose tip for missiles, in which the pores are filled with a material which melts at re-entry temperatures and flows to the surface to cool the tip.[1] Rounded pores also improve the

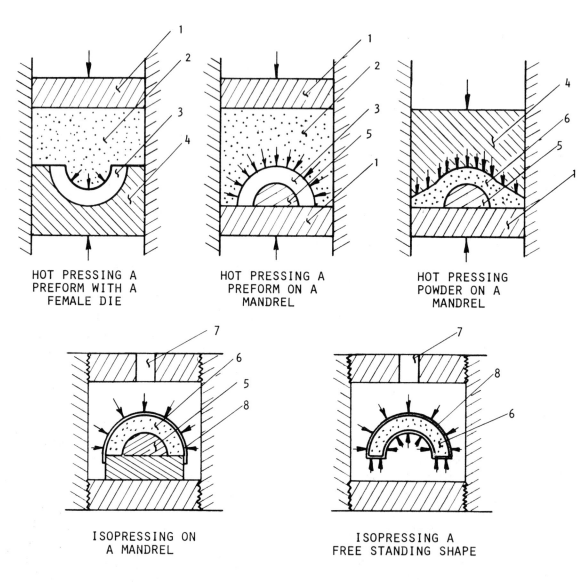

HOT PRESSING A
PREFORM WITH A
FEMALE DIE

HOT PRESSING A
PREFORM ON A
MANDREL

HOT PRESSING
POWDER ON A
MANDREL

ISOPRESSING ON
A MANDREL

ISOPRESSING A
FREE STANDING SHAPE

ARROWS INDICATE DIRECTION IN WHICH PRESSURE IS APPLIED.

1. PUNCH
2. GRAPHITE POWDER
3. PREFORM
4. DIE

5. MANDREL
6. POWDER
7. GAS PRESSURE INLET
8. ISOPRESSING CAN

FIGURE 1. Some methods for pressing short components.

dynamic properties of some materials, e.g., beryllium;[5] in this case, the pores are not interconnected.

Porous parts have long been made by pressing and sintering; now we have dem- onstrated that large parts can be made by (a) isopressing or by (b) plasma spraying and sintering. Hollow bodies having the shape and dimension of the part were made by the Union Carbide Corporation in Indianapolis to a controlled density of about 90% ± 2% using the latter process. An expendable substrate was coated with beryllium powder to wall thicknesses of .35 to 1.25". Then the substrate was removed by machining, and the porous preform was placed over a stainless steel mandrel for sintering. The mandrel was designed so that as it expanded the beryllium expanded also and then shrank due to sintering forces; at the end-point of the sintering, there was a slight interference fit between mandrel and part, which controlled shape and dimensions. During cooling from the sintering temperature, the stainless contracted away from the beryllium. At the point when the mandrel no longer supported the hollow part, the beryllium had regained sufficient strength to support its own weight. Wall thickness was controlled within about .002", and a .050" envelope was provided for final machining.

The same process which produced uniform density can also make parts having varying or graduated porosity;[2] several layers, each of different porosity, may be incorporated into the shape, or the porosity may change continuously from inner surface to outer. Porosity may be controlled from about 5% to 22%.

Parts of uniform porosity have been made by hot-isopressing, and, based on laboratory data, the hollow shape could be produced in full scale. And, there is some control over the selection of closed rounded pores vs. interconnected pores. For example, to make a density of 90%, beryllium powder was hot-isopressed between 600 and 1900 F (315–1040 C); at about 625 C (1150 F), a range of densities may be produced by controlling pressure, e.g., at 8000 psi a uniform density of 90% is achieved. Such porosity is uniform throughout the compact within less than two percent.

Both processes, plasma spraying and hot-isopressing, have their own advantages. Spraying offers better density control and is ideal for thin walls; however there is some possibility of contamination of metal purity, and there is substantial loss of input powder by overspray. Isopressing is not as well established; it is probably less costly; porosity must be uniform, not graduated.

Free Standing Shapes

The components discussed above were generally made by pressing the powder or preform against a die or mandrel. Such mechanical support is not essential, and it is possible to isopress a free standing shape, i.e., one in which the powder or preform is sealed in a thin walled container having the shape of the finished part. If the preform has closed pores, no container is required unless the pressurizing gas contains contaminants detrimental to the work piece.

However, the isopressing of unsupported shapes is much more liable to result in distortion and poorer dimensional control. For free standing shapes, powder loading is very important, and loading in vacuum offers advantages over air-loading followed by outgassing the container. Simple rectangular plates may be made economically, but unless they are isopressed against a flat support distortion may make the process noncompetitive with cutting plates from a large block. Hemispheres also may be isopressed free standing, but a die or mandrel is justified by the lower input weight required to produce an accurate blank.

Versatility of Isopressing

Isopressing offers a broader range of large shapes than is available from other metal shaping methods, possibly excepting casting. Cast shapes generally have poorer properties and geometries are limited to designs which avoid casting and shrinking problems.

Isopressing provides still greater versatility in that it accommodates many types of powders, including the irregular morphologies usually well suited to pressing and sintering and also the spherical powders made by rotating electrode or atomization method, which powders are more difficult to sinter. This is especially important in the case of titanium alloy powders which are usually alloyed prior to atomization; such powders have smooth surfaces and are not readily consolidated by cold pressing although hot pressing is quite practical.

Modifications to the isopressing method will produce a range of mechanical properties, since most powders will consolidate to full density at temperatures below the minimum necessary for best properties. Thus, one can attain high strengths using lower isopressing temperatures at some sacrifice of ductility (Table I); as a comparison, two grades of beryllium powder were isopressed at three temperatures into solid cylinders about 50 mm (2") diameter by 125 mm (5") long, with mechanical properties measured perpendiculary to the axis of the cylinder. Isopressing above the minimum needed for full densification is essentially a combining of the pressing step with a post-consolidation heat treatment.

Thus, isopressing provides near net shapes (a) using powders or powder mixtures which cannot be consolidated by

TABLE 1   Effect of Isopressing Temperature on Properties of Beryllium Compacts

HIP Temp.	Grade	U.T.S.	Y.S.	El.
760		68	43	4.6
915	CIP/HIP-1	69	43	4.9
1065		67.5	40	6.9
760		94	76.5	1.2
915	RR243	98.5	76.0	2.8
1065		94.1	71.9	3.6

other means, (b) in which mechanical properties can be controlled over a broad range, and (c) which are fully dense or which have controlled porosity, with some control over pore shape.

Summary

New techniques are being applied to producing near net shapes, work-pieces which have the desired shape and which require little machining for cleaning up and reaching final dimensions. The net shapes of titanium alloy in smaller sizes (under one foot) can be made close to final size. Other techniques developed for beryllium have made much larger net shapes, also close to final size. While both methods are suitable for hot-isopressing, the possibility of using conventional hot-presses for near net shapes has also been demonstrated, and the available production capacity is thus expanded manyfold. Also demonstrated is the practicality of making parts of controlled porosity when required for special properties.

References

1. Schwartzkopf, P., "Fabrication of Poly Porosity Microstructures," U. S. Patent 3,362,818
2. Taylor, T. A., "Low Density Sintered Beryllium,"

Union Carbide Corp. Project Report *42*, 16 June 1972

3. Pinto, N. P. and A. J. Martin, "High Purity Beryllium Powder Components," Powder Metallurgy, 1974, 17, 33

4. Pinto, N. P., J. P. Denny and G. J. London, "Isostatic Pressing of Beryllium Powder, Light Metals, 1974

5. Cooper, R. E., "Fracture Toughness of Beryllium: A Summary Report of the Present State of Knowledge," UKAEA-AWRE Report 017, 1972

6. Butcher, J. W. and Scott, V. D., "The Influence of Silicon and Other Elements in Controlling Interparticle Friction and Sintering of Beryllium Powder," Powder Metallurgy, *14*, 27, 1971

7. Stuart, W. I. and Price, G. H., "The High Temperature Reaction between Beryllia and Water Vapour," J. Nuclear Materials, *14*, 417–424, 1964

Manganese Steels for P/M Hot Forming

J. P. COOK*

ABSTRACT

Manganese steel powders easily oxidize during processing, however, high temperature sintering can reduce the oxides and yield material with impact properties and hardenability equal to wrought steels.

Introduction

The properties of P/M formed material increase as the inclusion content (oxygen content) decreases. The oxygen content in the final formed component is a function of the powder alloy used and the processing it receives. For example, nickel-molybdenum water atomized low alloy steel powder can be produced with a relatively low oxygen content (0.15–0.10% O_2) and this material is further refined with conventional sintering of preforms prior to hot forming. This refinement is possible owing to the fact that the oxides of Ni and Mo are relatively easy to reduce. This ease of processing along with good hardenability and mechanical properties has contributed to the fact that Ni–Mo low alloy steels are the principal materials being used in P/M hot forming.

Manganese and manganese-molyb-denum steels also offer the possibility of good hardenability and mechanical properties along with lower cost for powder forming applications. However, difficulties in manufacturing the powders and processing the preforms have held these steels off the market. The properties (mostly hardenability and impact strength) can be degraded by manganese oxidation during powder manufacture or during preform sintering. However, proper processing techniques can keep the oxygen content of these powders to relatively low levels as evidenced by the data for two 1500 steels (1.5% Mn) listed in Table 1. These data show one of the steels was oxidized during

* Hoeganaes Corporation, Riverton, N.J., U.S.A

TABLE 1 Oxygen Contents and Impact Properties of Two AISI 1500 Steel Powders

	Powder oxygen content %	Sintered oxygen content* %	Formed impact energy** ft-lbs.
Optimum Processing	0.20	0.16	26
Oxidized Processing	0.37	0.31	15

* Endogas—2050°F—30 min.
** Charpy V-Notch—Room temperature.

manufacture and preforms made with this steel had sintered oxygen contents over 0.30 percent. The impact properties of the oxidized steel are low while the other steel had much higher energy levels. However, both are much lower than wrought steel. Nevertheless the tensile properties of P/M formed steels are generally equal to wrought steels of equal hardness and limitations of an alloy's tensile strength are limited only by its hardenability.

Manganese steels can also be degraded by certain sintering conditions even if the powder has a very low oxygen content after manufacturing. Gas atomized powders generally have very low oxygen contents because they are protected from air during the melting and atomizing operation. However, Table 2 shows that after sintering in Endogas the gas atomized manganese steel preforms had an oxygen content equal to the water atomized manganese steel preforms. The gas atomized steel was degraded by the sintering treatment while the water atomized steel was improved. Again, the properties of these steels will be lower than that of wrought steels because of their high oxygen content.

Conventional wrought steel properties can only be duplicated when the inclusions or porosity content and oxygen contents are equal to those of wrought steels. Improvements in P/M formed impact proper-

ties can be achieved by increasing preform deformation while both impact properties and hardenability can be improved to the level of wrought steel by high temperature sintering.

High Temperature Sintering

High temperature sintering in an appropriate atmosphere reduced the manganese oxides as Fig 1 shows for an AISI 4025 steel (0.8% Mn, 0.25% Mo). Fig 1 also shows the oxygen content of preforms sintered in Endogas at 2050 F. Endogas was not employed at higher sintering temperatures due to the great difficulty of controlling the proper carbon potential with this gas at higher temperatures.

Powder preforms of AISI 4025 steels were sintered at 2050 F in Endogas and at 2300 F in dissociated ammonia then hot formed to full density by either repressing (no lateral deformation), upsetting (significant lateral deformation) or hot rolling (greatest lateral deformation). Room temperature impact energies of the Endogas sintered and *repressed* preforms were only 17 ft-lbs, while Fig 2 shows that impact energies

TABLE 2 Oxygen Contents of Gas and Water Atomized AISI 1500 Steels

	Powder oxygen content %	Sintered oxygen content* %
Gas Atomized	0.018	0.16
Water Atomized	0.20	0.16

* Endogas—2050°F—30 min.

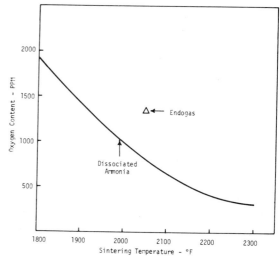

FIGURE 1 Oxygen Content Vs. Sintering Temperature— AISI 4025 Powder Preforms

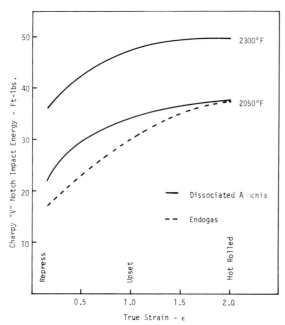

FIGURE 2 Impact Energy Vs. True Strain—P/M Hot Formed AISI 4025

increased sharply with increased deformation until 35 ft-lbs were reached with the *hot rolled* preforms.

Inclusions and porosity are detrimental to impact properties. Fig 3, a photomicrograph of the repressed sample, shows many inclusions and microporosity that define the prior particle boundaries. These are regions of weakness when impact loaded. The hot rolled sample, Fig 4, has the inclusions fragmented, aligned and elongated in the rolling direction, typical of wrought steel. This inclusion morphology and the better bonding across healed porosity in the hot rolled material provides a tougher structure.

Impact properties and hardenability are both improved by high temperature sintering. Fig 2 shows large increases in impact energies with increased sintering temperature (lower oxygen content). Preforms sintered at 2300 F and repressed had an impact energy of 36 ft-lbs while those

sintered at 2050 F had only 22 ft-lbs. Again, increased deformation improved the impact energy values significantly. Preforms sintered at 2300 F and repressed had an impact energy of 36 ft-lbs while hot rolled preforms had 50 ft-lbs and this equals wrought values for this steel. Fig 5 and 6 show that the repressed samples had micro-porosity while the hot rolled samples had little or none. Both samples had simi-

FIGURE 4 Hot Rolled P/M Hot Formed AISI 5025—Endogas Sinter—2050°F—Unetched—(42×)

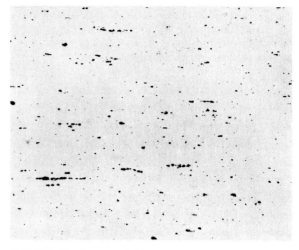

FIGURE 3 Repress P/M Hot Formed AISI 4025—Endogas Sinter—2050°F—Unetched—(42×)

FIGURE 5 Repress P/M Hot Formed AISI 4025—Dissociated Ammonia Sinter—2300°F—Unetched—(42 ×)

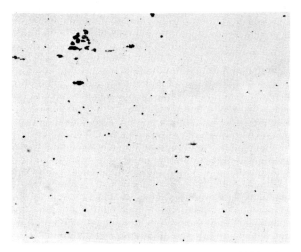

FIGURE 6 Hot Rolled P/M Hot Formed AISI 4025—Dissociated Ammonia Sinter—2300°F—Unetched—(41 ×)

lar oxygen contents and equal measured densities (7.87 g/cm³); however, hot rolling with its increased shear type deformation closed all the porosity while hot repressing did not. The hot rolled sample had wrought steel impact properties because high temperature sintering resulted in a cleanliness equal to wrought steel and rolling broke up the remaining inclusions and healed the porosity.

It is significant to note that the high oxygen content preforms benefited the most from increased deformation. This agrees with the work done by Eloff et al (1) where it was shown that increased deformation improved the impact properties of high oxygen content nickel-molybdenum steel preforms but had much less effect on preforms sintered at high temperatures to a low oxygen content. The purpose of a high temperature sintering is to reduce the number and size of inclusions and also clean the powder particle surfaces. This cleansing action facilitates particle welding when the preform is deformed thus removing a weak region and improving the impact properties.

Hardenability is dependent to a great deal on the alloying elements in solution. Oxidized alloying elements cannot contribute to hardenability, therefore high oxygen content P/M hot formed material will have lower hardenability than their wrought counterparts. High temperature sintering reduces oxides releasing alloying elements that dissolve to improve the hardenability of a P/M alloy to the same level of wrought steel. Endogas sintered P/M formed material of AISI 4025 steel had a high oxygen content (0.14) and was below the Jominy hardenability "H" band for 4027H steel while a P/M formed material sintered at 2300 F in dissociated ammonia had a low oxygen content (0.030) and was in the middle of the "H" band as shown in Fig 7. Fig 8 shows that the high temperature sintered material has approximately equal hardenability to commercially available nickel-molybdenum P/M steel.

Summary

Both the hardenability and impact properties of manganese steels can be degraded

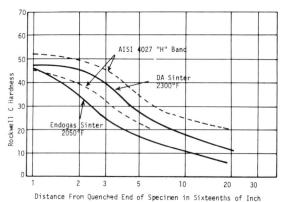

FIGURE 7 Jominy Hardenability Curves AISI 4025—Endogas Sinter—2050°F and Dissociated Ammonia Sinter 2300°F

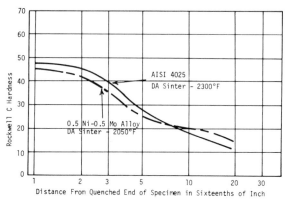

FIGURE 8 Jominy Hardenability Curves AISI 4025—Dissociated Ammonia Sinter—2300°F—And 0.5 Ni-0.5 Mo Alloy—Dissociated Ammonia Sinter 2050°F

by oxidation of the manganese during powder manufacturing or during preform sintering. The impact properties of oxidized material can be improved by increased deformation. However, high temperature sintering can reduce the oxides and yield material with impact properties and hardenability equal to wrought steels. Increased usage of manganese steels is expected when high temperature sintering becomes more commonplace since these P/M steels can offer savings over the presently available nickel-molybdenum steels.

Reference

1. Eloff, Peter C. and Guichelaar, Philip J., "Hot Formed Powder Preforms: The Relationships Among Deformation, Induction Sintering, and Mechanical Properties," *1972 Materials Engineering Congress*, Cleveland, Ohio, October 14–19, 1972.

Designing P/M Preforms for Forging Axisymmetric Parts*

C. L. DOWNEY** and H. A. KUHN***

ABSTRACT

Axisymmetric forged parts are considered to be composed of various sections, each undergoing one of the basic modes of deformation: upsetting (lateral flow), forward extrusion, backward extrusion. Experimental studies of each of these modes were carried out to determine the effects of preform size and shape on densification and fracture. Results of these studies are utilized to design the preform for a complex part consisting of a hub, flange, and rim. The technique can also be utilized to design preforms for other axisymmetric parts.

INTRODUCTION

P/M preforms contain 20 to 30% void space, and are thus susceptible to fracture during forging. Utilizing a preform shape very close to that of the final part (hot repressing) minimizes the amount of plastic deformation, and essentially eliminates the occurrence of cracking. Studies have shown, however, that the properties of forged P/M preforms increase as the amount of plastic deformation during forging increases.[1,2] This requires preforms of simple shape which then undergo substantial flow to reach the final shape. Fracture also becomes a distinct possibility during forging.

Current research in workability of fully-dense materials indicates that careful control of the deformation can reduce or eliminate the local stress and strain states that lead to fracture.[3,4] The technique is based on a fracture strain criterion. This approach is demonstrated through design of a preform for an axisymmetric part involving combined modes of flow.

EXPERIMENTAL PROCEDURE

Aluminum alloy powder (ALCOA 601 AB) was used in the study. Previous tests indicated that cold deformation of this material closely reproduces not only the deformation profile but the fracture strains of low alloy steel powder preforms forged at hot working temperatures.[5]

Compacts of aluminum alloy powder were treated as a model material. They were machined into various preform shapes, marked with grid lines, and forged to permit study of the basic modes of flow. Results of these tests were then used in

* Based in part on a thesis submitted in partial fulfillment of requirements for the Ph.D. in Metallurgical Engineering at Drexel University, Philadelphia, PA.

** Research Engineer, Cincinnati Incorporated, Cincinnati, OH.

*** Associate Professor of Metallurgical and Materials Engineering and Associate Professor of Mechanical Engineering, University of Pittsburgh, Pittsburgh, PA.

conjunction with a forming limit concept to design a preform for a part involving combined modes of flow. The prescribed preform, and several other preforms were forged to test the validity of the procedure.

BASIC MODES OF FLOW

The flow of metal in various sections of most axisymmetric forged parts can be considered to undergo one of three different modes of flow: upsetting, forward extrusion, or backward extrusion. Flow and fracture during upsetting of cylinders of P/M materials[6] and of fully-dense materials[3] have been thoroughly investigated previously. In the present study, formation of a cup shape (back extrusion), Fig 1a, and a flanged hub shape (forward extrusion plus lateral flow), Figure 1b, was examined.

In forming the cup shape, the P/M preform may be cup-shaped so that only densification occurs during the forming process. Since gross plastic deformation is beneficial to mechanical properties in the forged part, this option will not be considered in the present study. Two cup preform options that involve large deformation are shown in Fig 2. The preform fits against the die in Fig 2a so that the rim forms simply by flow up the die sidewalls. No fractures occurred in these preforms. In Fig 2b, the

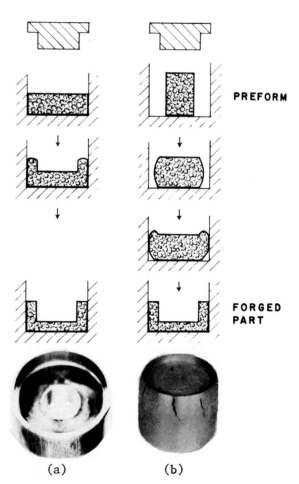

(a)   (b)

FIGURE 2 Two preform shapes for forging the cup shape. Preform (a) does not lead to cracking, while the rim cracks in preform (b) as metal flows around the punch corner.

preform is first upset and then back extrusion of the rim occurs. In this case, circumferential tensile strains due to expansion of material flowing around the punch corner lead to fracture of the rim.

In forming the flanged hub shape, Fig 1b, again the preform may have the same shape as the part so that forming simply involves densification. Three other options involving gross plastic deformation are shown in Fig 3. The first, Fig 3a, results in formation of the hub by extrusion. In this case, fracture occurred due to tensile

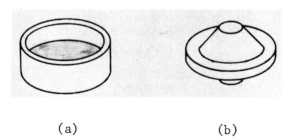

(a)   (b)

FIGURE 1 Two part shapes examined in this study: (a) cup; (b) flanged hub shape.

strains at the free surface of the hub resulting from friction between the die and preform. In the second preform option, Fig 3b, the preform fills the hub and the flange is formed by lateral flow. In this case, cracks occurred on the free expanding lateral surface because of tensile strains around the circumference. The third option, Fig 3c, involves formation of both the hub and flange sections by deformation. In this case, cracks formed at the hub surface when the draft angle was 45° or greater, while no cracks occurred when the angle was less than 45°.

FORMING LIMIT CRITERION

The strains on the surfaces of the parts depicted in Fig 2 and 3, as determined from grid lines, are shown in Figure 4. In addition, the fracture strain line for the preform material determined from upset tests, as described in previous work[1], is shown as the dotted line in Fig 4. For the preform 2b, the strains in the rim consist of circumferential tension and a small degree of vertical compression; in preform 2a no circumferential tension occurs. Consequently no cracks form in preform 2a while the rim cracks in preform 2b.

The strains in preform 3a are both tensile, leading to fracture. In the hub region of preform 3c, the strains are both tensile (leading to fracture) if the draft angle is greater than 45°, while they are both compressive (no fractures occurring) if the draft angle is less than 45°. In both preforms 3b and 3c, strains at the free expanding surface consist of large compressive and tensile components, similar to those in upsetting.

In all cases, as the strain paths cross the fracture line for the material, fracture oc-

curs. Therefore, use of the fracture strain line for the material, in combination with experimentally determined strain paths, provides a method for evaluation of the workability of various preform shapes.

PREFORM DESIGN

The characteristics of metal flow and fracture determined in the previous sections are utilized in the design of a preform consisting of a hub, flange, and rim (Fig 5). In order to develop a large amount of deformation in the material during forging, a flat ring preform will be considered. In addition, a flat ring preform is easier to compact, sinter, and handle with auto-

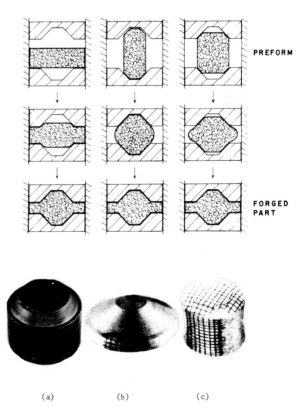

(a) (b) (c)

FIGURE 3 Three preform shapes for forging the flanged hub shape. Cracks occurred at various locations, depending on the preform shape.

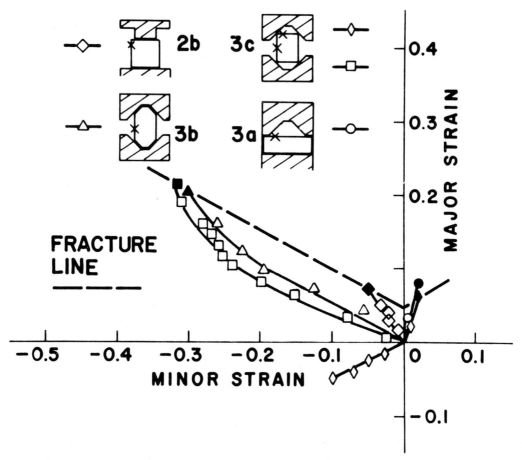

FIGURE 4 Representative strain paths for the fracture locations depicted in the preform shapes in Figures 2 and 3. The fracture strain line for the material, determined in Reference 1, is shown as the dotted line. As the strain paths cross the fracture strain line during progressing deformation, fracture occurs. The closed symbols represent fracture.

matic equipment than a preform in which the rim and hub are partially formed. The various possibilities for the ring preform are illustrated in Fig 6. The outside and bore diameters must be determined so that the part can be forged to full density without fracture.

In consideration of the outside diameter of the preform, Figure 2 indicates that when the preform fits against the die wall, fracture will not occur during formation of the rim. Thus, the preform O.D. should be as close as possible to the die wall diameter; this eliminates preforms 6b and 6d. For the material under consideration in the present work, the minimum tensile strain to fracture without the benefit of an attendant compressive strain is 0.05 (Fig 4). The maximum allowable clearance between the preform O.D. and die diameter, therefore, is 5%. It should be pointed out that fracture of the rim during back extrusion is entirely dependent on the degree of circumferential expansion of the rim mate-

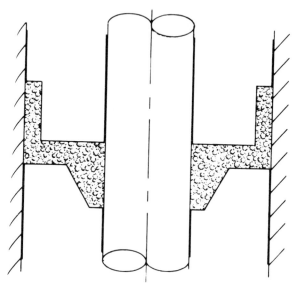

FIGURE 5 Cross-section of the part under consideration for powder forging.

rial and is therefore independent of the profile radius of the punch.

In consideration of the preform bore diameter, the results of preform 3a are recalled. Friction along the die surfaces resulted in tensile strains that led to fracture on the free surface. For the complex part in Fig 5, preform 6c will lead to similar results. Since the preform bore fits on the mandrel, friction along the mandrel and die surfaces will lead to tensile strains and fracture on the hub surface. A clearance between the preform bore diameter and the mandrel, preform 6a, will eliminate this problem. In addition, if the preform bore fits on the mandrel, the time of contact between the material and the mandrel will be increased, resulting in high operating temperatures of the mandrel. Coupled with the surface shear during stripping of the part from the mandrel, this will lead to excessive wear of the mandrel. A clearance between the preform bore diameter and mandrel will reduce the time of contact

and also the heat absorbed by the mandrel, thus reducing wear on the mandrel.

The clearance necessary between the preform bore and mandrel cannot be determined exactly. However, an excessively large clearance will result in a high, thin-walled ring preform and the shearing action of the punch may lead to cracking on the inside of the rim. For this reason, the preform bore diameter was taken midway between the mandrel and hub diameters.

CONCLUSION

To test the guidelines established in the previous section, the suggested preform design and three other designs expected to lead to cracking were tested. The preforms

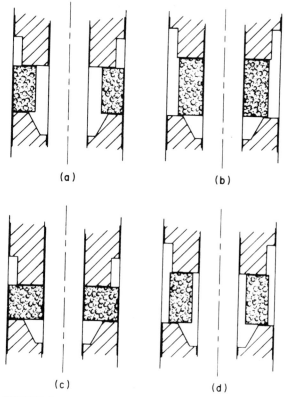

(a)   (b)

(c)   (d)

FIGURE 6 Possible configurations of the ring preform for forging the part shown in Figure 5.

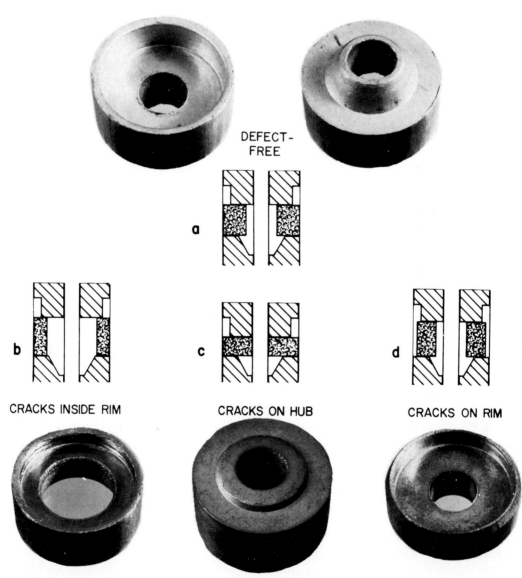

DEFECT-
FREE

a

b   c   d

CRACKS INSIDE RIM   CRACKS ON HUB   CRACKS ON RIM

FIGURE 7 Preform shapes tested and the parts forged from each shape. Preform (a) is an example of 6(a) with the bore midway between the mandrel and hub diameters. This preform leads to full-density, defect-free parts. Preform (b) is another example of 6(a) with a large bore. This leads to circumferential cracks on the inside rim. Preform (c) represents 6(c) and results in cracking of the hub. Preform (d) represents both 6(b) and 6(d) and leads to cracks at the top of the rim.

tested and the forged parts are shown in Fig 7.

Preform 7a has the prescribed dimensions, i.e. the O.D. fits against the die wall and the bore diameter is midway between the mandrel and hub diameters. Two views of a part forged from this preform are shown. No fractures occurred in the part.

Preform 7b is another example of a preform having the O.D. against the die wall, but the bore is very large and exceeds the hub diameter. During forging, the punch shears the thin layer of material at the bore causing circumferential cracks on the inside of the formed rim. Unlike the cracks that form at the top of a rim (Fig 2b), it is expected that those formed on the inside of the rim can be eliminated by using a large punch profile radius.

Preform 7c is equivalent to 6c, in which the preform bore fits against the mandrel. As predicted, cracking occurs on the hub.

Preform 7d is representative of both preforms 6b and 6d. In each case, there is a large clearance between the preform O.D. and the die wall. This leads to cracking at the top of the rim, as in Fig 2b.

Fig 7 indicates that the correct preform shape and dimensions can be determined from consideration of fracture during the basic modes of flow involved in forging the part. This technique does not take into account die wear. As further information on die wear and its relationship to metal flow is developed, however, this factor can be incorporated to develop P/M preform designs that optimize the forging process with respect to cracking, final part properties, and die wear.

ACKNOWLEDGMENT

The authors are indebted to Mr. Kal Buchovecky at the ALCOA Technical Center for generously supplying the powder and sintered compacts used in this study. This work was supported by the National Science Foundation, The Department of Defense through a THEMIS grant, and a grant from the ALCOA Foundation. The research was carried out while both authors were at Drexel University in Philadelphia, Pennsylvania.

REFERENCES

1. H. A. Kuhn and C. L. Downey, "How Flow and Fracture Affect Design of Preforms for Powder Forging," Int. J. Powder Met. and Powder Tech., *10*, 1974, pp. 59–66.
2. H. W. Antes, in Gear Manufacture and Performance, ASM, Metals Park, Ohio, 1974, pp. 271–293.
3. P. W. Lee and H. A. Kuhn, "Fracture in Cold Upset Forging—A Criterion and Model," Met. Trans., *4*, 1973, pp. 969–974.
4. H. A. Kuhn, P. W. Lee, T. Erturk, "A Fracture Criterion for Cold Forming", Trans. ASME, J. Eng. Materials and Tech., v. 95H, 1973, pp. 213–218.
5. C. L. Downey and H. A. Kuhn "The Application of a Forming Limit Concept to Design of Powder Preforms for Forging," presented at the ASME Winter Annual Mtg., Nov. 19, 1974, New York; to be published in Trans. ASME, J. Eng. for Industry.
6. H. A. Kuhn and C. L. Downey, "Material Behavior for Powder Preform Forging", Trans. ASME, J. Eng. Materials and Tech., v. 95H, 1973, pp. 41–46.

Progress in Gearmaking

P/M Hot Formed Gears

H. W. ANTES

THE main purpose of metalworking and metal fabricating is to change a mass of metal into a useful shape. The approach for making gears using powder metal (P/M) as a starting material is as follows. In conventional P/M processing, metal powder particles are compacted in a negative-shaped die at stresses that range from approximately 20 to 70 tons/in². The positive "green" powder compact is removed from the die, and sintered in a protective atmosphere at around 2000 to 2100°F (1100 to 1150°C) to develop a diffusion bond between the metal powder particles. Gears made by this process usually require little or no machining. The limitation of the process is that the resulting product contains an appreciable amount of porosity (5 pct or more), and thus lacks the mechanical integrity of higher density materials.

The P/M hot forming process was developed to realize the advantages of starting with a powder and yet provide a high integrity fully dense component. Processes currently being used to produce gears involve several steps. First, powder particles are compacted into a preform that may or may not have a geometry similar to that of the final component. Next, the preform is subjected to one or more steps of thermal processing to accomplish various objectives. Finally, the hot porous preform is densified as it is formed to the final shape. Cleaning or machining operations may be performed as well as heat treating before the component is ready for service.

This report describes some of the important details and reasons for specific types of processing in making gears by P/M hot forming. In addition, dimensional tolerance, surface finish, properties, and performance of P/M hot formed gears will be presented. This report will be limited to processes used for low alloy

H. W. ANTES is Senior Group Manger—R&D, Hoeganaes Corp., Riverton, N.J. This article appeared in "Gear Manufacture and Performance," the Proceedings volume for the Metalworking Forum on Gear Manufacture and Performance, held at Troy, Mich., on October 29 to 31, 1972, and sponsored by the ASM Mechanical Working and Forming Division.

steels. However, many of the procedures and concepts developed are applicable to other alloys.

POWDER FORMING PROCESSES

The basic process involves four steps:
1. Selecting metal powder
2. Fabricating preform
3. Forming preform to finished shape
4. Finishing operations

Three variations of the basic P/M Forming processes have evolved (Fig. 1). The first process (on the left in Fig. 1) utilizes a precision preform made by die pressing to a shape very similar to the final formed configuration. Therefore, the geometry and configuration of the preform is fairly well defined. The forming step in this process is more accurately described by the term "hot repressing."

Process 2 utilizes a flashless forming operation. In this process, the preform can vary from something relatively easy to design to probably the most difficult to design. In flashless forming, the preform weight must be controlled very accurately. If the preform is too light, the finished part will be undersized or less than full density. If the preform is too heavy, the component may be oversized, the die overloaded, or die failure may occur. In automated gear forming processes, the preform weight is usually controlled to ±0.5 pct.[1]

The third process (right hand side of Fig. 1) utilizes conventional closed die forming techniques. Weight of the preform does not have to be as carefully controlled as that in flashless forming; weight variation can be compensated by the amount of flash produced.

These three processes are used extensively for powder forming. The first process, "hot repressing," has been used predominantly where a fully dense gear is not required. Residual porosity in the finished component may range from almost 0 to 2 pct voids. Hot repressing pressures as high as 96.8 tsi have been reported.[2] The flashless and conventional closed die

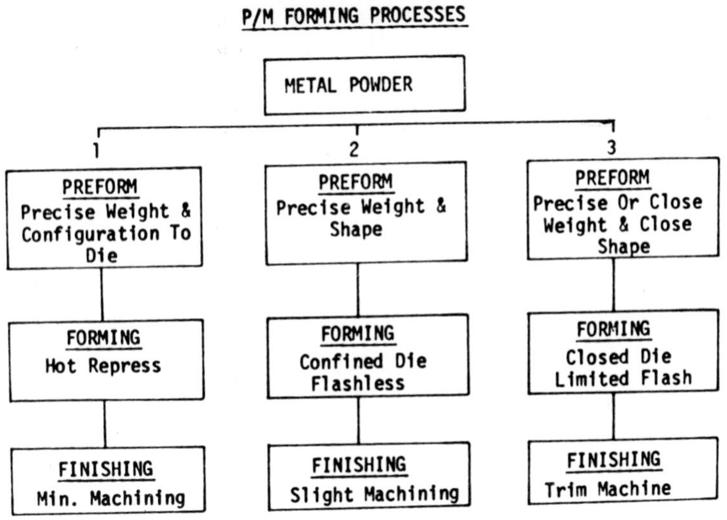

P/M FORMING PROCESSES

```
                    ┌──────────────┐
                    │ METAL POWDER │
                    └──────────────┘
          ┌──────────────┼──────────────┐
          1              2              3
  ┌──────────────┐ ┌──────────────┐ ┌──────────────┐
  │   PREFORM    │ │   PREFORM    │ │   PREFORM    │
  │Precise Weight│ │Precise Weight│ │Precise Or    │
  │& Configuration│ │& Shape       │ │Close Weight &│
  │To Die        │ │              │ │Close Shape   │
  └──────────────┘ └──────────────┘ └──────────────┘
          │              │              │
  ┌──────────────┐ ┌──────────────┐ ┌──────────────┐
  │   FORMING    │ │   FORMING    │ │   FORMING    │
  │ Hot Repress  │ │Confined Die  │ │Closed Die    │
  │              │ │Flashless     │ │Limited Flash │
  └──────────────┘ └──────────────┘ └──────────────┘
          │              │              │
  ┌──────────────┐ ┌──────────────┐ ┌──────────────┐
  │  FINISHING   │ │  FINISHING   │ │  FINISHING   │
  │Min. Machining│ │Slight        │ │Trim Machine  │
  │              │ │Machining     │ │              │
  └──────────────┘ └──────────────┘ └──────────────┘
```

Fig. 1—P/M forming processes.

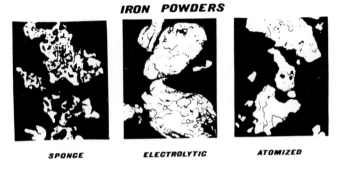

IRON POWDERS

SPONGE   ELECTROLYTIC   ATOMIZED

Fig. 2—Types of ferrous powders.

forming processes are used when full density is required. However, components made by these processes may not always be fully dense. The flashless process is used almost exclusively for making high-integrity, fully dense P/M hot formed gears.

The "confined-die flashless" process is somewhat of a combination of the other two processes. In this process, the component is made with a single blow or press on the preform. During compaction, lateral flow occurs, but no parting line flash is formed because confined or "trap" dies are used.

POWDER SELECTION

The principal types of ferrous powders available are electrolytic, sponge, and atomized (Fig. 2). Electrolytic powder is high purity material, but is expensive and will probably have limited use in P/M forming because of its high cost. Furthermore, the powder particles are somewhat rounded, and the green strength of this material is low. Sponge iron has an irregular type particle, and has good green strength even at low green density, which may be important for complex preform shapes. Sponge iron is inexpensive, but alloying must be achieved (for the most part) by mixing elemental powders. Furthermore, sponge irons may contain anywhere from 1 to 3 pct insoluble second-phase material. This may not be undesirable if less than full density is required in the final component. However, if full density is required and if maximum properties are to be

achieved, an atomized type powder will probably be preferred.

Atomized powder particles have a shape somewhere between that of sponge and electrolytic powders. Featuring good green strength and high compressibility, atomized powder can be made in a wide variety of alloy compositions by prealloying in the melt so that each powder particle is completely alloyed. These factors, coupled with the low cost and the relative cleanliness of atomized powder, dictate using an atomized product for most P/M forming applications.

PREFORM FABRICATION

The fundamental starting point for all preform design is the known weight of the final component plus that of the flash, if flash is produced. This information naturally fixes the weight of the preform. The problem of design is then resolved in the size of the preform (density) and the shape (amount and type of flow) that will facilitate making the component while minimizing die wear. The simplest example is for the hot repress forming process. The shape of the preforms is relatively fixed, since it must be approximately that of the final component so that it fits into the forming die. However, the size (height) or density may be varied.

It will be shown later, that flow is important in developing optimum properties in P/M forming. In hot repressing, the only freedom for flow is internal; no lateral (external) flow can occur because the preform touches the die wall. Therefore, the lower the preform density, the greater the amount of internal flow possible in hot repressing. However, low density preforms require higher forming pressures to achieve a given density.[3] Furthermore, lower density preforms are more prone to internal oxidation and decarburization on exposure to the atmosphere prior to forming or densifying.[4] The problem of internal oxidation of preforms will be covered later.

Preform design is more complicated for the flashless and limited-flash forming processes. In these processes, both preform size (density) and shape (amount of flow) must be determined. There are no hard-fast rules or formulas to use for preform design. However, some general guides for designing P/M preforms for closed or confined die forming have been presented.[3,5]

Consolidation

Green preforms have been made by both die compaction and isostatic compaction techniques (Fig. 3). Usually when die compacting is used, the powder is blended with a lubricant (e.g., 3/4-1 pct wax or stearate) to prevent die scoring. Rubber molds are used for isostatic compaction, and a lubricant is not required.

In comparing the two compacting processes, higher productivity and greater dimensional accuracy can be achieved with die compacting. However, isostatic compacting offers somewhat more freedom in preform design and more uniform densification. Furthermore, isostatically compacted preforms require no lubricant burn-off.

Sintering

Sintering of P/M hot forming preforms has somewhat different requirements than sintering conventional powder metal parts.[6] To summarize, the sintering operation must produce a relatively strong bond among the powder particles to permit handling and facilitate substantial flow or deformation without cracking. In addition, sintering must serve as a refining step to reduce oxygen content, provide the proper carbon content, and distribute alloying elements, resulting in a component with adequate heat treatability to provide good impact and fatigue resistance.

Sintering is usually accomplished with either gas fired or electrically heated sintering furnaces. Induction heating may be used for sintering too. In all instances, reducing atmospheres are used to provide reduction of oxides and preclude oxidation. Sintering temperatures are usually approximately 2000 to 2100°F (1100 to 1150°C) when conventional furnaces are used. Higher sintering temperatures are achieved more easily with induction sintering. Significant reduction in oxygen content may be realized with high temperature sintering (Fig. 4). The importance of low oxygen content in providing high impact strength is illustrated in Fig. 5.

An important advantage of high temperature sintering is that steels containing manganese and chromium as alloying elements may be refined to produce lower oxygen contents. Sintering at lower temperatures (2050°F) will not reduce the oxides of manganese and chromium.

Preforms containing lubricant are usually "burned-out" at 1000 to 1100°F (540 to 600°C) to remove the lubricant prior to induction sintering. Preforms without lubricant (isostatically compacted) may be induction sintered directly.

Post Sintering Treatments

After sintering, the preform may be cooled directly to the forming temperature, and hot formed. If this procedure is followed, it is important that transport time from the furnace to the forming equipment and the subsequent forming be as rapid as possible since the preform may be prone to surface and internal oxidation.[4] One technique used to increase the acceptable

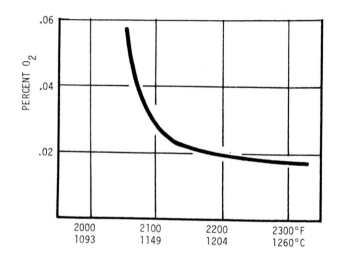

Fig. 4—In P/M preforms, oxygen contents drop as sintering temperatures rise. Material: 1.75 Ni-0.5 Mo steel.

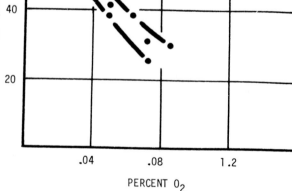

Fig. 5—Impact strength (Charpy V-notch) of sintered P/M parts drops as oxygen contents rise. Material: 1.75 Ni-0.5 Mo steel, Rc 25.

Fig. 3—Isostatically compacted preform (top left) and die compacted preform (top right) with corresponding P/M hot formed differential pinion gears.

transport time involves cooling the preform down to approximately 300 to 500°F (150 to 260°C), and then coating the preform with an aqueous suspension of graphite. The coated preform may be stored for short times prior to hot forming. Coated preforms are reheated to the forming temperature either by conventional atmosphere furnaces or by induction heating; most systems or processes use induction heating for the hot forming operation.

HOT FORMING

Porosity in P/M preforms is either partially or completely eliminated furing the hot forming operation. The efficiency of eliminating this porosity depends on the type of deformation. For example, it has been shown that lateral flow facilitates densification in contrast to repressing.[3] Other important factors that influence densification are forming temperature and forming pressure. These factors also affect the dimensional tolerance in the process.

Forming Temperature

The temperature range used for hot P/M forming is from about 1500°F (820°C) up to 2050°F (1120°C). The importance of increasing forming temperature in facilitating densification in P/M hot forming has been documented by many investigators.[7,8,9,10] Data of Fig. 6[7] are for single forming pressure—higher forming pressure provides higher density. Although higher forming temperature permits using lower forming loads, thermal shock and heat checking problems may be encountered and die life may be decreased for the higher temperature forming operations unless suitable die materials are used. Problems of die material selection have been discussed by Halter.[11]

Forming Pressure

Increasing the forming pressure also increases the formed density (Fig. 7).[12] The curve is valid only for a given component under a particular set of forging conditions.

It has been shown that material flow facilitates densification.[3] Making a given component by hot repressing will require higher forming pressure than if the same component is made by upsetting. With a pressure of 45 tsi, for example, low alloy steels can be formed to full density (2 in. × 2 in. × 5 in. component) by *upsetting* at 1800°F (980°C). However, if the same pressure is used for *repressing*, temperature must be increased to 2050°F (1120°C) to achieve full density. Combined effects of temperature and pressure are illustrated in Fig. 8.[13]

Repressing Vs. Upsetting

The principal purpose of hot forming P/M preforms is to provide a component with higher mechanical properties (realized as porosity is reduced). Now the question arises, "Will properties depend on the mode of deformation?" The answer is, "Yes." The properties will depend on the amount of *flow* received during the deformation. Results of a recent study demonstrate the magnitude of change that may be expected for Ni-

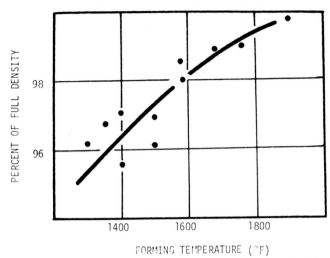

Fig. 6—High temperatures ease forming P/M parts to full density. Material: 0.5 Ni-0.5 Mo steel; forging pressure, 32 tsi.

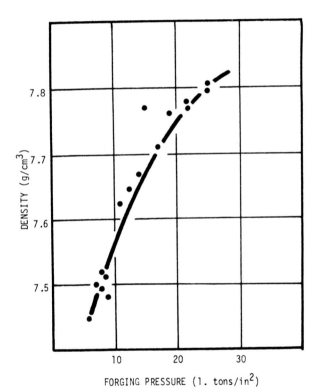

Fig. 7—Density increases with forging pressure. The part is a gear, 3 in. in diam.

Mo alloy steel preforms that were formed to full density by, (1) hot repressing, and (2) hot upsetting[14] (Fig. 9). In *hot repressing*, no external or lateral flow is experience. *Hot upsetting* (plane strain) provides considerable lateral flow (2 to 5 in.) in one direction. Much higher impact properties are obtained with the *hot upset* processed material.

Dimensional Tolerance

Hot repressing and confined die flashless hot forming (Processes 1 and 2 in Fig. 1) are both carried out in trap dies. Therefore, closer dimensional control is possible with these two processes than with conventional closed die forging where flash is produced. As a matter of fact, dimensional control in P/M hot forming approaches the dimensional control possible in the highly precise conventional P/M press-sinter components. Table I lists dimensional characteristics published by one parts manufacturer. Actual tolerances possible may vary from one parts manufacturer to another, and of course will depend on die wear and the process used. For example, tighter tolerances can be maintained with hot repressing P/M preforms than with the confined die flashless process owing to, among other factors, the lower die wear. However, the confined die process provides higher properties, as mentioned previously.

Another factor that contributes to better dimensional control and better surface finish is that oxidation or scaling is considerably less in P/M hot forming than in conventional forging. Surface finishes of 10 to 25 micro-inches have been achieved.

The typical automotive differential pinion offers a good example of the dimensional quality that can be maintained with P/M forming. Tolerance requirements for a part of this type are generally in the range of AGMA classification 8 to 10. Tests have shown that, with a single exception, all relevant aspects of the required quality are obtainable on a production basis.

Those factors that are met include pitch variation, tooth contact pattern, backlash, surface finish, and

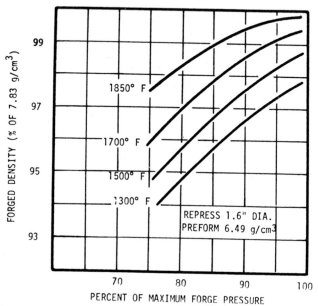

Fig. 8—Higher densities result when P/M parts are forged at higher temperatures and pressures.

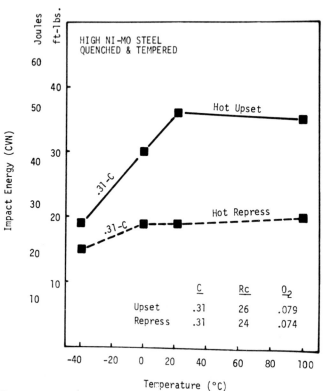

Fig. 9—Hot upsetting generally produces greater impact strengths than hot repressing. Material: 0.3 C-1.75 Ni-0.5 Mo steel.

Table I. Dimensional Characteristics of P/M Parts

Dimensional Characteristic*	Conventional P/M Sinterings Medium Density	P/M Sinterings Coined and Resintered High Density	P/M Sinterings Copper Infiltrated High Density	P/M Forgings (Fully Dense)
Outside diameters or widths transverse to pressing direction formed by die walls.	±0.0015 in./in.	±0.007 in./in.	±0.003 in./in.	±0.004 in./in.
Lengths formed by punches parallel to pressing direction.	±0.006	±0.003	±0.010	±0.012
Concentricity of cored holes to die wall dimensions.	0.004 tir	0.004 tir	0.008 tir	0.008 tir
True position of hole centers formed by core rods.	±0.005 in./in.	±0.005 in./in.	±0.010 in./in.	±0.010 in./in.
Squareness, die wall surfaces to punch face surfaces.	0.0015 tir	0.001 tir	0.003 tir	0.003 tir
Taper in direction of pressing.	0.002 in./in.	0.001 in./in.	0.010 in./in.	0.010 in./in.
Flatness transverse to pressing direction.	0.002 in./in.	0.001 in./in.	Up to 0.010 in./in.	Up to 0.015 in./in.
Surface roughness RMS in.	30-120	16-60	30-120	50-200
Allowance for burrs or flash.	0.002 in. typical at punch face surfaces	0.002 in. typical at punch face surfaces	Allow for 0.002 in. + rough spot at infil. location.	0.020 in. flash typical at punch face surface

*Conditions typical for parts conforming to mid-range of design rules but excluding extreme shape or density limits.

sizes. The one factor that may not be maintained on long production runs is the "bore to pitch-line" runout of 0.002 to 01003 TIR. Therefore, a pitchline chuck bore machining operation may be required.

The close dimensional control and good surface finish possible with P/M hot forming eliminates or minimizes machining or finishing operations. Reduced machining costs, material savings, reduced labor costs (owing to process automation) combined with high or superior performance makes P/M hot forming economically attractive for many gear applications

COMPONENTS AND PERFORMANCE

The first P/M hot formed component made in pilot production quantities was the differential pinion gear.[15] The confined die flashless process was used in this early work. The particular component and process were selected because of:
1. High potential volume
2. High strength requirements
3. Elimination of excessive material loss
4. Uniform shape and size
5. Tool costs not excessive

Since the initial formal report of this work in 1970, P/M hot forming has progressed; several companies have put one or more parts in limited or full production.[11] These parts include those made by hot repressing and by confined die flashless processes.

Hot Repressing

A beval gear that is in process is shown with the preform in Fig. 10. Although full density is not achieved in this component, mechanical integrity is more than adequate for the application in a garden tractor. The material is a nickel-molybdenum low alloy steel atomized powder.

Another area where hot repressing has been used extensively is for gear components made in Japan for motor bikes (Fig. 11). Recently, it has been indicated that the Japanese are concentrating on hot upsetting in confined dies because components for automotive applications currently being investigated require higher strength and toughness than those for motor bikes.

Flashless—Confined Die Upsetting

In the flashless process, the degree of lateral flow may be very little or considerable. In the automatic transmission input ring gear shown in Fig. 12, the amount of lateral flow is limited. The material is a sulfurized manganese-molybdenum low alloy steel, the teeth being machined in after hot forming. This component is replacing a casting, and the P/M formed material has 2 to 3 times the properties required for this application.

Probably the most publicized of all P/M hot formed components is the differential pinion gear shown in Fig. 3. This gear has been made from a variety of materials including Ni-Mo, Cr-Ni-Mo, and Mn-Mo low alloy steels. Most of the work has been concentrated on two Ni-Mo steels. The first contains approximately 1.7 pct Ni and 0.5 pct Mo, while the second contains less nickel, approximately 0.5 pct Ni and 0.5 pct Mo.

Fig. 10—Bevel gear; 127 g-2 in. diam. Preform on top; hot repressed gear on bottom.

Fig. 11—Hot repressed gears are used in some Japanese motor bikes.

Preform densities have ranged from about 6.0 to 6.6 g/cm^3 (24 to 16 pct porosity). Hot forming temperature ranged from 1500°F to 1800°F (815°C to 980°C). Maximum hot forming pressure for the pinion gear is

Fig. 12—Input ring gear for automatic transmission.

Table II. Fatigue Test Results for Gears

	Sinter				
Alloy	Temp, °F	Time (min)	B_{10}* Life	Lower Limit	Upper Limit
1.7 Ni-0.5 Mo	2100	3	2.92	1.56	5.47
1.7 Ni-0.5 Mo	2100	6	4.2	—	—
1.7 Ni-0.5 Mo	2350	1	2.21	1.05	4.60
1.7 Ni-0.5 Mo	2350	3	3.42	2.08	5.64
1.7 Ni-0.5 Mo	2350	6	2.8	—	—
0.5 Ni-0.5 Mo	2300	3	0.92	0.39	2.16
0.5 Ni-0.5 Mo	2300	6	0.74	0.49	1.11
4615 Bar Stock			0.76	0.49	1.79

*Estimates and two-sided 95% confidence limits for Weibull cumulative distribution function for differential pinion gears.

Table III. Impact Fatigue Tests on Differential Pinion Gears

	Sinter		
Alloy	Temp, °F	Time (min)	Cumulative Impact Energy
1.7 Ni-0.5 Mo	2100	3	20,400 in.-lbs.
1.7 Ni-0.5 Mo	2100	6	18,620
1.7 Ni-0.5 Mo	2350	1	17,760-18,620
1.7 Ni-0.5 Mo	2350	3	16,920-20,400
1.7 Ni-0.5 Mo	2350	6	16,100-23,220
0.50 Ni-0.5 Mo	2350	3	13,020-14,520
0.50 Ni-0.5 Mo	2350	6	12,300
4615 Bar Stock			13,020-16,100

usually 75 to 80 tons/in^2. The pinions can be made to essentially full density with slight porosity at the ends of the teeth. However, porosity is completely eliminated at the critically stressed areas, resulting in a high integrity component.

A variety of tests have been made on P/M hot formed pinion gears. These tests include accelerated abuse, durability, rock cycle, drop weight impact, and fatigue tests; results have been satisfactory. Comparing conventional machined gears with P/M hot formed gears by a dead weight impact test, investigators found the machined gear to have a strength of 2200 to 2600 inch-pounds, while the P/M formed gear had a strength of 2400 to 3600 inch-pounds.[16]

In a fatigue test especially designed to evaluate dif-

ferential gears in simulated operating conditions, the P/M hot formed gears performed better than or at least equal to machined gears (Table II).[17] Gears made for this study were also tested in a special impact fatigue test where a 20 pound drop hammer is dropped from a height of 6 in. onto a striker that is in contact with a single tooth of the gear. The drop hammer is raised 1 in. for each successive blow until the tooth fails. Results are shown in Table III.[17]

It was concluded from the above work that all of the lots of 4600 type (1.7 pct Ni) P/M forged pinions provided fatigue lives superior to those of conventionally cut pinions. In gears made from the lower Ni-Mo steel the B_{10} lives corresponded closely to those of conventional pinions. Possible explanations for the superior behavior of the P/M formed components include:

1. Finer grain size
2. Random orientation and equiaxed morphology of inclusions
3. Better surface finish

Other P/M hot formed gears that are being evaluated include side and ring gears for the differential (Fig. 13). It is interesting to note that the P/M formed ring gear appears to have an economic advantage even if the teeth have to be machined into the formed blank.

P/M hot forming of gears and other components is in limited production in Europe and England. A series of components that have been produced by one company is shown in Fig. 14.

P/M hot forming has made significant progress in the past few years, especially in the development of processing equipment and systems. Another area where

Fig. 13—Differential side and ring gears.

Fig. 14—A British company (GKN Ltd.) has made a variety of hot formed P/M components.

considerable progress has been made is development of information on the effect of fundamental processing variables. These factors, combined with continuing effort on alloy development, indicate that even greater progress will be made in the next two years in establishing a P/M forming industry.

REFERENCES

1. R. Halter: *Pilot Production System for Hot Forging P/M Preforms,* 1970 International P/M Conference, July, 1970.
2. Y. Takeya, T. Hayasaka, and Kamata: *Effect of Various Types of Iron Powder and Properties of High Density Fe-Based Alloys,* Brutcher Translation no. 7940.
3. H. Antes: *Cold and Hot Forging P/M Preforms,* SME Technical Report, EMR71-01, 1971.
4. J. P. Cook: *Effect of Air Exposure on Oxidation and Decarburization of Low Alloy Steel P/M Preforms,* Metals Engineering Congress, October, 1971.
5. H. Kuhn and C. Downey: "P/M Preform Design of Hot Forging," *Forging of Powder Metallurgy Preforms,* MPIF, 1973.
6. P. Eloff: *Sintering of Preforms,* MPIF Short Course on P/M Hot Forming, Philadelphia, September, 1972.
7. S. Kaufman and S. Mocarski: "The Effect of Small Amounts of Residual Porosity on the Mechanical Properties of P/M Forgings," *Forging of Powder Metal Preforms,* MPIF, 1973.
8. Kawakita, *et al:* "Properties of Forged Super-High Density Sintered Steel," *Forging of Powder Metal Preforms,* MPIF, 1973.
9. Huseby and Scheil: "Forgings from P/M Preforms," *Forging of Powder Metal Preforms,* MPIF, 1973.
10. Bargainnier and Hirschhorn: "Forging Studies of a Ni-Mo P/M Steel," *Forging of Powder Metal Preforms,* MPIF, 1973.
11. R. F. Halter: *Recent Advances in Hot Forming of P/M Preforms,* P/M 1973 International Meeting MPIF–Toronto, July, 1973.
12. P. K. Jones: *The Technical and Economic Advantages of Powder Forged Products, Forging of Powder Metallurgy Preforms,* MPIF, 1973.
13. S. W. McGee: *Design in Powder Metallurgy and Sinterforging,* SAE Mississippi Valley Section Winter Meeting, January, 1971.
14. H. Antes and Paul Stockl: *The Effect of Deformation on Tensile and Impact Properties of Hot P/M Formed Nickel-Molybdenum Steels,* P/M Joint Group Meeting, Eastbourne England; November, 1973.
15. G. Lusa: "Differential Gear by P/M Hot Forging," *Forging of Powder Metallurgy Preforms,* MPIF, 1973.
16. R. Halter: Cincinnati Inc., private communication.
17. P. C. Eloff and L. E. Wilcox: *Fatigue Behavior of Hot Formed P/M Differential Pinions,* P/M 1973 International Meeting MPIF, Toronto, July, 1973.

<u>POWDER FORGING FUNDAMENTALS</u>

by
Harry W. Antes
Senior Group Manager, R&D
Hoeganaes Corporation

A body subjected to a load will deform. The magnitude of the deformation
will depend on the intensity and the type of loading. The deformation may
be small and recoverable (elastic deformation) or large and permanent
(plastic deformation). Metalworking processes are concerned with plasti-
cally deforming metal under controlled conditions of flow to provide useful shapes.

During the mechanical working of conventional cast or wrought materials
there is essentially no change in volume of the material. This constancy of volume is one of the principal factors used in the development
of plasticity and metalworking theories. In contrast to the constancy
of volume for working conventional material, is the non-constancy of
volume or definite decrease in volume that occurs in working powder metal
(P/M) preforms. The magnitude of the change in volume that occurs during
deformation of P/M preforms depends on the preform density (porosity) and
the amount and type of deformation. Deformation, of course, is related
to the degree of straining while the load necessary to produce the deformation is related to the stress.

Since stress and strain are fundamental to the deformation of all materials
these quantities and their relation to one another will be reviewed.
Furthermore, application of these quantities to the fundamentals of flow
of <u>fully dense</u> and <u>porous materials</u> will be given and related to experimental results. Finally, the application of strain to predicting densification will be covered with results from experiments.

Some of the initial material is academic but it leads to mathematical concepts that have very practical significance. These concepts contribute
to the amount of deformation required and the loads or capacity of equipment required to form components. The details of the mathematics are
purposely omitted, but references are provided for those who desire an
understanding of the details. It is the object of this paper to put
into proper prospective the importance of understanding fundamentals
so that P/M Forging technology may be developed and applied; thus contributing to strengthening the foundations for a P/M Forging Industry.

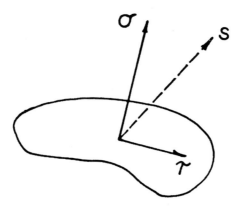

Figure 1. Total Stress S, its normal component σ and its shear component

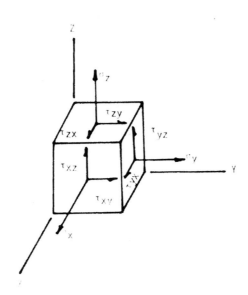

Figure 2. Components of total stress at a point relative to an x, y, z orthogonal system (Equal stresses must act on opposite faces of the unit cube for equilibrium)

Source: Technical Bulletin D211, Hoeganaes Corp., Oct 1971

STRESS

When a load "P" is applied to a body internal resistance or stresses "σ" are developed in the body that oppose the applied load. Now if the stresses in the body are uniformly distributed over a planar area that is perpendicular to the applied load then an equilibrium equation ($\Sigma F=0$) may be written:[1]

$$P = \int \sigma dA$$

$$P = \sigma A$$

$$\sigma = \frac{P}{A}$$ where σ = average stress
$\phantom{\sigma = \frac{P}{A}}$ P = applied load
$\phantom{\sigma = \frac{P}{A}}$ A = area perpendicular to applied load

The total stress "S" acting on a plane may not be perpendicular to the plane and usually for convenience the total stress is broken into a component perpendicular to the plane "σ" (a tensile or compressive stress) and a component in the plane "τ" (a shear stress) as shown in Figure 1. These components of the stress are broken-down further into components that are parallel to an orthoginal system of axes. Thus the total stress at a point for any general system of loading may be represented by the normal and shear components of stress shown in Figure 2. From Figure 2 it can be seen that there are 3 normal stresses (σ_x, σ_y and σ_z) and 6 shear stresses (τ_{xy}, τ_{yx}, τ_{xz}, τ_{zx}, τ_{yz}, and τ_{zy}). However, for isotropic polycrystaline cubic materials.

$$\tau_{xy} = \tau_{yx}$$
$$\tau_{xz} = \tau_{zx}$$
$$\tau_{yz} = \tau_{zy}$$

Therefore only 6 independent stresses have to be defined to describe completely the stress system (σ_x, σ_y, σ_z, τ_{xy}, τ_{xz} and τ_{yz}).

Frequently it will be found that the state of stress "S" (stress tensor) is written in a matrix form as shown below:[2]

$$S = \begin{vmatrix} \sigma_x & \tau_{xy} & \tau_{xz} \\ \tau_{yx} & \sigma_y & \tau_{yz} \\ \tau_{zx} & \tau_{zy} & \sigma_z \end{vmatrix}$$

It can be shown that it is always possible to orient the x, y and z axes in such a way that all the shear stresses are zero. The remaining normal stresses are called the "principal stresses" and are assigned subscripts of 1, 2 and 3 rather than x, y and z. Thus the stress tensor for the principal stresses is given by:

$$S = \begin{vmatrix} \sigma_1 & 0 & 0 \\ 0 & \sigma_2 & 0 \\ 0 & 0 & \sigma_3 \end{vmatrix}$$ where: $\sigma_1 > \sigma_2 > \sigma_3$

The total stress S may be separated into two parts as shown below:

$$S = S' + S''$$

where S' = Deviator part
S'' = Spherical part (hydrostatic or pressure part)

The spherical (pressure) stress tensor is given by:

$$S'' = \begin{vmatrix} \sigma_m & 0 & 0 \\ 0 & \sigma_m & 0 \\ 0 & 0 & \sigma_m \end{vmatrix}$$

where $\sigma_m = \dfrac{\sigma_x + \sigma_y + \sigma_z}{3}$ = mean stress

This part of the stress causes volume changes and contributes to the apparent ductility of a material during deformation. The effect of this stress on ductility at fracture is shown schematically in Figure 3. As shown in this figure the more negative the mean stress the greater the apparent ductility.

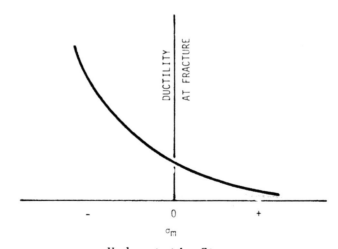

Hydrostatic Stress

Figure 3. Effect of hydrostatic or mean stress on apparent ductility

The sum of the three normal stresses:

$$\sigma_x + \sigma_y + \sigma_z \quad \text{or} \quad \sigma_1 + \sigma_2 + \sigma_3$$

is one of the invariants of stress and will therefore have a constant value regardless of the orientation of the x, y, and z axes. There are two other important stress invariants that are relatively simple expressions in terms of the principal stresses. The three invariants are given below:

$$I_1 = \sigma_1 + \sigma_2 + \sigma_3 = \sigma_x + \sigma_y + \sigma_z$$

$$I_2 = \sigma_1\sigma_2 + \sigma_2\sigma_3 + \sigma_3\sigma_1 = \sigma_x\sigma_y + \sigma_y\sigma_z + \sigma_z\sigma_x - \tau_{xy}^2 - \tau_{yz}^2 - \tau_{zx}^2$$

$$I_3 = \sigma_1\sigma_2\sigma_3 = \sigma_x\sigma_y\sigma_z + 2\tau_{xy}\tau_{yz}\tau_{zx} - \sigma_x\tau_{yz} - \sigma_y\tau_{zx} - \sigma_z\tau_{xy}$$

These invariants are the coefficients of the cubic equation for stress.

$$\sigma^3 - I_1\sigma^2 + I_2\sigma - I_3 = 0$$

Solving the cubic equation for σ gives the three principal stresses. Other important aspects of the invariants will be discussed later with reference to flow theories for both fully dense and porous powder preforms.

The deviator stress S' is determined by subtracting the mean stress from each of the normal components of the total stress.

$$S' = S - S'' = \begin{vmatrix} [\sigma_x - \sigma_m] & \tau_{xy} & \tau_{xz} \\ \tau_{yx} & [\sigma_y - \sigma_m] & \tau_{yz} \\ \tau_{zx} & \tau_{zy} & [\sigma_z - \sigma_m] \end{vmatrix}$$

The deviator stress, S', is responsible for flow and plastic deformation of conventional fully dense materials.

STRAIN

The important types of strain in deformation are:

1. normal strain
2. shear strain

There may also be rigid body rotation; however, this is not related to deformation. There are two types of normal strains

1. engineering strain = e
2. true strain = ε

The engineering strain is defined as the change in length per unit original length or gauge

$$e = \frac{\Delta L}{L_o} = \frac{L - L_o}{L_o} \qquad \text{where: } L = \text{deformed length} \\ L_o = \text{original length of gauge}$$

Engineering strain is satisfactory for small strain (i.e. $e < 0.1$); however, for large strains like those encountered in plastic flow and metalworking, the change in length, ΔL may be very large. If the engineering strain (e) is measured at various stages of the deformation process and then all of the individually measured strains added, the sum will not equal the value of e calculated from the beginning and final values of the gauge length. To overcome this problem a concept of <u>true strain</u> was developed and is used in metalworking problems. True strain (ε) is defined as:

$$\varepsilon = \frac{L_1-L_0}{L_0} + \frac{L_2-L_1}{L_1} + \frac{L_3-L_2}{L_2} + \ldots \ldots$$

$$\varepsilon = \int \frac{dL}{L}$$

$$\varepsilon = \ln \frac{L}{L_0}$$

It can be shown that:

$$\varepsilon = \ln (1+e)$$

Shearing strain, γ, is defined as the change in right angle that occurs during deformation. That is, if a square grid is scribed on the surface of a specimen and if after deformation the angles of the grid have changed from 90°, then during deformation shear occurred on this plane.

The total strain tensor E may be written in matrix form as was the stress tensor:

$$E = \begin{vmatrix} \varepsilon_x & \gamma_{xy}/2 & \gamma_{xz}/2 \\ \gamma_{yx}/2 & \varepsilon_y & \gamma_{yz}/2 \\ \gamma_{zx}/2 & \gamma_{zy}/2 & \varepsilon_z \end{vmatrix}$$

Like stress, the strain may be divided into a <u>deviator component</u> and a <u>spherical or hydrostatic component</u>.

$$E = \underset{\text{Deviator}}{E'} + \underset{\text{Hydrostatic}}{E''}$$

where:

$$E'' = \begin{vmatrix} \varepsilon_m & 0 & 0 \\ 0 & \varepsilon_m & 0 \\ 0 & 0 & \varepsilon_m \end{vmatrix}$$

and $\varepsilon_m = \dfrac{\varepsilon_x + \varepsilon_y + \varepsilon_z}{3}$

The elastic relationships between the stress and strain tensors are given by:

$$S' = 2GE'$$
$$S'' = 3KE''$$

where: G = shear modulus
         K = bulk modulus

The deviator part of the strain is associated with deformation in conventional full density materials while the hydrostatic part of the strain is associated with a volume change. However, in the plastic deformation of fully dense materials there is essentially no volume change; therefore

$$\Delta = \varepsilon_x + \varepsilon_y + \varepsilon_z = 0$$

or $\quad \Delta = \varepsilon_1 + \varepsilon_2 + \varepsilon_3 = 0 \qquad\qquad$ where $\quad \Delta$ = change in volume/unit vol.

In the case of P/M preforms volume changes during deformation (density increases) and the hydrostatic components of stress (S") and strain (E") that had no effect on yielding for conventional fully dense materials become very important for the flow of P/M preforms.

YIELDING AND FLOW

Metalworking theory is concerned with prediction of the stresses acting during metal deformation and therefore, the forces that must be applied.[3] These forces or working loads determine the capacity of the equipment required. Since metalworking is concerned with plastic deformation or yielding then a yield criterion is required as the first step to the development of a theory. Yield criteria for fully dense materials were developed many years ago (Tresca - 1865 and von Mises - 1913).[4,5] Yield theories for porous materials have also been presented but these have been more recent.[6,7]

Fully Dense Materials

Since the principal mechanism of flow in metals is by slip, which is a shearing mechanism, it is logical that flow occurs when the maximum shearing stress reaches a critical value. This is Tresca's Maximum-Shear-Stress Criterion. Stated mathematically, yielding occurs when:

$$\sigma_o \ \lessgtr \ \sigma_1 - \sigma_3$$

$$\frac{\sigma_1 - \sigma_3}{2} = \tau \text{ max.} \qquad \text{where} \quad \begin{aligned} \sigma_o &= \text{yield strength in tension} \\ \sigma_1 &= \text{maximum principal stress} \\ \sigma_3 &= \text{minimum principal stress} \\ \tau \text{ max.} &= \text{maximum shear stress} \end{aligned}$$

This yield criterion completely ignores any contribution that the intermediate principal stress (σ_2) has on yielding. The von Mises Distortion Energy Criterion includes the intermediate principal stress. This theory states that only the deviator part of the stress (S') is responsible for yielding and therefore yielding will occur when the deviator strain energy per unit volume reaches a critical value. Stated mathematically, yielding occurs when:

$$\sigma_o \ \lessgtr \ \frac{1}{\sqrt{2}} \ [(\sigma_1 - \sigma_2)^2 + (\sigma_2 - \sigma_3)^2 + (\sigma_3 - \sigma_1)^2]^{1/2}$$

Where σ_o = yield strength in tension or compression
σ_1, σ_2 and σ_3 = principal stresses

150

It has been found that the von Mises distortion energy theory fits experimental results better than the Tresca Maximum-Shear-Stress Theory (See Figure 4). Using the von Mises yielding criterion and the fact that no volume change occurs during deformation; two principal techniques are used to determine the stresses (or loads) required for metalworking. These techniques are:

1. Slip-line field theory
2. Load bounding techniques

Detailed information and application of these techniques are given in References 3 and 8. Although these techniques have been successful when applied to metalworking problems for full density material, they cannot be applied without modificiation to the deformation of P/M preforms.

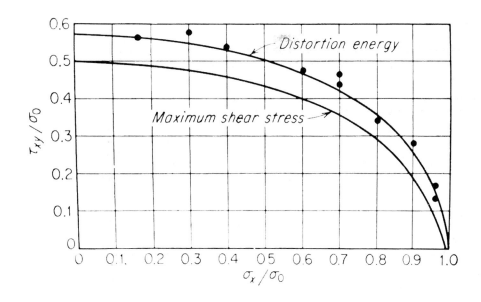

Figure 4. Comparison between maximum-shear-stress theory and von Mises distortion-energy theory. Actual data points shown by the dots. [1]

P/M Preforms

The von Mises criterion for yielding is really a function of the second invariant of the stress deviator [2], I_2':

$$I_2' = -1/6 \ [(\sigma_1 - \sigma_2)^2 + (\sigma_2 - \sigma_3)^2 + (\sigma_3 - \sigma_1)^2]$$

Therefore the von Mises yield condition may be written: $\sigma_0 = (-3I_2')^{1/2}$

As stated previously the hydrostatic part of the stress does not affect yielding in fully dense materials, but it does affect yielding in P/M preforms. Therefore, any yield condition for P/M preforms must include some function of the hydrostatic component of stress which will be some function of:

Source: Technical Bulletin D211, Hoeganaes Corp., Oct 1971

$$\sigma_1 + \sigma_2 + \sigma_3 = I_1 : \text{the first invariant of stress.}$$

and in addition must contain some function of I_2' the second invariant of the deviator stress.

The Mohr-Coulomb yield criterion as mentioned by Suh[6] and Kuhn and Downey[7] involve both of the required invariants, I_1 and I_2'.

Thus for P/M preforms:

$$\text{Yielding} = f (I_1, I_2')$$

But:

$$I_1 = [3(I_2 - I_2')]^{1/2}$$

Therefore:

$$\text{Yielding} = f_2 (I_2, I_2')$$

It was proposed by Kuhn and Downey that yielding for P/M preforms may be expressed as:

$$\text{yield function} = [-3I_2' + (1-2\nu)I_2]^{1/2} \underline{\hspace{4cm}}$$

Where ν = Poisson's ratio

Poisson's ratio for full density materials is equal to 0.5 for plastic deformation. This is a requirement if volume is to remain constant. In all the cases of porous P/M preforms Poisson's ratio is less than 0.5 but increases to a limiting value of 0.5 as the density approaches the maximum value.[9] This means that volume decreases during deformation. The yield function proposed by Kuhn and Downey indicates that as the density increases ($\nu \rightarrow 0.5$) the hydrostatic component part of yielding, $(1-2\nu)I_2$ contributes less and less to yielding. When full density is reached ($\nu = 0.5$) the hydrostatic part of yielding vanishes and yielding depends only on I_2', von Mises yielding, that applies to full density materials.

By combining the Kuhn and Downey yield function with the normality flow rule of St. Venant (See Ref. 7) the following equations were developed.

1. For repressing powder preforms

$$\sigma_{re} = [(1-\nu)/(1-\nu-2\nu^2)]^{1/2} \quad \sigma_{uc}$$

2. For plane strain compression where $\varepsilon_z = 0$

$$\sigma_x = \sigma_{uc}/(1-\nu^2)^{1/2}$$

Where: σ_{re} = stress required for repressing
σ_{uc} = yield stress in uniaxial compression
ν = Poisson's ratio
σ_x = Axial compressive stress required for plane strain compression

152

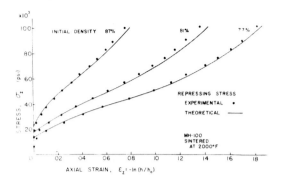

Figure 5. Comparison of theoretical and experimental results in repressing sponge iron powder compacts.[7]

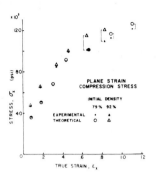

Figure 6. Comparison of theoretical and experimental stress in plane strain compression of sponge iron powder.[7,10]

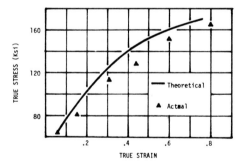

Figure 7. Comparison of theoretical and experimental stress in plane strain compression of modified 4630 atomized steel powder

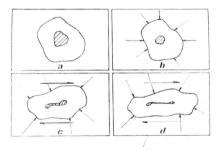

Deformation of Voids

Figure 8

Source: Technical Bulletin D211, Hoeganaes Corp., Oct 1971

These equations are analogous to those developed by using slip-line field theory for fully dense materials. The equations may be used to predict flow curves for more complex type loadings (i.e. repressing and plane strain compressing) from simple uniaxial compression flow curves, provided that Poisson's ratio as a function of density is known.

The utility of these equations is illustrated by the plots shown in Figures 5 and 6. It can be seen from these figures that there is good agreement between the experimental results and the theoretical predictions.

Application of this theory has been made to 4630 type steel P/M preforms. A plot of the theoretical prediction and actual data is shown in Figure 7. It should be noted that data for Poisson's ratio (ν) for 4640 preforms as a function of density were not available. The theoretical curve was developed assuming that ν was the same for 4640 preforms as that given in Reference 9 for sponge iron. The error in this assumption may contribute to the differences observed in Figure 7. Another contribution to the difference may be that in the uniaxial compression tests for 4640, lubrication was not perfect and slight barreling occurred. The above equations were developed for frictionless (perfect lubrication) deformation. The interesting aspect of the data shown in Figure 7 is that the theoretical prediction is high or a safe prediction and actual values will be lower than this. This is the same approach that is taken in slip-line field theory and upper load bounding techniques used for full density materials. That is, they predict a safe, higher than actual stress or loads required for the deformation. This means that if the metal forming equipment is capable of providing the predicted loads then the material will deform.

DEFORMATION AND DENSIFICATION

The application of a compressive type load to a P/M preform causes a decrease in volume and thus an increase in density. The amount of density increase depends on type of loading, magnitude of the loads preform density and the type of material. The type of loading is important if maximum efficiency of densification is to be achieved.

Voids and Inclusions

One unique difference in deforming P/M preforms is that a large volume of voids are present (up to 30 or 40%). Therefore, consideration must be given to the manner that voids deform during loading. In Figure 8a the shaded area represents a void in an unloaded body. Applying a hydrostatic type load (S") will cause the volume to decrease; however, as shown by Bockstiegel the pressure needed to close the void completely will be much too high for any practical process and even may be infinite, depending on the model that is chosen.[12]

Although hydrostatic loading contributes to yielding and densification, a practical deformation process for completely eliminating voids must be one where conventional shear or flow type deformation (E') occurs. Deforming with shear and hydrostatic loading could produce void closure as illustrated in Figure 8c and d.

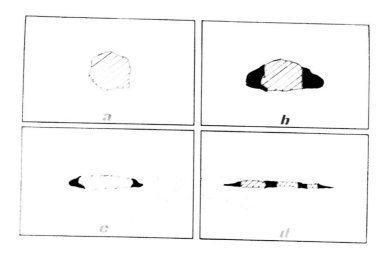

Figure 9 - Deformation of Inclusions

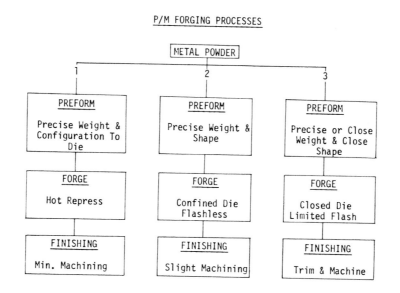

Figure 10. Three processes being used for P/M Forging

Source: Technical Bulletin D211, Hoeganaes Corp., Oct 1971

In addition to void closure the problem of what happens to inclusions during deformation must be considered. An inclusion (shaded area) in an unloaded material is shown in Figure 9a. If the relative strength of the inclusion to the matrix is high, the inclusion may not deform and cracks or voids may open at the matrix-inclusion interface, as illustrated by the black regions in Figure 9b, or the inclusion itself may crack without appreciable deformation. A high hydrostatic stress (S") helps to minimize these types of cracks. If the inclusion is not significantly stronger than the matrix, then the inclusion may be deformed and elongated in the direction of maximum straining. There is a tendency for voids to form at the ends of deformed inclusions too (Figure 9c). Increasing the deformation beyond the point shown in "c" may cause the inclusion to fracture and leave voids between the fragments (Figure 9d). These phenomena have been observed in forged P/M preforms[11].

One factor that may be very important in determining the extent of inclusion deformation would be forging temperature. The strength of the matrix material may be more sensitive to temperature than various types of inclusions. Therefore the ratio of inclusion strength to matrix strength would change with temperature. In any event, minimizing inclusions in the P/M forging with clean powder and proper precautions to prevent oxidation during processing will minimize the degrading effect on the properties of P/M forgings.

P/M Forging and Simple Types of Deformation

P/M Forging processes that have been used most frequently are outlined in Figure 10. Each of the processes requires three stages: preform manufacture, forging and finishing operations. The process on the left (#1) in Figure 10 can best be described as a hot repressing type forging process. It has been used principally by people with a P/M background. The process on the right (#3) is a conventional closed die forging process where limited flash is produced. In this process the preform is hit only once (or as few times as possible) in one die impression as opposed to conventional forging where multiple die impressions and multiple blows in each impression are used to produce the final forging. This process has been used principally by people with a forging background. The final process, the one shown in the center (#2) of Figure 10, borrows from both P/M and conventional forging. This process provides a flashless product because the forging is made in a confined or trap-die by a single blow or press on the preform.

It was mentioned previously that the type of deformation affects the rate of densification. Deformation tests in the laboratory are usually confined to:

1. Simple repressing in a closed die (uniaxial strain)
2. Plane strain compression (biaxial strain)
3. Uniaxial Compression (triaxial strain)

The above laboratory processes are listed in decreasing rate of densification per unit amount of compaction strain. Actual densification data on iron, carbon steel, low alloy steel and stainless steel preforms have been presented for the three types of deformation listed above.[10,11,13] These types of deformation, singularly or combined, simulate the actual deformations encountered in P/M Forging. Densification as a function of preform

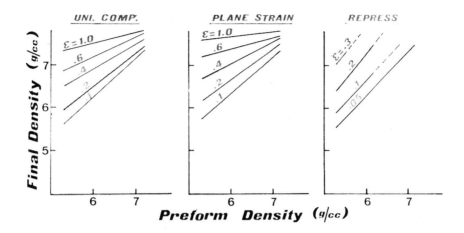

Figure 11 - Densification - Atomized Fe Preform

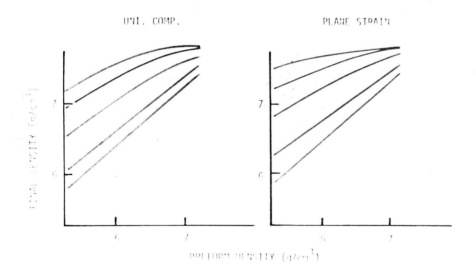

Figure 12. Densification of 4620, 4630 and 8620 atomized steel preforms. Each curve represented the fixed amount of axial compression true strain (ε) as shown.

Source: Technical Bulletin D211, Hoeganaes Corp., Oct 1971

density, with amount of strain as the parameter, is shown in Figures 11 and 12. An interesting aspect of the curves in Figure 12 is that the single set of curves are representative of several low alloy steels. These curves also represent the deformation-densification of carbon steel preforms in the range of 0 to 0.40% carbon. These results indicate that a common denominator for relating densification for various types of materials and geometries may be strain, or more precisely the sum of the normal strains.

Since, $\epsilon_x + \epsilon_y + \epsilon_z = \Delta$: change in volume/unit volume, then densification may be calculated from the strains. For the case of plane strain deformation $\epsilon_z = o$. Therefore, $\Delta = \epsilon_x + \epsilon_y$. Plane strain compression tests were made on 1-1/4 x 1/2 x 1/2 inch P/M preforms. Compaction was in the 1-1/4" (x) direction with lateral flow in one of the 1/2" (y) directions. The strains were calculated by:

$$\epsilon_x = \ln \frac{x_f}{x_o}$$

$$\epsilon_y = \ln \frac{A_z/x_f}{y_o}$$

Where: $x_o = 1\text{-}1/4"$
A_z = area of z face
A_z/x_f = average deformed length in y direction
$y_o = 1/2$ inch

Δ, was determined from:

$$\Delta = \epsilon_x + \epsilon_y$$

Density was then calculated as follows:

total change in volume = $\Delta \times V_o$ where: V_o = original volume of preform

Calculated density = $\dfrac{Wt.}{V_o - V_o\Delta}$ Wt.= Weight of preform

Measured density was determined by using an Archimedian technique. Calculated and measured densities are compared in the table on the next page:

Material	Preform Density	Range of ε_x From	To	Range of % Δ_ρ From	To	Average % Δ_ρ
1020	5.3	.08	.79	0.7	6.6	3.5
	6.2	.07	.94	-1.2	2.9	0.8
	7.2	.06	.81	-0.6	1.6	0.3
1030	5.3	.08	.79	0.3	5.4	3.1
	6.2	.07	.90	0.5	4.3	2.6
	7.2	.06	.76	-3.1	-0.7	-1.6
1040	5.3	.08	.76	0.2	5.4	1.7
	6.2	.07	.76	-1.3	2.6	0.1
	7.2	.06	.68	-2.4	0.0	-1.5
4620	5.3	.08	.96	-2.0	3.5	0.7
	6.2	.06	.80	-2.5	3.4	0.1
	7.2	.05	.73	-1.6	0.3	-0.7
4630	5.3	.07	.59	-0.5	4.6	2.2
	6.2	.06	.76	-2.0	-0.3	-1.1
	7.2	.06	.73	-1.4	1.5	0.3
8620	5.3	.08	.61	0.3	3.4	1.9
	6.2	.07	.86	-3.9	-0.7	1.8
	7.2	.05	.78	-3.4	-0.7	1.9

The above data indicate that densities calculated from deformation
stains agree well with actual measured densities. Although the range
in variation in the differences in densities were approximately 5 to
6% maximum, most of the data were in agreement to within 2-3%.

This implies that at least for simple shapes (e.g. rectangular prisms, cylinders, etc.) preform density and dimensions may be calculated so that in closed
die (trap die) flashless forging, essentially full density can be achieved
when the preform flows to the die walls. This minimizes the amount of repressing required in the final steps of deformation.

Although the results shown here are for simple types of deformation and
simple geometries, it represents a good beginning. This work is being
extended and laboratory tests are in progress that will contribute to
furthering the understanding of deformation-densification of P/M preforms.
However, this effort represents only one of many interrelated phases that
contribute to a fuller understanding of P/M Forging. Other important
phases for consideration of P/M Forging are included in this brochure.

REFERENCES

1. Dieter, G. E., Jr., _Mechanical Metallurgy_, McGraw-Hill Book Co., 1961.

2. Hoffman, O. and Sachs, G., _Introduction to The Theory of Plasticity for Engineers_, McGraw-Hill Book Co., 1953.

3. Rowe, G. W., _Principles of Metal Working_, Edw. Arnold Ltd., London, 1965.

4. Tresca, H., C.R. _Acad. Sci._, Paris, vol. 59, p. 754-756, 1964.

5. vonMises, R., "Mechanik der festen Korper im plastisch – deformablen Austand," Nachr. Ges. Wiss. Gottingen, Math-phys. Klasse, 582-592, 1913.

6. Suh, N. P., "A Yield Criterion for Plastic, Frictional, Work Hardening Granular Materials," Int. J. Powder Metallurgy, 5, 1, p. 69, 1969.

7. Kuhn, H.A. and Downey, C.L., "Deformation Characteristics and Plasticity Theory of Sintered Powder Materials," Int. J. Powder Metallurgy, 7, 1, p. 15, 1971.

8. Avitizur, B. _Metal Forming Processes and Analysis_, McGraw-Hill Book Co., 1968.

9. Kuhn, et.al., "Deformation Characteristics of Iron Powder Compacts," 1970 International Conference on P/M.

10. Antes, H. W., "Cold Forging Iron and Steel Preforms," 1970 International Conference on Powder Metallurgy to be published.

11. Antes, H. W., "Cold and Hot Forging P/M Preforms," SME International Meeting, Philadelphia, April 1971, to be published as a special SME report.

12. Bockstiegel, G., "A Study of the Work of Compaction in Powder Pressing," Powder Metallurgy International, vol. 3, No. 1, 1971.

13. Moyer, K. H., "Cold Forging of 316L Powder Preforms," Westec Conference ASM & SME, Los Angeles, March 1971.

POWDERS FOR FORGING

By
Cornelius Durdaller
Director of Research
Hoeganaes Corporation

Powder preform forging (P/M Forging) is a new process, a combination of powder metallurgy and forging. Powders now used for P/M Forging are essentially those used for conventional powder metallurgy practice. There is a need for powders developed specifically for P/M Forging. They cannot be developed in a vacuum, but need to be done in close conjunction with the development of the P/M Forging process - the powder and the process are intimately related.

POWDER-PROCESS-PROPERTIES

The mechanical properties of a part made from powder depend not only upon the powder (or powders) used and the density to which it is made, but also upon the manner in which it is processed. This is clearly the case in conventional powder metallurgy (P/M) practice. The pressing, sintering and post-sintering conditions all affect the mechanical properties independently of the powders used and the final density of the part. Through the years the interrelation among metal powders, the processes used to make parts and the properties of these parts has become well-established in the P/M industry.

The commercially available powders suit the industry's needs well. The properties of these powders reflect the compromise made to combine the optimum economic powder manufacturing process with the part manufacturing process requirements and the minimal properties necessary for the final part.

The same situation holds true for the powders to be used for P/M Forging. If anything, the influence of process and properties on the kind of powders used will be greater. A P/M Forging powder, since it must be made into a preform, must combine the properties of a conventional P/M powder with those suitable for producing a satisfactory forging. New developments in preform manufacture and the forging process would be expected to require powders different from those in common use today. The economics of the powder manufacturing processes restricts the range of useful powders that can be produced; this, naturally, must be considered by those developing the preform manufacturing and the forging processes.

The primary thrust of P/M Forging development has been the reproduction of conventional forged properties in the forged powder part. The mechanical properties and structures of conventional forgings are well-known, in contrast to the more limited knowledge of standard P/M properties and struc-

Source: Technical Bulletin D211, Hoeganaes Corp., Oct 1971

Fig. 1a – Iron Powder Plus
Graphite

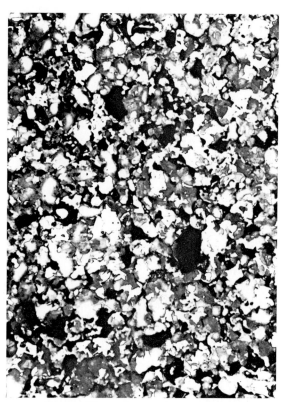

Fig. 1b – Iron Powder Plus
Copper Powder

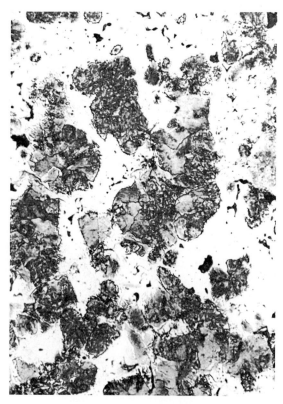

Fig. 1c – Iron Powder Plus
Nickel Powder Plus Graphite

Fig. 1d – Iron Powder Semi-
Alloyed With Copper, Nickel
and Molybdenum Powders Plus
Graphite

tures. The P/M forged parts will have to reproduce conventional forged properties and structures. Any differences must be shown to be harmless.

P/M Forging also offers the opportunity for achieving mechanical properties greater than those of conventional P/M parts, but less than those of conventional forgings. This can be accomplished by forging the powder preforms to something less than full density. The powder requirements may very well depend upon the properties desired in the final part in these cases. And the choice of different powders may be broader owing to less of a demand to reproduce specific structures.

POWDER - THE POSSIBILITIES

Commercially available metal powders are made by three processes: atomization, electrolysis and solid-state reduction of metal oxides. The great majority of the iron powder used for conventional P/M parts is made by the solid-state reduction of iron oxides, although the usage of atomized iron powder is growing rapidly. Small quantities of atomized low alloy heat treatable steel powders have begun to become available to P/M Forging. These prealloyed powders are finding only minor use in conventional P/M applications.

These commercial iron powders are designed to be used most effectively for conventional P/M part manufacture. Each of the three types has particular advantages in P/M applications, but all three must meet certain requirements to make them useful in these applications. Each, even when mixed with other powders, must flow readily enough to fill the die cavity fast enough to permit economical production rates. Each must have a particle size small enough to give a satisfactory surface finish to the part and to give sintering rates rapid enough to obtain sufficient strength in the part. Each must have compressibility high enough to permit a reasonable combination of press capacity and part size while keeping die wear as small as possible. And each must have sufficient strength to withstand the pressing and handling operations while still green, even when the powder is mixed with lubricants and graphite.

The strength and hardness of conventional P/M parts can be increased by three methods: increasing the part density, sintering for longer times or at higher temperatures and by mixing other powders with the base iron powder. Owing to limitations on press size and sintering furnaces, the latter method is the one normally chosen. Typical mixes include graphite, copper, nickel and molybdenum powders. These powders alloy with the iron to varying degrees during processing. Figure 1 illustrates typical P/M structures.

Alloying is accomplished by diffusion during sintering. Only in the case of the graphite addition is a homogeneous alloy formed which, except for the pores, is identical in structure to plain carbon steels. The others are inhomogeneous and the structures are unique to powder metallurgy.

Source: Technical Bulletin D211, Hoeganaes Corp., Oct 1971

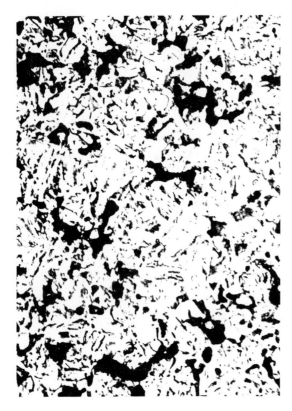

Fig. 2 – Sintered Microstructure
of AISI 4640 Steel Powder

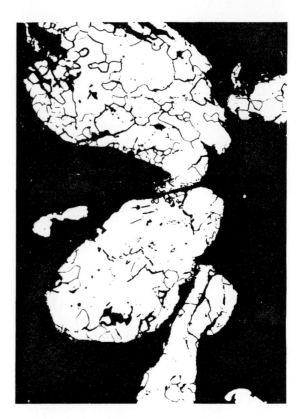

Fig. 3 – Electrolytic Iron
Powder

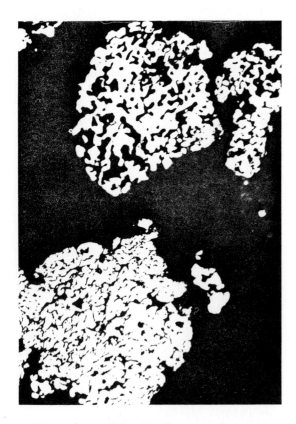

Fig. 4a – Finely Porous Sponge
Iron Powder

Fig. 4b – Coarsely Porous Sponge
Iron Powder

If powder preforms are to be used to reproduce conventional steel structures, this method of mixing powders cannot be used. Except for carbon,
the powders themselves must be homogeneous alloys. Figure 2 illustrates
the homogeneous structures obtained using atomized prealloyed powders.
The carbon, since it can alloy completely if temperatures in excess of
1800° F are used, can be either premixed or prealloyed. There is little
doubt that prealloyed homogeneous powders will find widespread use in P/M
Forging. The question only is the extent that conventional P/M structures
made by powder mixing will be acceptable.

1. Commercial Iron Powders

    The three types of commercial iron powders have different character-
    istics and these characteristics give each particular advantages and
    disadvantages for use in conventional powder metallurgy and in P/M
    Forging.

    a. Electrolytic Iron Powder (Figure 3) - This powder is the purest
        and most compressible of the three types. Because of this, it
        is used in applications that require high purity (e.g. magnetic
        uses) or high density when these properties obtained justify
        the higher cost of the electrolytic powders.

        The electrolytic powders are very dense and can be produced at
        high apparent densities. They also have poor green strength
        which, when combined with low sintered strengths in all but high
        density part applications (greater than 6.8 g/cm^3) limits their
        use markedly.

        In a practical sense, prealloyed electrolytic powders cannot be
        manufactured. This fact, when combined with their high cost
        (about three times the cost of the other types) argues strongly
        against any widespread use of electrolytic iron powders in P/M
        Forging. If used at all, it will be in applications requiring
        their special combination of purity, density and compressibility.

    b. Solid-State Reduced Iron Powder (Figures 4a and 4b) - The purity
        of these so-called "sponge iron" powders is dependent upon the
        purity of the raw material used to make it; no refining is possible
        because they are never melted during their manufacture. The den-
        sity of the particles depends upon the temperature at which the
        iron oxide raw material is reduced: finely porous powder is pro-
        duced at low temperatures (Figure 4a); coarse pores result from
        high temperature reduction (Figure 4b).

        Although these powders are characterized by having low to medium
        apparent density (0.8 to 2.4 g/cm^3) and compressibility (5.8 to
        6.4 g/cm^3), they can be altered somewhat by further processing.
        The practical limit, however, is approximately 2.8 g/cm^3 in ap-
        parent density and 6.6 g/cm^3 in compressibility while still keep-
        ing useful P/M properties. These powders have good green strength
        and have a wide variety of sintered properties that make them ex-

extremely useful for conventional part production. For low to medium density part production (to about 6.6 g/cm³) they are unexcelled in standard part manufacture. Above this density, they compete with atomized iron powder.

The fine internal porosity and residual impurities of the sponge iron powders will limit their usage in P/M Forging. It is extremely difficult, if not impossible in a practical sense, to remove this fine porosity. The residual impurities, although of academic interest only in parts made even 95% dense, have a detrimental influence on dynamic properties once the porosity of the part is reduced to a small, but as yet undefined level. This level of porosity and inclusions is thought to be about 1 to 2%.

There is no practical way to produce prealloyed sponge iron powders. Some semi-alloyed powders are produced but they are far from the level of homogeneity required to reproduce conventional steel structures (See Figure 1d). If sponge iron powders are to be used in P/M Forging, they will be used in premixed or semi-alloyed form and their usefulness will depend upon the inherent advantages sponge iron powders have in preform manufacture (a conventional P/M practice) and whatever strength or economic superiority that can be demonstrated at reasonably high forged densities. This information is being developed now and it appears that sponge iron powders will find application in P/M Forging, but probably not at very high density, i.e., densities at which the dynamic properties of conventional forgings can be reproduced.

c. <u>Atomized Iron Powder</u> (Figure 5) - Since atomized powder is made from liquid metal, it has a flexibility neither of the other two kinds of commercial powders have. The purity is a matter of choice of raw material and melting practice. A broad range of particle shape and

Fig. 5 - Atomized Iron Powder

size is possible through variations of the atomizing parameters. It is the only method of the three that can make prealloyed powders. Atomized powders are therefore being widely evaluated for use in P/M Forging.

Atomized iron powders are characterized by high apparent density and high compressibility. They are normally used for higher density parts (greater than 6.6 g/cm^3) where their strength becomes comparable with that of sponge iron parts and where their poorer green strength becomes less of a problem. The powder particles are dense, containing very little porosity. The normal commercial atomized iron grades are pure enough to be used in some magnetic applications.

2. Powder Developments For Forging Applications

The first steps in producing a powder specifically made for P/M Forging have already been taken. The availability of atomized iron powders on a commercial scale suitable for conventional P/M practice has made it possible to think seriously about using powder forged parts to replace conventional forged parts. Just five or ten pounds of powder forged parts per car would take more powder than most powder producers could manufacture just a few years ago. In addition, in the beginning at least, conventional forged parts are only going to be replaced in any significant quantity if the conventional forged structure can be reproduced by the P/M forging process. The only way to do this is with homogeneous powder. And the only practical way to make a homogeneous powder is by atomizing it.

The other significant step in powder development taken in recent years has been the clear demonstration that superalloy compositions can be made into powder and then successfully fabricated into fully dense parts capable of replacing the conventional cast and forged superalloy parts in many critically demanding high temperature applications. The superalloy example is especially instructive because it illustrates the advantages of P/M forging dramatically.

Nickel-base superalloys are highly complex both chemically and structurally. Controlled additions of aluminum and titanium are used in these alloys to give them high temperature strength. Both these elements are very strong oxide formers and, for them to remain effective in the atomized alloy, no more than 50 parts per million (0.0050%) of oxygen can be added during processing from melt to part. This is an extraordinarily restrictive specification, especially when a well annealed iron powder contains at least 1000 parts per million of oxygen. These superalloy powders have been made and successfully fabricated into parts within the specification limits.

The point of the superalloy example is that practically any alloy, certainly all the AISI low alloy heat treatable steels, can be made successfully into powder and processed into parts. The only question is how much cost is economical to add to part making process by using powder

as the starting material. Superalloys cost several dollars per pound
more as powder, but give advantages that justify the added costs.
These advantages include:

> Increased yield from the forging operation

> Increased forgeability

> More predictable property response during heat treatment

> Broader range of alloys can be used for forging

> New Alloys can be developed specifically from the powder
> approach

The economic question in P/M Forging is a critical one. The answers to
it will certainly affect the kinds of powders that will be available.

3. Directions for Forging Powder Development

The initial powders used for P/M forging were conventional P/M powders.
A great deal of work has been done taking shapes similar to conventional
P/M parts and forging them to higher densities. Concurrently some prealloyed powders were included in the development effort to reproduce conventional forged structures; these included some AISI grades of low alloy
heat treatable steels and stainless steels. Since the powder production
process was based on a conventional P/M product, these AISI alloys reproduced atomized iron powder characteristics. The successful powders ex-
cluded all alloying elements which oxidized more readily than iron. The
next generation of powders for forging will reflect more closely the needs
of the P/M Forging process.

a. Chemical Composition – From a technical point of view every alloy
steel can be produced by atomization. Most AISI compositions can
be made by water atomization the cheapest atomization method.
The problem with many of the compositions is that they contain
alloying elements that will oxidize during the post-atomizing
operations. These operations through the preform sintering step
have been designed to process iron powder; the forging steps expose a porous preform not a solid material. Oxidation would therefore be a problem unless precautions are taken to avoid it.

Standard AISI alloys containing chromium and manganese, the two major
alloying elements which oxidize more readily than iron, have been
made and hot forged to full density under controlled conditions without any significant oxidation. It seems certain that sufficient protection can be given these alloys during the critical P/M Forging
steps to reproduce these results. The use of chromium and manganese
is essential to the production of cheaper low alloy heat treatable

steel powders. The use of prealloyed atomized powders containing these elements will make it easier to protect them from oxidation than if mixtures of the elemental powders were used.

In addition to standard AISI compositions for the low alloy powders, the powder approach makes it possible to use other elements or alloy types that are not common to conventional forging practice. Elements that make hot working difficult and are avoided in conventional steelmaking for that reason can be used for their contribution to hardenability in P/M Forging because the hot working temperatures can be reduced. P/M methods are ideal for creating dispersion-strengthened alloys or for creating alloys from materials normally incompatible like the cermet materials.

Finally, the practice of adding carbon as graphite must be reevaluated. Graphite additions are useful in conventional powder metallurgy because elevated temperatures are necessary for sintering and the graphite will react to carburize the iron at these temperatures. Carbon, as prealloyed in the powder, will make it possible to eliminate the sintering step if sufficient bonding can be done during hot forging, i.e., the highest temperature the powder needs to see is the forging temperature.

b. <u>Powder Characteristics</u> - The characteristics required of powders for conventional P/M are not those necessarily required for a P/M Forging powder. Particle size and distribution, compressibility, green strength and apparent density all should be reexamined to determine what the optimum ranges of each are for P/M Forging.

Although most powder manufacturing processes cannot produce any <u>particle size distribution</u> desired, the size distribution being typical of a process, some adjustment is possible especially if the average particle size is varied. All the reasons for using the fine P/M particle sizes are not valid for P/M Forging. Surface finish and particle bonding are more a function of forging conditions than particle size. A coarse powder, for example, would give a void structure in a preform that would be easier to eliminate than a fine powder. Coarse powder may very well minimize the contamination and oxidation problems.

The <u>compressibility</u> of a powder is important for conventional powder metallurgy because it determines the density or the size of a part possible. Although the final density of a powder forged part does depend to an extent on the density of the preform when a specific forging pressure is used, it is only one of many controls available. The preform density, and therefore the powder compressibility, is a variable that can be manipulated to optimize the P/M Forging process. The lower density preforms, however, are more susceptible to oxidation from exposure during forging and extra

Source: Technical Bulletin D211, Hoeganaes Corp., Oct 1971

protection must be used. Lower compressibility powders would be interesting because it would give a much greater latitude in choosing powder composition, especially carbon content.

Sufficient green strength is required to hold the part (or preform) together until a metallurgical bond can be made between the powder particles by a thermal treatment, be it sintering or hot forging. Conventional P/M uses additions of lubricant and graphite as standard practice. Both seriously decrease the green strength, requiring rather large initial powder green strength to give the green part enough strength to withstand the handling operations. If both the lubricant and graphite additions were not used, powders with much lower green strength would be useful. More regular particle shapes could be used. Higher apparent densities, and therefore faster die filling and simpler tooling, would be possible.

The sum total of this discussion of possible directions for forging powder development is not to say that these directions will be followed, but rather that they are interesting. More importantly, it is to say that there is a great deal more flexibility in the properties of a powder for forging than there is in a conventional P/M powder. And, finally, an attempt has been made to indicate that there must be developments in the forging process to take advantage of the powder possibilities.

FORGING POWDERS AND ALLOYING ELEMENTS

Alloying elements are added to atomized steel powder for the same reasons they are added to conventionally produced steel: for solid solution strengthening, increased hardenability, increased toughness, increased corrosion resistance, etc. Powders however, owing to their methods of manufacture and fabrication into finished parts, are affected by alloying elements in ways not common to conventional steelmaking. These effects must be considered when using powders for P/M Forging.

1. Alloying Elements And Powder Manufacturing Methods

Conventional metalworking starts with the casting of large ingots. This process protects the alloying elements from oxidation, the largest problem being getting the elements in solution and preventing excessive segregation of the elements during solidification. The segregation problem is minimized in atomized powders owing to the rapid solidification of extremely small droplets, each typical of the melt composition. The use of powders effectively solves most of the segregation problems encountered in conventional ingots.

The use of alloying elements in manufacturing atomized powders creates two additional factors not present in conventional steelmaking. The first factor is concerned with alloying elements changing the character of the powder produced. For a given set of manufacturing variables, many of the powder characteristics, e.g. particle size and shape, apparent density, green strength, sintering behavior, etc., are a function of the amount and kind of alloying elements present. The manufacturing variables are under control and, within rather broad limits, undesirable variations in powder characteristics can be avoided.

The second factor is concerned with the reaction of the alloying elements during processing. The proper functioning of alloying elements of interest in this discussion, ignoring the behavior of carbon, depends upon their remaining in solution in the base iron. They can be removed from solution by oxidation and powders are subjected to potentially oxidizing conditions many times during their processing: during atomization, annealing, sintering and forging or other thermo-mechanical processing. The products of oxidation can prevent proper bonding among the powder particles and their presence can result in poor mechanical properties of the final part. Naturally, the loss of alloying elements diminishes whatever effect they were originally added to accomplish.

2. Oxidation of Alloying Elements

The powder manufacturer has two of these potentially oxidizing conditions to consider. The others concern the users of the powder - the part manufacturers. The water atomizing process is highly oxidizing to most metals. The liquid metal droplets, however, are subjected to these conditions for an extremely short period of time. Examination of the as-atomized particles reveals that the oxidation occurring during atomization is superficial and is primarily iron oxide for normal AISI heat treatable compositions even if some of the alloying elements form more stable oxides than iron. There is simply not enough time for the alloying elements to oxidize. When cold, the temperature is too low to permit this atomic motion to take place.

Iron and iron-base powders are normally annealed to reduce any remaining oxides, fix the carbon content and soften the powder. In the normal temperatures used for annealing (1500 to 2100° F), the reduction of iron oxide is accomplished easily in readily available furnace atmospheres, e.g. a hydrogen or dissociated ammonia atmosphere with a dew point of 75° F is suitable. Metal oxides less stable than iron oxide, e.g., Cu_2O, NiO, CoO and MoO_3 of interest to steelmaking, reduce more easily than the iron oxide. Those more stable, e.g. Cr_2O_3, MnO, V_2O_3, SiO_2, TiO_2 and Al_2O_3, are more difficult to reduce; some impossible when using practical furnace atmospheres. The problem in annealing low alloy heat treatable powders is not to reduce oxides because an insignificant amount of alloying element oxide is formed during atomization, but rather to prevent oxidation. It has proven possible to anneal powders containing manganese and chromium in normal production furnaces without oxidation of these alloying elements. Elements that form more stable oxides, e.g. silicon, titanium and aluminum, have proved very difficult to protect from oxidation during annealing using typical production atmospheres. The properties of commercially water atomized and annealed low alloy heat treatable powders will be discussed later to illustrate that they can be produced without significant oxidation during the manufacturing process.

Source: Technical Bulletin D211, Hoeganaes Corp., Oct 1971

Oxidation during the part manufacturing stages is a real problem and processes must be designed to prevent oxidation from occurring. If the iron can be protected, copper, nickel, cobalt and molybdenum can be safely used without fear of oxidizing them. Since the initial processes developed were based on iron powder metallurgy, the first heat treatable low alloy powders contained only these alloying elements in significant amounts.

However, since chromium is a common addition for deep hardening steels and since it, in common with manganese, is among the cheapest of alloying additions, the broadest and most economical P/M Forging processes should be able to use powders containing both these alloying elements.

3. Alloying Elements and Powder Compressibility

The concept of compressibility is unknown in conventional steelmaking, although the deformability of steel is an important material property for metalworking. Compressibility is a measure of the degree by which powders deform under a given load. It is related to softness of the metal itself and, other variables held constant, the softer the powder, the higher the compressibility. Other factors, like powder porosity, shape and size affect the compressibility. These are functions of the powder manufacturing method and are under the control of the powder manufacturer.

In general, the higher the compressibility of a powder, the more desirable it is for P/M manufacturing operations. And the purer the iron powder, the higher will be its compressibility. The final density of conventional

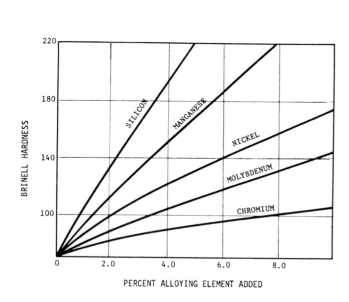

Fig. 6 - Effect of Alloying Elements on the Hardness of Iron

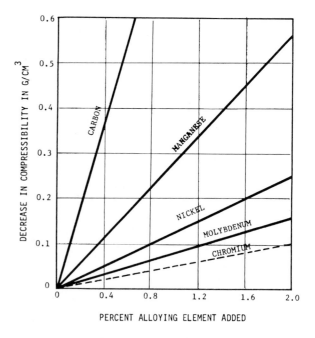

Fig. 7 - Effect of Alloying Elements on the Compressibility Of Iron Powder

P/M produced parts is primarily determined by the compaction process. Since higher compressibility powders permit larger and higher density parts to be made without changing the size of compacting presses, P/M part manufacturers historically have tried to retain as much of the compressibility of the iron powder as possible. The use of powder mixes instead of fully alloyed powders reflects this desire for highest possible powder compressibility.

Although it is desirable to have as high a compressibility as possible in powders for P/M Forging, it is not the overriding consideration it is for conventional P/M part manufacture. Reproduction of standard steel structures requires the use of prealloyed powders and prealloyed powders have lower compressibility.

The addition of alloying elements to iron harden the iron normally by solid solution strengthening. Figure 6 illustrates this effect for some common alloying elements. Carbon and nitrogen, although potent solid solution strengtheners, are not included because they are soluble in iron only to a limited extent at room temperature. They strengthen iron through transformation processes; the transformation by which carbon strengthens iron is, in fact, the basis for the extraordinary combination of strength and ductility that is obtained by heat treating steels.

The effect of alloying elements on the compressibility of iron powder follows the same pattern as the alloying elements affect the hardness of iron: the compressibility decreases as the amount of alloying elements in the iron increases. Figure 7 illustrates this effect for the alloying elements of present interest for steel powders for forging applications.

The general trends illustrated in Figure 7 are accurate, but should only be considered as relative trends because different manufacturing techniques may change the magnitude of the alloying element's effect on compressibility. Using atomized iron as a base iron powder, the following compressibility ranges might be expected for some standard AISI compositions (See page 2-19)

AISI Series*	<0.020% C	0.20% C	0.40% C	Cost of Alloying Elements $/Ton **
1000	6.69/6.60	6.50/6.41	6.32/6.23	1.20/1.80
1500	6.48/6.39	6.29/6.20	6.11/6.02	2.70/3.30
4000	6.64/6.58	6.45/6.39	6.27/6.21	10.32/15.18
4100	6.59/6.50	6.40/6.31	6.22/6.13	12.73/18.87
4300	6.42/6.30	6.23/6.11	6.05/5.93	60.65/76.64
4600	6.44/6.33	6.25/6.14	6.07/5.96	54.87/69.18
8600	6.57/6.44	6.38/6.25	6.20/6.07	22.35/37.09

*Based on standard AISI composition ranges
**Based on the following prices - Mn $200/ton,
Cr $840/ton
Ni $2700/ton
Mo $4460/ton

Source: Technical Bulletin D211, Hoeganaes Corp., Oct 1971

The choice among the different kinds of steel powders depends first of all upon its meeting the material specifications for the application. If the material specifications permit some latitude in powder composition, the powder user can base his choice on the cost of the powder and the cost of processing (resistance to oxidation during processing, powder compressibility, etc.) the powder to the finished part. The two elements easiest to use, nickel and molybdenum, are expensive, although both have only a moderate effect on the powder compressibility. Some decrease in cost can be obtained by lowering the nickel content or replacing part of the nickel with a less expensive element like copper.

Manganese is the cheapest of the alloying elements. However it forms an oxide more stable than iron and, except for carbon, has the most drastic effect on compressibility. Its use would require special care to prevent oxidation during processing. Chromium does not appear to have a very strong effect on the powder compressibility and this, when combined with its strong contribution to hardenability, makes it a very desirable alloying element. However, since chromium can form a stable coherent oxide on the powder particles if the powder is subjected to oxidizing conditions, it like manganese, would require special care during processing. Both oxides cannot be easily reduced using normal processing atmospheres and temperatures.

Carbon causes a drastic decrease in compressibility. Its use, however, would offer some protection from oxidation to the alloying elements in the powder and can make the P/M Forging process simpler by avoiding the necessity for admixing graphite and the additional treatments required to put this graphite into solution in the iron.

FORGING PROCESS - THE UNKNOWN QUANTITY

The forging process is an unknown quantity. It occupies a position of critical importance in determining the character of the powders which will be available for P/M Forging. There are several variations of P/M Forging possible (Figure 8), each incorporating a preform manufacturing, forging and a finishing step. Each has similar requirements for powder and any developments in one would, in general, be applicable to the others. The individual processes, however, may be most efficient at different preform sizes and shapes and at different temperatures of forging. If this is true, they may also require different types of powders.

The two most critical parts of the forging process for the powder are preform manufacturing and forging of the preform. To a very large degree, once the preform manufacturing and forging processes are chosen, the limits of the useful powders have been defined. The full potential of the forging powders will not be reached unless the potential of both these processes have been explored fully.

1. The Preform Manufacturing Process

The preform manufacturing process involves blending of the powder with any additions that are desired, compacting the powder into the preform shape

174

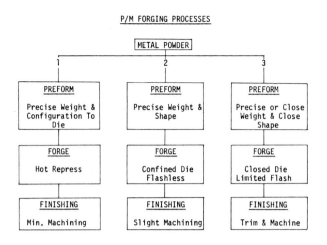

FIGURE 8 - P/M FORGING PROCESSES

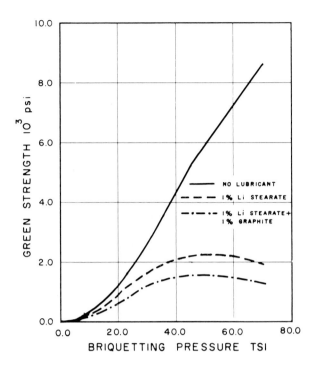

Fig. 9 - Effect of Admixed Lubricant And Graphite on the Green Strength of Atomized Iron Powder

and sintering the preform for the forging operation. Each of these needs further evaluation as part of a P/M Forging operation, not strictly as a conventional P/M operation. Preform manufacture starts with powders; what follows remains to be determined.

a. Powder Blending - The conventional use of powder blending is mixing the alloy ingredients and the lubricant with the iron powder. The alloy ingredients, other metal powders and graphite, when added in this manner have a number of detrimental effects. They introduce chemical heterogeniety into the finished part because the powders can never be mixed perfectly and the practical sintering times and temperatures used are not sufficient to permit complete interdiffusion of the different elements. The use of graphite requires that the powder mix be heated to at least 1800° F to put the carbon into solution; it seriously decreases the ability of the powder to flow freely and substantially lowers the green strength. (Figure 9)

The lubricant addition serves no purpose except to aid the compaction of the powders into the green shape. The lubricant decreases die wear, makes ejection of the part easier and acts to distribute the pressure throughout the part while it is being pressed. Once the part is pressed the lubricant must be removed. The removal of it is a source of many serious problems and, for P/M Forging, problems that will limit its range of application. The lubricants are volatile and burn off at resonably low temperatures (less then 800° F). The products of the burning normally deposit on the furnace walls and, to avoid contaminating the sintering furnaces, the lubricant burn-off must either be done in a separate furnace or in an extra zone put on the sintering furnace for this purpose.

Source: Technical Bulletin D211, Hoeganaes Corp., Oct 1971

If the green parts are large (greater than a few pounds) or dense (above 6.8 g/cm^3), complete lubricant removal becomes extremely difficult and the entrapped lubricant either remains as an impurity in the part or erupts during sintering to cause imperfections in the parts. It is almost certain that in large parts (greater than five to ten pounds), the lubricant will not be admixed with the powder if the mechanical properties of the part are at all critical. In addition to the difficulties in removing the lubricant after compaction, admixed lubricant also restricts the free flowing ability of the powder and decreases the strength of the green part. (Figure 9)

The compromises made in conventional P/M practice to achieve the highest possible compressibility (admixed metal powders and graphite) and to extend die life (admixed lubricant) reflected the techniques available for part making. Final density is reached primarily in the green compacting step, therefore, requiring the highest possible compressibility in the powder and limiting any alloying to after the green part is made. The green part itself must be compacted to close size and shape tolerances, requiring complex and expensive tooling and, generally, an admixed lubricant. The powders used must be irregular to give the requisite strength to the green part when the powder is mixed with graphite and lubricant. Fine particle sizes are required to give sintering rates fast enough to achieve satisfactory mechanical properties under economical sintering conditions.

These compromises are not preordained in P/M Forging. The final part density is not reached after the green compaction step. The structural integrity of the part is not obtained during sintering, but rather during forging. The final size and shape of the part is not controlled by the green compaction step and additions to the powder mix, but by the forging operation. The entire preblending process, i.e., the alloying metal powder, graphite and lubricant additions, may be eliminated by the proper choice of the preform compacting and sintering steps and the final forging practice. Blending can just be used to control apparent density and to insure a uniform distribution of particle size.

Preform Compaction - The subject of preform compaction has already been introduced and partially discussed because the subjects of powder blending and compressibility have been discussed. The design of the powder preform depends very much on the forging process. The preform design, i.e., its size, shape, weight, density and design tolerances, limits the methods that can be used to manufacture it.

Simple shapes with generous size tolerances give the most flexibility in choosing the preform compaction method. The use of die wall lubrication or isostatic pressing gets rid of the admixed lubricant and the problems associated with using it. These problems have already been discussed. It is worth mentioning that preform weight control and speed of die cavity filling are critical production factors and the removal of the admixed lubricant permits more precise preform weight control and faster die filling.

A final word should be said about the response of different powder types to compaction pressures because, even if the materials were identical, they would behave differently owing to their morphology and porosity content. A given void volume made up of fine porosity is more difficult to remove than the same void volume composed of coarse porosity. Coarse, dense, regular-shaped powders would therefore be easier to densify than fine, porous, irregular powders and, provided their lower green strength and sintering activity could be tolerated, they would be more desirable for both the preform compaction and preform forging operations.

c. <u>Preform Sintering</u> - The P/M Forging processes now being used all include a sintering step after preform compaction and before the forging operation. The sintered preforms are then normally cooled to room temperature and reheated to the forging temperature.

This procedure must consider two problems. The first is preventing the oxidation of the alloying elements. The second is insuring that the admixed graphite burns completely enough to achieve the carbon level desired in the part. The level of unreacted graphite and combustion products should not be high enough to affect the mechanical properties of the part. If the part contains admixed lubricant, a similar problem exists: the lubricant must be removed before forging because its presence will prevent metal-metal bonding and thus deteriorate the mechanical properties of the final part.

The prevention of oxidation is a serious concern for all the alloying elements, including those which can be reduced using temperatures and atmospheres typical of the industry. Oxidation localizes the element and this heterogeniety leads to a variation of properties throughout the part. If the oxide remains unreduced in the final part, the additional problem of inclusions presents itself.

An alloying element in dilute solution in iron is more difficult to oxidize than if it existed as the pure metal or as a rich master alloy. It would be easier to prevent oxidation using prealloyed powders rather than mixtures of elemental powders. This fact also explains why oxidation-prone elements like chromium and manganese can be treated using practical atmospheres and temperatures without oxidizing them, while the same conditions could not reduce their oxides.

Sintering of parts made of alloy steel powders, in fact any thermal treatment of these parts, must seriously confront the oxidation problem, especially when these parts pass through lower temperatures at which oxidation can more easily occur. Rapid heating and cooling techniques, part coatings (especially carbon-base coatings) and highly protective atmospheres would all be helpful and should be considered to prevent oxidation.

2. Preform Forging

Most of the powder considerations for the forging process have already been discussed. The forging process is concerned with reaching the desired final density of the part and how the deformation of the preform is done to reach the density. And because of this, the deformability and the oxidation characteristics of the preform and the material making it up are an important part of the forging operation.

The preforms must be protected from oxidation if the forging operation is done at elevated temperatures. There is experimental evidence that shows that carbon-containing preform coatings are protective against oxidation. The lower the density of the preform, the more serious the oxidation problem becomes. Oxidation, in general, becomes worse with increasing time and temperature of exposure although the oxides become less stable and carbon more protective toward the metal as the temperature is raised. Exposure studies are being done to detail these pheonomena.

The deformability of a preform is a function of many variables, e.g. forging temperature, preform design and the composition and characteristics of the powder making up the preform. In general, the more highly alloyed a steel is, the more difficult it is to deform. It has previously been noted that coarse, regular and dense powder would probably be the most desirable from the single standpoint of ease of densification. Experimental investigations are determining the extent of the influence of these variables upon the P/M Forging processes.

PROPERTIES OF FORGED POWDERS

It is difficult to talk about the properties of forged powders without discussing at the same time how they were forged. The properties of a part forged from powder are a direct function of the process used to forge it. The quality of the powder is important, if not indispensable to a successfully forged part, but the proper forging process must be used to realize the properties. Figure 10 illustrates the variation of impact strength with final forged density for constant density iron powder preforms cold forged by upsetting. Impact strength, as all dynamic propeties, is especially sensitive to forging conditions and its value is not just determined by part density. The curve represented in Figure 10 would be changed by using a preform of different density, by forging the preform differently and by changing the iron powder. The proper choice of each, however, will insure reproducible properties and, if desired, properties entirely equivalent with forgings of similar composition.

A series of AISI alloys were manufactued under commercial conditions to evaluate the quality of alloy powders as manufactured. The powders were compacted, sintered and forged under highly protective conditions to prevent any contamination of the powders during these operations. The forging, although done under protective conditions, was severe enough to produce typically good forged properties from the powders. The compositions chosen were AISI 1040, 1540, 4040, 4140, 4340, 4640 and 8640. These alloys have the following ranges of compositions:

Alloy	% C	% Mn	% Cr	% Ni	% Mo
AISI 1040	0.38/0.43	0.60/0.90	--	--	--
AISI 1540	0.36/0.44	1.35/1.65	--	--	--
AISI 4040	0.38/0.43	0.70/0.90	--	--	0.20/0.30
AISI 4140	0.38/0.43	0.80/1.05	0.90/1.20	--	0.08/0.15
AISI 4340	0.38/0.43	0.65/0.85	0.70/0.90	1.65/2.00	0.20/0.30
AISI 4640	0.38/0.43	0.70/0.90	--	1.65/2.00	0.20/0.30
AISI 8640	0.38/0.43	0.75/1.00	0.40/0.60	0.40/0.70	0.15/0.25

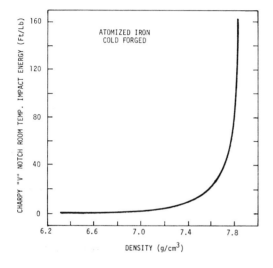

Fig. 10 - Impact Strength of Cold Forged Atomized Iron Powder As A Function of Density

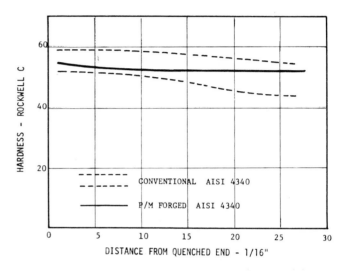

Fig. 11 - Hardenability of P/M Forged AISI 4340 Steel Powder

Silicon should be avoided in these powders because of its pronounced tendency to oxidize during the manufacturing processes. Since AISI 4340 is the most heavily alloyed material and contains both chromium and manganese which oxidize more readily than iron, it will be discussed in detail. The other low alloy powders gave very similar results to those discussed for the 4340 powder.

The powder was forged into bars 99.9+% dense by compacting the powder in steel cans at room temperature. The cans were sealed, vacuum sintered and hot pressed before the final hot forging. They were then hot forged at 2100° F using a multiple-blow air hammer technique.

The hardenability of the material was tested using the standard Jominy End-Quench technique (Figure 11). Despite the fact that the powder forged steel had the fine grain size typical of powder products (ASTM #8 vs ASTM #4 for conventional steels) and would therefore be expected to exhibit poorer hardenability than conventional steels, the curve for the powder forged steel falls well within the hardenability band for AISI 4340.

Source: Technical Bulletin D211, Hoeganaes Corp., Oct 1971

The powder forged 4340 bars were austenitized at 1550° F oil quenched and tempered at 750° F for one hour to a hardness of RC 40. The mechanical properties of these bars were:

Mechanical Property	Powder Forged	Conventional
Ultimate Tensile Strength in psi	203,000	191,000
Yield Strength (0.2% offset) in psi	183,000	175,000
Elongation in % (2" gauge length)	12	15
Reduction in Area – in %	48	54
Impact Strength (32° F) "V"-Notch Charpy in ft-lbs.	30	30
Hardness (Rc)	40	40

The mechanical properties of the powder forged bars compare very well with those of conventionally processed AISI 4340. All the properties are in the range expected for this steel. The finer grain size of the powder forging could well account for higher tensile and yield strengths and the slightly smaller ductility. In the light of so much of the information available of properties of powder forged steels, the identical values of the impact strength for the differently produced steels is very encouraging.

The dynamic properties of the powder forged steels, i.e., impact and fatigue strength, have typically been reported as lower than conventional forged steels, while the tensile properties, even for powder forgings not at full density, normally are equivalent to conventional steels. This is an important distinction and one that bears critically upon the P/M Forging processes.

In contrast to the tensile properties, the dynamic properties are extremely sensitive to material impurities like inclusions and voids. There are also indications that these properties are sensitive to the degree of deformation given the preform to reach its final density. The void removal and the preform deformation are properly subjects of a forging discussion and will not be covered here. The inclusions, however, are very much concerned with the powder used and the manner in which the powder is handled through all the steps of P/M Forging.

The good impact properties of the powder forged AISI 4340 can be directly related to the cleanliness of the structure (Figure 12). The structure is typical of a clean conventionally forged steel. The inclusions themselves are generally products from the melting practice, very few from the powder manufacturing practice. Naturally, the care exercised during the forging prevented any inclusions from forming during this operation. The forging of the powder preform can be done so that internal oxidation does take place. When this happens (Figure 13), severe deterioration of the dynamic properties takes place.

A successful P/M Forging process must control the oxidation of the powder preform from the compaction through the final forging step. In general, any oxidation that occurs must be less than that which would affect the dynamic properties or the hardenability if the properties of conventional forgings are desired. This level of oxidation has yet to be determined and remains one of the major pieces of design information required to put the technology of P/M Forging on a firm commercial basis.

Satisfactory powders for P/M Forging can now be made. This discussion has attempted to show this. But, more importantly, an attempt was made to place the powders within the complete forging process. The powder, the process used to forge the powder into a part and the properties of the finished part are all intimately related. All three must be evaluated together when considering the P/M Forging processes.

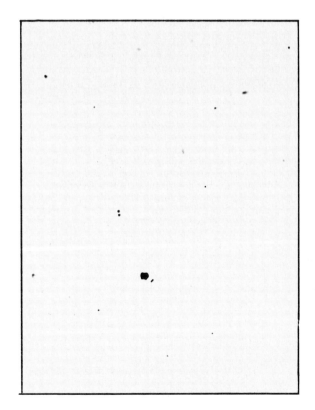

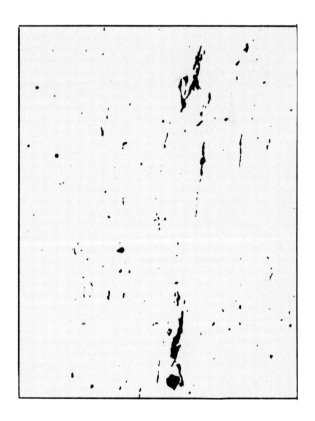

Fig. 12 – P/M Forged AISI 4340 Steel Powder Forged Using Protective Conditions

Fig. 13 – P/M Forged AISI 4340 Steel Powder Forged Using Non-Protective Conditions.

Source: Technical Bulletin D211, Hoeganaes Corp., Oct 1971

The Effects of Heat Treatment and Deformation on the Homogenization of Compacts of Blended Powders

R. W. HECKEL AND M. BALASUBRAMANIAM

Homogenization of compacts of blended powders is reviewed in terms of the influence of variables which are known to affect the process. Mathematical models which describe major process variables are discussed and evaluated with experimental data. Consideration is given to the effects of both heat treatment and deformation on the rates of homogenization in one-phase and multiphase binary systems. Guidelines are presented which will assist other workers in the use of the mathematical models as predictive tools.

THE fabrication of alloys by powder metallurgy processing may be carried out through the use of prealloyed powders or by diffusional homogenization of compacted blends of powders. The latter method must involve additional processing to achieve a suitably mixed blend of powders and to bring about the required interdiffusion of the elements in the compacted blend. However, the homogenization processing route often provides the advantage of minimal compaction difficulties compared with the prealloy route, since one or more of the powders may be a soft, elemental powder. In addition, diffusion processing offers the potential for providing microstructures with "tailored" amounts of interdiffusion.

The early treatment of the problem of homogenization in powder compacts by Rhines and Colton[1] has provided considerable insight into a) the analysis of the appropriate diffusion problem for one-phase systems, and b) the method of applying this analysis to experimental data. They pointed out that both the deviation of particle composition from the mean composition and the distance between adjacent centers of maximum and minimum concentration are important parameters in the homogenization process. Their data obtained on Cu-Ni compacts of various compositions and particle sizes, using electrical resistivity measurements to assess the progress of homogenization, showed that the amount of interdiffusion varies a) as the square of the distance between the maximum and minimum concentration centers; and b) as the exponential of the reciprocal of the absolute homogenization temperature. Rhines and Colton were successful in showing that the attainment of a given degree of homogeneity can be expressed as:

$$\frac{X^2}{\tilde{D}t} = k \qquad [1]$$

where:

X is the distance between adjacent centers of maximum and minimum in concentration

R. W. HECKEL, on leave of absence from the Department of Metallurgical Engineering, Drexel University, Philadelphia, Pa., is Visiting Professor, Department of Metallurgy and Materials Science, Carnegie-Mellon University, Pittsburgh, Pa. M. BALASUBRAMANIAM is Graduate Student, Department of Metallurgical Engineering, Drexel University.

This paper is based on an invited talk presented at a symposium on Homogenization of Alloys, sponsored by the IMD Heat Treatment Committee, and held on May 11, 1970, at the spring meeting of The Metallurgical Society of AIME, in Las Vegas, Nev.

t is the time of homogenization

$\tilde{D}$ is the interdiffusion coefficient ($\tilde{D} = D_0 e^{-Q/RT}$)

k is a constant which describes the degree of homogenization

Rhines and Colton also determined that X was a function of both the mean composition of the compact and the particle size of the powder. Thus, by showing that Eq. [1] provided a description of the process, these workers demonstrated that the basic nature of compact homogenization may be defined in terms of normal diffusion parameters. The studies of Rhines and Colton, therefore, laid the groundwork for subsequent studies, both theoretical and experimental.

Mathematical models of the homogenization process in one-phase systems have been described by a number of authors. The major variations among these studies center about the assumed geometry of arrangement of the particles in the powder compact. Chevenard and Wache[2] and Duwez and Jordan[3] have modeled the alternate plate (planar) geometry where the plates of different original composition were of equal thickness [$\bar{C}$ (mean composition) = 0.50].* Cubic arrays of

---

*Compositions throughout this paper will be given in atom fraction.

---

particles have been considered by Duwez and Jordan,[3] Weinbaum,[4] Gertzriken and Feingold,[5] and Raichenko and Fedorchenko[6,7] with several of these models considering compositions other than $\bar{C}$ = 0.50. Raichenko,[8] Fisher and Rudman,[9] and Heckel[10] have based mathematical models upon the concentric-sphere geometry where the minor constituent particles are assumed to be isolated in a continuous matrix of the major constituent. This geometry permits the coupling of the particle diameter of the isolated constituent l_0 with the mean composition of the compact, $\bar{C}$, by:

$$\bar{C} = (l_0/L)^3 \qquad [2]$$

where L is the diameter of the sphere-shell composite. Eq. [2] assumes constant molar volume in the alloy system of interest and a mixture of elemental powders.

Compact homogenization in two-phase binary systems has been analyzed by Tanzilli and Heckel.[11-13] Their studies include both stages of the process; a) the solution of the unstable second phase, followed by b) the homogenization in the remaining phase. The homogenization process in binary three-phase systems is currently under investigation.[14]

A wide variety of experimental techniques have been

applied to studies of homogenization in powder compacts. These have included quantitative metallography, electrical resistance measurements, electron microprobe analysis, and X-ray measurements (both compositional line broadening and quantitative determination of amounts of phases) and are discussed in Ref. 15.

Past experimental and analytical studies on powder homogenization[15] have shown that the process is influenced by a large number of variables which include:

Powder variables
  i) particle size of minor constituent, l_0
  ii) particle size of major constituent
  iii) composition of minor constituent, C_B
  iv) composition of major constituent, C_A
  v) mean compact composition, $\overline{C}$
  vi) degree of mixing
  vii) particle geometry
  viii) compact porosity (both interparticle porosity and Kirkendall porosity developed during interdiffusion)

Phase equilibrium variables (multiphase systems)
  i) two phase systems
     $C_{\alpha\beta}$, the concentration of solute in α in equilibrium with β
     $C_{\beta\alpha}$, the concentration of solute in β in equilibrium with α
  ii) for multiphase systems, in general, the number of solubilities is $2(p-1)$, where p is the number of phases

Diffusion variables
  i) interdiffusion coefficients
     $\tilde{D}_\alpha$ for one-phase systems
     $\tilde{D}_\alpha, \tilde{D}_\beta$ for two-phase systems
     (in general, p coefficients are necessary)
  ii) compositional dependence of interdiffusion coefficients
  iii) surface diffusion contributions to the homogenization process

Processing variables
  i) temperature, T, affects both the phase equilibrium and diffusion variables
  ii) time of homogenization, t
  iii) deformation processing conditions, including amount and method of deformation

The homogenization models mentioned previously have, in general, been formulated so that the effects of most of the process variables may be considered analytically. The principal exceptions include degree of mixing, particle shape, compact porosity, compositional dependence of interdiffusion coefficients, surface diffusion contributions, and deformation processing conditions. Past experimental studies have shown that, in the range of normal powder processing conditions for low-porosity compacts, the effects of particle shape, compact porosity, and compositional dependence of interdiffusion coefficients are small compared with effects of variables that can be accounted for by mathematical models.

It is the purpose of this paper to provide an overall appraisal of the mathematical models for the analysis of powder compact homogenization. This will be carried out by comparing both new and previously presented data with the predictions of a complete mathematical analysis of homogenization using the concentric-sphere (spherical), concentric-cylinder (cylindrical), and parallel-plate (planar) geometries. These geometries were chosen for comparison since they provide fairly good descriptions of the possible environments surrounding a particle in a compact and are flexible enough to permit the greatest number of variables to be considered. In addition, the cylindrical and planar models can be assumed to describe, in the limit, the geometry of compacts that have been extruded and rolled, respectively. This will allow the analysis to be extended to consider the effects of deformation processing on powder homogenization.

MATHEMATICAL MODELS

The diffusion analysis of homogenization processing must consider a finite geometry problem if the solution is to be applicable to situations approaching complete homogeneity (*i.e.*, long times). The most direct method for treating such problems involves the use of numerical methods and computer techniques.[16]

The mathematical analysis given in this section of the paper will be presented for the general situation where $0 < C_A < C_B < 1.0$. However, in order to condense the graphical presentation of the results of calculations based upon the various models (Figs. 2 through 5, 7 and 8), the composition scales were normalized from zero to unity [*i.e.*, $C_A = 0$ and $C_B = 1.0$] to remove the effects of the C_A and C_B variables. Thus, strictly interpreted, these figures are directly applicable only to compacts of elemental powders (pure A and pure B). However, the figures may be applied to the more general problem [*i.e.*, $0 < C_A < C_B < 1.0$] by converting actual compositions (usually only $\overline{C}$ is required) to normalized compositions by $(C - C_A)/(C_B - C_A)$, where C is the composition to be normalized. In addition, it should be remembered throughout the paper that equations taken from graphs summarizing calculations also contain compositions in normalized form.

One-Phase Systems

The concentric-sphere geometry considers a particle of the minor constituent of radius $l_0/2$ in the compacted powder blend to be surrounded by a continuous matrix of the major constituent which is assumed to have lost its particulate form. Perfect distribution of minor constituent particles is also assumed in order to apply the diffusion solution for the sphere-shell composite of radius $L/2$ to the behavior of the entire compact. If the composition of the minor and major constituent particles are initially C_B and C_A, respectively, the concentration-distance profiles in the composite will be as shown in Fig. 1.* Numerical solution of the diffusion

---
*For $C_A = 0$ and $C_B = 1.0$, Eq. [2] may be used for $\overline{C}$, as shown in Fig. 1. However, for the general situation where $0 < C_A < C_B < 1.0$, $\overline{C} = [l_0^3(C_B - C_A)/L] + C_A$.
---

problem in the concentric-sphere model to yield concentration-distance profiles from Fick's Law was carried out in the manner described previously[11] under the initial and boundary conditions shown in Fig. 1.

The concentration-distance profiles resulting from the numerical solution to the concentric-sphere geometry problem were related to the homogenization process in terms of the "degree of interdiffusion" parameter, F, defined previously by Fisher and Rudman.[9]

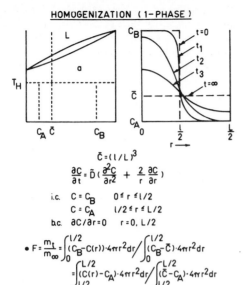

HOMOGENIZATION (1-PHASE)

$$\bar{C} = (1/L)^3$$

$$\frac{\partial C}{\partial t} = \tilde{D}\left(\frac{\partial^2 C}{\partial r^2} + \frac{2}{r}\frac{\partial C}{\partial r}\right)$$

i.c. $C = C_B$ $0 \leq r \leq l/2$

    $C = C_A$ $l/2 \leq r \leq L/2$

b.c. $\partial C/\partial r = 0$ $r = 0, L/2$

$$\bullet \; F = \frac{m_t}{m_\infty} = \int_0^{l/2}(C_B - C(r))\cdot 4\pi r^2 dr \Big/ \int_0^{l/2}(C_B - \bar{C})\cdot 4\pi r^2 dr$$

$$= \int_{l/2}^{L/2}(C(r) - C_A)\cdot 4\pi r^2 dr \Big/ \int_{l/2}^{L/2}(\bar{C} - C_A)\cdot 4\pi r^2 dr$$

$$\bullet \; C_{max} = C_{r=0} \; ; \quad C_{min} = C_{r=L/2}$$

Fig. 1—Representation of the one-phase, concentric-sphere model in terms of the phase diagram and concentration-distance profiles as a function of time. $\bar{C}$ expression valid for $C_A = 0$ and $C_B = 1.0$.

(A similar parameter was originally proposed by Rhines and Colton).[1] F provides a measure of the ratio of the mass transferred across the spherical surface at radius $l_0/2$ after a diffusion time, t, to that transferred at $t = \infty$. Fig. 1 gives the mathematical form of F. In addition, the range of compositions existing in the sphere-shell composite was determined by the values of concentration at $r = 0$ and $r = L/2$, $C_{max} = C_{r=0}$ and $C_{min} = C_{r=L/2}$, respectively, as shown in Fig. 1.

The cylindrical and planar diffusion problems were also solved by numerical methods by considering the problem described in Fig. 1. For the cylindrical model:

$$\frac{\partial C}{\partial t} = \tilde{D}\left(\frac{\partial^2 C}{\partial r^2} + \frac{1}{r}\cdot\frac{\partial C}{\partial r}\right) \qquad [3]$$

$$F = \int_0^{l_0/2}[C_B - C(r)]\cdot 2\pi r dr \Big/ \int_0^{l_0/2}(C_B - \bar{C})\cdot 2\pi r dr$$

$$= \int_{l_0/2}^{L/2}[C(r) - C_A]\cdot 2\pi r dr \Big/$$

$$\int_{l_0/2}^{L/2}(\bar{C} - C_A)\cdot 2\pi r dr \qquad [4]$$

$$\bar{C} = [l_0^2(C_B - C_A)/L^2] + C_A \qquad [5]$$

For the planar model, where X is the distance perpendicular to the plates and $X = 0$ at the midpoint of the plate of the minor constituent:

$$\frac{\partial C}{\partial t} = \tilde{D}\,\frac{\partial^2 C}{\partial X^2} \qquad [6]$$

$$F = \int_0^{l_0/2}[C_B - C(X)]dX \Big/ \int_0^{l_0/2}(C_B - \bar{C})dX$$

$$= \int_{l_0/2}^{L/2}[C(X) - C_A]dX \Big/ \int_{l_0/2}^{L/2}(\bar{C} - C_A)dX \qquad [7]$$

$$\bar{C} = [l_0(C_B - C_A)/L] + C_A \qquad [8]$$

The calculated results for the degree of interdiffusion, F, for the spherical, cylindrical, and planar geometries are presented in Fig. 2 as a function of the variables $\tilde{D}$, t, l_0, and $\bar{C}$. It should be noted that the results given in Fig. 2 for the spherical model agree with previously reported calculations[10] in the range of $\bar{C} = 0.50$, and represent somewhat more accurate values than previously reported[10] for lower values of $\bar{C}$.

The range of compositions (C_{max} and C_{min}) are presented in Fig. 3 (also as a function of $\tilde{D}$, t, l_0, and $\bar{C}$). The two branches of the curves for any given value of $\bar{C}$ define the limits of compositions which exist for any value of $\tilde{D}t/l_0^2$. The composition scale in Fig. 3 should be considered as normalized between $C_A = 0$ and $C_B = 1$.

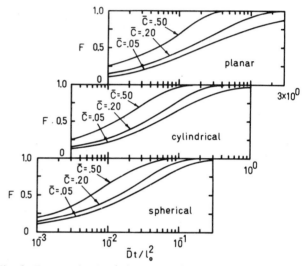

Fig. 2—Homogenization behavior of alternate plate (planar), concentric-cylinder (cylindrical), and concentric-sphere (spherical) models in terms of the degree of interdiffusion, F, as a function of normalized homogenization time, $\tilde{D}t/l_0^2$, for various mean compositions, $\bar{C}$, in one-phase systems.

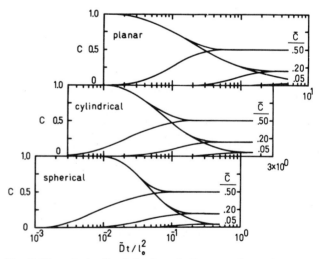

Fig. 3—Homogenization behavior of alternate plate (planar), concentric-cylinder (cylindrical), and concentric-sphere (spherical) models in terms of C_{max} (upper branch) and C_{min} (lower branch) as a function of normalized homogenization time, $\tilde{D}t/l_0^2$, for various mean compositions, $\bar{C}$, in one-phase systems.

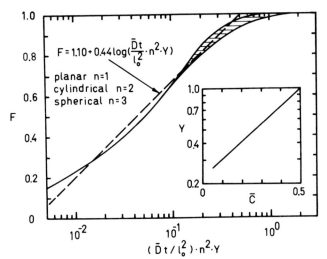

Fig. 4—Normalized degree of interdiffusion, F, as a function of $(\tilde{D}t/l_0^2)n^2 Y$ which collapses all of the F vs t curves to essentially a single curve describing most of the major homogenization variables in one-phase systems.

Consideration of variables like geometry and $\bar{C}$ on the behavior of F in Fig. 2 led to the empirical normalization of the effects of these parameters in terms of $(\tilde{D}t/l_0^2)n^2 Y$ as shown in Fig. 4. The value of n is determined by geometry and Y is a function of $\bar{C}$ as shown in Fig. 4. In general, the parameter $(\tilde{D}t/l_0^2)n^2 Y$ allows most of the curves in the range of mean compositions considered $(0.05 \leq \bar{C} \leq 0.50)$ to be superimposed on a single curve with a slight spread in the range of $F > 0.70$. In the range of F values of interest $(0.20 < F < 0.90)$ the curve in Fig. 4 may be closely approximated by the equation:

$$F = 1.10 + 0.44 \log_{10}[(Dt/l_0^2)n^2 Y] \qquad [9]$$

Either Eq. [9] or the curve given in Fig. 4 provides a description of the finite homogenization problem and defines most of the significant variables involved. In addition, Fig. 4 provides the basis for consideration of deformation processing effects associated with changes in geometry.

Deformation processing by rolling transforms the diffusive flow geometries from spherical to planar; extrusion transforms the diffusive flows from spherical to cylindrical. The extent of these transformations is a function of the extent of the deformation. Such deformation processing brings about two significant geometrical effects relative to homogenization rates: a) the reduction in minor constituent particle size, l_0, accelerates the homogenization, and b) the change in geometry [spherical ($n = 3$) to cylindrical ($n = 2$) in extrusion or spherical ($n = 3$) to planar ($n = 1$) in rolling] retards the homogenization [both effects are seen in the $(\tilde{D}t/l_0^2)n^2 Y$ parameter in Fig. 4]. Obviously, both effects occur simultaneously during deformation. The reduction in minor constituent particle size will be dominant for large reductions. It should be noted though that for large extrusion reductions the l^2 effect must offset the n^2 change of 9 to 4 (increase in homogenization time by a factor of 2.27). For large rolling reductions, the l^2 effect must offset the n^2 change from 9 to 1 (increase in homogenization time by a factor of 9.0).

The effects of deformation can be analyzed by con-

sidering the simultaneous changes in l and n brought about by the alteration in diffusion geometry. If R represents the fractional reduction of l_0, the initial minor constituent particle diameter, caused by the deformation and it is assumed that the state of homogeneity is

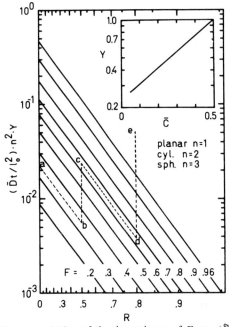

Fig. 5—Representation of the dependence of F on $(\tilde{D}t/l_0^2)n^2 Y$ as a function of total reduction, R, in the original minor particle size, l_0, as a means of assessing the influence of geometry changes due to deformation on the progress in one-phase systems. A hypothetical homogenization-deformation sequence is shown by the dashed line; 0-a, b-c, and d-e represent homogenization steps; a-b and c-d represent deformation steps. Incremental values of $(\tilde{D}t/l_0^2)n^2 Y$ (b-c and d-e) are added to the $(\tilde{D}t/l_0^2)n^2 Y$ values obtained after deformation (b and d).

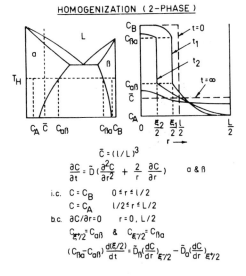

Fig. 6—Representation of the two-phase, concentric-sphere model in terms of the phase diagram and concentration-distance profiles as a function of time. $\bar{C}$ expression valid for $C_A = 0$ and $C_B = 1.0$.

unaffected by the deformation (*i.e.*, F remains constant), the plot of iso-F curves as a function of R shown in Fig. 5 is obtained (The values of F at $R = 0$ were obtained from Fig. 4). For example, the rolling of alternate plates (planar model; no variation in n due to deformation) by a reduction of 90 pct ($R = 0.90$) results in a reduction of l_0 to $l_0/10$ [*i.e.*, $l_0(1 - R)$] and an effective reduction in homogenization time to 0.01 times the homogenization time with no deformation, since the effect of particle size enters the analysis as a squared term. Thus, the curves in Fig. 5 are two orders of magnitude lower in value at $R = 0.90$ than at $R = 0$. Concurrent changes in n must be considered when this variable also changes as a function of R. The dashed line a-b-c-d-e in Fig. 5 shows the method for considering a multiple sequence of homogenization and deformation treatments. It should be pointed out that values of R have been defined in terms of the reduction in minor particle dimension; actual reduction of the dimension of the particle may be significantly less than the axial deformation reduction of the overall powder compact.

Evaluation of predictions based upon Figs. 2 through 5 will be undertaken in subsequent sections of this paper.

Two-Phase Systems

Mathematical models for the two-phase homogenization problem based upon the description provided in Fig. 6* have been presented previously for the spheri-

---

*For $C_A = 0$ and $C_B = 1.0$, Eq. [2] may be used for $\overline{C}$, as shown in Fig. 6. However, for the general situation where $0 < C_A < C_B < 1.0$, $\overline{C} = [l_0^3(C_B - C_A)/L] + C_A$. For cylindrical and planar problems, Eqs. [5] and [8] are applicable.

---

cal, cylindrical, and planar models.[11-13] The treatment differs from that of the one-phase problem in that Fick's Law must be applied to both the α and β phases, and the flux balance equation must be used to define the velocity (and, therefore, position) of the moving interface between the α and β phases, as shown in Fig. 6. The results of these previous numerical calculations have been presented in terms of the normalized position of the α:β interface, ξ/l_0, and the normalized departure of the composition at $L/2$ from the mean composition, $(C/\overline{C})_{L/2}$. The composition at $L/2$ represents, in most instances, the maximum departure from the mean composition in the α phase. The values of ξ/l_0 have been shown to be primarily dependent upon $\widetilde{D}_\alpha$, t, l_0, $C_{\alpha\beta}$, and geometry. To a first approximation,

$$\xi/l_0 \cong f_1\left[(\widetilde{D}_\alpha t/l_0^2) \cdot C_{\alpha\beta}^y\right] \qquad [10]$$

$$(C/\overline{C})_{L/2} = f_2[(\widetilde{D}_\alpha t/l_0^2) \cdot \overline{C}^y] \qquad [11]$$

where $y = 1.10$, 1.40, and 2.10 for the spherical, cylindrical, and planar geometries, respectively. Some range in ξ/l_0 and $(C/\overline{C})_{L/2}$ values was found to exist relative to Eqs. [10] and [11] due to the effects of the minor variables in the model shown in Fig. 6. Furthermore, the geometry parameter, y, was not able to account for the various geometries completely in the values of $(C/\overline{C})_{L/2}$.

These previously developed calculations for ξ/l_0 and $(C/\overline{C})_{L/2}$ were adapted to analyze the effects of deformation in the same manner as those used for the one-phase analysis. The treatment of the ξ/l_0 data as a function of R is shown in Fig. 7. The bands for ξ/l_0 = 0.50 and zero reflect the magnitude of the effects of minor variables that are not considered in the model (*i.e.*, $\widetilde{D}_\beta$, $C_{\beta\alpha}$, and $\overline{C}$). The treatment of the $(C/\overline{C})_{L/2}$ data as a function of R is shown in Fig. 8. The bands for $(C/\overline{C})_{L/2}$ = 0.10 (three lower bands) and 0.90 (three upper bands) reflect the magnitude of the effects of minor variables that are not considered in the model

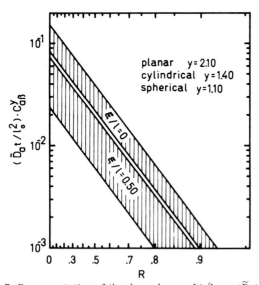

Fig. 7—Representation of the dependence of ξ/l_0 on $(\widetilde{D}_\alpha t/l_0^2)\overline{C}_{\alpha\beta}^y$ as a function of total reduction, R, in the original minor particle size, l_0, as a means of assessing the influence of geometry changes due to deformation on the progress of homogenization in two-phase systems. The use of this diagram is the same as that indicated for Fig. 5.

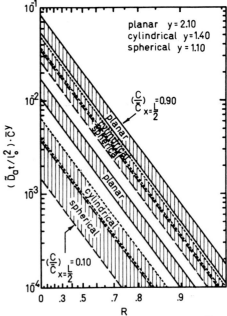

Fig. 8—Representation of the dependence of $(C/\overline{C})_{L/2}$ on $(\widetilde{D}_\alpha t/l_0^2)\overline{C}^y$ as a function of total reduction, R, in the original minor particle size, l_0, as a means of assessing the influence of geometry changes on the progress of homogenization in two-phase systems. The use of this diagram is the same as that indicated for Fig. 5.

$(i.e., \tilde{D}_\beta , C_{\alpha\beta}$, and $C_{\beta\alpha})$. Figs. 7 and 8, as in the analysis of the one-phase homogenization analysis, are based upon a continuous change in the geometry parameter, y, with reduction, R. The reduction is in terms of the size of the minor constituent particles size which may not be as great as the overall deformation of the powder compact.

DATA ANALYSIS

A direct comparison between mathematical model predictions and actual experimental data is necessary to provide a meaningful evaluation of the models. Rhines and Colton[1] suggested that the spectrum of lattice parameters in a partially homogenized powder compact could be determined by X-ray diffraction measurements and analyzed to provide data on the distribution of compositions in the alloy. They were unable to apply this technique due to the unavailability of experimental equipment. Rudman[17] later developed this technique by correcting X-ray diffractometer peak profiles (line broadening analysis) and using known lattice parameter data as a function of composition to provide information in the form of number of unit volumes having a given composition, N_C, as a function of composition, C. Furthermore, Rudman was able to show that N_C vs C data could be converted to a concentration-"effective distance" profile which described all of the regions in the sample which contributed to the diffraction peak in terms of a single mean concentration gradient. This was accomplished through the use of the effective penetration, y_C:

$$y_C = \int_{C_B}^{C} N_C dC \Big/ \int_{C_B}^{C_A} N_C dC \qquad [12]$$

Heckel[10] has shown that the instrumental broadening inherent in the diffraction peaks can be "removed" from the peak profiles, giving more accurate data.

The complete analysis of the peak profiles for a typical X-ray compositional line broadening experiment is shown schematically in Fig. 9. The upper portion of the figure shows the raw data in the form of diffracted X-ray intensity, I, as a function of diffraction angle, θ, for three different stages of homogenization in a one-phase system. N_C is shown to be proportional to the raw I vs θ and the correction functions, $f_1(\theta)$, discussed by Rudman[17] and Heckel,[10] for an alloy system where the composition variation with lattice parameter, and thus diffraction angle, is known $[C = f_2(\theta)]$. The formulation for y_C shown in Fig. 9 may be used therefore, to obtain the composition-effective distance profiles shown at the bottom of Fig. 9 as a function of homogenization time from X-ray diffraction scans.

Rudman's analysis,[17] using the y_C parameter, is applicable to any experimental technique which provides data which are sensitive to the spectrum of compositions existing in the inhomogeneous material. Fig. 10 shows the manner in which electron microprobe scans across the surface of a specimen may be analyzed to give composition-effective distance profiles.

The effective penetration analysis of Rudman[17] may also be applied to the concentration-distance profiles resulting from the numerical solution of the diffusion problems discussed previously, Figs. 1 and 6. This analysis is shown in Fig. 11 for the one-phase, spheri-

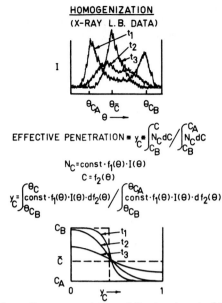

Fig. 9—Schematic representation of the treatment of X-ray diffraction data by the compositional line broadening analysis. Also shown is the analysis of the composition data in terms of the Rudman effective penetration parameter, y_C.

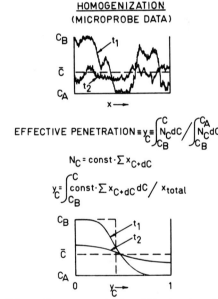

Fig. 10—Schematic representation of the treatment of electron microprobe scan data by the Rudman effective penetration parameter, y_C, to yield concentration-distance profiles.

cal-geometry model. Thus, the C vs y_C plots obtained from experimental data and the mathematical models may be compared directly. Degree of interdiffusion data (F values) for one-phase systems may be obtained from C vs y_C plots by:

$$F = \int_0^{l'/2} (C_B - C_y)dy \Big/ \int_0^{l'/2} (C_B - \bar{C})dy$$

$$= \int_{l'/2}^{L'/2} (C_y - C_A)dy \Big/ \int_{l'/2}^{L'/2} (\bar{C} - C_A)dy \qquad [13]$$

where $l'/2$ is the value of y_C at the compositional dis-

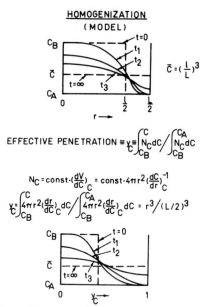

HOMOGENIZATION
(MODEL)

EFFECTIVE PENETRATION $\equiv y_C \equiv \int_{C_B}^{C} N_C dC \Big/ \int_{C_B}^{C_A} N_C dC$

$N_C = \text{const} \cdot \left(\frac{dV}{dC}\right)_C = \text{const} \cdot 4\pi r^2 \left(\frac{dC}{dr}\right)_C^{-1}$

$y_C = \int_{C_B}^{C} 4\pi r^2 \left(\frac{dr}{dC}\right)_C dC \Big/ \int_{C_B}^{C_A} 4\pi r^2 \left(\frac{dr}{dC}\right)_C dC = r^3/(L/2)^3$

Fig. 11—Schematic representation of the treatment of concentration-distance profiles from the concentric-sphere model to yield concentration-effective distance profiles by the Rudman analysis.

continuity at $t = 0$, and $L'/2$ is the y_C value which is equivalent to $L/2$. Eq. [13] has the added advantage that it is independent of geometry for either models or experimental data.

EXPERIMENTAL PROCEDURE

Compacts of blended Ni-Cu powders were used in the present investigation to obtain data for homogenization in a one-phase system; Ni-W powders were studied to obtain two-phase system data.* Descriptions of

*Thus, for the purpose of analyzing the data to be presented in this paper, the normalized model calculations are directly applicable since $C_A = 0$ (pure Ni) and $C_B = 1.0$ (either pure Cu or pure W).

the powders used are given in Table I. The powders were mixed in approximately 200 g batches in a twin-shell blender for times long enough to achieve a high level of uniformity of particle distribution as determined by qualitative examination of the powder mix and of metallographically prepared compact sections. Mixing times usually ranged between 2 and 20 hr.

Details of the compaction and processing conditions are given in Tables II through V. Compacts were formed in a standard, floating-action, transverse-rupture-bar die; compact dimensions were 0.50 by 1.25 by approximately 0.25 in. thick. Green densities were generally in the range from 80 to 90 pct of theoretical density, depending upon compaction pressure, type of powder, and mean compact composition. All diffusion treatments were carried out in a hydrogen atmosphere. Most of the deformation treatments in the homogenization-deformation sequences given in Tables III and V were carried out by unidirectional rolling (straight rolling) at room temperature; the amounts of deformation shown in these tables represent the total thickness reduction by all deformation treatments (i.e., the percentage total reduction in thickness of the original compact at the given stage in the sequence). In some instances, the original reduction was carried out by

Table I. Powders Used in the Present Investigation

Designation	Manufacturer	Size (Mesh)*	Shape
Cu-A	U. S. Bronze C-133	−200/+270	Irregular
Cu-B	U. S. Bronze C-113	−200/+270	Equiaxed
Cu-C	U. S. Bronze C-113	−140/+200	Equiaxed
Ni-A	Glidden F-210	−140/+200 31 pct −270/+325 38 pct −325/+400 31 pct	
Ni-B	Glidden F-210	−140/+200 31 pct −200/+270 38 pct −270/+325 31 pct	
Ni-C	Glidden F-210	−400	
Ni-D	Inco 123	4 to 7 μ	
W	Linde	−140/+200	Spherical

*Percentages in mesh ranges given in weight percent.

open-die, press forging at room temperature in order to minimize fracture of the compacts during the initial deformation treatments.

Evaluation of the homogeneity of the Ni-Cu compacts was carried out primarily by X-ray compositional line broadening using the Rudman analysis[17] as modified and used previously by Heckel.[10] (Electron microprobe data on selected samples were used to verify the X-ray data). CuK_α radiation was used to monitor the {311} diffraction peak (90 deg < 2θ < 93 deg, depending on the degree of homogeneity). The peak profiles were corrected for instrumental broadening by the Stokes method[19] using the computer program of DeAngelis and Schwartz.[20] These corrected profiles were analyzed by the method described previously, Fig. 9, to yield values of F, C_{max}, and C_{min} as reported in Tables II and III. The C_{max} and C_{min} values were assessed at $y_C = 0.02$ and 0.98, respectively, rather than $y_C = 0$ and 1.00, to avoid the scatter inherent in analyzing the "tails" of the diffraction peaks.

The two-phase Ni-W compacts were evaluated for the normalized diameter of the tungsten-rich phase ξ/l_0, by comparison of the amount of the phase at $t = 0$ with that present after the homogenization process. It has been shown[15] that:

$$\xi/l_0 \cong \left(\frac{V_V}{V_{V_0}}\right)^{1/3} \qquad [14]$$

where V_{V_0} and V_V are the volume fractions of the tungsten-rich phase determined by quantitative metallography before and after the homogenization treatment, respectively. Values of ξ/l_0 determined in this manner are given in Table V.

RESULTS

One-Phase System (Ni-Cu)

The as-compacted structure of the Ni-Cu compacts was, in general, similar to that shown in Fig. 12 for Ni-0.20 Cu. Although many isolated copper particles (dark etching) may be seen (thus supporting the concentric-sphere) model, numerous copper particles are contiguous with other copper particles. The frequency of occurrence of Cu:Cu contacts decreased with increasing copper content, and decreasing nickel particle size as would be expected on a statistical basis.

The progress of homogenization in Ni-Cu compacts

Table II. Summary of Homogenization Treatments for Ni-Cu Alloys

Specimen	Powders Cu	Powders Ni	Compact Pressure, kpsi	T, °C	t, hr	$\tilde{D}t/l_0^2$ *	Density, g/cc Green	Density, g/cc Sintered	F	C_{max}†	C_{min}†
						Mean Composition ($\overline{C}$) = 0.20 Cu					
1	B	A	80	850	1.0	1.25×10^{-3}	8.05	7.10	0.37	0.99	0
2	B	A	80	850	2.0	2.50×10^{-3}	8.05	7.10	0.40	0.99	0
3	B	A	80	1050	0.42	3.50×10^{-3}	7.91	6.93	0.43	0.98	0
4	B	B	80	950	1.0	3.90×10^{-3}	–	–	0.40	0.98	0
6	B	A	80	1050	1.0	8.50×10^{-3}	7.87	6.97	0.59	0.88	0
7	A	C	100	950	5.0	1.95×10^{-2}	7.83	6.80	0.67	0.67	0
9	A	C	100	950	10.0	3.90×10^{-2}	7.83	6.95	0.78	0.56	0.06
10	B	A	80	1050	5.0	4.25×10^{-2}	7.90	6.83	0.79	0.55	0.02
11	A	C	100	950	20.0	7.80×10^{-2}	–	–	0.80	0.54	0.08
12	B	A	80	1050	10.0	8.50×10^{-2}	7.76	6.81	0.84	0.45	0.03
13	A	C	100	950	31.0	1.20×10^{-1}	7.91	7.48	0.85	0.48	0.05
14	B	A	80	1050	20.5	1.75×10^{-1}	–	–	0.87	0.44	0.09
15	B	D	80	850	1.0	1.25×10^{-3}	7.26	7.16	0.54	0.95	0
16	B	D	80	850	2.0	2.50×10^{-3}	7.30	7.24	0.57	0.95	0
17	B	D	80	850	3.0	3.75×10^{-3}	–	–	0.60	0.94	0.01
18	B	D	80	950	1.0	3.90×10^{-3}	–	–	0.55	0.95	0
19	B	D	80	1050	1.0	8.50×10^{-3}	–	–	0.55	0.94	0
20	B	D	80	950	3.0	1.17×10^{-2}	–	–	0.57	0.93	0
22	B	D	80	1050	10.0	8.50×10^{-2}	7.39	7.42	0.79	0.50	0.03
						Mean Composition ($\overline{C}$) = 0.30 Cu					
23	B	D	80	850	1.0	1.28×10^{-3}	7.15	7.06	0.39	0.98	0
24	B	D	80	950	1.0	4.26×10^{-3}	–	–	0.45	0.97	0
25	B	D	80	1050	1.0	1.07×10^{-2}	–	–	0.52	0.93	0.01
26	B	D	80	1050	10.0	1.07×10^{-1}	7.30	7.29	0.78	0.65	0.09
27	B	D	80	1050	17.5	1.86×10^{-1}	–	–	0.82	0.65	0.21
						Mean Composition ($\overline{C}$) = 0.40 Cu					
28	C	D	100	950	1.0	2.50×10^{-3}	–	–	0.42	0.98	0
29	C	D	100	950	3.0	7.50×10^{-3}	–	–	0.42	0.98	0
30	C	D	100	1050	3.0	2.00×10^{-2}	7.75	7.24	0.57	0.89	0.02
31	C	D	100	1050	10.0	6.67×10^{-2}	7.73	7.01	0.76	0.70	0.10

*$\tilde{D}$ values for mean alloy composition obtained from reference 18; l_0 (for the minor constituent (Cu) obtained from the average mesh size (Cu-A and Cu-B(−200/+270), $l_0 = 65 \mu$; Cu-C(−140/+200), $l_0 = 89 \mu$.

†C_{max} and C_{min} obtained at $y_C = 0.02$ and 0.98, respectively, rather than $y_C = 0$ and 1.0 in order to eliminate scatter resulting from errors involved in treating the "tails" of the X-ray diffraction profiles.

proceeded as shown in Fig. 13 for a short-time treatment ($F = 0.55$) and in Fig. 14 for a long-time treatment ($F = 0.84$). The development of Kirkendall porosity, as analyzed by Fisher and Rudman,[21] is observed in Fig. 14 and in the density data given in Table II.

Comparison of the F and composition range (C_{max} and C_{min}) data for the Ni-Cu compacts with the predictions afforded by the concentric-sphere model showed that model predictions of homogenization were low for short-time treatments and high for long-time treatments. The comparison for the Ni-0.20 Cu experimental data with the model is given in Figs. 15 and 16; data for other values of $\overline{C}$ exhibited similar trends. Several different effects appear to bring about the discrepancy between the model and the data:

a) Surface transport in addition to volume transport appears to explain why the fine nickel powder matrix gives rise to high values of F at low values of $\tilde{D}t/l_0^2$. That is, the effective value of $\tilde{D}$ should contain a surface diffusion contribution during the initial stages of homogenization when the amount of original particle surface area is large (especially for fine powders). This effect has been discussed by Fisher and Rudman.[9]

b) The variation in $\tilde{D}$ as a function of composition (almost an order of magnitude higher in copper-rich alloys than in nickel-rich alloys[18]) should initially give a large decrease in C_{max} with increasing $\tilde{D}t/l_0^2$ when compared with the model since the model assumes a value of $\tilde{D}$ characteristic of the mean composition, $\overline{C}$.

c) The Cu:Cu contacts increase the effective value of l_0 causing the homogenization rate to be slower than predicted by the model for large values of $\tilde{D}t/l_0^2$. The concept of an increasing effective l_0 with increasing F has been considered previously for copper-rich Cu-Ni compacts.[10]

d) Localized variations in mean composition within the compact increase the difference between C_{max} and C_{min}. This was borne out by sampling studies on seventy-five different 1 by 1 mm areas (about 100 particles) on a polished section of an as-compacted specimen, Fig. 12. The variation in $\overline{C}$, measured by quantitative metallography (point counting), was between 0.10 copper and 0.30 copper for these areas. Concentric-sphere model values of C_{max} for $\overline{C} = 0.30$ and C_{min} for $\overline{C} = 0.10$ are approximately the same as the data values in Fig. 16 for $\tilde{D}t/l_0^2 \cong 10^{-1}$.

The type of correlation suggested by Fig. 4, where the measured and predicted F values may be compared on a single plot containing the variables $\tilde{D}$ (as a function of temperature), t, l_0, and $\overline{C}$, was also employed in the data analysis. The Ni-Cu data of the present investigation (0.20 Cu ≤ $\overline{C}$ ≤ 0.40 Cu) were supplemented by the

Cu-Ni data of a previous study[10] (0.11 Ni $\leq \overline{C} \leq 0.52$ Ni). The model treatment for the Ni-Cu data assumes a copper particle surrounded by a shell of nickel; the Cu-Ni data assume a nickel particle surrounded by a shell of copper. A summary of all of these data along with their comparison with the concentric-sphere model is presented in Fig. 17. Although the data scatter is relatively large as a result of powder processing variables mentioned earlier that are not taken into account in the model, it should be noted that the effects of the variables T, t, l_0^2, and $\overline{C}$ on the F values are correlated fairly well in terms of the parameter $(\tilde{D}t/l_0^2)n^2 Y$ as predicted by the model. For example, F as a function of t for the data shown in Fig. 17 would have a maximum spread of over $2\frac{1}{2}$ orders of magnitude in time for $0.60 < F < 0.80$. Thus, it may be concluded that, in general, the concentric-sphere model adequately describes the relative effects of processing temperature (through the volume interdiffusion coefficient), particle size of the minor constituent, processing time, and mean compact composition. The principal exceptions to this statement were discussed previously in relation to Figs. 15 and 16. In addition, it is noteworthy that the data for compacts with high values of $\overline{C}$ are lowest in the data band in Fig. 17, and *vice versa*. This effect is presumably due to the high probability of having contiguous minor constituent particles for large $\overline{C}$ blends and, thus, to a

Table III. Summary of Homogenizatin-Deformation Treatments for Ni-0.20 Cu Alloys

Specimen	Compact Pressure, kpsi	Homogenization T, °C	t, hr	Deformation† Total Reduction By Rolling, Pct		F	C_{max}‡	C_{min}‡
32	100	950	10.0 +	79	+			
		950	10.0			0.89	0.44	0.12
33	100	950	10.0 +	76	+			
		950	1.0 +	95	+			
		950	9.0			0.92	0.49	0.16
34	80	1050	10.0 +	47	+			
		1050	5.0			0.88	0.45	0.13
35	80	1050	10.0 +	73	+			
		1050	5.0			0.90	0.42	0.15
36	80	850	1.0 +	24	+			
		1050	0.33 +	47	+			
		1050	0.17 +	79	+			
		1050	0.17			0.63	0.75	0.0
37		Specimen 36	+	95	+			
		1050	1.0			0.76	0.60	0.01
38		Specimen 37	+	98	+			
		1050	0.33			0.82	0.48	0.04
39	80	850	0.5 +	15	+			
		1050	0.5 +	38	+			
		1050	0.5 +	69	+			
		1050	0.5 +	95	+			
		1050	2.0			0.82	0.48	0.02
40	100	1050	0.17 +	34	+			
		1050	0.08			0.48	0.90	0.0
41		Specimen 40	+	81	+			
		1050	0.25			0.65	0.68	0.0
42		Specimen 41	+	95	+			
		1050	0.25			0.76	0.54	0.01

*Cu-A and Ni-A powders used for all specimens.

†Original reduction on specimens 36, 39, and 40 carried out by open-die, press forging rather than rolling.

‡C_{max} and C_{min} obtained at $y_C = 0.02$ and 0.98, respectively, rather than $y_C = 0$ and 1.0 in order to eliminate scatter resulting from errors involved in treating the "tails" of the X-ray diffraction profiles.

Table IV. Density Changes in Specimens 40, 41, and 42 as a Function of Homogenization-Deformation Processing Sequence

	Processing Homogenization T, °C	t, hr	Deformation Total Reduction By Rolling, Pct	Density g/cc	Density Pct of Theoretical
		as compacted		8.05	90.3
	1050	0.17		7.16	80.3
			34*	8.78	98.5
Specimen 40	1050	0.08		8.46	94.9
			81	8.70	97.6
Specimen 41	1050	0.25		8.42	94.5
			95	8.63	96.8
Specimen 42	1050	0.25		8.50	95.0

Control Samples (no deformation processing)

		as compacted		8.12	91.1
	1050	0.17		7.12	79.8
	1050	0.08		7.29	81.7
	1050	0.25		7.15	80.2
	1050	0.25		6.87	77.1

*First deformation by open-die press forging rather than rolling.

Table V. Summary of Homogenization Treatments (With and Without Deformation) for Ni-0.06 W Alloys*

Specimen	Homogenization T, °C	t, hr	Deformation Total Reduction By Rolling, Pct	Volume Fraction of Unstable Tungsten-Rich Phase t = 0	After t	ξ/l
43	1207	16	—	0.0542	0.0458	0.945
44	1207	23	—	0.0542	0.0427	0.924
45	1207	40	—	0.0542	0.0231	0.750
	1207	16	+ 44			
46	1207	1	+ 67			
	1207	6		0.0542	0.0309	0.829
	Specimen	46	+ 94			
47	1207	17		0.0542	0.0106	0.580

*Ni-D powder using a compaction pressure of 98.4 kpsi.

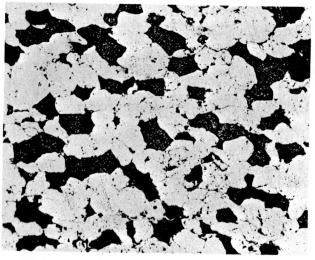

Fig. 12—As-compacted structure of Ni-0.20 Cu compacts using the Cu-B and Ni-A powders and a compaction pressure of 80,000 psi. Magnification 110 times.

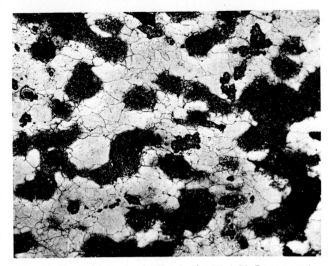

Fig. 13—Partial homogenization in the Ni-0.20 Cu compact shown in Fig. 12 after a heat treatment of 45 min at 1050°C (F = 0.55). Magnification 110 times.

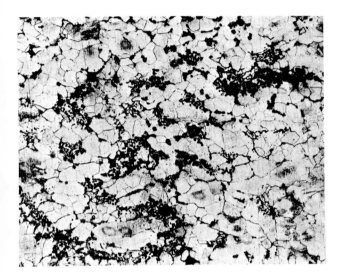

Fig. 14—High degree of homogenization in the Ni-0.20 Cu compact shown in Fig. 12 after a heat treatment of 10 hr at 1050°C (F = 0.84). Magnification 107 times.

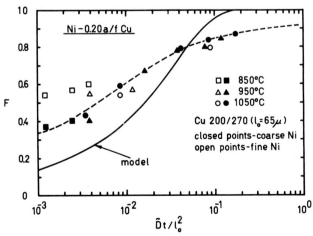

Fig. 15—Comparison of experimental F data on Ni-0.20 Cu compacts with model predictions. Open points were obtained from compacts fabricated from Ni-D powder.

larger effective particle size giving rise to slower kinetics than would be expected.

Deformation treatments by rolling at room temperature were successful in altering the diffusive flow geometry in the Ni-0.20 Cu compacts. Fig. 18 shows the microstructure of a specimen (no. 41) that had undergone a sequence of homogenization and deformation treatments (total reduction of 81 pct). The microstructure indicates that a transition from spherical flow (as in the original compact) to almost planar flow has been effected by the rolling. Because of the relatively large amounts of inhomogeneity (F = 0.65) existing in this specimen, the copper-rich regions are readily attacked during metallographic preparation. Subsequent processing, by deformation and homogenization, of the specimen shown in Fig. 18 resulted in the improved homogeneity exhibited metallographically in microstructure shown in Fig. 19 (specimen no. 42; Fig. 19 (specimen no. 42; F = 0.76).

The deformation treatments not only altered the diffusion geometry, but also succeeded in improving the density of the alloys. The density data presented in Table IV indicate that a gradual approach to theoretical

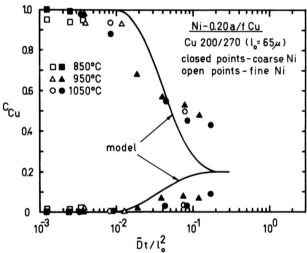

Fig. 16—Comparison of experimental C_{max} and C_{min} data on Ni-0.20 Cu compacts with model predictions. Open points were obtained from compacts fabricated from Ni-D powder.

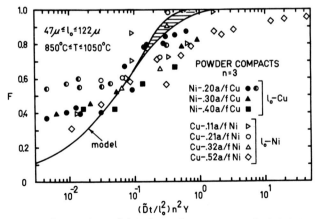

Fig. 17—Comparison of data from the present study (Ni-Cu) and a previous study[10] (Cu-Ni) with model predictions to show the normalizing effect of the parameter $(\tilde{D}t/l_0^2)n^2Y$.

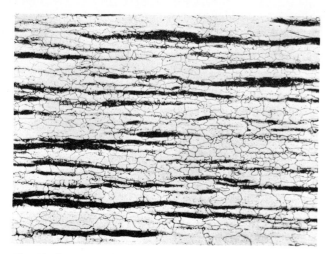

Fig. 18—Structure of a Ni-0.20 Cu compact given a homogenization-deformation sequence resulting in a total of 81 pct deformation and $F = 0.65$ (specimen 41). Plane of polish is perpendicular to the plane of the sheet and parallel to the rolling direction. Magnification 110 times.

Fig. 19—Structure of a Ni-0.20 Cu compact given a homogenization-deformation sequence resulting in a total of 95 pct deformation and $F = 0.76$ (specimen 42). Plane of polish is the plane of the sheet. Magnification 110 times.

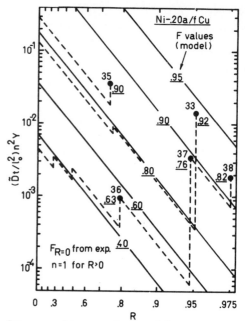

Fig. 20—Summary of homogenization-deformation sequence data and comparison with the predictions of the model. Data points are marked by specimen numbers and experimentally determined F values.

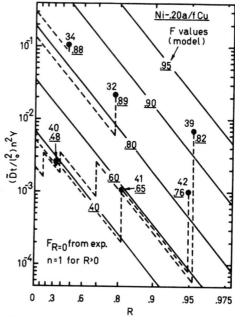

Fig. 21—Summary of homogenization-deformation sequence data and comparison with the predictions of the model. Data points are marked by specimen numbers and experimentally determined F values.

density is achieved by the deformation treatments. However, it should be noted that each homogenization treatment lowers the density, presumably due to the formation of Kirkendall porosity.[21]

The homogenization-deformation sequences for the Ni-0.20 Cu specimens, Table III, were analyzed in the manner suggested by Fig. 5. However, the actual data on homogenization of compacts (without deformation) as shown in Fig. 15 (dashed line) were used as the F values for $R = 0$ in the development of the model curves for F as a function of R. To facilitate the display of the data, it was assumed that $n = 1$ for all deformation treatments and that R was the same as the deformation of the overall compact. Figs. 20 and 21 give both the model predictions and the actual data. Each experimental determination of F is indicated as a data point described by the specimen number and the value of F. The experimental values of F generally are higher than the predicted values of F for low R

values and *vice versa* for high R values. The former effect is due to the restriction that $n = 1$ even for low values of R when the geometry is still very close to being spherical. The low experimental values of F at large R are believed to result from the fact that R values were assumed to be the same as the deformation of the compact. The data appear to show that the deformation in the copper-rich regions is less than that of the overall compact. This effect is magnified by the extreme sensitivity of model predictions for $R > 0.80$.

The data from these deformation studies have shown that the reductions of greater than 80 to 90 pct (the amount usually necessary to reduce the porosity to less than 5 pct) may be used to reduce markedly the time necessary to get to a relatively homogeneous alloy ($F > 0.80$). Typically, it was found that total homogenization time was reduced to 10 to 20 pct of that needed for compacts which received no deformation processing. More effective use of the deformation processing would probably result from extrusion (with R relative to the radial strain) since the reduction in n value from 3 to 2 is more favorable than the reduction from 3 to 1, as in rolling.

Two-Phase System (Ni-W)

Homogenization of compacts in the two-phase, Ni-W system with $\overline{C}$ values in the nickel-rich solid solution field have already been shown to be in accord with concentric-sphere models[11-13, 15] developed using numerical techniques as discussed relative to Fig. 6. The data obtained in the present investigation of the effects of deformation (by rolling) on the homogenization rates have been given in Table V in terms of the normalized size of the unstable tungsten-rich phase, ξ/l_0, obtained by quantitative metallography.

The effects of deformation in enhancing the rate of homogenization in Ni-W alloys are minimal as shown in Table V. Specimens 44 and 46 received the same total homogenization time at 1207°C with the latter being rolled (a total of 67 pct). The amount of homogenization (reduction in ξ/l_0) for the deformed alloy may be seen in Table V to be greater than for the undeformed compacts, but the difference is relatively small. Increasing the amount of deformation to 94 pct (compare specimens 45 and 47) results in more rapid homogenization, but the enhancement in rate is much less than would be predicted from Fig. 7 if R is assumed to be the same as the compact reduction. Metallographic studies showed that the deformation of the tungsten-rich phase during rolling was minimal, especially when compared with the deformation by rolling of the copper-rich areas in the Ni-0.20 Cu compacts. The only slight improvement in homogenization rates by rolling the Ni-W compacts is attributed to the inability of the deformation processing to effect a significant reduction in the thickness of the hard, tungsten-rich phase. Thus, in the use of Figs. 5, 7, and 8 in the analysis of deformation processing effects, the value of R can be assumed to be the same as the overall compact deformation only as the limiting case; hard minor constituent particles will contribute to poor model predictions using this assumption. If accurate predictions are required, the relationship between particle deformation and compact deformation must be known.

DISCUSSION

Evaluation of the concentric-sphere models of homogenization for undeformed powder compacts contained in the present and prior studies[9, 10, 13-15] indicates their usefulness in the prediction of the relative effects of major process variables in both one-phase and multiphase systems. This is substantiated in the present study by the condensation of homogenization data for both nickel-rich and copper-rich alloys in the isomorphous Ni-Cu system into a single band of data when plotted as F vs $(\tilde{D}t/l_0^2)n^2Y$ as shown in Fig. 17. The breadth of the band is postulated to result from the effects of a number of minor variables that are known to affect homogenization rates, but are not included in the model. Available homogenization data[13, 15] on the Ni-W compacts (two-phase) led to the same conclusion: ξ/l_0 and $(C/\overline{C})_{L/2}$ data may be normalized to a single, relatively narrow data band when plotted vs the parameters $(\tilde{D}_\alpha t/l_0^2)C_{\alpha\beta}^y$ and $(\tilde{D}_\alpha t/l_0^2)\overline{C}^y$, respectively, which contain the major variables in the process.

Absolute predictions of the degree of homogeneity in undeformed compacts in either one-phase or multiphase binary systems using the concentric-sphere models generally give estimates which are too low when the degree of homogeneity is low, and too high when the degree of homogeneity is high. The exception to this appears to be in ξ/l_0 measurements in two-phase systems which appear to agree well with the concentric-sphere model.[13, 15] The relatively rapid initial rates of homogenization are most easily associated with the contribution of surface diffusion in addition to volume diffusion, since a) the effect is operative usually at short homogenization times [when the original matrix particles (major constituent) retain their identity], and b) the effect is enhanced by increased particle surface area (decreased matrix particle size).

The progressive slowing of the rate of homogenization relative to the predictions of the model with increasing time is ascribed to mixing effects. The concentric-sphere model implicitly assumes a perfect distribution of minor constituent particles in a continuous matrix. The statistical nature of the mixing problem, on the other hand, indicates that such an array represents a limiting situation and that contiguous minor constituent particles are to be expected. The probability of contiguous particles should increase with increasing mean composition, leading to larger departures from the model. Available data are in support of this. Furthermore, it is to be expected on statistical grounds that groupings of particles in a compact will depart from the mean compact composition, independent of whether contiguous minor particles occur. The mean composition of the groupings will obviously approach the mean composition of the compact as the group size increases. It may therefore be concluded that the treatment of the mixing problem in homogenization using the concentric-sphere models should be approached from the standpoint of either considering an increasing effective particle size with increasing time,[10] or considering compacts to contain a range of mean compositions and using the models to consider the behavior of alloys within the range.

A technological solution to the mixing problem has been achieved by Lund, et al.[22,23] These investigators studied the homogenization of compacts fabricated from a variety of different powders coated with nickel. They observed improved homogenization rates compared with compacts made from blended powders, since each of the coated powder particles had approximately the same composition as the overall compact. In effect, each particle was the sphere-shell composite of the concentric-sphere model.

Consideration of the effects of geometry changes due to deformation on the homogenization rates may be carried out only through consideration of both a) the reduction in the smallest dimension of the minor constituent particles (effect of R on l_0), and b) the alteration of the geometry of the diffusive flows (from spherical toward cylindrical in extrusion; from spherical toward planar in rolling). Under conditions where there are large amounts of deformation [$i.e.$, where cylindrical ($n = 2$) and planar ($n = 1$) geometries are developed by deformation] and the reduction of the minor constituent particles may be approximated by the axial reduction of the compact, homogenization rates should approach the model predictions given in Figs. 5, 7, and 8 where R is the compact reduction. If the particle reduction may not be approximated by the compact reduction, one must determine the relationship between these variables to use the model. The deformation models also indicate that for a given axial reduction of the compact, extrusion is a more effective deformation treatment to promote homogenization (in subsequent heat treatments) than rolling, since the cylindrical geometry is more favorable than the planar geometry.

SUMMARY

The homogenization behavior of powder compacts has been described in terms of powder, phase equilibrium, diffusion, and processing variables. It has been shown that mathematical models, assuming specific geometries, can be formulated using numerical methods to describe the effects of the major variables in the homogenization process. Experimental data obtained from X-ray compositional line broadening, electron microprobe analysis, and quantitative metallography have been used to evaluate the models. These data confirm the predicted effects of the major variables, namely, volume interdiffusion coefficient (the coefficient of the stable phase in two-phase systems), particle size of the minor constituent, time, mean composition, and, for two-phase systems, the stable phase solubility. The principal effects which may influence the homogenization process, but which are not included in the models are a) the contribution of surface diffusion, b) the departure from ideal mixing, and c) the effects of nonuniform particle sizes.* The effects of deforma-

---

*Uniform size models have, however, been shown to be applicable to non-uniform size problems under certain circumstances.[13,24]

---

tion of the homogenization process includes both the reduction in particle dimension and the alteration of diffusive flux geometry. A method for assessing the effects of deformation on homogenization rates is developed from the mathematical models and is shown to be applicable in limiting cases.

ACKNOWLEDGMENTS

The authors gratefully acknowledge the support of this research through a Department of Defense grant under Project THEMIS. The authors also express their appreciation to H. Markus and F. I. Zaleski of the Frankford Arsenal who contributed to discussions during the course of the research and who supplied the tungsten powder used in the two-phase homogenization study. Discussions with R. L. Coble of M. I. T. on the topic of mixing and R. A. Tanzilli of the General Electric Co. and R. D. Lanam of Drexel Univeristy on the modeling and data analysis are also greatly appreciated.

REFERENCES

1. F. N. Rhines and R. A. Colton: *Trans. ASM*, 1942, vol. 30, p. 166.
2. P. Chevenard and X. Wache: *Rev. Met.*, 1944, vol. 41, p. 353.
3. P. Duwez and C. B. Jordan: *Trans. ASM*, 1949, vol. 41, p. 194.
4. S. Weinbaum: *J. Appl. Phys.*, 1948, vol. 19, p. 897.
5. S. D. Gertzriken and M. Feingold: *Zh. Theor. Fiz.*, 1940, vol. 10, p. 574.
6. A. I. Raichenko and I. M. Fedorchenko: *Dokl. Akad. Naul. Ukr. SSR*, 1958, vol. 3, p. 255.
7. A. I. Raichenko and I. M. Fedorchenko: *Dokl. Akad. Naul. Ukr. SSR*, 1958, vol. 8, p. 835.
8. A. I. Raichenko: *Fiz. Metal i Metalloved.*, 1961, vol. 11, p. 49.
9. B. Fisher and P. S. Rudman: *J. Appl. Phys.*, 1961, vol. 32, p. 1604.
10. R. W. Heckel: *Trans. ASM*, 1964, vol. 57, p. 443.
11. R. A. Tanzilli and R. W. Heckel: *Trans. TMS-AIME*, 1968, vol. 242, p. 2313.
12. R. A. Tanzilli and R. W. Heckel: *Trans. TMS-AIME*, 1969, vol. 245, p. 1363.
13. R. A. Tanzilli: Ph.D. Thesis, Drexel University, 1970.
14. R. D. Lanam, Ph.D. Thesis Research, Drexel University.
15. R. W. Heckel, R. D. Lanam, and R. A. Tanzilli: *Advanced Experimental Techniques in Powder Metallurgy, Perspectives in Powder Metallurgy*, Vol. 5, J. S. Hirschhorn and K. H. Roll, eds., Plenum Press, 1970.
16. M. L. James, G. M. Smith, and J. C. Wolford: *Analog and Digital Computer Methods in Engineering Analysis*, International Textbook Co., 1964.
17. P. S. Rudman: *Acta Cryst.*, 1960, vol. 13, p. 905.
18. L. C. C. daSilva and R. F. Mehl: *AIME Trans.*, 1951, vol. 191, p. 155.
19. A. R. Stokes: *Proc. Phys. Soc. London*, 1948, vol. 61, p. 382.
20. R. J. De Angelis and L. H. Schwartz: *Acta Cryst.*, 1963, vol. 16, p. 705.
21. B. Fisher and P. S. Rudman: *Acta Met.*, 10, 1962, vol. 10, p. 37.
22. J. A. Lund, T. Krantz, and V. N. Mackiw: *Progr. Powder Met.*, 1960, vol. 16, p. 160.
23. J. A. Lund, W. R. Livine, and V. N. Mackiw: *Powder Met.*, 1962, vol. 10, p. 218.
24. D. L. Baty, R. A. Tanzilli, and R. W. Heckel: *Met. Trans.*, 1970, vol. 1, p. 1651.

RBO: A New Approach to Lubricant Removal

By A. P. CREASE JR.

Rapid burn-off system practically eliminates lubricant residue problems. It also replaces the preheat furnace in a sintering line. The advance could open the door to much larger sintering furnaces.

A NEW DEVICE for removing lubricants from P/M parts — the result of four years' study at Drever Co. —shows promise of making a major contribution to the future of powder metallurgy. We call it the RBO (for rapid burn-off).

The need for something new became acute about five years ago when new high-production sintering furnaces with capacities up to 800 lb/h (363 kg/h) were placed in operation. The rapidly expanding use of Acrawax was also posing major lubricant residue problems in conventionally constructed furnaces. The burnout and cleanout interval on the high-production furnaces was ranging between three to five days, causing considerable lost production time and cost.

Idea — The original concept of the RBO was to remove a major percentage of the lubricant from P/M parts prior to entering the con-

Mr. Crease is sales manager, Drever Co., Huntingdon Valley, Pa.

ventional preheat or burn-off furnace and sintering system, thereby increasing the time interval between burnouts and furnace cleanouts.

As development work and experience with the unit progressed, it was learned that the RBO could completely replace the preheat or burn-off furnace in a sintering line.

Description — The RBO is a gas-tight heating unit directly coupled to the furnace atmosphere system and can presently be located ahead of the sintering furnace (Fig. 1). Within its active heating chamber it heats the parts passing through it to a temperature of 1000 to 1500 F (540 to 815 C) in a very short time. Heat is supplied by special gas-fired radiant burners which fire directly into the chamber.

Typical heating times to reach the desired temperature range from 3 min for small flat parts to 15 min for 1000-ton press parts weighing 14 lb (6 kg). In general, the

unit heats parts being processed five to six times as fast as conventional heating methods.

A curtain between the RBO and the sintering furnace atmosphere system helps to keep burn-off products from mixing with the furnace atmosphere.

In the RBO, the conveyor belt travels over an open grid through which liquid residues (such as Acrawax) can pass to a collection pan underneath which then drains the lubricant in liquid form to an outside collection point.

In normal operation the burner combustion system fires on the reducing side so that no oxygen can touch the parts being heated. In the event that a burner panel clean-up becomes necessary, the air-to-gas ratio can be changed to oxidizing to quickly remove hydrocarbon residues.

Results — Work to date on our test unit has been primarily conducted on iron, iron-carbon, iron-copper-carbon, and low-alloy parts.

The majority of these parts contained Acrawax C lubricant, but parts containing zinc stearate, lithium stearate, Polymist A-12, and mixtures of these have also been processed successfully. Figure 2 shows typical temperature profiles.

Processing experience has been as follows:

Strength: Where transverse rupture strength tests have been made, parts processed in the RBO were equal to or stronger than those conventionally sintered.

Growth or shrinkage: No significant difference in growth or shrinkage has been experienced between parts processed in the RBO and those conventionally processed.

Oxygen: The oxygen content of parts after sintering in endothermic gas has been extremely low when using RBO. We believe that this is due to the isolation of the majority of the burn-off products from the furnace atmosphere.

Decarburization: We have been unable to detect decarburization caused by the RBO. Decarb depths in parts sintered with the RBO and with conventional systems are approximately the same.

Popcorning: We have been unable to blow up any part processed in the RBO. Even parts processed at 7.3 to 7.55 g/cm^3 were free of any physical defects caused by the rapid heating.

Density: Parts have been processed in the RBO from 6.2 to 7.55

Fig. 1 — Rapid burn-off unit is located in enclosed portion of photograph. In this instance, it is installed on a 12 in. (305 mm) conveyor belt furnace. It has an hourly capacity of 344 lb (155 kg); loading is 20 lb/ft^2 (9 kg/m^2); effective RBO heating length is 17 in. (422 mm).

RBO Can Eliminate the Need for Preheat or Burn-Off Furnaces in a Sintering Line

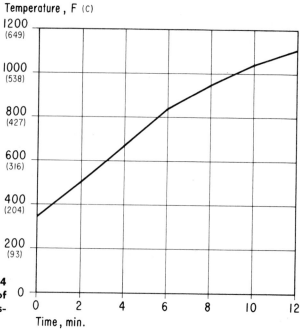

Fig. 2 — Following charts show typical temperature profiles. Temperatures were taken from thermocouples embedded in the center of the mass of parts. Smaller parts normally heat faster.

A round part 5 1/4 in. in diam by 1 3/4 in. thick (134 by 45 mm), weighing 6 lb and having a density of 6.8 g/cm³. Material: iron + 1% Acrawax. Atmosphere: endothermic +37 F (+3 C).

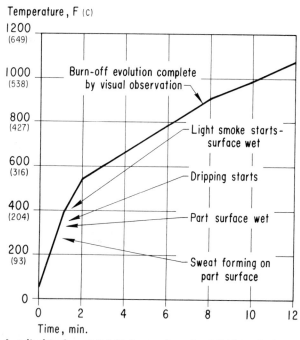

A cylindrical part 3 1/2 in. in diam by 2 3/4 in. high (89 by 70 mm), weighing 3 lb 5 1/2 oz (2 kg 156 g). Material: iron + 1% Acrawax. Atmosphere: endothermic +34 F (+1 C).

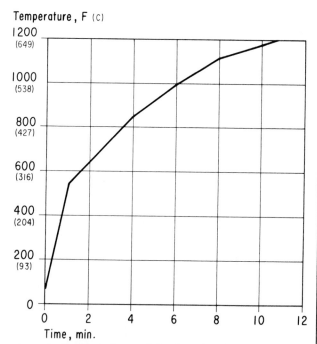

Irregularly shaped part 4 in. long by 4 1/2 in. wide by 1 3/4 in. thick (102 by 114 by 45 mm), weighing 2 lb 1 1/2 oz (0.91 kg 44 g). Material: iron + 1 1/2% Acrawax. Atmosphere: endothermic +46 F (+8 C).

g/cm³ density. Even at the high densities, no visual surface defects have been detected.

Part size: The RBO has processed parts ranging from about 1 oz up to 14 lb (29 g to 6 kg).

Temperature control: The RBO does not utilize an automatic temperature controller. You just turn it on.

Stacking: Under regular production conditions, parts may be stacked two and three layers deep with separator screens to a net loading of 15 to 20 lb/ft² (7 to 9 kg/m²) — where furnace conveyor allows.

Oxidation: Iron and iron carbon

parts do not pick up oxygen in the RBO unit. Manganese-bearing alloys and other oxide-sensitive materials may pick up some oxygen if their temperature is allowed to go too high. The same is true for stainless steels. Test work to determine threshold oxidation temperatures of these materials is scheduled in the near future.

Advantages — The primary contribution of the RBO is that it is a highly efficient and effective means of removing and disposing of compacting lubricants in vapor and liquid form in the sintering operation. The combination of high dew point, high CO_2 content, and the added volume of waste products of combustion, when added to the furnace atmosphere, combine to effectively dispose of lubricant burn-off products.

Another advantage is that the unit is very small. A 17 in. (422 mm) effective heating length can serve a 12 in. (305 mm) wide mesh belt furnace producing 344 lb (156 kg) of parts per hour at 20 lb/ft^2 (9 kg/m^2) loading. This compares with an 8½ ft (3 m) long conventional burn-off furnace length. The shorter length of the RBO reduces the stress on the mesh belt and therefore allows heavier belt loading with resultant higher output capacity from a given sintering furnace.

Because of its efficiency in disposing of lubricants, the RBO now opens the door to much larger sintering furnaces than have ever before been considered possible. This is particularly important in the area of P/M hot forming where furnace capacities of 1500 to 2000 lb/h (677 to 907 kg/h) are needed in many cases to balance the forging press output.

A furnace equipped with an RBO burn-off system has a much cleaner, more consistent atmosphere in the sintering and cooling zones because the atmosphere is not heavily contaminated with lubricant burn-off products.

Disadvantages — The RBO is presently available as a gas-fired unit only. Since gas is in short supply in many areas, many sintering companies cannot make use of it. An electrically heated unit is currently undergoing development, but many problems remain to be solved before it can be operational.

The unit is presently available only for use on furnaces which are equipped with the Convecool high-convection cooling system. An atmosphere directional control system is used with the convection cooler which can offset the higher atmosphere pressures in the RBO chamber. ⊕

FURNACE ATMOSPHERES

By
Cornelius Durdaller
Director of Research
Hoeganaes Corporation

To those persons who work in any of the diverse areas of the powder metallurgy industry, the use of furnace atmospheres is a part of their daily experience. Many of the P/M processes occur at elevated temperatures where chemical reactions are extremely rapid; the fact that powders have very large specific surfaces and that P/M parts are porous makes these chemical reactions even faster and, therefore, more critical to control.

When powder or a powder part is put into a furnace at an elevated temperature, the atmosphere it is placed in participates actively in whatever happens in the furnace. Furnace atmospheres in powder metallurgy normally do one or a combination of the following:

> Oxidize or Reduce
> Carburize or Decarburize
> Nitride or Denitride
> Remain Neutral.

Which they do depends on the atmosphere composition, the temperature, the pressure and the material in the furnace.

The general field of gas-solid reactions at elevated temperatures (high temperature chemical metallurgy) is one that is particularly amenable for study and one, owing to its importance to steelmaking, that has received a great deal of effective study. As a result, much of what happens between gases and solids at higher temperatures is understood well. And being understood, the reactions can be controlled. The powder metallurgist, be he a part maker, equipment manufacturer or powder producer, has this knowledge available to him. The growth of powder metallurgy depends upon how well he avails himself of this knowledge, extends it and puts it into practice.

BASIC COMPONENTS OF FURNACE ATMOSPHERES

The discussion in this section is limited to only the particular combination of gases in each heading. It is understood that realistic atmospheres contain many different gases which interact with one another and the material in the furnace in a complex manner. The basic trends discussed are correct and will help in understanding the practical atmospheres better.

1. Hydrogen-Water Vapor (H_2-H_2O)

   The relative amounts of hydrogen and water vapor determine whether the atmosphere is oxidizing or reducing. When a pure metal oxide (MeO) is placed in a pure hydrogen atmosphere, the oxide and the hydrogen will react to form the metal and water (or water vapor), the relative amount of water formed is dependent upon the temperature and pressure of the system and the relative stability of the metal oxide. The following chemical reaction describes the process:

$$MeO + H_2 \rightleftarrows Me + H_2O. \tag{1}$$

The stability of the oxide, MeO can be described, at any one temperature and pressure, as the ratio of hydrogen to water vapor when the reaction is at equilibrium. The smaller the H_2/H_2O ratio, the more stable the oxide is. The ratio can also be stated in terms of the dewpoint of the hydrogen atmosphere (Figure 1).

If, for example, the equilibrium H_2/H_2O ratio for this chemical reaction at a given temperature and pressure is 10^4 (10,000 parts H_2 for every part of H_2O) or a dewpoint of $-40°$ F, any metal present will turn completely to oxide if the dewpoint of the hydrogen is <u>maintained</u> greater than $-40°$ F. If the dewpoint is held below $-40°$ F, the oxide will completely transform to metal. Oxidation at dewpoints higher than $-40°$ F, reduction below. In practice, if oxidation is to be prevented (or oxide reduced) using hydrogen as a furnace atmosphere, the dewpoint of the atmosphere in contact with the metal has to be maintained lower than the equilibrium dewpoint. This requires sufficient flow to prevent local building up of water vapor owing to leakage of air into the furnace or oxide reduction. Metal oxides become less stable as the temperature is raised: the equilibrium H_2/H_2O ratio gets smaller, equilibrium dewpoint larger.

The oxidation of iron and reduction of iron ore are critically important in powder metallurgy. These reactions can be described as follows:

$$Fe_3O_4 + 4H_2 \rightleftarrows 3Fe + 4H_2O \qquad \text{below } 1030° \text{ F} \tag{2}$$

$$Fe_3O_4 + H_2 \rightleftarrows 3FeO + H_2O \tag{2a}$$
$$\text{above } 1030° \text{ F}$$
$$FeO + H_2 \rightleftarrows Fe + H_2O. \tag{2b}$$

The equilibrium curves for these reactions are given in Figure 2 as a function of temperature in an H_2-H_2O atmosphere. FeO is unstable below $1030°$ F; alpha iron (carbon free ferrite) is stable below $1670°$ F, gamma iron (carbon-free austenite) stable above. Since the total pressure in the system is fixed, choosing a temperature and water vapor percentage will describe what is present in the system. Note that as the temperature is increased, the oxides get less stable, e.g., at a constant water vapor percentage of 30%, increasing the temperature through $1110°$ F will reduce Fe_3O_4 to FeO, past $1390°$ F will reduce FeO to Fe.

The general question of stability of oxides and furnace atmospheres will be discussed in further detail later in this presentation. Hydrogen also acts as a decarburizing atmosphere. This will be discussed along with methane.

Source: Technical Bulletin D174, Hoeganaes Corp., Dec 1972

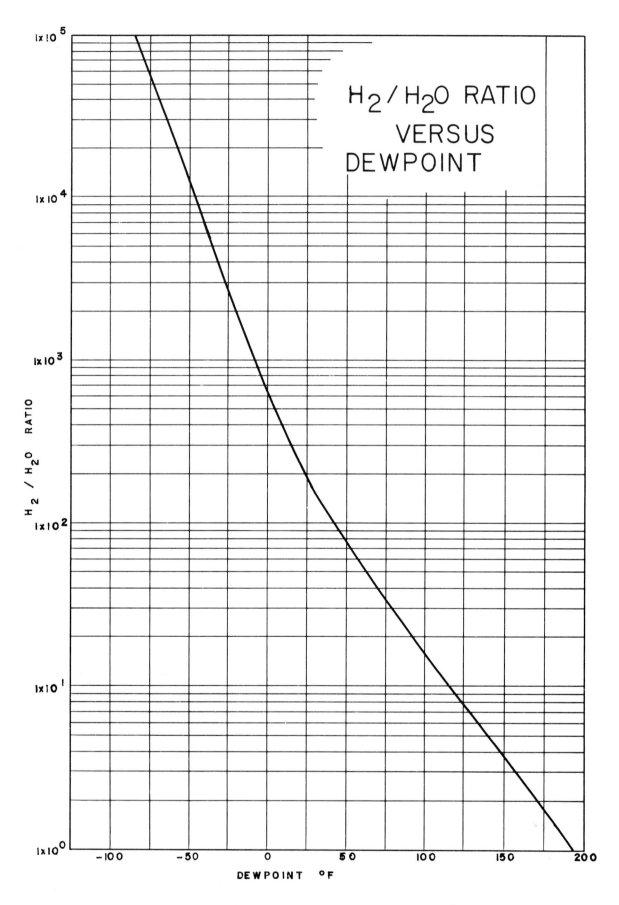

Figure 1 – Relation Between Dew Point and H_2/H_2O Ratio

FE-O-H EQUILIBRIUM

Figure 2

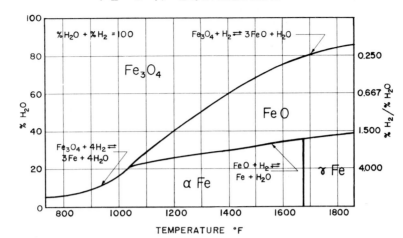

FE-O-C EQUILIBRIUM

Figure 3

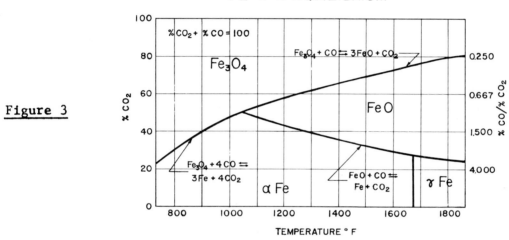

FE-O-C EQUILIBRIUM DIAGRAM

Figure 4

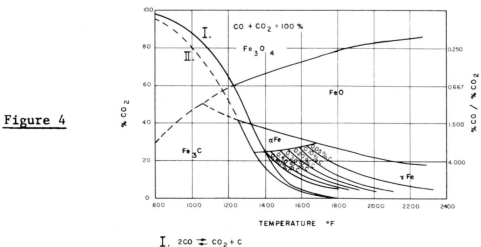

I. $2CO \rightleftharpoons CO_2 + C$

II. $Fe_3C + CO_2 \rightleftharpoons 3Fe + 2CO$

Source: Technical Bulletin D174, Hoeganaes Corp., Dec 1972

201

2. Carbon Monoxide and Carbon Dioxide

Mixtures of carbon monoxide and carbon dioxide can also be oxidizing or reducing.

$$MeO + CO \; \rightleftarrows \; Me + CO_2. \tag{4}$$

In the same manner that the H_2/H_2O can be used to describe the stability of the oxide in an hydrogen atmosphere, the carbon monoxide to carbon dioxide ratio can describe it in a carbon monoxide atmosphere. The smaller the CO/CO_2 ratio, the more stable the oxide.

Carbon monoxide is a common reducing agent for metal oxides. When used to reduce iron oxides, the following reactions occur:

$$Fe_3O_4 + 4CO \; \rightleftarrows \; 3Fe + 4CO_2 \qquad \text{below } 1030°\text{ F} \tag{5}$$

$$Fe_3O_4 + CO \; \rightleftarrows \; 3FeO + CO_2 \tag{5a}$$

$$\text{above } 1030°\text{ F}$$

$$FeO + CO \; \rightleftarrows \; Fe + CO_2. \tag{5b}$$

The equilibrium curves for these reactions are given in Figure 3 as a function of temperature in a CO-CO_2 atmosphere. The same general description given for Figure 2 is valid for Figure 3.

Figure 3, however, does not completely describe what occurs once the oxides are reduced. A CO-CO_2 atmosphere is also a carburizing-decarburizing atmosphere for iron. Once the oxide is reduced, the presence of carbon monoxide can carburize the iron as carbon entering into solution in the alpha or gamma iron or as iron carbide (Fe_3C).

$$2CO \; \rightleftarrows \; C \text{ (in } \alpha \text{ or } \gamma \text{ or as pure carbon)} + CO_2 \tag{6}$$

$$3Fe + 2CO \; \rightleftarrows \; Fe_3C + CO_2. \tag{7}$$

Figure 4 superimposes these two equilibria upon Figure 3, including the lines of constant carbon content in the gamma (now austenite) iron. Several features of this figure are interesting. As temperature increases the carburizing effect of carbon monoxide becomes less, i.e., to keep a constant carbon content in the austenite, greater amounts of carbon monoxide must be added as the temperature increases. At sintering temperatures, slight changes in the amount of carbon dioxide or the temperature can cause a marked change in the carbon content of the austenite. This emphasizes the difficulty in controlling the carbon content of P/M parts during the sintering process, especially when cooling from the sintering temperature.

3. Hydrogen and Methane

Hydrogen will react with carbon in solution in iron (or as Fe_3C) to form methane (CH_4). This accounts for the decarburizing action of hydrogen on steel. The decarburization reaction is enhanced by the presence of water vapor (increasing the dewpoint). Below 1300° F hydrogen has no pronounced decarburizing effect. The reaction can be described by either of two reactions.

$$C \text{ (in solution)} + 4H \rightleftarrows CH_4 \qquad \qquad (8a)$$
$$C \text{ (in solution)} + 2H_2 \rightleftarrows CH_4 \qquad \qquad (8b)$$

$$Fe_3C + 2H_2 \rightleftarrows 3Fe + CH_4. \qquad \qquad (9)$$

Reaction (8a) is a normal carburizing-decarburizing reaction. The dissociated hydrogen comes from whatever water vapor is around. The oxygen released reacts with the carbon in the steel to form carbon monoxide. Figure 5 describes the equilibria in an iron-methane-hydrogen system as a function of temperature. In contrast to the $Fe-CO_2-CO$ system, an increase in temperature increases the carburizing effect of methane (as well as its reducing effect). The amount of methane required is very small, i.e., small additions of methane into a furnace atmosphere will result in either soot deposition (at low temperatures – less than 1400 to 1500° F) or carburization of iron at higher temperatures.

Other hydrocarbons, ethane (C_2H_6), propane (C_3H_8) and butane (C_4H_{10}), are also used as furnace additions for carburization. The soot forming tendency of the hydrocarbons is proportional to the amount of carbon atoms in the basic hydrocarbon molecule. Therefore, the use of these higher hydrocarbons will give a greater soot forming potential than methane when used in a furnace atmosphere. Again, the lower the temperature the more pronounced the sooting problem becomes.

4. Nitrogen (Ammonia)

Pure molecular nitrogen is an inert atmosphere. Its use as a protective atmosphere depends upon the impurity gases present in the nitrogen. The impurity gases act as they would if the nitrogen was absent.

Nitrogen is used as a hardener for iron and metal nitrides are, in general, stable compounds. The action of nitrogen on iron is almost identical to that of carbon. Nitrogen, however, will not react with the iron if it exists as molecular nitrogen. Atomic nitrogen will.

The most common source of atomic nitrogen comes from the thermal decomposition of ammonia.

$$2NH_3 \rightleftarrows 2N + 3H_2 \qquad \qquad (10)$$

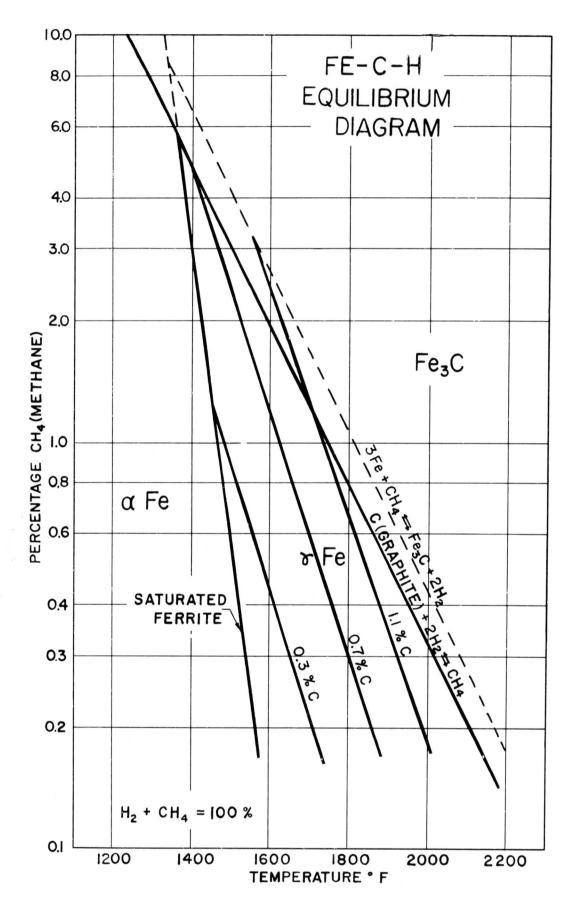

Figure 5 – Fe-C-H Equilibrium

This reaction is greatly accelerated when the ammonia comes in contact with the surface of hot iron or steel. Nitriding is a commercial practice used to surface harden steel in the same manner as carburizing. Very small amounts of ammonia in a furnace atmosphere can cause nitriding; if the odor of ammonia can be detected (about 50 ppm or 0.0050%), the danger of nitriding exists. As the furnace temperature increases, nitriding is accelerated; it becomes pronounced above 900° F.

5. Other Inert Atmospheres

Argon, helium and vacuum are other common inert atmospheres. Inert is a relative term because, unless these atmospheres are totally pure, they are not inert and, in practice, all contain impurity gases. The effects of the impurity gases depend upon the kind of impurity gases, the amount of them, the chemical reactions between these gases and the material under consideration, the stability of the reaction products and the temperature and pressure of the system.

For example, if pure nickel was placed in an argon atmosphere at 1600° F, and if a partial pressure of oxygen in that system was maintained at approximately 8×10^{-9} atmospheres, the nickel would be completely oxidized to nickel oxide (NiO). This oxygen potential corresponds to an H_2/H_2O ratio of 3×10^3 (dewpoint of -26° F) in a hydrogen atmosphere or a CO/CO_2 ratio of almost 2×10^4 in a carbon monoxide atmosphere.

MIXTURES OF BASIC FURNACE ATMOSPHERES

The discussion to this stage has dealt with simple atmospheres, the gaseous reactants and products being simply related to one another. Normally, furnace atmospheres contain combinations of these simple atmospheres, the components of which are related to one another in more complex ways because they interact with each other. Many possibilities of simple atmosphere combinations exist, but one is sufficient to illustrate the nature of this interaction.

The equilibrium among carbon monoxide, carbon dioxide, hydrogen and water vapor is called the Water Gas Reaction and can be written as:

$$CO + H_2O \rightleftarrows CO_2 + H_2. \tag{11}$$

At any constant temperature and pressure where these species exist, the ratio among these gases is constant:

$$K = \frac{P_{CO_2} \times P_{H_2}}{P_{CO} \times P_{H_2O}} \quad \text{or} \tag{12}$$

$$\frac{P_{H_2}}{P_{H_2O}} = K \frac{P_{CO}}{P_{CO_2}}. \tag{12a}$$

Source: Technical Bulletin D174, Hoeganaes Corp., Dec 1972

The value of the constant, k, is a function of temperature and is given as a function of temperature in Figure 6. At a temperature of 1525° F this constant equals one and the H_2/H_2O ratio equals the CO/CO_2 ratio. Increasing the temperature decreases this constant and therefore moves the reaction (11) to the left. The significance of this reaction is that gas components cannot be added independently of one another. The ratio of H_2/H_2O cannot be fixed independent of the ratio of CO/CO_2 and for any given ratio of one the ratio of the other is a function of temperature.

The carbon content of steel can be regulated by the amount of carbon monoxide in the furnace atmosphere (water vapor can only decarburize by itself). If the amount of hydrogen and the combined amount of carbon monoxide and carbon dioxide is held constant in a furnace atmosphere, the carburizing-decarburizing action of the atmosphere can be controlled by the water content. If P_{h_2} = A (constant) and $P_{co} + P_{co_2}$ = B (constant), then from equation (12a)

$$P_{h_2o} = \frac{A}{K} \quad \frac{B - P_{co}}{P_{co}} . \qquad (13)$$

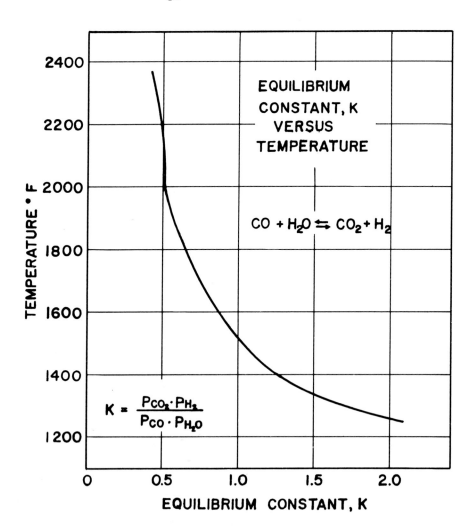

Figure 6 - Temperature Dependence of Water Gas Reaction Equilibrium Constant

Figure 7 illustrates the effect of dew point (water vapor content) on the equilibrium carbon concentrations in steel as a function of temperature for an atmosphere containing 40% H_2 and CO_2 + CO = 20%. A constant water vapor content becomes increasingly decarburizing as the temperature is increased.

Reactions (1) and (4) describe the reduction of metal oxides or oxidation of metal using hydrogen-water vapor and carbon monoxide-carbon dioxide. For iron, the reactions,

$$FeO + H_2 \rightleftarrows Fe + H_2O \qquad\qquad (2b)$$

$$FeO + CO \rightleftarrows Fe + CO_2 \, , \qquad\qquad (5b)$$

have a different dependence on temperature (See Figures 2 and 3). As temperature increases, the H_2/H_2O ratio in equilibrium with FeO decreases (FeO will reduce at higher dewpoints); the CO/CO_2 ratio increases. An atmosphere with a constant water content becomes more reducing as temperature is increased (more oxidizing as the temperature is decreased). The reverse is true for an atmosphere with a constant carbon dioxide content. This has practical consequences for furnace atmospheres.

Since many furnace atmospheres are produced by reacting a fuel gas with air over a wide range of air/gas ratios, a wide range of compositions are available. Lean atmospheres (high air/gas ratios) have low CO/CO_2 ratios

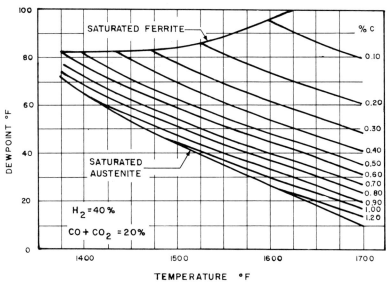

CARBON CONTENT OF AUSTENITE AS A
FUNCTION OF TEMPERATURE & DEWPOINT

Figure 7 - Variation of Carbon Content of Austenite with Dewpoint and Temperature for an Atmosphere Containing a Fixed H_2 and CO_2 + CO Content

Source: Technical Bulletin D174, Hoeganaes Corp., Dec 1972

and tend to be oxidizing to iron at normal working temperatures. The reducing potential of hydrogen is overwhelmed because so little of it is present. At lower air-gas ratios, the amount of CO_2 decreases, the CO/CO_2 ratio increases, but now the H_2/H_2O ratio becomes more important and the atmosphere has a greater potential for oxidation on cooling. Great care must be exercised in choosing the proper air/gas ratio for atmosphere production to avoid these problems. Very lean atmospheres (air/methane ratios of 7 or greater) require the removal or pronounced reduction of CO_2 and H_2O to prevent oxidation of iron above about 800 to 900° F.

COMMERCIAL FURNACE ATMOSPHERES

Commercial atmospheres are of various types, of varying purities. Control of water vapor and carbon dioxide, as already discussed, are critical to prevent oxidation and to maintain carbon potentials. These are normally controlled external to the basic atmosphere producing equipment. Water vapor can be reduced to a +40° F dewpoint by refrigeration, to a -50° F dewpoint by adsorbent tower dehydration. Lowering the dewpoint also lowers the carbon dioxide content. For example, in an endothermic atmosphere, lowering the dewpoint from a +70°F dewpoint to 0° F at the normal operating temperature of the generator, lowers the carbon dioxide content from 0.75% to 0.05%. Carbon dioxide can also be removed through high pressure scrubbing.

1. Hydrogen

    Hydrogen, owing to its excellent reducing potential at reasonable dewpoints, is an excellent furnace atmosphere for iron powder and parts. It can even be used to sinter stainless steels provided that the dewpoint of the atmosphere is maintained at about -40° F (2050° F). Care must be taken to prevent oxidation on cooling. It cannot be used for heating carbon-containing iron parts because of the marked decarburizing tendency of this atmosphere if water vapor is present. In practice however, the products from the powder or part reactions with the furnace atmosphere (primarily CO and CO_2) sometimes may stay in the furnace and establish a proper carbon potential to prevent decarburization.

    Hydrogen is normally produced in two ways: electrolysis and cracking of natural gas. The electrolytic process gives the highest purity. The only significant impurity is water vapor and it can be removed almost entirely by refrigeration, use of desiccants and molecular sieves; the latter will remove water vapor beyond the limits of normal detecting devices.

    Cracking of natural gas with steam gives hydrogen, carbon monoxide, carbon dioxide and water vapor as products. The carbon monoxide can be converted to carbon dioxide. The carbon dioxide and water vapor can be removed by means already discussed. The result is a reasonably pure Hydrogen (98%).

The cost of hydrogen is high. The equipment is expensive and, in the case of the electrolytic process, the cost of electricity is also considerable. Hydrogen is therefore limited to critical applications, e.g., stainless steel (where no nitriding is desired) and magnetic parts, except where local conditions (e.g., cheap electrical power) makes it competitive. Table I lists common commercial atmospheres and an estimate of the cost of each.

2. Dissociated Ammonia

A cheaper source of hydrogen is produced by heating anhydrous liquid ammonia in the presence of an iron or nickel catalyst.

$$2NH_3 \rightleftharpoons N_2 + 3H_2 \; . \tag{14}$$

The reaction is endothermic and therefore requires heat. The decomposition starts at about 600° F, increasing in rate as the temperature increases. Ammonia dissociators are normally operated at 1650 to 1800° F; the operating temperature is limited primarily because of equipment life considerations.

The gas produced by the dissociation is 75% hydrogen and 25% nitrogen. Dewpoints of less than -100° F are possible through careful operation of the equipment. The major impurity in this atmosphere is undissociated ammonia which, even in minute traces, is nitriding toward steel. This ammonia can be removed by bubbling the gas through water but, in doing so, the dewpoint of the gas becomes that of the water temperature.

TABLE I

COST OF COMMON COMMERCIAL ATMOSPHERES

Atmosphere	Cost[1]	Principal Cost Variable
Bottled Hydrogen Hydrogen[2]	$6-12/1000 ft^3 $0.86/1000 ft^3	Quantity Purchased Cost of Electricity
Dissociated Ammonia[3]	$1.50-5.00/1000 ft^3	Cost of Ammonia Quantity Manufactured
Exothermic	$0.15-0.20/1000 ft^3	Cost of Fuel Gas
Endothermic	$0.40-0.50/1000 ft^3	Cost of Fuel Gas
Exothermic-Endothermic	$0.35-0.45/1000 ft^3	Cost of Fuel Gas

1. Based on 1964 costs given by the American Society for Metals.

2. Cost based upon a generator capacity of 1000 ft^3/hr.

3. Cost for 1000 ft^3/hr generator is $2.69/1000 ft^3.
   Cost for 2000 ft^3/hr generator is $1.53/1000 ft^3.

If a low dewpoint gas is required, this water vapor must be removed. Ammonia in the gas is normally caused by operating the dissociator beyond its capacity or by variations of the dissociation temperature. Care in the operation of the equipment and proper maintenance will insure complete dissociation of the ammonia.

Dissociated ammonia is the most common hydrogen atmosphere used in furnaces. It costs about one-fifth of bottled hydrogen produced electrolytically and, for most applications, performs equally well.

Dissociated ammonia is also used to produce the so-called "burned" or combusted ammonia atmospheres. These are high nitrogen content (80 to 99+% N_2, balance H_2) atmospheres formed by burning normal dissociated ammonia. The reaction is very complete and less than one part per million of oxygen remains. The water vapor resulting from the burning is removed by dehydration.

Burned ammonia is an extremely pure atmosphere. It is also expensive and is therefore limited to applications in which a very pure inert (or slightly reducing) atmosphere is needed. In the P/M industry it is used in processing refractory metal powders. Normally, if a nitrogen-rich atmosphere is required, and carbon monoxide, carbon dioxide and methane impurities are not harmful, nitrogen-base atmospheres formed by burning mixtures of air and natural gas are used (see below).

3. Atmospheres Based Upon Combustion of Air-Natural Gas Mixtures

A wide variety of furnace atmospheres based upon mixtures of nitrogen, carbon monoxide, carbon dioxide, methane and water vapor are formed by the partial or full combustion of natural gas (or other hydrocarbon fuels such as propane, butane and kerosene). The degree of combustion is controlled by the amount of air entering into the process, normally called the air/gas ratio. Low ratios limit the combustion the most and result in a gas strongly reducing toward iron oxide. High ratios result in neutral or even oxidizing atmospheres. Table II lists some common furnace atmospheres, their composition, air-gas ratios and fuel requirements.

TABLE II

CHARACTERISTICS OF COMMON FURNACE ATMOSPHERES BASED UPON
AIR-NATURAL GAS COMBINATION

| Atmosphere | Nominal Composition | | | | Air/Gas[1] Rates | Fuel Req.[2] ft^3 of CH_4 |
	% N_2	% CO	% CO_2	% H_2		
Exothermic	86.8/71.5	1.5/10.5	10.5/5.0	1.2/12.5	9.0/6.0	120/155
Endothermic	45.1/39.8	19.6/20.7	0.4/0	34.6/38.7	2.6/2.5	190/200[3]
Exothermic-Endothermic	63.0/60.0	17.0/19.0	--	20.0/21.0	7.5/7.2	120/220

1. Based on methane (CH_4) as the fuel gas.

2. To produce 1000 ft^3 of atmosphere.

3. Requires 250 ft^3 of heating gas/1000 ft^3.

a. <u>Endothermic-Type Atmospheres</u> - These are manufactured by reacting a heated mixture of air and a hydrocarbon gas (air/gas ratio of approximately 2.5 for methane) in the presence of a nickel catalyst. The reaction can be approximated by:

$$2CH_4 + O_2 + 3.8N_2 \rightarrow 2CO + 4H_2 + 3.8N_2. \tag{15}$$

A pure methane gas would ideally result in 20.4% CO, 40.8% H_2 and 38.8 N_2 if complete combustion took place. In a "rich" endothermic gas this is close to being true. In practice, some methane, water vapor and carbon dioxide are present. This gas is strongly reducing toward iron oxide, even more so if the water vapor is further reduced. Control of the water vapor content (Figure 8), carbon dioxide content (Figure 9) or the methane content will control the carbon potential of the atmosphere.

If another hydrocarbon gas is used, the air/gas ratio must be varied accordingly to obtain complete combustion, i.e., to obtain a "rich" endothermic gas. For the case of propane:

$$2C_3H_8 + 3O_2 + 11.4N_2 \rightarrow 6CO + 8H_2 + 11.4N_2. \tag{16}$$

The air/gas ratio in this case is 7.2. If combustion is complete the gas composition using propane is 23.6% CO, 31.5% H_2 and 44.9% N_2, containing more carbon monoxide and nitrogen, and less hydrogen than if methane were used. The two atmospheres, although somewhat different in composition, do act similarly as furnace atmospheres.

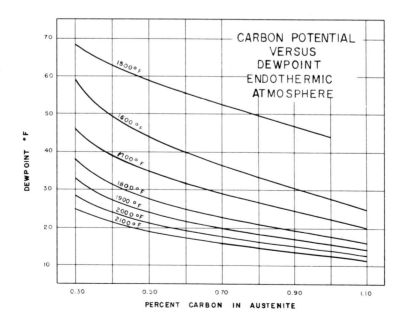

Figure 8 - Endothermic Atmosphere: Control of Carbon Potential By Water Vapor Control

Source: Technical Bulletin D174, Hoeganaes Corp., Dec 1972

Another type of endothermic gas, richer in hydrogen, can be prepared by reacting a hydrocarbon gas with steam:

$$CH_4 + H_2O \rightarrow CO + 3H_2. \tag{17}$$

This atmosphere would be used if a higher hydrogen content is desired, e.g., oxide reduction.

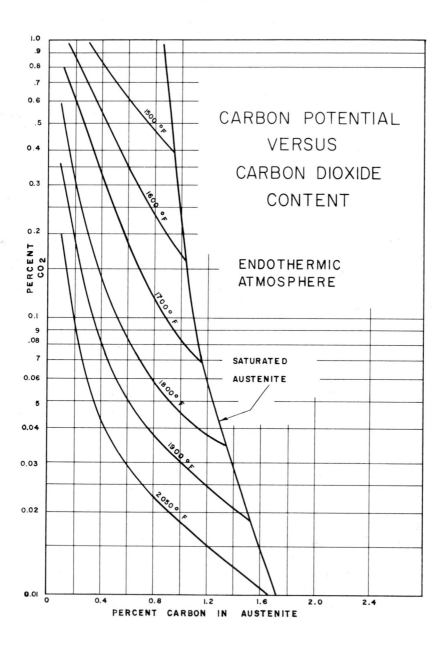

Figure 9 - Endothermic Atmosphere: Control of Carbon Potential By Carbon Dioxide Control

The production of endothermic-type atmospheres requires heat and becomes more efficient as the temperature is raised (thus the name endothermic). The generators are usually run at about 1900° F. Endothermic atmospheres are excellent for iron and steel products because they can offer protection from oxidation and hold useful carbon potentials at normal operating temperatures. They cannot be used if a non-carburizing atmosphere is desired, e.g., annealing of stainless steels which readily form chromium carbides.

b. <u>Exothermic-Type Atmospheres</u> - These atmospheres are prepared by the partial combustion of hydrocarbons in air (18). The reaction is self-supporting and generates heat. This heat, in the presence of a catalyst, will crack any of the hydrocarbon not burned into hydrogen and carbon monoxide (19). Two basic reactions take place:

$$CH_4 + 2O_2 \rightarrow 2H_2O + CO_2 + heat \qquad (18)$$

$$2CH_4 + O_2 + heat \rightarrow 4H_2 + 2CO \qquad (19)$$

The relative amounts of nitrogen, hydrogen, carbon dioxide, carbon monoxide and methane are determined by the air/gas ratio which, if methane is used as the fuel gas, ranges from 9 for a "lean" exothermic gas to 6 for a "rich" one. "Rich" means higher contents of carbon monoxide, hydrogen and methane, lower contents of nitrogen, carbon dioxide and water vapor.

The exothermic atmospheres are among the cheapest of the common furnace atmospheres. The generation equipment is cheap and easy to maintain. The nature of the atmosphere can be changed from rich to lean easily by simply varying the air/gas ratio. Although this atmosphere is not a strongly reducing atmosphere for iron, the water vapor content can be reduced low enough to prevent oxidation on cooling. The relatively high carbon dioxide contents makes high temperature oxidation a problem. Normally, this gas is strongly decarburizing and cannot be used when decarburization must be avoided. However, a rich exothermic atmosphere with a substantial amount of the water vapor and carbon dioxide removed can have a useful, controlled carbon potential. This atmosphere cannot be used for protecting metals which form more stable oxides than iron owing to its relatively high oxidizing potential.

c. <u>Exothermic-Endothermic Type Atmospheres</u> - This class of atmospheres is prepared by reforming an exothermic gas. The product is more reducing than exothermic gas, but less than endothermic gas. The exothermic-endothermic atmosphere can be used to replace exothermic and endothermic atmospheres in almost all their applications. It can hold a high enough carbon potential to be used as carburizing and carbonitriding atmospheres.

The reforming part of the production of this gas is simply a stage added after the exothermic gas has been manufactured and dried. The dried gas is heated and reacted with an additional amount of hydrocarbon gas in the presence of a catalyst to form carbon monoxide and hydrogen from the carbon dioxide:

$$CH_4 + CO_2 \rightarrow 2CO + 2H_2. \tag{20}$$

Although potentially useful and flexible (a generator may produce exothermic or exothermic-endothermic atmospheres), the exothermic-endothermic atmosphere requires more expensive generators and is more costly to manufacture than exothermic gas, although not as costly as the richer endothermic gases. Some installations have proved to be economical when the generators were modified to also produce an endothermic atmosphere.

4. Oxidizing Atmospheres

Oxidizing atmospheres are used in powder metallurgy for two purposes: lubricant burn-off and steam treating. The lubricant, which is mixed with the powder for better compaction, is volatile and will burn in air. It is desirable to perform burning before any sintering takes place to avoid rupturing the part and to keep the reaction products from contaminating the furnace. The lubricant burn-off is normally done between 600 and 900° F in any available oxidizing atmosphere, be it air or waste gas at the end of the furnace. The oxidizing reaction will sometimes form a superficial oxide on the powders. This oxide will be reduced during the sintering cycle. The thin layer of porous iron left on the particles after reduction can accelerate sintering and result in better mechanical properties of the part.

Steam treating is performed to form a thin, protective and tenacious oxide coating on a sintered part. This oxide reduces the porosity of the part, while it gives them increased corrosion resistance, wear resistance and compressive strength. Steam treating is done in the temperature range of 650 to 1200° F by introducing steam into the furnace once the temperature of the parts are above the boiling point of water. Rusting will occur if the temperature is below 212° F.

Another atmosphere limitation during steam treating is that the air in the furnace must be purged if the steam treating is done above 800° F. This will prevent the formation of non-protective oxides.

5. Vacuum

Vacuum, the absence of atmosphere, is also thought to be an inert atmosphere. The word "inert" has to be understood with caution: One metal's inert atmosphere is another metal's poison.

Although it has been said that <u>vacuum</u> is the ideal inert atmosphere, it is hardly inert in the sense of nothing happening to material when placed in a vacuum. Desorption of gases takes place; evaporation takes place (the more perfect the vacuum, the more likely will evaporation be); oxidation or reduction can take place.

Evaporation is a serious problem when using metals at elevated temperatures in a vacuum. All materials evaporate; the practical concern is the case when there is net evaporation, i.e., when there is a net loss of material. When the pressure acting on a material becomes less than the vapor pressure of the material, net evaporation takes place. For a pure material, the vapor pressure is defined as that pressure at which the gas phase exists in equilibrium with solid (or liquid). It is a function of temperature, increasing with increasing temperature.

Metals and alloys, which can be heated in a protective atmosphere under normal pressure without any danger of vaporization or loss of alloying elements by vaporization, can vaporize when heated to the same temperatures in a vacuum.

Some of the more common elements are listed below in an approximate order of their susceptibility to evaporation when heating in vacuum. The rating is based upon whether the element <u>can evaporate before it can melt at the particular pressures listed.</u> The melting point of each at one atmosphere pressure is given in parenthesis. One torr is equivalent to 1 mm or 10^3 microns of Hg pressure and to 0.132×10^{-2} atmospheres.

<u>Element</u>

Increasing Susceptibility to Vaporization →

Element	Vaporization
Chromium (3430° F)	
Boron (3650–4750° F)	will vaporize below <u>10 torr</u>
Magnesium (1202° F)	will vaporize below <u>1 torr</u>
Lead (621° F)	
Zinc (787° F)	will vaporize below <u>10^{-1} torr</u>
Cadmium (321° F)	
Manganese (2273° F)	
Iron (2802° F)	will vaporize below <u>10^{-2} torr</u>
Molybdenum (4760° F)	
Titanium (3120–3480° F)	
Beryllium (2340° F)	
Nickel (2651° F)	
Copper (1981° F)	will vaporize below <u>10^{-3} torr</u>
Silver (1761° F)	
Platinum (3224° F)	will vaporize below <u>10^{-4} torr</u>
Gold (1945° F)	will vaporize below <u>10^{-5} torr</u>
Aluminum (1220° F)	will vaporize below <u>10^{-6} torr</u>

Source: Technical Bulletin D174, Hoeganaes Corp., Dec 1972

Vacuums of 10^{-5} torr are within the limits of commercial vacuum furnaces. The following pumping equipment is used:

<u>Rotary Pumps</u> to 10^{-1} torr
<u>Mechanical Booster Pumps</u> to 10^{-2} torr
<u>Oil Diffusion Booster Pumps</u> 10^{-2} to 10^{-3} torr
<u>Fractionating Oil Diffusion Pumps</u> 10^{-3} to 10^{-5} torr

The limit for this equipment is between 10^{-6} to 10^{-7} torr. It is possible, therefore, to vaporize almost all of the above listed elements in commercial equipment without exceeding their melting point (provided the temperature can be approached using the equipment). Chromium, lead, zinc and magnesium are troublesome elements to handle when heating in vacuum.

Vacuum furnaces can protect many metals from oxidation and are therefore useful. The effective dewpoint at 10^{-3} torr is -105° F, -130° F at 10^{-4} torr. At this level of dewpoint, the vacuum would be protective to most metals at elevated temperatures (above 1200 to 1500° F).

6. Inert Atmospheres

Argon, helium and nitrogen are usually considered to be inert atmospheres. Only argon and helium are inert in the sense they do not react chemically. Nitrogen does, although the stability of the N_2 molecule normally makes it unreactive.

Argon is made by selective evaporation of liquid air, helium by selective evaporation of liquid natural gas. Their commercial purity is typically 99.9%. The impurity gases must be low enough to prevent undesired reactions between these gases and the material in the furnace. Argon and helium are rarely used as furnace atmospheres because of their high cost, although they are used for heat treating special grades of stainless steel and titanium when it is critical to avoid gas reactions. The major application of argon and helium is their use as the protective atmospheres in shielded arc welding.

<u>CONTROL OF FURNACE ATMOSPHERES</u>

The subject of control of furnace atmospheres has already been introduced by showing that the properties of an atmosphere are a function of the concentration of its components. Control of the water vapor content of hydrogen (Figure 2) or the carbon dioxide content of carbon monoxide (Figure 3) will control the oxidizing or reducing potential of the gas. Control of water vapor content (Figures 7 and 8), carbon dioxide content (Figures 4 and 9) or methane content (Figure 5) will determine the carbon potential. Control of a furnace atmosphere has four parts:

Determining which component gas to control
Obtaining a sample of the furnace atmosphere
Analyzing the furnace atmosphere
Feed-back

1. Determination of Control Gas Component

The choice of which component gas to control is determined once the purpose of the operation and the most economical furnace atmosphere to perform it is determined. Some choice can be made, e.g., both carbon dioxide and water vapor content can control the carbon potential in many atmospheres. The choice between them many times depends upon the kinds of analytical and control equipment available and the level of concentration of the component gas.

2. Sampling The Furnace Atmosphere

Obtaining a representative furnace atmosphere sample at the most meaningful point of the process is a critical step in controlling the properties of the atmosphere. The necessity for a representative sample speaks for itself, although it should be mentioned that the sampling process should introduce no impurities into the sample. The sample must be taken at a point where changes in the gas analysis will reflect changes in the material in the furnace. For example, it would serve no useful purpose to measure the composition of stagnant gas in the furnace because it would not contact the material in the furnace, nor would it reflect changes in the gas that is flowing through the furnace.

A more subtle aspect of the location of the sample taking equipment is concerned with control of the critical step in the process. When using hydrogen to protect against oxidation, the dewpoint of the gas is critical. But water vapor becomes more oxidizing as the temperature is decreased, being limited only when the temperature is low enough to slow the rate to prevent a significant oxidation from taking place. A dewpoint of +100° F will not oxidize iron at 1800° F, yet at 200° F a dewpoint of −10° F will. The most desirable place to control the dewpoint would be at the point where the temperature is just high enough to permit significant oxidation. The furnace atmosphere will not be oxidizing in a practical sense if the dewpoint here is held just below that necessary for oxidation.

The same sort of reasoning holds for carbon control. The carbon potential of a fixed gas composition is a function of temperature. If, for example, the carbon potential of an endothermic atmosphere is neutral to a steel at the maximum furnace temperature, it will be strongly carburizing when the steel is cooled. If the carbon potential were held at some point during the cooling cycle to just permit carburization to the desired level, then a uniform carbon content can be held in the steel. The gas composition will decarburize at the maximum furnace temperature, carburize back again on cooling. The carburization reaction, once past this point in the cooling cycle, would be too slow to affect the surface composition of the steel. Control at this critical point in cooling will put the controls where they are most needed and since the carbon potential can be maintained more exactly at lower temperatures, the accuracy of the carbon control is better.

3. Furnace Atmosphere Analysis

There is a number of different kinds of devices and equipment made for furnace atmosphere analysis. The choice of which to use depends upon what component is to be measured, the concentration of the component, the time required to perform the measurement and what the measurement is to be used for. The analysis can be as simple as finding the approximate composition to insure burning, making corrections by turning a coarse valve, or as complicated as requiring an accurate analysis in parts per million immediately for a complex, continuous, automatic feed-back control system. The devices and equipment chosen, obviously, depend upon the controls necessary for the proper conduct of the operation.

a. Orsat Analysis - This method of analysis has been around for years and basically works by bubbling the atmosphere sequentially through a series of solutions which selectively absorb one of the gaseous species at a time. Simple units measure oxygen and carbon dioxide contents, more complex ones will measure oxygen, carbon monoxide, carbon dioxide, hydrogen, methane, ethane and nitrogen (by difference).

Orsat analysis is cheap and very simple. Its accuracy is sufficient to coarsely control a furnace atmosphere, i.e., check for massive furnace leaks or generator breakdowns. It takes a long time to perform (30 minutes for a complete analysis) and cannot easily be made part of an automatic control system. In addition Orsat analysis cannot measure small (less than 1%) concentrations accurately and therefore cannot be used for accurate carbon potential control.

b. Gas Chromatography - Orsat analysis is gradually being replaced by gas chromatography because it is faster, can measure almost any gas species and can do it more accurately. Like the Orsat analysis it is a discontinuous analytical method, i.e., a sample is taken and then the sample is analyzed; continuous measurements of gaseous species cannot be done.

The chromatograph operates by separating the gaseous species by adsorption or partition. The thermal conductivity or ionization characteristics of each species is measured after separation. The conductivity and potential of each species is unique to it and the intensity can be calibrated to give the concentration of the species. The gas must be dried before separation and the water vapor content must be done by another operation. Gas chromatography can measure gaseous species in concentrations as low as parts per million.

Although gas chromatography is intermittent, it can be used as part of any automatic control system. The primary disadvantages of this analytical method are its inability to give continuous analysis and, therefore, continuous control, and the fact that the gas chromagraph is a complicated device. Skill, knowledge and experience are required to operate, calibrate and maintain the equipment and to interpret the raw data from it.

c. <u>Dew Point Measurement</u> - There are several methods of dewpoint measurement, some of which can be used as part of an automatic control system. Automatic systems based on dewpoint measurements, although relatively cheap to purchase, are extremely difficult to maintain. They have to be kept free from dirt and the sampling lines must be free from any moisture. Constant maintenance and additional equipment are required to do this. Dewpoint apparatus are difficult to calibrate and need additional calibration if the hydrogen content of the atmosphere varies.

The Dew Cup, Fog Chamber and Chilled Metal or Mirror methods are all based upon the phenomenon that, when air is saturated with water vapor, the water vapor will condense. The temperature at which the condensation takes place is, by definition, the dewpoint of the atmosphere - a direct measurement of the concentration of water vapor in the atmosphere.

The <u>Dew Cup method</u> requires skill and, since it does depend on the skill of the operator, it is not useful for the careful, consistent analysis needed for close atmosphere control. The method simply consists of passing the furnace atmosphere across a polished cup filled with acetone. The cup is cooled by placing dry ice in it and when the moisture in the atmosphere condenses on the cup, the dewpoint is reached.

The <u>Fog Chamber</u> is a portable instrument, simple to use and is based on the principle that, at a constant temperature during the adiabatic expansion of a gas, the pressure change to cause condensation is related to the water vapor content of the gas. The fog chamber instrument subjects the gas sample to varying adiabatic pressure changes. The pressure that just causes visible condensation in the fog chamber is recorded. The dew point is found by using a series of curves (for different ambient temperatures) relating dew point to pressure change. The fog chamber can measure a wide range of dewpoints, from below $-100°$ F to ambient temperature. It, like the dew cup method is discontinuous and therefore not suitable for automatic control.

The <u>Chilled Mirror</u> technique uses a photoelectric cell to record the intensity of light from an illuminated mirror placed in a sample of the furnace atmosphere. The temperature of the mirror is regulated by heating and cooling devices. The amount of water condensed on the mirror affects the reflectivity of the mirror; the photoelectric cell records this intensity and can feed back information to the heating and cooling devices to change the temperature of the mirror. When the reflected light intensity is that associated with the onset of condensation, the temperature of the mirror is stablized and is a direct measurement of the dew point. This technique is continuous and can be used for continuous automatic control of atmosphere dew point.

Source: Technical Bulletin D174, Hoeganaes Corp., Dec 1972

Since the chilled mirror technique depends upon the reflectivity of a polished surface, it is critical to insure that any changes in reflectivity are owing to water condensation or evaporation, not to the deposition of contaminants. In practice, contamination is a major problem and special precautions must be taken to avoid contamination of the mirror surface, a difficult task when natural gas base atmospheres are used.

The Chilled Metal technique is based upon the completion of an electrical circuit between two platinum electrodes. The electrical circuit is open when the temperature of the electrodes is above the dewpoint. When the electrode temperature reaches that of the dewpoint, the water vapor condenses from the atmosphere and closes the electrical circuit. This temperature is recorded and the dewpoint is measured. The electrodes must be heated to evaporate the water before another dewpoint measurement can be taken.

Another method that is suitable for continuous automatic dewpoint control is based on the hygroscopic nature of Lithium Chloride. A saturated solution of lithium chloride in water is placed in a sample of the furnace atmosphere and heated to a temperature at which the evaporation potential of the solution just equals the water absorbing potential of the lithium chloride. This point can be measured by changes in the electrical characteristics of the solution. An electrical signal is fed back to the heating and cooling circuit to permit the solution temperature to remain at the dewpoint when the dewpoint of the atmosphere the solution sees is changing. This continuous methods is increasing in popularity, especially for use in automatic dewpoint control devices. Its main drawback is that the solution is contaminated by ammonia.

d. Infrared Absorption Analysis

Many gas molecules absorb infrared radiation. When they do, each different molecule absorbs characteristic wave lengths of radiation; the amount of radiation absorbed is directly related to the amount of that molecule in the atmosphere being tested. Infrared analysis uses this characteristic to analyze gaseous constituents.

Infrared analysis can measure most of the gaseous species in furnace atmospheres, e.g. carbon monoxide, carbon dioxide, ammonia, methane and water vapor. It cannot, however, measure the elemental gases (e.g. hydrogen and oxygen) because they do not absorb radiation in the infrared range. Infrared absorption can measure small concentrations accurately (less than 0.1%), do it precisely ($\pm$1%) and quickly (in a few seconds).

In practice, infrared analysis equipment is of two types: Negative filtering and positive filtering. The two methods both measure the amount of any particular constituent by differences in the infrared absorption between the sample gas and a non-absorbing reference standard gas. Both methods use a detector gas composed of only the particular constituent being analyzed, e.g. pure carbon monoxide detector gas if measuring the carbon monoxide content of the sample gas. However, the two methods use different techniques to measure the differences.

The positive filtering method measures the difference in heating of the detector gas when it is subjected to the infrared source radiation after it passes through the standard gas and after it passes through the sample gas. Since no radiation characteristic of the constituent is absorbed by the standard gas, the source radiation that passes through the standard gas will heat the detector gas more than when it passes through the sample gas. The difference in heat energy is a direct measure of the amount of constituent gas in the sample gas. The negative filtering method works on the same general principles, except that it measures the difference in intensity of the source radiation after it passes through the standard gas and after it passes through the sample gas. In this case, the difference in intensity of the wave lengths characteristic of the constituent gas is a direct measure of the amount of constituent gas in the sample gas.

Infrared analysis is an excellent method to monitor and control furnace atmospheres. If only a single constituent is measured, the method can give continuous control. If several constituents are measured, each analysis must use a different detector gas; the change of detector gas can be done in less than a minute. The method is sensitive, stable and accurate, ideal for furnace control. It is typically used to control the carbon potential by measuring carbon monoxide, carbon dioxide and natural gas contents. Infrared analysis equipment is expensive and relatively complicated. The services of a skilled technician are required to maintain and repair them. It is difficult to recognize when one is not operating correctly and constant care must be given the analysis and control equipment to be sure it does operate correctly.

e. Other Methods of Analysis

Other Methods of gas analysis exist for special applications. The Hot-Wire Analyzer offers a continuous method of measuring the carbon potential of furnace atmospheres by measuring the change in resistance of a fine steel wire caused by any change in its carbon content while in the furnace atmosphere. It is quite suitable for use as part of a continuous control system. The hot-wire method can be used at temperatures ranging from 1450 to 1860° F for carbon contents of 0.15 to 1.15% and will give an accuracy of $\pm0.05\%$ in the carbon analysis. The use of this method is limited because the steel wires are very fragile and the measuring instruments are very sensitive to the sort of contaminents associated with commercial furnace practice.

Source: Technical Bulletin D174, Hoeganaes Corp., Dec 1972

The Flame-Temperature Analyzer measures the temperature of endothermic gas flames under carefully controlled conditions of burning. The temperature of the flame is directly related to the dew
point of the gas and, thus, is a direct measure of the dew point.
Since the temperature is measured with a thermocouple, this method
can be made into a control system. It requires frequent recali-
bration because the flame temperature is also sensitive to the
composition of the endothermic gas.

Mass Spectroscopy is used for analyzing the constituents present
in a vacuum. It can measure the amount of the constituent and
also identify what it is. It is not limited to the species it
can measure, although low molecular weight molecules (below 50)
are difficult to distinguish from one another. The analysis is
rapid and very accurate for low concentrations.

There are several kinds of mass spectrometers; the two most common
are the deflection and time-of-flight. The deflection type separate the different species by their deflection in a constant magnetic field; the lighter the species, the greater the deflection.
The time-of-flight separates the species by the different times it
will take them to travel a fixed distance under a constant accelerating force; species of low molecular weight travel faster.

Mass spectrometers are expensive, very complex and the data they produce are very difficult to interpret. They require a very high level
of technical ability to maintain them and to use them for analysis.
They perform a unique function and are necessary when vacuum analysis
is required.

Nothing has been said about the object of all this analysis: The composition of the product. Many methods are available for analyzing the
material when it comes out of the furnace and the choice depends on what
methods are convenient and what kind of property is important. For example, the metallographic structure is important for steel because it
is the combined carbon and the degree of uniformity of the structure
that must be controlled; therefore, metallographic analysis is required.
Other analytical methods include all the variations of chemical analysis,
hardness measurements, tensile strength measurements, magnetic or electrical properties, etc. The property desired in the product must be related to the furnace variables if control over these properties is desired
by exercising control over the furnace variables.

3. Feed-Back: Completing the Central Circuit

Furnace atmosphere control requires that something is done with the information gathered by correctly choosing the proper species to control,
by carefully sampling the atmosphere to insure that it is representative
of the process and by performing an accurate analysis to know how much
of the species is present in the sample. Using this information, in
the context of the properties of the material coming out of the furnace,
to control the concentration of the species in the furnace atmosphere
closes the feed-back control cycle.

The methods of doing this are too numerous to discuss in a general review of this sort. The devices are as plentiful as the different kinds of analytical equipment. The numbers being limited only by the ingenuity of the persons using the equipment. They can range from manual regulation to highly sophisticated electronic control systems. The important point is to use the control information in the most effective manner consistent with the object of the operation. Changes in the furnace atmosphere should be made at the generator in a manner dictated by the extent of control required by the operation itself.

THE RELATIVE STABILITY OF OXIDES

All metals, except gold, form stable oxides in air at normal temperatures and pressures. The relative stability of the oxides is reflected in the ease or difficulty in reducing the oxides, or in protecting the metals from oxidation by using furnace atmospheres. The science of thermodynamics permits an exact description of the stability of metal oxides in terms of practical variables, i.e., temperature, pressure and gas composition.

The measure of stability used by thermodynamics is called the free energy and the free energy difference between reactants and products for a chemical reaction is a measure of the tendency for the reaction to take place. This free energy difference, or free energy change (ΔG) can be measured experimentally. For the reaction:

$$Me + O_2 \rightarrow MeO_2, \tag{21}$$

if the free energy change is negative, the oxide is more stable than the metal and oxygen; if positive, the metal and oxygen is more stable: the oxide will not form. If the free energy change is zero, all three will exist together in equilibrium.

The standard free energy change (standard, in this case, meaning pure metal and pure oxide) for reaction (21) can be written as:

$$\Delta G° (T) = RT \log_e P_{O_2} (eq), \text{ where} \tag{22}$$

$\Delta G° (T)$ is the standard free energy change for the reaction; (T) means that it is a function of temperature

R is the gas constant = 1.987 calories/degree Kelvin/gram mole

T is the absolute temperature in degress Kelvin

P_{O_2} (eq) is the equilibrium partial pressure oxygen for the reaction at temperature T.

If P_{O_2} (eq) = 1, or an atmosphere of pure oxygen is required for equilibrium at a specific temperature, $\Delta G° (T) = 0$ and metal, oxide and the pure oxygen atmosphere co-exist. This is the limiting case for a stable oxide, for if

Source: Technical Bulletin D174, Hoeganaes Corp., Dec 1972

$\Delta G°$ (T) is positive using pure oxygen, the oxide will not form under the given conditions of temperature and pressure. The smaller the value of the oxygen partial pressure required to achieve equilibrium, the more stable the oxide is (the more negative the free energy change).

In the same manner, the free energy change for the formation of a metal oxide can be described in terms of the H_2/H_2O and CO/CO_2 ratios.

$$Me + H_2O \rightarrow H_2 + MeO \tag{23}$$

$$\Delta G° \text{ (T)} = -RT \log_e (P_{h_2}/P_{h_2o}) \text{ eq, or}$$

$$\Delta G° \text{ (T)} = -RT \log_e (H_2/H_2O) \text{ eq} \tag{24}$$

$$Me + CO_2 \rightarrow CO + MeO \tag{25}$$

$$\Delta G° \text{ (T)} = -RT \log_e (P_{co}/P_{co_2}) \text{ eq, or}$$

$$\Delta G° \text{ (T)} = -RT \log_e (CO/CO_2) \text{ eq} \tag{26}$$

A plot of $\Delta G°$ (T) versus temperature for various metal-metal oxide reactions will give, visually, the relative stability of the metal oxides as a function of temperature. Since $\Delta G°$ (T) is a direct function of P_{O_2} (eq), (H_2/H_2O) eq and (CO/CO_2) eq for the metal-metal oxide reactions, the plot will also give these as a function of temperature for each reaction.

This plot is available and is called the <u>Richardson Free Energy Chart</u> (Figure 10). It can be used to rapidly estimate the relative stability of many oxides, the equilibrium partial pressure of oxygen and the equilibrium H_2/H_2O and CO/CO_2 ratios for the metal-metal oxide reactions, all over a wide range of temperatures. This information is critical for control of furnace atmospheres and those persons working with furnace atmospheres should become skillful in the use of the chart.

1. <u>Relative Stability of Oxides</u>

Each line on the Richardson Chart represents the variation of $\Delta G°$ (T) for a particular metal-metal oxide reaction with temperature. The lower down on the chart the line is, the more stable will be the oxide. For example, at 500° C the oxides have the following order of stability:

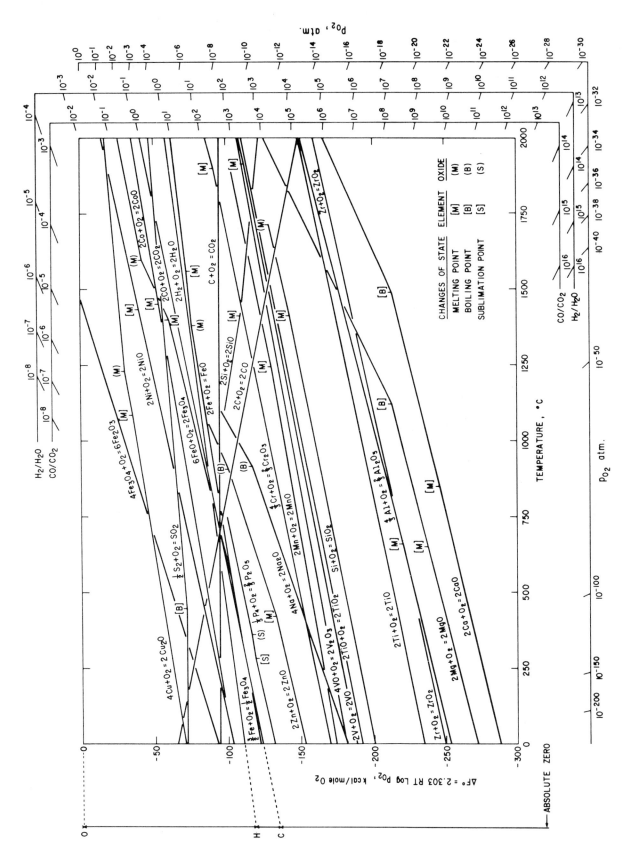

Fig. 10 – The free energies of formation of oxides for the standard states: pure elements, pure oxides and gases at 1 atm.

$$Cu_2O$$
$$Fe_2O_3$$
$$NiO$$
$$CoO$$
$$SnO_2$$
$$Fe_3O_4$$
$$P_2O_5$$
$$Cr_2O_3$$
$$MnO$$
$$V_2O_3$$
$$SiO_2$$
$$TiO_2$$
$$Al_2O_3$$
$$UO_2$$
$$MgO$$
$$CaO$$

When two lines cross, as they sometimes do, the two metal oxides are equivalently stable at the temperature of intersection. Since the stability of each oxide varies differently with temperature, the relative stability of the oxides will vary with temperature.

2. Equilibrium Partial Pressure of Oxygen

The equilibrium partial pressure of oxygen for each metal-metal oxide equilibrium can be determined at any temperature by drawing a straight line between two points and intersecting this line with the (P_{O_2} atm) scale. The point of intersection with this scale, which runs along the bottom and right hand side of the Richardson Chart, is the equilibrium partial pressure of oxygen. The two points that determine the first line are the "0" point at the upper left-hand side of the chart and the point of intersections of the temperature and the individual metal-metal oxide free energy line.

For example: at 1000° C the standard free energy change for the Ni-NiO reaction is -62 kilo calories/gram mole. A line drawn through this point on the Ni-NiO line and the "0" point will intersect the (P_{O_2} atm) scale at 10^{-10} atmospheres. At this partial pressure of oxygen and at 1000° C, both nickel and nickel oxide coexist. At smaller partial pressures only pure nickel will exist at 1000° C; at larger partial pressures, only the oxide will exist. As the temperature increases, the equilibirum partial pressure of oxygen increases.

3. Equilibrium H_2/H_2O Ratio

The equilibrium H_2/H_2O ratio is found in a manner similar to that used to find the equilibrium partial pressure of oxygen. A line is drawn from the "H" point (below the "0" point) through the point of intersection of the temperature and the individual metal-metal oxide free energy line until it meets the (H_2/H_2O) ratio scale. The scale is just to the left of the (P_{O_2} atm) scale. The point of intersection of this line with the (H_2/H_2O) scale will give the equilibrium H_2/H_2O ratio.

For example: At 1000° C, the point of intersection for the Ni-NiO reaction with the (H_2/H_2O) ratio scale is approximately 1×10^{-2}. At 1000° C, therefore, the equilibrium H_2/H_2O ratio is 1×10^{-2}. For ratios greater than this at 1000° C, only pure nickel can exist; for smaller ratios, only the oxide can exist. Therefore, to anneal nickel in hydrogen at 1000° C, the H_2/H_2O ratio of the gas must be greater than 1×10^{-2} to prevent oxidation. At higher annealing temperatures, smaller H_2/H_2O ratios can be used (higher dew points - see Figure 1).

4. <u>Equilibrium CO/CO_2 Ratios</u>

The equilibrium CO/CO_2 ratio is the intersection of the line drawn through the "C" point (just below the "H" point) and the point of intersection of the temperature and the individual metal-metal oxide free energy line with the (CO/CO_2) ratio scale. The scale is just to the right of the (H_2/H_2O) scale.

For example: For Ni-NiO at 1000° C, the point of intersection, and therefore the equilibrium CO/CO_2 ratio, is approximately 1×10^{-2}. Nickel will oxidize at 1000° C in a carbon monoxide atmosphere if the ratio is smaller than this. NiO will reduce in a carbon monoxide atmosphere at 1000° C at greater CO/CO_2 ratios. The equilibrium CO/CO_2 ratio decreases with increasing temperature.

5. <u>Limitations of the Richardson Free Energy Chart</u>

The Richardson Free Energy Chart was constructed using certain assumptions and these assumptions should be understood by those who intend to use it. The principles that such charts as these are based upon are fundamental to metallurgy and any serious practitioner should become familiar with them. Any textbook of physical chemistry or physical metallurgy will give this information and also give a more thorough description of the use of charts like the Richardson Chart.

a. The Richardson Chart is based upon thermodynamics, not kinetics. In other words, the chart will only describe what will happen, not how long it will take. Hydrogen and oxygen will react to form water explosively at room temperature. They may never do so if the reaction is not started by an external means.

b. The free energy changes are those for the reactants and products in their standard states, i.e., pure metal and pure oxide. The oxidation of alloying elements in solution in steel would require the free energy to be corrected for deviations from the standard state of the elements. The chart does give a valuable estimate quickly, and that is its great value.

c. The chart assumes that the oxides are compounds having a fixed composition. This is not true for many oxides, e.g. FeO.

d. The chart does not consider the possibility that intermetallic compounds may be formed between the reactants and products.

e. And, finally, there is no consideration given to the possibility that the reactants and products may be distributed among different phases, e.g. in solution in solid or liquid phases.

SECTION VI:
Powder Mixing and Blending

PROBLEMS OF POWDER MIXING AND BLENDING

By Henry H. Hausner and John K. Beddow

INTRODUCTION

There are more problems involved in mixing of two or more powders than are usually recognized.

Powder mixing is a complex process involving a great number of variables. In the following, some of these variables and their effects on the degree of mixing will be discussed briefly. The bibliography will then give more information and will, hopefully, stimulate further studies in the mixing process.

The term "mixing" of powders refers to "the thorough intermingling of powders of two or more materials." The term "blending" refers to "the thorough intermingling of powders of the same composition."* In the literature concerning the science and technology of powders, both terms are frequently used interchangeably. In the following short discussion of the problems, the term "mixing" will be used exclusively. For the characterization of the mixture the terms "mixedness" or "degree of mixedness" are frequently used.

"MIXEDNESS"

An ideal mix, i.e., one in which all the particles of each material are distributed uniformly among all the particles of each other material in the mix, is impossible to achieve. The laws of nature insist upon a certain amount of non-uniformity even if the powders were mixed under the best possible conditions for an infinite period of time. This degree of mixing, i.e., that occurring under the best possible conditions at an infinite time, is a useful concept because it provides a point of reference to describe how well any mix is mixed compared to the best degree of mixing that can be really obtained. The degree of approach or "mixedness" (M) uses this concept. When a mix is completely non-uniform, $M = O$; when the mix is equivalent to that mixed for infinite time, $M = 1$. For mixes in between the two extremes, $0 < M < 1$.

MIXER VARIABLES

Table 1 and Fig. 1 list and show a variety of different types of mixing equipment. A multitude of different mixers or blenders is available from the industry. The selection of the best type of mixer for a given powder is still extremely difficult and the question "which is the best mixer?" can by no means be answered in a general way. Tests must be made in each case.

TABLE 2
VARIABLES IN THE MIXING PROCESS

1. Type of mixer (see Table 1)
2. Volume of the mixer V_m
3. Actual dimensions of mixer
4. Interior surface of mixer A
5. Volume of powder in the mixer V_p
6. Volume ratio of the component powders V_{p1}, V_{p2}--

7. Characteristics of the powders (see Table 3)
8. Rotational speed of mixer
9. Mixing temperature
10. Mixing time
11. Mixing medium (gaseous or liquid)

*ASTM designation B243 (1958)
"Standard Definitions of Terms used in Powder Metallurgy"

TABLE 1
VARIOUS TYPES OF POWDER MIXERS
(Types A thru J illustrated Fig. 1)

A Cylindrical mixer: horizontal rotating
B Cylindrical mixer: inclining axis
C Hexagonal or octagonal mixer
D Rotating cube
E Double cone
F Pyramid mixer
G Twin shell tumbler mixer
H Y-Cone mixer

I Mixer with internal baffles
J Mixer with helical plates
K Vibratory mixers
L Ribbon blender
M Paddle blender
N Vertical rotating auger
O Continuous mixers

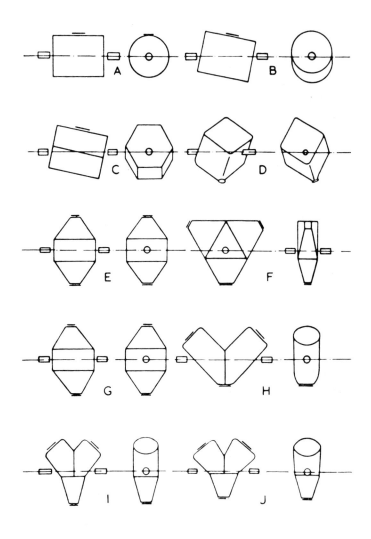

Fig. 1 Diagrammatic outlines of various powder mixers.

Source: Technical Report BM173, Hoeganaes Corp., Dec 1970

There are many variables involved in mixing. Table 2 shows some of the most important variables and the mixing process. Mixers can be characterized by their volume. However, two cylindrical mixers of identical volume will perform entirely differently when their length to diameter ratios differ. Although they may be of identical volume, the interior surface area of the two mixers will be different, thereby changing the frictional effect and the movement of the particles will therefore be quite different. It is the interior surface of the mixer which gives movement to the powder mass. No final conclusion about the behavior of a production model of a mixer can be drawn from experimental work with a small scale counterpart because the interior surfaces of the large and small models are different. In spite of the great importance of the ratio mixer volume to interior surface in the mixing process, hardly any investigation concerning this effect can be found in the literature.

SIZE OF LOAD

For every type and size of mixer the optimum powder charge will vary. The ratio of the mixer volume to the volume of the powder is of great importance as shown in a schematic way in Fig. 2.

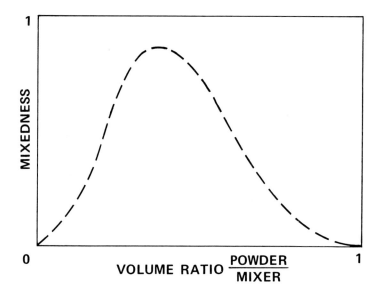

Fig. 2 Effect of the ratio: powder volume to mixer volume on mixedness.

There is always an optimum amount of powder in a given mixer — below and above this optimum amount the mixedness is inferior. This optimum amount however, depends on the type of powders to be mixed and on many other factors which affect the frictional behavior of the powder.

POWDER CHARACTERISTICS

The powders to be mixed may differ in many respects. Table 3 lists the most important characteristics of a powder. If the powders differ very much in density, segregation of the heavier powder may occur because gravitational forces may be stronger than the frictional forces. A change in the type of mixer and in the mixing procedure may sometime be helpful to avoid this segregation.

Powders of identical materials and identical particle size and particle size distribution, may mix in entirely different ways when the particle shape differs. Metal powders with oxidized surfaces flow faster and mix differently than identical metal powders with pure metallic surfaces. All these differences in the mixing behavior are due to the different friction conditions in the powders.

TABLE 3
CHARACTERISTICS OF A POWDER PARTICLE

A. Material characteristics
    1) Structure
    2) Theoretical density
    3) Melting point
    4) Plasticity
    5) Elasticity
    6) Purity (impurities)

B. Characteristics due to the process of fabrication
    1) Density (porosity)
    2) Particle size (particle diameter)
    3) Particle shape
    4) Particle surface area
    5) Surface conditions
    6) Microstructure (crystal grain structure)
    7) Type and amount of lattice defects
    8) Gas content within a particle
    9) Adsorbed gas layer
   10) Amount of surface oxide
   11) Reactivity

TABLE 3a
CHARACTERISTICS OF A MASS OF POWDER

    1) Particle characteristics (see Table 3)
    2) Average particle size
    3) Particle size distribution
    4) Average particle shape
    5) Particle shape distribution
    6) Specific surface (surface area per 1 gram)
    7) Apparent density
    8) Tap density
    9) Flow of the powder
   10) Friction conditions between the particles

FRICTION

Friction occurs between the particles in a powder mass. This friction and the manner of mixing determine the movement of powder particles. Studies of friction in a powder mass are a necessity for a better understanding of mixing problems.

During mixing, friction occurs between the powder particles and between the powder and the mixer wall. Friction causes an increase in temperature. If this increase in temperature occurs in an oxygen containing atmosphere, and if some of the powders to be mixed oxidize, the mixed powders will contain particles with an oxide film. This is highly undesirable in some cases, especially when the oxide film affects further processing of the mixed powder.

If friction between powder particles is a factor of prime importance as claimed above, the temperature during mixing should also play an important part in the process because the friction coefficient between most materials increases with increasing temperature. The flow of a powder is not as good at elevated temperature as it is at a lower temperature. If it should be desirable to improve the flow or movement of powder particles during mixing, a lowering of the mixing temperature will also lower friction and may in some cases improve the mixing.

ROTATIONAL SPEED

The rotational speed of a mixer greatly affects the mixedness. Increasing the speed up to a certain point is useful in shortening the mixing time. However, there is for practically all types of mixers and powders, an optimal speed. Above this speed, the mixedness decreases because the rotational forces are then too strong and the powders will not move away from the wall of the mixer.

MIXING TIME

The optimum mixing time can hardly be determined in advance without some testing and, at the present time, it is almost impossible to draw any intelligent conclusions from the optimum time for one powder to that of another powder in the same mixer, as long as our knowledge of the friction in a powder mass is still so limited.

Mixing time strongly affects the quality of the mix. Some powders are well mixed in some equipment after a relatively short time whereas, when using other powders and/or other equipment, the mixing time has to be considerably prolonged to achieve the same degree of mixedness, and in other cases a prolonged mixing time results in an even lower degree of mixedness. This is shown in Fig. 3a, b, c. Theoretical considerations do not as yet permit predictions of the optimal mixing time.

Figs. 3a, 3b and 3c show
  Mixedness vs. mixing time for
  three different types of mixers.

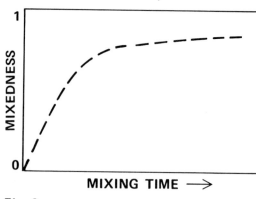

Fig. 3a

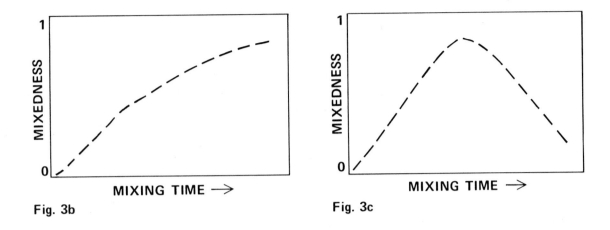

Fig. 3b

Fig. 3c

Fig. 4 shows the correlations between mixing time and mixedness for a mixer operated at various speeds from 16 rpm to 80 rpm, and Table 4 shows that in this case the total number of revolutions decreases with the increasing mixing speed. This shown for a degree of mixedness of 0.4 and 0.6 respectively.

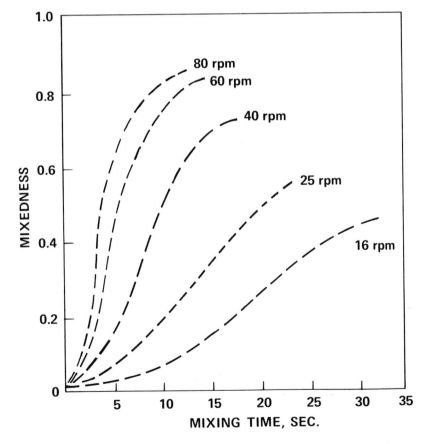

Fig. 4 Degree of mixing vs. mixing time for five speeds of rotation.

Source: Technical Report BM173, Hoeganaes Corp., Dec 1970

TABLE 4
NUMBER OF REVOLUTIONS OF THE MIXER
TO OBTAIN IDENTICAL DEGREES OF MIXING

Degree of mixing = 0.4		
Mixing speed rpm	Mixing Time min.	Total number of revolutions
80	3	240
60	4	240
40	8	320
25	17	425
16	30	480
Degree of mixing = 0.6		
80	5	400
60	7	420
40	12	480

Material: red and white sand
Mixer: horizontally rotating cylinder

ATMOSPHERE IN MIXERS

Powders can be mixed in air or in other gaseous atmosphere, or in a liquid. Mixing in a liquid usually results in a better mix because of the lubricating action of the liquid. When mixing in a liquid is not permissible, the selection of the gas atmosphere should be carefully evaluated. The powder may react with the gas, changing the frictional characteristics thus affecting the degree of mixing.

POWER CONSUMPTION

The power consumption for the mixing process depends on the size of the mixer, the amount of powder to be moved in the mixer, the type of powder as well as on the rotational speed of the equipment. This is shown in Fig. 5 for 3 mixers of identical design but different diameters (a > b > c) which are filled with approximately 50% of limestone powders of approximately 1 mm particle size. With increasing rotational speed, the power consumption goes up. However, when the speed is high enough, all particles tend to rotate with the inner wall of the mixer, the power consumption goes down but no mixing occurs.

POWDER MOVEMENT

The powders are in many cases in a continuous rolling movement during which a kind of diffusion occurs when the particles from one powder diffuse into the mass of another powder (micromixing). An additional movement of the powder occurs when the powder mass is separated into parts such as occur in double cone or Y-type mixers (macromixing). A similar effect of additional movement occurs also in the cylindrical mixers with baffle plates, which divide

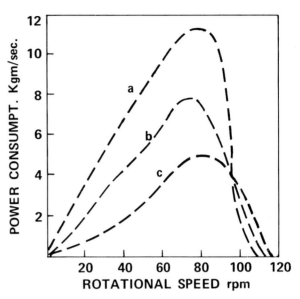

Fig. 5 Power consumption vs. rotational speed cylindrical mixers of various diameters.

Diameters a = 35 cm, b = 30 cm, c = 25 cm
(50% of mixer filled with powder)

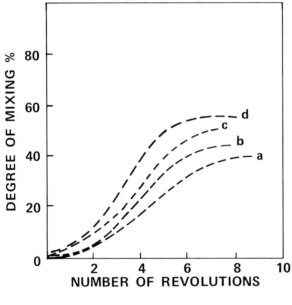

Fig. 6 Degree of mixing vs. number of revolutions for various conditions of baffling.
a. no baffle plates
b. height of baffle plate = 0.66R
c. height of baffle plates = 0.5R
d. height of baffle plates = 0.33R

Radius of cylindrical mixer = R
Rotational speed = 25 rpm

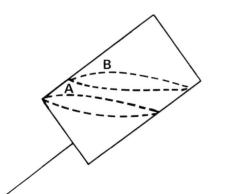

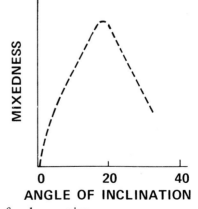

Fig. 7 Mixedness vs. angle of inclination of a drum mixer.

Source: Technical Report BM173, Hoeganaes Corp., Dec 1970

the powder mass into several parts of different particle movements. The ratio of the height of the baffle plates in relation to the height of the powder to the diameter of the rotating cylinder determines to some extent the rate of mixing. This is shown in Fig. 6 for a cylindrical mixer rotated at 25 rpm and containing particles of approximately 1 mm in diameter. The height of the baffle plates varies between 0.33 to 0.66 R, where R is the radius of the cylinder. Baffles in the mixer can shorten the mixing time and improve the degree of mixedness.

Horizontally rotating cylindrical mixers are frequently used to improve mixing. The axis of some mixers may be inclined. The example shown in Fig. 7 indicates that the mixedness depends on this angle of inclination and that there is an optimum angle under which the best mixture can be obtained. This optimum angle varies with the dimensions of the mixer, with the rotating speed and with the type of powders to be mixed; it depends on the friction in the powder. The authors of some experimental work using such a mixer rotating at an angle, state also that filling the mixer to position A will work very well, whereas a larger amount of powder (position B) would hardly mix at all.

MIXING PROBLEMS

There are many special problems involved in the mixing process: some of them are listed in Table 5.

TABLE 5
PROBLEMS OF POWDER MIXING

1. Filling the powder into the mixer
2. Determination of optimum amount of powder
3. Changes in particle size distribution during mixing
    a. Grinding action, resulting in finer particles
    b. Agglomeration, resulting in coarser particles
4. Changes in particle surfaces (oxidation)
5. Abrasive action, resulting in the addition of impurities
6. Determination of optimum mixing time
7. Extraction of mixed powder from the mixer
8. Problems of sampling
9. Evaluation of mixedness

The determination of the optimum amount of powder in the mixer depends on the type and dimensions of the mixer, but to some extent also on the characteristics of the powder.

One has to keep in mind further that the particle shape of the powders may change during mixing. A kind of abrasive self-grinding may occur especially in brittle particles, and result in a finer powder after mixing. However, the opposite may olso occur and the very fine powders may form agglomerates of larger sizes during mixing. One never should take it for granted that the particle size distributions before and after mixing are identical. Another kind of abrasive action may take place between the powder and the inner surface of the mixer whereby fine particles which break off the wall surface are added to the powder in form of finely divided impurities, an effect which sometimes results in a different behavior of the powder in further processing.

DISCHARGING

The problems of extraction of the mixed powders from the mixer are also quite unsolved. It can happen that during extraction of the mix, segregation or some other kind of de-mixing may occur and that tests may reveal results which are not due to the mixing procedure but to the extraction procedure and are therefore not conclusive. In any evaluation of mixing equipment, the problems of extraction play a major role. In close connection with the extraction problems are also those of sampling.

MIXING OF METAL POWDERS

In the following, examples are given to indicate how much a higher or lower degree of mixing may affect the behavior of metal powders in further processing. The mixing of metal powders with lubricant powders is widely applied in powder metallurgy and the optimal mixing procedure depends on the type of lubricant — all the other mixing variables being equivalent.

Fig. 8 shows the effect of mixing time on the apparent density of 316L stainless steel powders plus 1% of various types of lubricants. Maximum apparent densities can be obtained by mixing the 316L powder for 45 to 60 minutes with an amide wax or some stearates. Mixing the same alloy powder with stearic acid results in maximum apparent densities when mixed for only 15 minutes, after which time the apparent density of the mix again decreases. Moreover, the flow of the five powder mixes shown in Fig. 8 varies with the mixing time.

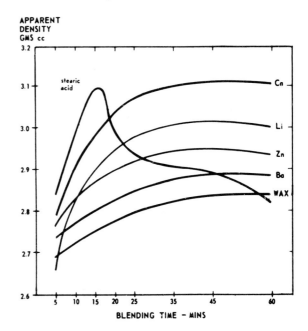

Fig. 8 Apparent density of 316L - stainless steel powder mixed with 1% of various lubricants as affected by mixing time.

Fig. 9 indicates how much the apparent density of these mixtures depend on the amount and type of the lubricant to be mixed with the alloy powder and on the mixing time.

The effect of mixing time on the pressure required to eject the compact from the die shown in Fig. 10.

Source: Technical Report BM173, Hoeganaes Corp., Dec 1970

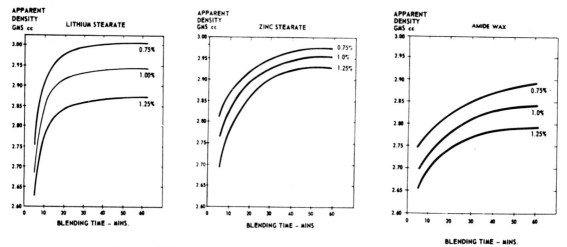

Fig. 9 Effect of mixing time on the apparent densities of
316 L - stainless steel powder mixed with 3 amounts
of 3 lubricants.

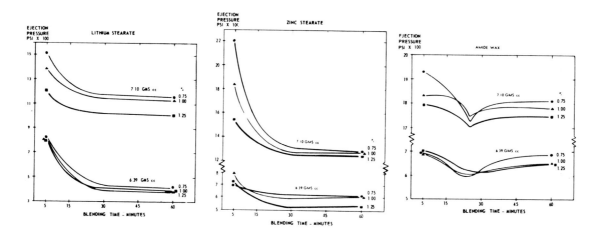

Fig. 10 Effect of mixing time on the ejection pressure
of compacted 316L - stainless steel powder
mixed with 3 amounts of 3 lubricants.

The type and amount of lubricant and the time for mixing the alloy powder with the lubricant both affect the green strength of the ejected part as shown in Fig. 11. In an indirect way the mixing time also affects the properties of the sintered part.

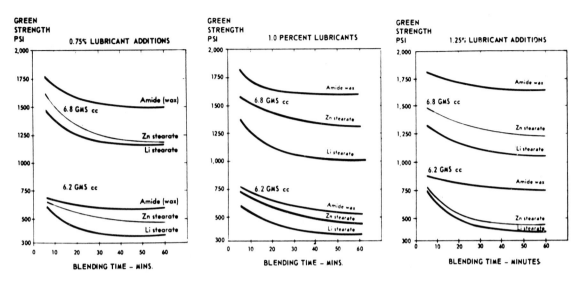

Fig. 11 Effect of mixing time on the green strength of 316L - stainless steel mixed with 3 amounts of 3 lubricants and compacted to two different green densities.

SUMMARY

Although the above refers to just a few problems of powder mixing, it has been shown that the mixing process is more complex and complicated than normally assumed. It has been shown further that satisfactory mixing conditions can be obtained by modifications (sometimes minor) of the process or the equipment.

Source: Technical Report BM173, Hoeganaes Corp., Dec 1970

SECTION VII:
Tooling for P/M

P/M—Profitable Tooling for Powdered Metal Parts

Profit or loss in P/M depends on the tool material selected. Selection, in turn, depends on tool design, tool construction, and subsequent processing—the subjects of this article

ROBERT KUNKEL

Ford Motor Company

TOOLS, USED IN THE PRODUCTION of powdered metal parts, perform a variety of operations, such as compacting, coining, sizing, and hot forming. In the early days of the powdered metal industry, when only bronze powders were compacted, little was required of tooling, other than high wear resistance. With the introduction of iron and steel powders—and, more recently, refractory metal and superalloy powders—tool materials are required to have the additional feature of high i m p a c t resistance. Additionally, there is an increasing demand for high density compacts, a requirement that imposes even heavier loads on the compacting tools. And, with recent activity in the

hot forming of powdered metal parts, wear resistance at elevated temperatures has been added to the growing list of tooling requirements.

Wear Resistance or Impact Strength

The characteristics of wear resistance and high impact strength are difficult to find in one material. Usually, a tool material has one or the other—but not both.

Powdered metals are by their nature abrasive. Moreover, high pressures—often in excess of 35 tons per square inch —are required to cold bond the powder particles. Some powders, such as brass and bronze, compact at the relatively low

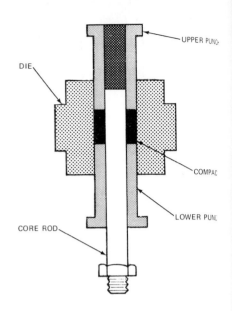

1. COMPACTING tool set used to form a bushing from powdered metals.

◀ *TOOLS REQUIRED to produce the gear compact shown in foreground. The tools are upper and lower punches, core rod, and die with carbide insert.*

pressures of 20 to 25 tons per square inch, and are not very abrasive. In contrast, powders such as iron and nickel alloys require compacting pressures of 35 to 50 tons per square inch, and are highly abrasive. Ceramic powders that require compacting pressure of 6 to 20 tons per square inch are extremely abrasive.

Selecting a Tool Material

Because the range of powder characteristics is broad the selection of tool materials is a complex engineering decision based upon a thorough examination of the application. There are many instances in which a compromise choice must be made. Either wear or impact resistance must sometimes be sacrificed to select an optimum material for the application. For example, if a highly abrasive powder, such as tungsten carbide, is to be compacted, a tool material having high wear resistance must be selected. This means that impact strength must be sacrificed to obtain the more desirable feature of high wear resistance. In other applications, a high impact strength material might be selected at the sacrifice of wear resistance.

In no instance, however, should performance reliability be sacrificed. Material cost is a factor that must be considered, but more important is the tool's performance and reliability. Thus, the tool engi-

neer must also analyze the tool design to determine load deflection curves, while giving due consideration to wear resistance and impact strength.

The final choice of a tool material is also influenced by the P/M processes by which the part will be made. A compacted part which will be subsequently cold or hot formed can be pressed at lower pressures in the briquetting stage. This means that there will be lower tool loads and less abrasive wear during this operation. At low compacting pressures, there is also less danger of galling from powder particles. In this case, an abrasion-resistant tool material can be selected without sacrificing tool reliability. Selection of this type of material may also result in lower tool costs.

A part that is to be compacted only, w i t h no subsequent operations performed, requires tooling capable of withstanding high compacting pressures. In this case, since the final density must be attained during compacting, wear resistance must be sacrificed to obtain greater impact strength.

Coining o p e r a t i o n s impose high compressive loading on the tooling, because of the high pressures required to cold form a part. There is relatively little wear on the tooling because the powders are sintered prior to coining. Because they are not loose, as in the compacting

operations, they cannot become lodged in the tool clearances to cause wear. Coining tool materials are usually selected for their high impact strength. Since tool loss is due more to breakage than to wear in coining, superior impact resistance becomes an even more important factor.

Sizing operations require lower pressures than those used for compacting or coining. However, the operation is comparatively abrasive, especially if the ironing method of sizing is used. In sizing, wear resistance is the most important factor in tool material selection.

When tool materials are selected for hot forming, it must be remembered that the high temperatures involved adversely affect impact strength and wear resistance. Since no material possesses the ideal combination of high wear resistance and high impact strength, various design techniques are used to minimize the effects of wear and impact on tools. Floating core arrangements are perhaps the most common devices to lessen tool wear. However, a detailed discussion of the various devices used is beyond the scope of this article.

Tool Construction

Not only do the methods used to actuate the tools affect the selection of tool materials; so do the methods of tool construc-

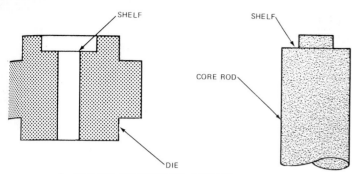

2. *STRESS CONCENTRATIONS occur at tooling shelves used to form levels in the compact.*

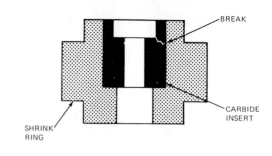

3. *INSUFFICIENT SUPPORT of the die insert will lead to breakage at the tooling shelves.*

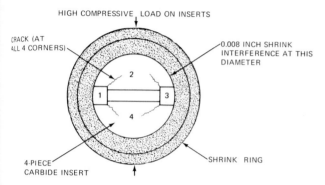

4. *CARBIDE DIE INSERTS cracked by the high stresses set up as the shrink ring cools.*

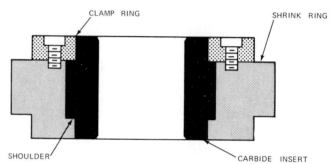

5. *CLAMP RING die construction used to prevent overstressing of the carbide insert.*

tion. The effects of three elements—the tool actuating mechanisms, tool construction, and tool use—are inseparable when considering a tool material. Each of the elements interact to affect the selection of tool materials.

Die Materials

A simple compacting tool set, used to form a compact having the configuration of a bushing, is shown in *Figure 1*. The die, which forms the outside diameter of the compact, is subjected to considerable wear, since the compact walls are in intimate contact with the die. Therefore, wear resistance is a vital factor in selecting a die material. However, if compact levels are formed on tool shelves, as shown in *Figure 2*, consideration must also be given to the material's ability to withstand deflection, and to its impact resistance.

The most common method of die construction is that of placing a die insert in a shrink ring adapter. The shrink ring supports the insert at compacting loads of up to 50 tons per square inch. In many cases, the shrink ring can be designed to carry the bulk of this load. The die insert can then be made of a highly wear-resistant material, because the impact load will be borne by the shrink ring.

Carbides are the obvious first choice as die insert materials. Although carbides are brittle, their wear resistance is high. In applications in which steel dies produce 200,000 compacts, a carbide die may produce 2,000,000.

Although the composition of carbides can be altered slightly to improve their impact strength, suitable design of the shrink ring is the most important factor in establishing the amount of support given to the die insert. Where design limitations restrict the amount of support that can be given, the die insert should be made of a wear-resistant steel. In some instances, pressing on a shelf die, *Figure 3*, causes breakage of the carbide because of insufficient support beneath the shelf. In such cases, the use of D2, M2, D7, M43, or T15 steels as the insert material provides the necessary impact resistance.

Impact resistance of carbide can be increased by increasing the cobalt in the basic material. However, as impact resistance increases, wear resistance declines. D2 is less wear resistant than D7, but its impact resistance is greater. In designs in which the die insert can be adequately supported by the shrink ring adapter, a carbide insert should be used. If there is insufficient support, a steel insert should be used. All grades of carbide exhibit practically the same wear resistance in compacting operations. Steels vary considerably in their wear resist-

ance, however. The wear resistance increases in the steels previously listed from D2 low to T15 high. When resistance to impact is more important than resistance to wear, a steel insert should be used. The relative importance of these two characteristics should then be evaluated to determine the most suitable steel for the application.

In all applications, die inserts made from wear-resistant, brittle materials must be supported by shrink rings. The greater the support, the more reliable the tool in withstanding the loads applied. In forging applications, it is not uncommon to preload the case to its ultimate applied load in order to adequately support the die insert. This same practice can be followed when developing compacting tools.

A case history illustrates the importance of carefully considering all factors when choosing compacting tool materials. A fabricator recently ordered a rectangular cavity die comprising four carbide inserts held in place with a shrink ring, as shown in *Figure* 4. The amount of shrink interference was left to the tool maker's discretion. In this instance, an interference of 0.002 inch per inch was used. Since the insert diameter was four inches, the total interference amounted to 0.008 inch. Carbide was chosen as the die insert material because of the requirement for

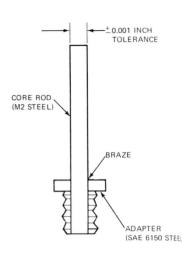

6. *CORE ROD made of an M2 steel blank brazed to a low alloy adapter.*

HIGH-TONNAGE PRESS used to produce large, multilevel compacts. These compacts, made from powdered iron, require abrasion-resistant tooling.

high wear resistance.

Unfortunately, the stresses created by the shrink ring on cooling caused both of the large insert sections to crack. The first thought was to make the insert from a solid block of carbide, and cut the cavity by electrical discharge machining. However, it was realized that there was no guarantee that a solid carbide insert wouldn't crack also—particularly in view of the fact that carbide does not flex as readily as steel. A second approach is to change the die insert material to steel. In this case, M43 or T15 steel would be suitable. If a carbide insert is mandatory, die construction can be changed so that the inserts are held in place with clamp rings and shoulders, as shown in *Figure 5*. In this design, a small press or shrink interference would be used to hold the inserts laterally. The shoulders and clamp ring prevent vertical movement. In this type of die, steel inserts can also be used.

Core Materials

Cores are subject to the same abrasive wear as dies, but unfortunately they're much more fragile, hence careful selection of core materials is essential.

If a core rod diameter is small in proportion to the surface in contact with the compact wall, a tool material that will

not separate under tension loads must be used. Carbide would be a poor material choice for such a core rod. Inexpensive core rods can be made from M2 steel blanks brazed into a low alloy steel (SAE 6150) adapter, as shown in *Figure 6*. M2 blanks finish ground to ± 0.0001 inch tolerance and having a high quality surface finish are available for these parts. Since M2 rod is readily available, other steels offer few advantages. The cost to produce a core from M43 or T15 would be three or four times the cost of making it from brazed M2 rod.

Although M2 steel does not have the wear characteristics of carbide, several M2 cores can be made for the price of one carbide core. Thus, the cost of the two types tends to equalize. The same type of construction can also be used with carbide rods. However, since the carbide has poor resistance to tension loads, it is necessary to use a free-floating core mechanism to prevent core breakage during ejection. When a free-floating mechanism is used, another satisfactory method of core rod construction is butt brazing a carbide tip onto a steel shank.

When cores are made of solid M2 or D2 steel, surface coatings are sometimes applied to increase the rod's wear resistance. Nitriding or hard-chrome plating increases wear resistance up to ten times.

Since hard chrome can be renewed, the same core rod can be reused many times. Nitriding is applicable only on core rods that are free of sharp edges. Surface strains set up in the tool steel by nitriding cause sharp edges to pit or flake.

Carbide flame plating can be used to provide a highly wear-resistant surface on a tough base material, such as SAE 6150 steel. Additionally, recent advances have been made with processes such as Chromalizing. This is a combination plating and heat treating process in which hard chrome is diffused into a base core material. This process improves wear resistance, and minimizes, or eliminates entirely, the danger of flaking or chipping.

Punch Materials

The factor of impact resistance is more important than wear resistance in selecting punch materials. However, because loose powder particles wedge into the tool clearances, wear resistance cannot be completely ignored. Thus the ideal material in punch fabrication is one that offers high impact resistance without too great a sacrifice in wear. Latrobe Steel's Kon-cor is suitable in this application because it has wear characteristics similar to M2 plus twice the impact strength.

Good punch design requires the cor-

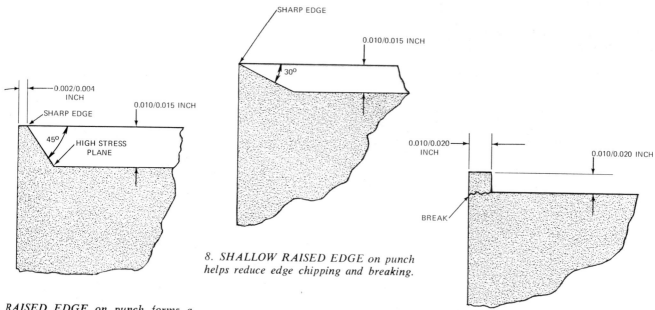

RAISED EDGE on punch forms a tap to eliminate burrs on the compact.

8. SHALLOW RAISED EDGE on punch helps reduce edge chipping and breaking.

9. HIGH STRESS concentrations in this design cause excessive breaking and chipping.

ners to have generous radii. Heels should be thick as possible with a large bearing surface to evenly distribute heavy loads. Stress risers at the heel rim edges should be relieved.

Many fabricators favor L6, H13, or A2 steel as punch materials. These steels are selected primarily for their ability to resist impact.

If the face of the punch has projections, they are subject to the same wear conditions as dies. In these instances, more wear-resistant steels—M2 or D2—are better choices as punch materials. Carbide faces cannot be used on punches, as carbide cannot withstand the high forces encountered.

Tool clearances between punches and dies are dimensioned for optimum operating conditions. Since these clearances are necessary for the proper functioning of the tool set, there is no way to completely eliminate the burr formed by powder compacting into the clearance. Fabricators often specify raised edges on punches to provide a burr trap, as shown in *Figure* 7. The purpose of these edges is to prevent the burr from rising above the part face to cause operating interference of the part when it is assembled. The edges are natural stress risers on the punch, and punch maintenance is high because the sharp knifelike edges break

and chip after a number of cycles. This problem can often be alleviated by good tool design. *Figure* 8 shows a redesigned chamfer which wears to unusable limits before it chips or breaks.

Recently , a fabricator had a problem with the rectangular chamfer design in *Figure* 9. The stresses encountered were so great that chamfer breakage occurred at least once in every eight-hour shift. Tool rework and replacement costs averaged $13,000 per year. without adding the additional costs of press downtime and setup time. The end item specification for the burr trap a l l o w e d a 0.010/0.020 x 0.010/0.020 inch envelope. The geometry of the chamfer could be anywhere in this rectangular boundary. A simple change in design to the chamfer type shown in *Figure* 8 eliminated the problem of breakage. After this redesign, an inspection program was initiated to prevent production of undersized parts because of worn burr traps.

Adapter Material

The adapters used to secure the tooling elements to the press are generally made from low alloy steels with good fatigue properties. SAE 6150 steel, heat treated to a hardness of R_C 42-45, is the most commonly used steel for this purpose. In

instances where hard wear plates are used, they should be totally enclosed and supported to protect personnel against flying fragments if they should break.

Deflection is perhaps the most significant parameter to consider in selecting a tool steel for adapters. Since steel, whether hard or soft, deflects the same amount under unit loads, steel selection is secondary to adapter design. The adapter should be designed to have as large a cross section as possible in order to support the applied load with minimum deflection. It should also have an appreciable amount of hardness to raise the level of load at which permanent set occurs in the material. However, the hardness must be kept low enough to prevent danger to the operator should overload stresses occur.

The design and construction of the tooling used for the production of powder metal parts is as important as the tool material selected. In line with this, there are two rules to follow when designing tooling for powdered metals. First, work closely with tool suppliers—the people who can often provide answers to difficult problems. Second, don't be afraid to experiment with the new materials that are continually being developed. In many cases, they can provide simple—and inexpensive—answers to your more difficult tooling problems. ◄

SECTION VIII:
Impregnation

Fig. 1-Parts to be impregnated are loaded into a simple dip rack which is lowered into the vacuum tank. The rack is lifted out of the tank to remove impregnated parts.

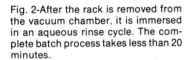

Fig. 2-After the rack is removed from the vacuum chamber, it is immersed in an aqueous rinse cycle. The complete batch process takes less than 20 minutes.

Impregnation expands P/M applications

Vacuum impregnated P/M parts are opening up major new application opportunities for P/M parts.

Impregnated P/M components can be successfully plated and painted. They can be used in high pressure hydraulic and pneumatic applications, and they can be more easily machined.

These benefits are being realized from a new vacuum impregnation process using anaerobic resins. The method is far more effective than impregnation systems which use polyester resins or steam oxiding, according to Loctite Corp., developer of the vacuum process.

Cures without air. Anaerobic resins produced by Loctite have excellent locking, sealing, retaining and bonding properties. The same characteristic that makes the resins valuable for these functions, that they cure in the absence of air, is important to P/M impregnation.

In the vacuum system, parts (figure 1) are dipped into a chamber, air

is removed, and the ultra-thin resin fills the natural porosity of the powdered metal parts (see figure 2).

Parts are then dipped into rinses and set aside to cure. The total process can be completed in as little as 20 minutes per batch. Although vacuum equipment designed specifically for anaerobic impregnation is available from Loctite, vacuum equipment built for other resins can be easily converted to the anaerobic material.

Pressure tightness. The greatest potential for impregnated P/M parts is for pressure-tight components such as valves, pumps, meters, compressors, brake pistons and hydraulic systems (see figure 3).

One major P/M fabricator which was quick to realize the potential for pressure applications is Burgess-Norton Mfg. Co. It is impregnating a number of parts including a hy-

draulic end plate for John Deere; a hydraulic motor valve plate for Eaton's Char-Lynn Div.; and hydraulic parts for Century Engineering. In each instance, impregnation removed porosity from parts which otherwise had the structural strength of steel. According to Doug Davis of Burgess-Norton, prior to anaerobic impregnation parts such as the John Deere hydraulic end plate required 23 grams of sealant per part at a cost of $.01 per gram. The former process resulted in a high percentage of "leakers" at 2,500 psi (175.8 kg/cm²), but with anaerobics there have been zero defects.

Machinability. Parts for guns, micrometers and compressors are currently being impregnated because of the resulting improvement in machinability. Sinterbond Corp. experience is typical. According to

Edward Enos, "We increased tool life by switching from steam oxided parts to anaerobic impregnation. Steam oxiding hardens part surfaces, making machining difficult. Impregnation seals the parts without affecting hardness, prolonging tool life up to ten times." The process was particularly useful on a Carlyle Compressor Co. valve plate (see figure 4) which had to be both pressure tight and machined to make a groove for an O-ring.

Machinability is increased by eliminating the chattering that develops as the tool jumps across pore openings. Although the resin does not replace machining oils, it does help lubricate the machining process.

Surface finishing. P/M parts now being impregnated to prepare them for plating, bluing and painting include gun components, pole pieces, decorative automotive parts and outboard motor parts. Filling the parts with hardened resin prevents entrapment of other fluids which would later leach out and ruin surface finishes. Related benefits include improved structural strength, and internal corrosion is virtually eliminated.

Gun parts to be blued are being impregnated by Olin Corp. According to Dick Mathews of that company, such parts include gun sights, triggers and trigger guards.

Applications. Although iron-based components are the most common form of P/M parts being impregnated, the process is equally effective with aluminum, brass, bronze and stainless steel.

Anaerobic impregnation offers P/M fabricators more parts applications and the possibility of subcontracting impregnation with inhouse equipment, since vacuum impregnation is equally useful for sand and die castings. In addition, experience has shown that the equipment does not require skilled operators. **PM**

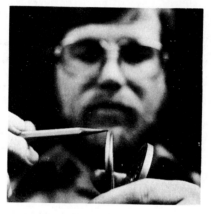

Fig. 3-Typical impregnated P/M parts. Burgess-Norton hydraulic components (left) are pressure tight. Three Sinterbond parts (upper right) are both pressure tight and easier to machine. Impregnated Olin Corp. parts (lower right) are more easily blued.

Fig. 4-Tools which machine the groove in the O.D. of this P/M part last 10 times longer on impregnated parts than they did on steam oxided parts.

GUIDELINES FOR IMPREGNATING P/M PARTS

1 — Plan for a density range between 80% — 90% theoretical, or 6.2 to 7.0 g/cc for iron parts. This is the best range for maximum penetration without bleeding.

2 — Heat treat prior to vacuum sealing. Anaerobic sealants are limited to upper temperatures of 400F (204.4C). Also, remove quenching oils prior to impregnation by baking, annealing or vapor degreasing.

3 — The optimum time for sealing is immediately after sintering. Clean, open pores aid penetration. Tumbling, burnishing and machining tend to smear surfaces and block the sealant entry. Also, fluids used in these operations can penetrate the pores and inhibit impregnation.

4 — P/M parts can be coined, sized and repressed after vacuum sealing. Volume changes of up to 2% can be tolerated without difficulty.

SECTION IX:
Sizing - Coining

PRESSING AND SIZING FACTORS AFFECTING THE PRECISION AND MECHANICAL CHARACTERISTICS OF SINTERED PARTS*

M. Eudier†

All stresses, and particularly those due to friction, which occur during pressing and sizing operations have consequences on the dimensions, on internal stresses, and eventually on micro-cracks. This friction can have an influence on the variations of the density from one point to another but, when used in a proper way, it can help the movement of punches and increase homogeneity. Three different types of re-pressing are considered—drawing, sizing, and coining—and examples are given to show their influence on mechanical properties and especially on the fatigue limit.

THE purpose of this paper is to show the influence of pressing and repressing parameters on the quality of the finished products. Certain defects are mentioned which, happily, are not often encountered in small parts, but are of greater importance in large parts and also in parts which have to work under fatigue conditions.

This paper in certain respects is a continuation of the study of compaction techniques published in this journal.[1]

Powder Mixes

Apart from the ideas that we have already expressed on the evacuation of air during the pressing operation it must be said that for large parts this problem is very important even when using such a very common powder as the NC100.24 Höganäs powder mixed with a stearate. With a mechanical press running at 600 parts/h, when the upper punch enters the die it often happens that a cloud of stearate is observed all around the punch, showing the high speed of evacuation of the air. Though the compacts seem to be sound, one can see in the

* Manuscript received 11 November 1974. Contribution to a Symposium on 'Factors Affecting the Uses of PM Products', held in Eastbourne on 28–30 October 1974.
† Société Metafram, Beauchamp, France.

finished products preferential alignments of porosities which are not cracks but which reduce the mechanical properties. The usual remedy is to slow down the press, and especially to reduce the speed of the press at the beginning of compression. There are other possibilities since the main parameter for the evacuation of air is the size of the porosities between grains. This size is proportional to the value of the mean grain size given by Fisher's apparatus which measures the diameter of the holes and not the particle diameter. What is important is the variation of the pore size as a function of the particle size. The flow of a gas according to Poiseuille's law is proportional to r^4 ($r=$ radius of pore) for one pore. As the number of the pores at the surface is proportional to r^2, the permeability for a given compact is also proportional to r^2. This means that by doubling the size of the particles the permeability is multiplied by 4. Hence for large parts it is good practice to use a coarser powder, which can be made by removing the finer particles of the powder. A better solution is to use a comparatively old method of lubrication which was used in Germany and then by Husqvarna and consists in mixing 10% stearic acid with 10% of the powder at a temperature above the melting point of the acid and afterwards mixing this to the remainder of the powder. This method is especially useful when finer powders like nickel are used. Then, all the fine powder must be mixed with the molten stearic acid. By this method agglomerates of the finer powders with the basic powder are made which have the proper size.

Lubricants other than stearic acid may be used if they melt before decomposition.

Compression

As has been explained previously,[1] the manufacture of a cup (Fig. 1) is one of the best examples of the difficulties which are encountered during compression. Apart from elastic considerations

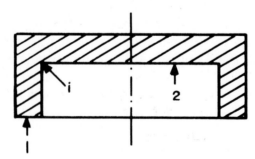

FIG. 1. Compacted part in the shape of a small cup (schematic). The arrows indicate the action of two punches.

which show theoretically that the part cannot be produced without internal stresses, let us consider how it can be produced at least without cracks. The idea is that at point i there will always be the least possible displacement of powder. The part is composed first of a disc, the centre of which is pressed between the upper punch and a central lower punch 2 (Fig. 1), and secondly of a ring which is pressed between the outer lower punch 1 and the upper punch.

If we suppose that punch 2 and the die are immobile, the pressure in i by the upper punch is, as a first approximation, equal to the pressure received by the upper surface of punch 2 (and equal to the pressure applied by the upper punch if the disc is thin). If the same pressure is applied to punch 1, the pressure by this punch at the level of i will be much lower because of the friction forces between punch 2 and die. Thus there will be a rapid gradient of pressure just below the level of i which will cause a crack. The first idea is to increase the pressure on punch 1 but this pressure can be very high if the ring has a thin wall, so that the density at the lower part of the ring approaches the maximum possible. We can also try to diminish friction on the walls of the ring. This is not possible with punch 2, since the displacement between the two punches is fixed by the depth of fill. The only possibility is to move the die upwards during compression. The maximum effect is obtained when the die goes up simultaneously with punch 1. If it moved faster the result would be no better, since the dynamic coefficient of friction is not higher than the static one.

When using a press which has no lower moving member the same results can be obtained by having the die and punch 1 immobile during compression and allowing punch 2 to descend. The pressure under punch 2 must start from zero and increase as a function of the displacement of the punch. The value of this pressure (and especially its maximum) must depend on the friction of the lateral periphery of the punch against the powder of the ring. As an example, for the part shown in Fig. 2, which is pressed under a load of 400 N/mm², the final force, without considering the friction on punch 2, would be 3 MN (300 tons). When friction with a coefficient of 0·12 is taken into account, 0·8 MN (80 tons) is the maximum value. In practical tests the force was increased linearly after the movement of punch 2 from 0·2 tons to the maximum value and the experiment showed that the parts were satisfactory when the maximum applied loads were 0·3–1 MN (30–100 tons).

Another solution, which is satisfactory for the production of small quantities, is to sand-blast the periphery of punch 2 for a height slightly less than the height of the compacted part. This causes an increase of the friction coefficient from 0·12 to 0·15 for a CLA

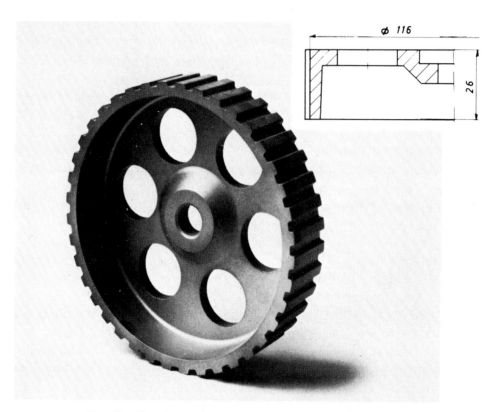

FIG. 2.   Specimen of part pressed under 400 N/mm² load.

roughness of 0·3 μm, and there is no more need for any pressure under punch *2*. Even when operating with a satisfactory pressure, the results are better if the punch is depolished in the vicinity of point *i*.

After compression, when the die is withdrawn downwards, the elastic springback of the compact is such that it no longer touches punch *2* and its outer diameter is larger than that of the die. When the die returns to the filling position, it pushes up the compact which can then be expelled.

We think that this example gives a few ideas which are sufficient to avoid cracks in most pressings.

Fig. 3 shows two parts with which similar problems are encountered.

Re-pressing, Coining, and Sizing

Re-pressing is often needed after a presintering operation when a high density is necessary combined with good tensile or fatigue properties. When a part has walls with a high height-to-thickness ratio, friction can be so strong that the place where, after pressing, the

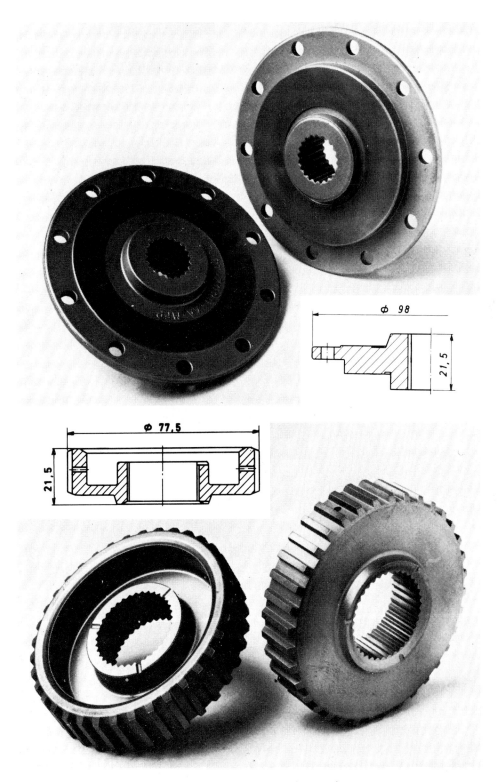

FIG. 3.   Specimen parts (see text).

density is the lowest is not densified at all. After sintering there are then wide differences of properties from one point to another.

Friction is largely dependent on the wall thickness in the die as compared with the wall thickness of the presintered part. It is obvious that the effect is worst when the second is larger than the first. On the other hand, when a comparatively large clearance exists between the part and the several components of the tool (die, core rods, and auxiliary punches) the densification of the portions which have the lowest density can occur before the walls of the part touch the components of the tool. After applying a high pressure, the friction is very low and the density of the part is homogenized. The part shown in Fig. 4 has been re-pressed by this method and a density of 7·1 has been achieved with a pressure of 600 N/mm², starting from a density of 6·5 after pressing.

FIG. 4.   Specimen part repressed at 600 N/mm² after presintering.

The method is very difficult to apply for such a part. If for example we have a uniform clearance of 0·3 mm, and if we consider the region between two holes, we will observe that the volume of material between these is much less than between these and the outer diameter. Thus, after pressing, the density between the holes will be less than that of the outer regions near the walls. This is not what is desired in the finished part; and it can also result in the production of cracks (due to unequal pressures) during re-pressing. Several solutions are possible. One is to mould the part with oval holes, which is not easy. The easiest way is to vary the height of the part as a function of the wall thickness. The lower punch must then have depressions in places where the wall thickness is smaller.

When the wall thickness varies in an easier way as in Fig. 5, then the correction entails varying the clearance from one point to another. The subject of Fig. 5 is a part made of copper with a final density of 97–98% of theoretical density.

FIG. 5.   Specimen part in copper, repressed to ~98% of theoretical density.

The good results obtained are not entirely explained by the previous considerations. Friction is due to the contact of the part with the tool and this contact and the pressure which is produced is a function of the lateral plastic deformation of the part (plastic Poisson's ratio). In order to calculate this effect, we have studied the plastic deformation of presintered parts. Fig. 6 gives the value of the plastic Poisson's ratio as a function of the relative reduction of height for an initial density of 30%.

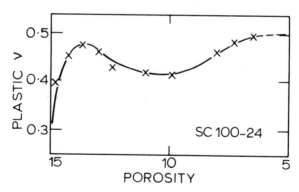

FIG. 6.   Plastic Poisson's ratio as a function of height reduction, initial density 30%.

We were surprised to observe that this ratio has a minimum, which means that after a certain initial lateral deformation the rate of increase of deformation diminishes. In order to explain this fact, we supposed that the pores are deformed and that flat pores, being easier to flatten, give a low lateral deformation. Micrographs seem to confirm this hypothesis. This is why the final density increases comparatively more than expected as a function of lateral clearance.

When the clearances are calculated properly the lateral pressures produced by repressing are homogeneous.

We now look at a different aspect of the techniques of plastic deformation after sintering or presintering—the sizing problem, taking the very simple example of the sizing of a hole without a die.

Up to now the consequences of erroneous operation were either a lack of homogeneity in density or the presence of cracks; and these defects are usually easily detected. When sizing after final sintering or heat-treatment internal stresses are produced which are difficult to detect and which adversely affect the properties.

Let us consider the sizing of a hole 30 mm dia. in a flat disc 50 mm dia. made of a steel with an elastic limit of 400 N/mm². The core rod does not produce any sizing effect unless its diameter is ~60 μm larger than the hole. To obtain a good surface finish inside the hole the diameter has to be 20 μm larger and 30 μm have to be added to allow for variations of dimensions after sintering.

After sizing with a cylindrical rod 110 μm larger than the hole, the external diameter is 50 μm greater than before. This shows that a tensile stress of 150 N/mm² has been created at the outside diameter. For the steel under consideration the fatigue limit is 250 N/mm². If the part is a gear, fatigue strength at the root of the teeth is then roughly halved and tensile strength reduced by 30%.

The great advantage of such a sizing operation is that final variations of the dimensions of the hole are due only to differences in the elastic limit of the material and to the variation of this elastic limit with the work hardening (consolidation curve). Under good conditions this represents a possible variation of 5 μm but as has been explained the operation may have an adverse effect on mechanical properties. The danger is very much lessened if the end of the core rod has the form of an olive and if the height of the part is sufficient to localize the deformation so that the stresses transmitted through the part are smaller.

If we now suppose that there is a die outside the disc and that we only apply the necessary pressure to enter first the part into the die and then the core rod, the stresses in the part will depend on the outer diameter of the disc before sizing as compared with the inner diameter of the die. When the part is larger than the die it is possible to reduce or even suppress the stresses at the outer diameter after sizing. Then, with the above example, the hole will be larger than the core rod and for a similar polishing action will require a core rod 10–20 μm smaller than the final dimension of the hole.

This method is better for reducing the residual stresses but there will be greater variation in the hole dimensions since variation of the outer diameter will cause variations in the springback of the hole. Tolerances smaller than 25 μm are difficult to maintain, and the residual internal stresses will probably vary, though they will be smaller than in the preceding method.

Another solution is to apply a high pressure (coining) to the part after entering into the die and around the core rod in order to reach the plastic range throughout the part. Then all stresses are nearly equalized; and having withdrawn the part from the die internal stresses are minimized but the metal is work-hardened. Elongation before rupture is then reduced with a consequent risk of brittleness or ease of crack propagation. It is however safe then to anneal, with the advantage that the dimensions are changed very little.

The inconvenience of this method is that for comparatively high strength steels, the applied pressure must also be high. If reduced pressure is used, one can incur the cumulative disadvantages of all the methods. It is possible to calculate to a good approximation the different forces which must be applied. The main parameters which are necessary are Young's modulus, Poisson's ratios (elastic and plastic), and the curves of plastic deformation as a function of the loads in tension and in compression.

Conclusions

A sound method of evaluating the influence of pressing operations on the quality of sintered mechanical parts would have been to establish a correlation between the different parameters of pressing operations and the quality factors (mechanical properties and dimensional variations). This paper, however, confines itself to those parameters which have given some trouble in our production plants.

Reference

1. M. Eudier, 3rd Europ. Powder Met. Symp. 1971, *Powder Met., Conf. Supp. Part II*, **1971,** 452.

SECTION X:
Machining

MACHINABILITY OF SINTERED IRON

By Ernst Geijer and Roy B. Jamison

INTRODUCTION

The practical machinability of sintered metal powder parts has to be judged from tool wear and surface finish. Neither property is in itself conclusive or independent of the other. Both have to be taken into consideration. The machinability, measured as tool wear with corresponding surface finish, has been investigated for a large number of materials.

The machinability of parts from straight iron and graphite containing mixes can be considerably improved by sulfur additions. While close to optimum improvement in machinability can be obtained on 0.5% combined carbon containing parts through an addition of 0.5% sulfur, it seems that for both straight iron and 0.85% combined carbon containing parts a sulfur content of 1.0% is required. Since hydrogen in the sintering atmosphere desulfurizes the surface of the part through the formation of hydrogen sulfide, only the sulfur content at the point of actual machining should be taken into consideration, and a necessary adjustment should be made of the average sulfur content of the mix.

Copper additions to the mix composition generally improves machinability, though enough copper to provide free unalloyed copper in the sintered part generally is required for really good machinability. The amount required, therefore, depends both on cooling rate and combined carbon content. Parts requiring 0.85% combined carbon can be made machinable through the addition of 10% copper. Larger copper additions give even better results. The 0.85% combined carbon and 10% copper containing material has very high sintered strength and hardness.

Parts with low copper contents (2.5%), with or without combined carbon, can be made more machinable, through an 0.5% sulfur addition without noteworthy change in sintered properties. If the copper content is increased to 10%, the sulfur addition will cause a large dimensional change after sintering.

Phosphorous additions to carbon containing mix compositions cause excessive tool wear. Phosphorous is much less detrimental in the presence of copper.

Lead improves the machinability of both straight iron and combined carbon containing materials without changing the sintered properties. Suitable alternatives are copper or sulfur.

Finally, it can be said that, when it can be used, an increased nose radius on the cutting edge and/or a lower feed rate will considerably improve surface finish.

We thank Mr. Robert Campbell of the Spring Garden Institute in Philadelphia for performing the actual machining and for valuable advice.

TEST CONDITIONS

The number of variables present when determining the machinability of sintered iron powder parts is so great that, by necessity, only a limited number of them could be investigated.

All work has been done on turning, since it is one of the most commonly used machining operations and relatively easy to perform under controlled test conditions. The lathe used for the machining had variable cutting speed, feed rate and depth of cut. The profilometer used for measuring surface finish had a .0005 in. diameter penetrator and recorded the average surface roughness in micro inches (average root mean square deviation from a line drawn halfway between "the tops and bottoms" of the rough machined surface).

For all mix compositions, the same lot of Ancor MH-100 sponge iron was used. Sulfur was added by prealloying the iron powder with 0.5% or more. When less than 0.5% sulfur was desired, straight iron powder and the 0.5% sulfur containing powder were mixed proportionally. 0.75% stearic acid was used in all mixes for a die lubricant. 0.85%, 1.25% and 1.40% graphite was added to give 0.5%, 0.85%, and 1.10% sintered combined carbon respectively.

The test specimens used were short sleeves with an inner diameter of 1.5 in., an outer diameter of 2.5 in. and height of 0.75 in. All specimens were pressed to give a sintered density of 5.8 g/cc. Specimens with a sintered density of 6.2 g/cc were also tested. No significant differences were found in machining properties. The test specimens were sintered in dissociated ammonia for 45 min. at 2050° F (dewpoint - 30° F). The cooling rate after sintering was relatively high since only one sleeve was sintered at a time and, after sintering, was pushed directly into the coolest section of the cooling zone.

In order to eliminate the effect of any surface decarburization or desulfurization, all machining was done 0.100 in. to 0.250 in. from the outer surface of the test specimens. Several cuts were taken on each specimen and uniform results were obtained throughout this range. Figure 1 shows the increase in carbon and sulfur contents and the improvement in surface finish as the distance from the surface increases. Naturally, it is only the analysis of the material in the actual cut that determines the machining properties and not the average analysis of the part.

To duplicate the machining done by industry today, the depth of cut was standardized at only .005 in. on the radius, the feed rate (except when it has been used as a variable) was .005 in. using a carbide cutting edge with a 1/32 in. nose radius. The cutting speed was 500 surface feet per minute (SFM), using a carbide edge with the following data:

A standard indexable type tool holder.

A precision ground, triangular carbide insert, cast iron grade.

Back and side-rake 0°

End and side-relief 5°

Source: Technical Bulletin 125D, Hoeganaes Corp., Oct 1968

Lead angle (side cutting edge) 15°

Nose radius 1/32 in.

All machining was made dry, without coolant.

SINTERED PROPERTIES OF TEST SPECIMENS

The different specimens tested contained, except for iron, combined carbon, sulfur, copper, phosphorous and lead in different compositions. To check the physical properties, transverse rupture strength bars were made of all used mix compositions, and sintered under the same conditions as the test specimens. The mix compositions and sintered properties are listed in figure 2.

MICROSTRUCTURES OF TEST SPECIMENS

Figures 3 to 9 show the microstructures of some of the more interesting mix compositions, as follows:

Fig.	Mix composition, %			
	Comb. Carb.	Copper	Sulfur	
3	.5	—	—	Ferrite and pearlite.
4	.85	—	—	Only pearlite, no ferrite or cementite visible.
5	1.1	—	—	Pearlite and cementite.
6	—	—	.5	Ferrite and islands of ironsulfide.
7	.5	—	.5	Ferrite, pearlite and ironsulfide.
8	—	10	—	Ferrite and free copper (no difference could be observed in the microstructure between slow and fast cooling rate).
9	.5	10	—	Ferrite, pearlite and free copper. 0.5% combined carbon determined chemically.

All specimens have been etched 6 seconds in 1% Nital. Lens magnification 1,000 X.

IRON-CARBON MATERIALS
(Figs. 3, 4, 5, 10, and 18)

Figures 3, 4 and 5 are the microstructures that refer to this group.

Figure 10 shows the surface finish for varying combined carbon contents at increasing feed rates (new cutting edges). The highest combined carbon contents give the best surface finish. However, they all are quite poor. The feed rate does not influence the finish to any extent. When the finish is poor at low feed rates, it generally remains constant as the feed rate increases, even to the extent that the finish at the highest feed rate is better than theoretically possible. (The curve for the theoretical finish is included in Fig. 10). A finish better than the theoretical is possible since the ridges between the valleys, formed for every revolution, on poorly machining materials is torn off and never fully developed.

Figure 18 shows tool wear and the corresponding surface finishes versus machining time for varying combined carbon contents. Tool wear increases rapidly with the combined carbon content and the surface finish is consistently poor.

There is an indication in figure 18 that in some cases the finish improves slightly up to a certain machining time. This can be explained by the fact that, as the cutting edge wears, its radius flattens out, and it acts like one with a larger nose radius and gives a better finish (compare figure 34). Consequently, the finish after a certain machining time can also be better than that with a new cutting edge (zero machining time) at a corresponding feed rate.

IRON-SULFUR MATERIALS
(Figs. 6, 11, and 19)

Figure 6 is the microstructure that refers to this group.

Figure 11 shows the surface finish for varying sulfur contents at increasing feed rates (new cutting edges). The surface finish improves rapdily for increasing sulfur contents at low feed rates (up to .006 inch/rev). As all hydrogen containing gases will cause some surface desulfurization, the sulfur content at the actual machining area has to be considered, not the average sulfur content of the part.

Figure 19 shows tool wear and the corresponding surface finishes versus machining time. Tool wear decreases rapidly with increasing sulfur content, although it is realtively low even with zero sulfur content. Surface finish improves slowly with low sulfur additions. It is still quite poor at a 0.5% sulfur addition, but very good at a 1.0% addition. To reach this high sulfur content at the surface of a part probably requires a practically hydrogen free sintering atmosphere.

From the table in figure 2, it can be seen that the sintered strength is slightly higher at intermediate sulfur contents than at 0 or 1.0%. The dimensional change increases slightly with increasing sulfur content.

Source: Technical Bulletin 125D, Hoeganaes Corp., Oct 1968

IRON-CARBON-SULFUR MATERIALS
(Figs. 7, 12, 13, 20, and 21)

Figure 7 is the microstructure that refers to this group.

Figures 12 and 13 show the surface finish for 0.5% to 0.85% combined carbon and varying sulfur contents at increasing feed rates (new cutting edges). The surface finish improves rapidly for increasing sulfur contents at low feed rates.

Figures 20 and 21 show tool wear and the corresponding surface finishes versus machining time. Tool wear decreases rapidly with increasing sulfur content for both combined carbon contents. Figure 20 (0.5% combined carbon) shows that the lowest tool wear has practically been reached at 0.5% sulfur and that surface finish improves rapidly with increasing sulfur content. Figure 21 (0.85% combined carbon) shows that the tool wear, in this case, does not decrease significantly until the sulfur addition reaches 1.0%. Again, this can be hard to reach in the surface of a part sintered in hydrogen containing atmosphere. The surface finish improves rapidly with increasing sulfur content.

From the table in figure 2 can be seen that the sintered strength is slightly higher at intermediate sulfur contents and that the dimensional change increases with increasing sulfur contents.

IRON-COPPER MATERIALS
(Figs. 8, 14, 22, and 23)

Figure 8 shows the microstructure that refers to this group. A 10% copper addition means that free copper is still present after sintering.

Figure 14 shows the surface finish for varying copper contents at increasing feed rates (new cutting edges). The surface finish is vastly improved by a small copper addition at low feed rates. The best obtainable finish has practically been reached at a 2.5% copper content.

Figure 22 shows tool wear and the corresponding surface finishes versus machining time. Tool wear reaches an optimum at 10% copper content and then decreases rapidly. 18% copper content gives very low tool wear. Iron can, at the sintering temperature used (2050°F), dissolve about 8% copper. At the high cooling rate used, very little copper had time to precipitate, in spite of the fact that the solubility of copper in iron at room temperature is very low. As copper dissolved in iron increases hardness (see table in figure 2) it is natural that tool wear should increase as long as the percentage of dissolved copper increases. At higher copper contents, when saturation is reached and free copper appears, the tool wear decreases. The optimum tool wear was probably reached at a copper content less than 10%, at the very point where free copper started to appear.

The surface finish has improved considerably at 2.5% copper content and reaches an optimum at 10%. If the tests had been run longer, the 18% copper would probably have given an acceptable surface finish long after the cutting edge used for the 10% copper had worn out, since the difference in tool wear is so tremendous.

Sintered hardness, strength and dimensional change reach a maximum at an intermediate copper content.

Figure 23 shows the importance of cooling rate on copper containing parts, both for tool wear and surface finish. The slow cooling rate has not been obtained through slow cooling of the furnace, but only through increasing more than 10 times the load of the tray used in the laboratory sintering furnace. Tool wear decreased and the surface finish deteriorated to such an extent that the machinability strongly resembles that of straight iron.

Sintered hardness and strength decrease rapidly with slower cooling, while dimensional change remains practically the same. A slightly slower cooling rate than the "fast cool" in figure 23 is probably obtained in most production furnaces and is beneficial for tool wear.

IRON-CARBON-COPPER MATERIALS
(Figs. 9, 15, 24, and 25)

Figure 9 shows the microstructure that refers to this group.

Figure 15 shows the surface finish for varying combined carbon and copper contents at increasing feed rates (new cutting edges). The surface finish improves rapidly with increasing copper content at both 0.5% and 0.85% combined carbon, especially at low feed rates.

Figures 24 and 25 show tool wear and the corresponding surface finishes versus machining time. Tool wear reaches an optimum at about 2.5% copper content instead of at 10% as in figure 22 (no combined carbon). This is probably due to a decreased solubility of copper with an increasingly pearlitic structure. In figure 24 (0.5% combined carbon) 10% copper content does not give much lower tool wear than 5%, while in figure 25 (0.85% combined carbon) 10% copper content gives a tool wear low enough to approach the 18% copper content. However, the difference in tool wear between the 10% and 18% additions is still appreciable.

The surface finish in both figures 24 and 25 is vastly improved by low copper additions.

The table in figure 2 shows, as with iron-copper materials without combined carbon, an optimum sintered strength at 5% copper content and also an optimum hardness and dimensional change at intermediate copper contents. This is true at both 0.5% and 0.85% combined carbon.

IRON-COPPER MATERIALS WITH
SULFUR, PHOSPHOROUS AND
COMBINED CARBON
(Figs. 26 to 31)

All figures show tool wear and the corresponding surface finishes versus machining time.

Figure 26 shows the influence of sulfur, phosphorous, and combined carbon on materials containing 2.5% copper. A 0.5% combined carbon content results in the shortest tool life and a 0.5% sulfur content in the longest.

Surface finish is poor and is about the same in all cases.

Sulfur does not improve sintered strength, but increases the dimensional change slightly. The largest increase in sintered strength is obtained with 0.5% combined carbon and the largest decrease in dimensional change with 0.5% phosphorous.

Figure 27 shows the influence of sulfur, phosphorous, and combined carbon on a material containing 10% copper. A 0.5% phosphorous content results in the shortest tool life and a 0.5% sulfur content in the longest.

The surface finish is good for materials with 10% copper, with or without 0.5% sulfur.

The material containing 0.5% combined carbon is the only one that shows considerably improved sintered strength and hardness. While combined carbon and phosphorous strongly decreases dimensional change, sulfur causes a considerable increase.

Figure 28 shows again the influence of sulfur additions on materials with 2.5% and 10% copper contents. Tool wear decreases especially for 10% copper content, while surface finish remains the same. The increase in dimensional change when adding sulfur to a material containing 10% copper is, however, so large that it is mostly of theoretical interest.

Figure 29 shows the influence of sulfur and phosphorous on a material containing 2.5% copper and 0.5% combined carbon. Phosphorous causes tremendous tool wear while sulfur decreases tool wear significantly(better than a material containing only 2.5% copper). The surface does not change appreciably.

Combined carbon content (0.5%) causes a large increase in sintered hardness and strength, while the presence of sulfur or phosphorous results in a modest further increase. The dimensional change remains about the same for all four materials.

Figure 30 shows the influence of additions of sulfur and phosphorous on a material containing 10% copper and 0.5% combined carbon. Phosphorous causes increased tool wear, while the presence of sulfur results in a very low tool wear. The surface finish, however, deteriorates seriously with the sulfur content.

Phosphorous gives increased sintered hardness and strength, while sulfur causes a tremendous increase in dimensional change.

Figure 31 shows the influence of sulfur on a material containing 10% copper and 0.85% combined carbon. As should be expected, tool wear decreases, surface finish remains the same, sintered hardness and strength decrease slightly and dimensional change increases tremendously.

IRON WITH PHOSPHOROUS AND
COMBINED CARBON
(Figs. 16 and 32)

Figure 16 shows the surface finish for phosphorous containing materials with and without combined carbon at increasing feed rates (new cutting edges). The surface finish improves with phosphorous and even more so with combined carbon.

Figure 32 shows tool wear and the corresponding surface finishes versus machining time. Tool wear increases heavily in the presence of phosphorous and further with combined carbon. The tool wear for the phosphorous-combined carbon material seems prohibitive in spite of the good surface finish.

The sintered strength of the phosphorous-combined carbon material is quite high.

IRON WITH LEAD AND
COMBINED CARBON
(Fig. 33)

Figure 33 shows tool wear and the corresponding surface finishes versus machining time. Tool wear decreases in the presence of lead both with and without combined carbon. The decrease, however, is not outstanding. Sintered properties remain unchanged.

VARYING NOSE RADIUS ON
CUTTING EDGES
(Figs. 17 and 34)

Figure 17 shows the theoretically possible surface finish at varying feed rates for a cutting edge with a 1/32 in. nose radius. Also included is a standard steel (B-1112). Because of the small nose radius the theoretical finish deteriorates at increasing, though modest, feed rates. Figure 34 shows the influence of varying nose radii for increasing feed rates on a 0.5% sulfur containing material. The surface finish reported here for 1/32 in. nose radius very closely resembles the theoretical finish in the previous diagram (figure 17). Figure 34 shows that a larger nose radius considerably improves surface finish at higher feed rates, caused by the fact that at a constant feed rate a larger nose radius gives smaller ridges at machining. Consequently a larger nose radius is recommended for a machining operation when the part design permits.

VARYING CUTTING SPEED
(Fig. 35)

Figure 35 shows the influence of cutting speed on surface finish at varying feed rates (new cutting edges). The material contains 0.5% combined carbon. The surface finish improves considerably with increasing cutting speed at lower feed rates. The higher the cutting speed, the closer the surface finish approaches the theoretical finish (figure 17). However, tests run at different cutting speeds to constant tool wear (Taylor-curves) on different materials show that when the cutting speed is doubled, the tool wear increases 4 to 10 times compared to the wear at the lower cutting speed.

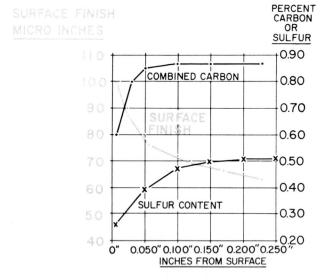

Figure 1 — SURFACE DECARB. AND DESULF— THEIR INFLUENCE ON SURFACE FINISH (NEW CUTTING EDGES) — 0.85% COMBINED CARBON + 0.5% SULFUR CONTENT.

SINTERED PROPERTIES

Figure 2   45 MIN. AT 2050°F, CRACKED NH₃— TRANSVERSE RUPTURE STRENGTH BARS.

MIX— COMBINATION				SINTERED PROPERTIES AT 5.8g/cc (SINTERED DENSITY)		
% COMB. CARBON	% Cu	% S	% MISC.	HARDNESS R_B	TRANSV. RUPTURE STRENGTH PSI	DIM. CHANGE FROM DIE SIZE INCH/INCH
—	—	—	—	−40	38,000	−.001
—	—	.1	—	−35	45,000	−.001
—	—	.25	—	−35	48,000	−.001
—	—	.5	—	−35	44,000	0
—	—	1.0	—	−40	38,000	+.001
.5	—	—	—	+30	65,000	0
.5	—	.1	—	+30	64,000	+.001
.5	—	.25	—	+30	67,000	+.002
.5	—	.5	—	+40	69,000	+.002
.5	—	1.0	—	+35	64,000	+.003
.85	—	—	—	+50	76,000	+.001
.85	—	.1	—	+50	80,000	+.002
.85	—	.25	—	+50	83,000	+.003
.85	—	.5	—	+55	85,000	+.004
.85	—	.75	—	+55	92,000	+.005
.85	—	1.0	—	+50	81,000	+.006
1.1	—	—	—	+45	71,000	+.002
—	2.5	—	—	0	55,000	+.004
—	5	—	—	+25	63,000	+.016
—	10 FAST COOL		—	+30	63,000	+.023
—	10 SLOW COOL		—	0	47,000	+.021
—	18	—	—	+20	54,000	+.013
.5	2.5	—	—	+50	79,000	+.004
.5	5	—	—	+65	85,000	+.005
.5	10	—	—	+65	76,000	+.003
.5	18	—	—	+55	66,000	−.005
.85	2.5	—	—	+65	90,000	+.004
.85	5	—	—	+70	98,000	+.003
.85	10	—	—	+70	85,000	0
.85	18	—	—	+65	64,000	−.006
—	2.5	.5	—	−5	55,000	+.005
—	10	.5	—	+30	63,000	+.030
.5	2.5	.5	—	+55	84,000	+.005
.5	10	.5	—	+60	74,000	+.014
.85	10	.5	—	+60	78,000	+.013
—	—	—	.5P	+10	54,000	−.005
—	2.5	—	.5P	+20	70,000	+.002
—	10	—	.5P	+35	65,000	+.009
.5	2.5	—	.5P	+65	85,000	+.003
.5	10	—	.5P	+70	90,000	+.004
.85	—	—	.5P	+60	100,000	+.001
—	—	—	2Pb	−50	36,000	−.001
.85	—	—	2Pb	+45	77,000	+.002

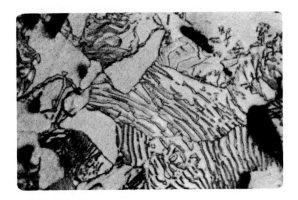

See page 258
for mix
composition tabulation
of these microstructures.

Figure 3

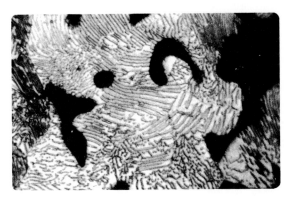

Figure 4

Figure 7

Figure 5

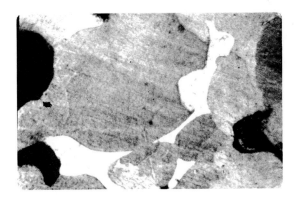

Figure 8

Figure 6

Figure 9

Source: Technical Bulletin 125D, Hoeganaes Corp., Oct 1968

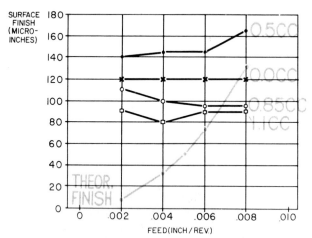

Figure 10 — SURFACE FINISH VS. FEED
FOR NEW CUTTING EDGES

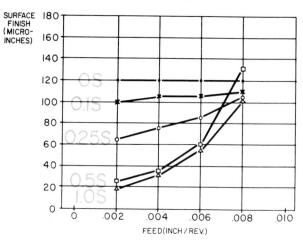

Figure 11 — SURFACE FINISH VS. FEED
FOR NEW CUTTING EDGES

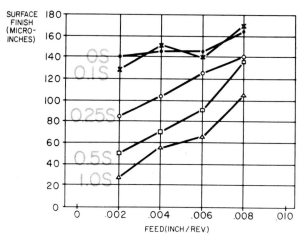

Figure 12 — SURFACE FINISH VS. FEED
FOR NEW CUTTING EDGES

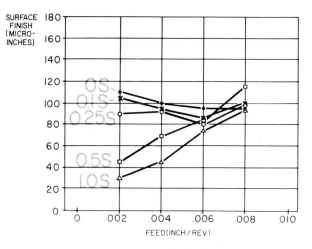

Figure 13 — SURFACE FINISH VS. FEED
FOR NEW CUTTING EDGES

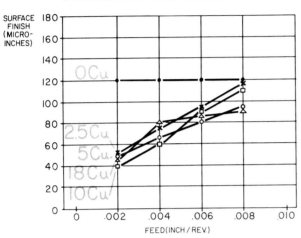

Figure 14 — SURFACE FINISH VS. FEED
FOR NEW CUTTING EDGES

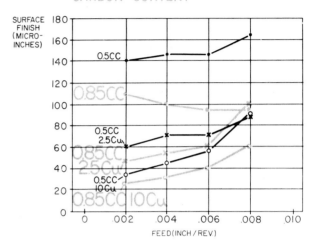

Figure 15 — SURFACE FINISH VS. FEED
FOR NEW CUTTING EDGES

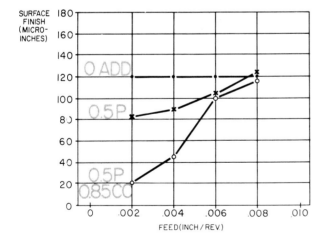

Figure 16 — SURFACE FINISH VS. FEED
FOR NEW CUTTING EDGES

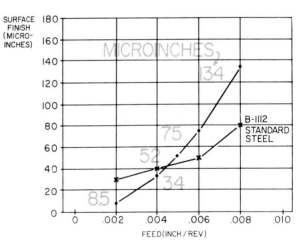

Figure 17 — SURFACE FINISH VS. FEED
FOR NEW CUTTING EDGES

Source: Technical Bulletin 125D, Hoeganaes Corp., Oct 1968

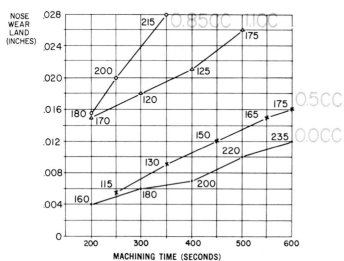

Figure 18 — TOOL WEAR VS. MACHINING TIME — DATA ON CURVES
REPRESENT SURFACE FINISH IN MICRO INCHES

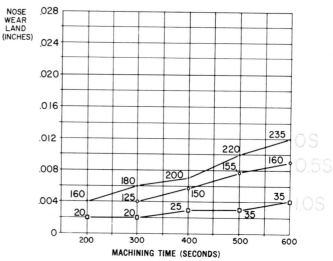

Figure 19 — TOOL WEAR VS. MACHINING TIME — DATA ON CURVES
REPRESENT SURFACE FINISH IN MICRO INCHES

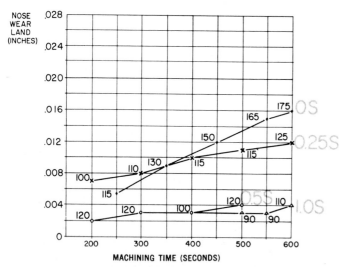

Figure 20 — TOOL WEAR VS. MACHINING TIME — DATA ON CURVES
REPRESENT SURFACE FINISH IN MICRO INCHES

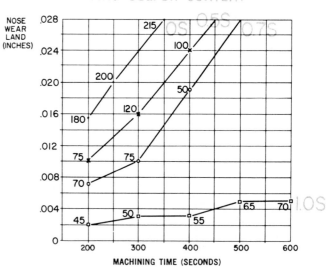

Figure 21 — TOOL WEAR VS. MACHINING TIME — DATA ON CURVES
REPRESENT SURFACE FINISH IN MICRO INCHES

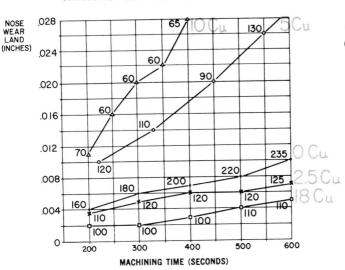

Figure 22 — TOOL WEAR VS. MACHINING TIME — DATA ON CURVES
REPRESENT SURFACE FINISH IN MICRO INCHES

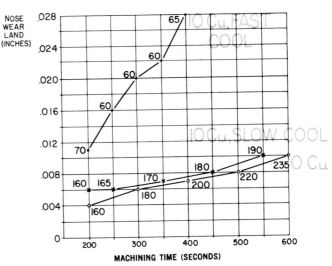

Figure 23 — TOOL WEAR VS. MACHINING TIME — DATA ON CURVES
REPRESENT SURFACE FINISH IN MICRO INCHES

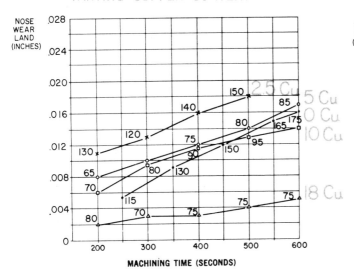

Figure 24 — TOOL WEAR VS. MACHINING TIME — DATA ON CURVES
REPRESENT SURFACE FINISH IN MICRO INCHES

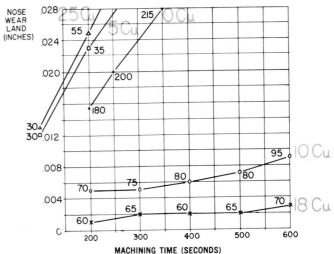

Figure 25 — TOOL WEAR VS. MACHINING TIME — DATA ON CURVES
REPRESENT SURFACE FINISH IN MICRO INCHES

Source: Technical Bulletin 125D, Hoeganaes Corp., Oct 1968

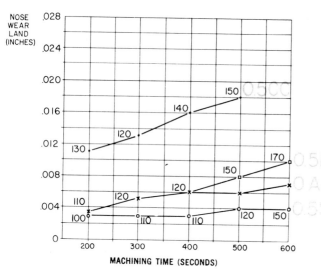

Figure 26 — TOOL WEAR VS. MACHINING TIME — DATA ON CURVES
REPRESENT SURFACE FINISH IN MICRO INCHES

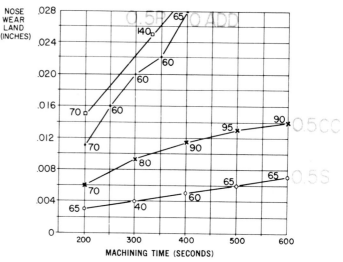

Figure 27 — TOOL WEAR VS. MACHINING TIME — DATA ON CURVES
REPRESENT SURFACE FINISH IN MICRO INCHES

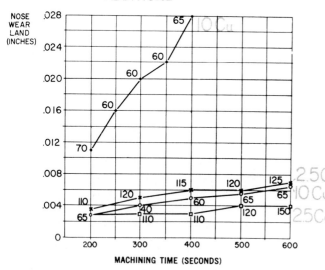

Figure 28 — TOOL WEAR VS. MACHINING TIME — DATA ON CURVES
REPRESENT SURFACE FINISH IN MICRO INCHES

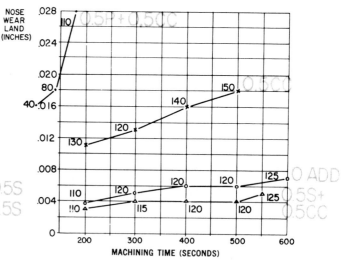

Figure 29 — TOOL WEAR VS. MACHINING TIME — DATA ON CURVES
REPRESENT SURFACE FINISH IN MICRO INCHES

Figure 30 — TOOL WEAR VS. MACHINING TIME — DATA ON CURVES
REPRESENT SURFACE FINISH IN MICRO INCHES

Figure 31 — TOOL WEAR VS. MACHINING TIME — DATA ON CURVES
REPRESENT SURFACE FINISH IN MICRO INCHES

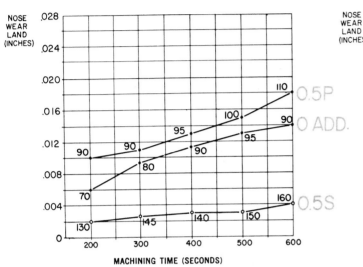

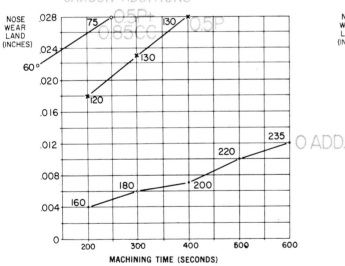

Figure 32 — TOOL WEAR VS. MACHINING TIME — DATA ON CURVES
REPRESENT SURFACE FINISH IN MICRO INCHES

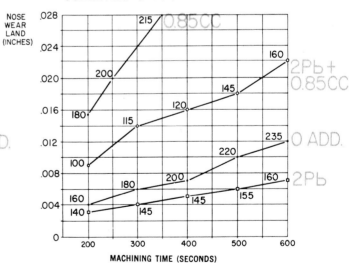

Figure 33 — TOOL WEAR VS. MACHINING TIME — DATA ON CURVES
REPRESENT SURFACE FINISH IN MICRO INCHES

Source: Technical Bulletin 125D, Hoeganaes Corp., Oct 1968

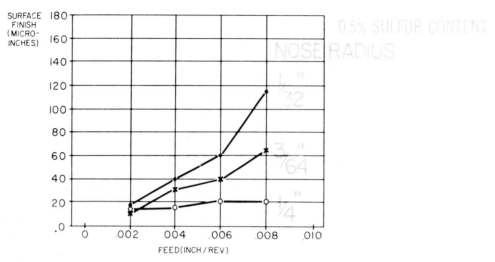

Figure 34 — VARYING NOSE RADII ON NEW CUTTING EDGES

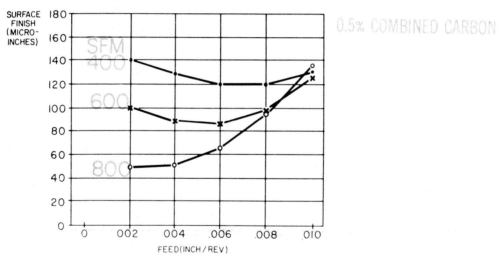

Figure 35 — VARYING CUTTING SPEED ON NEW CUTTING EDGES

MACHINING P/M COMPONENTS

By
Gerald L. Pearson
Manager of Technical Service
Hoeganaes Corporation

Before discussing machining of P/M parts, it must be recognized that variations in machinability may be caused by differences in sintering conditions and by different typical compositions used for their machinability.

One of the major problems that we find after examining parts sent to us is carburization or decarburization to various degrees. Both of these conditions have been found even in large production runs. When variations of this nature occur, it is virtually impossible to establish a standard machining procedure and still obtain good tool life, surface finish, and a constant production rate.

As a result of compacting tool wear, the sintering atmosphere, or furnace conditions, are often altered to obtain a desired dimensional change. In the case of carbon containing materials under such changing conditions, the combined carbon levels will change and machinability will vary. Machinability will also be affected if the parts are either over or under sintered.

Another factor which affects machinability of the material is density. As parts increase in density, they begin to approach the same machinability as wrought and cast parts. This is principally due to the porosity reacting as a series of interrupted cuts with the tooling being intermittently in contact then out of contact with the metal. To partially eliminate or minimize this condition, many parts are copper infiltrated to fill the voids, or impregnated with a polyester resin or waxes. In many instances, due to subsequent treatments, if it is not possible to use these methods then you must resort to either machining dry or alternatively impregnating with the oil then re-oiling with the required lubricant. Another method is to impregnate with water and use a soluble oil coolant again followed by degreasing, or a burn-off operation to remove the moisture. Still another approach is to cool the cutting edge either with an air stream of with a water soluble wax (i.e. Johnson Lubracool) vaporized with a Mistic Mist spray unit.

There have been in the past numerous recommendations that sulfurized base cutting lubricants can be used for machining iron base parts. This is quite satisfactory provided the parts are not to be plated. This applies not only to electroplating, but electrolysis plating, such as the Cannogin process.

In order to understand some of the problems encountered in machining, we should also consider some common sintered compositions and their characteristics as machined.

SINTERED COMPOSITIONS

1. Straight Iron - is the most difficult material to machine because of its softness and tendency to tear easily in all machining operations. One method of fabrication for improved machinability without changing the composition, is to repress or otherwise densify the area to be machined. In repressing or densifying the part, a degree of coldworking occurs, thus obtaining better machinability. As an example, pole pieces are produced with a projection on the tang section which during repressing is sized to the same thickness as the rest of the part. This coining increases density and cold works the area to be drilled and tapped so that tearing of the threads, common in low density press, is reduced.

2. Iron Carbon - To provide a better chip breaking effect, a combined carbon of 0.2/0.4% is used by a number of fabricators. However, much better machinability can be obtained by the use of materials containing sulfur, copper, lead or phosphorus and copper. The effects of varying combined carbon on tool wear versus machining time are shown in Figure 1. From this figure it appears that the carbon free material would be the best material for tool wear, however, excessive galling occurs in drilling, tapping, etc., and the tool must be changed for this condition not for wear.

3. Iron-Sulfur and Iron-Carbon-Sulfur Materials - In the case of iron-sulfur, the surface finish improves slowly with low sulfur additions. The finish is poor at 0.5% sulfur addition, but very good at a 1% addition. In order to maintain this high content on the surface of a part, a sintering atmosphere which is essentially free of hydrogen is required.

    For higher strength structural material, the iron-carbon-sulfur mixes are used. The best results in actual production were obtained for 0.5% sulfur and a combined carbon of 0.35 to 0.5%. (Figures 2, 3 and 4)

4. Iron-Copper Materials - Normally this composition is used for bearings with the copper content varying from 2.5% to possibly as high as 20%. The extent of machining would normally be either I.D. grooving, or O.D. turning to obtain a configuration which normally cannot be pressed in the part. For these operations programmed machining is ideal, with the rough bore being performed at one feed and speed, and the finished bore being automatically changed to obtain the maximum pore surface. In O.D. surface turning where open pore surfaces are not required, the main concern is to obtain the highest productivity, with satisfactory surface finish, and minimum tool wear.

5. Iron-Copper-Carbon - The addition of copper or carbon increases the machinability of iron to different degrees and when both additions are made machinability is improved still further. As the copper percentage is increased, the material approaches that of free machining steels.

When copper infiltrating is used, the best machining condition is obtained, which is low tool wear, increased production rate, and better surface finish.

It should be pointed out that the majority of machine parts do not fall in a category of copper infiltrated, but have a composition range of 2.0/5.0% copper with the combined carbon being in the ranges of 0.5/0.8%. (Figure 5)

6. Miscellaneous Additions – Frequently, other elements such as lead, copper, bronze, phosphorus and molydisulfide are added to the iron powder to obtain better machinability. These additions result in smoother surface finishes on machine parts.

   The use of lead is somewhat limited due basically to difficulties in furnace control and blending of the atmosphere.

   Phosphorus is the least used of the additions, and is virtually eliminated due to high tool wear.

   Molydisulfide is only used with iron-copper-carbon materials where solid lubrication is required for the part in its applications. However, it does help in prolonging tool life.

7. Because material compostions have so many other major and minor affects on machinability, we have included in this section our brochure "Machinability of Sintered Iron."

MACHINING

1. Turning and Boring – It has been found that carbide tools with a sharp nose point work best. Typical specifications are:

   ### General Turning Tool

   Material – Grade C-4 Carbide
   Side Cutting Angle   -5 degrees
   End Cutting Angle   -15 degrees
   Back Rake   -10 degrees
   End Relief   -10 degrees
   Side Relief   -10 degrees
   1/32" nose radius for roughing cuts
   Sharp point for finishing cuts

Cutting speeds of 150 to 325 surface feet per minute are satisfactory. Maximum depth of cut should be about .005" and feeds of .001" to .003" per revolution give the best results.

2. Drilling – Carbide or HSS drills with a low right hand helix angle will help prevent the drill from digging in. The cutting edges should also be dubbed to reduce the axial rake. Speeds of 70 sfm for high speed steel and up to 200 sfm for carbide drills with a mechanical feed will

get optimum performance. For a 1/8" diameter hole, a feed rate of .003" per revolution has been found satisfactory. Feed rates can be increased proportionally to .010" per revolution for a 1" diameter hole.

3. <u>Tapping</u> - Tapping is easily accomplished by using two-fluted carbide taps for holes less than 1/2" in diameter, or three-fluted taps for holes over 1/2" diameter.

    Spiral pointed taps are most desirable because they throw the chips ahead and prevent them from driving into the pores. If difficulties are encountered, the relief of the tap can be increased to nearly twice that used for conventional ferrous materials.

4. <u>Milling</u> - Milling is generally difficult because of the tendency of the material to smear. To minimize the smearing, it is recommended that dead sharp helical tool cutters with an axial rake be used so that the chips are sheared on an angle. Most parts can be machined with the same cutters as are used on cast iron or alloys of low tensile strength. High speed steel cutters, should be run about 70 sfm, whereas, carbide cutters can be run as high as 300 sfm. For roughing cuts, the feeds should be about .002" to .005" per tooth, whereas, finishing feeds should be finer and in the order of .001" to .002" per tooth.

5. <u>Shaping</u> - Shaping is, of course, very similar to turning or boring in that a single point tool is employed. It is important that the operator makes sure that the tool does not drag on the return strokes which will tend to mar the finish.

6. <u>Reaming</u> - Reaming is definitely not recommended if porosity is to be maintained because of the tendency to smear the bearing surface. However, for structural parts reaming is satisfactory. A left-handed helical reamer with a right-hand periphery cut is recommended. Allowances vary from .002" and can be as much as .005" depending on the size of the hole. Use a standard floating holder and a feed of 25 to 50 feet per minute.

7. <u>Ball Sizing or Burnishing</u> - Ball sizing or burnishing of holes maintains open pores, providing that proper sintered dimensions are held. Normally, with carbon material, the sizing stock should not exceed .002". To hold close tolerances, it is a must to have a least three sizes of balls or burnishing punches having a difference of .0002" in diameter. This is to allow for spring back from the sintered size.

8. <u>Broaching</u> - Broaching is not advised if porosity is to be maintained. It is recommended that at least .015" of stock should be removed. Standard draw broaching is recommended to obtain best tolerances and finish. The best results are obtained when the broach has at least 16 or more cutting edges and a minimum of six burnishing surfaces.

9. <u>Grinding, Honing and Lapping</u> - These operations are usually a finishing operation for heat treated materials, and only a small degree of pore closure occurs, however, if sintered non-heat treated parts are honed, the honing stock should be held to less than .0014".

VARYING COMBINED CARBON

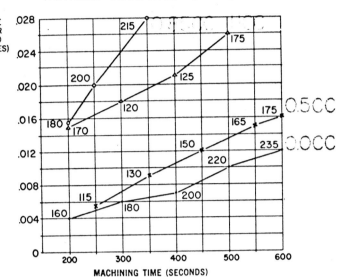

— TOOL WEAR VS. MACHINING TIME — DATA ON CURVES
REPRESENT SURFACE FINISH IN MICRO INCHES

Fig. 1

VARYING SULFUR CONTENT

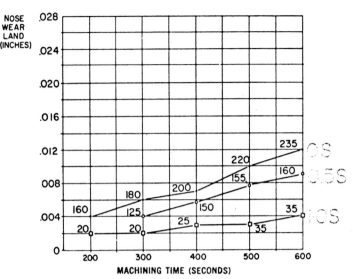

— TOOL WEAR VS. MACHINING TIME — DATA ON CURVES
REPRESENT SURFACE FINISH IN MICRO INCHES

Fig. 2

0.5% COMBINED CARBON
VARYING SULFUR CONTENT

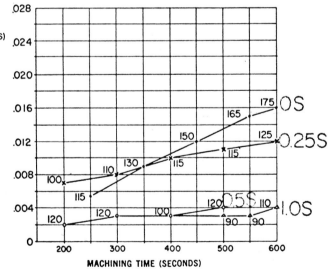

— TOOL WEAR VS. MACHINING TIME — DATA ON CURVES
REPRESENT SURFACE FINISH IN MICRO INCHES

Fig. 3

0.85% COMBINED CARBON
VARYING SULFUR CONTENT

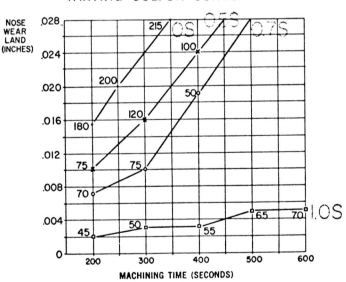

— TOOL WEAR VS. MACHINING TIME — DATA ON CURVES
REPRESENT SURFACE FINISH IN MICRO INCHES

Fig. 4

277

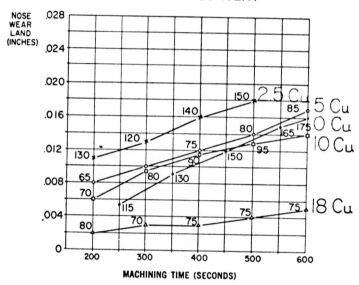

0.5% COMBINED CARBON
VARYING COPPER CONTENT

— TOOL WEAR VS. MACHINING TIME — DATA ON CURVES
REPRESENT SURFACE FINISH IN MICRO INCHES

Fig. 5

SECTION XI:
Properties and Applications

Progress in Gearmaking

Ferrous-Based Powder Metallurgy Gears

WALTER E. SMITH

THE powder metallurgy process is one of the most flexible metalworking processes for the production of gears. This process is capable of producing close tolerance gears with strengths to 160,000 psi at economical prices in volume quantities. Spur, helical, bevel, face, spur-helical and helical-helical gears are produced by P/M techniques. The process is particularly attractive when the gear contains depressions, through holes, levels, or projections. Examples of gears used in the appliance, business machine, automotive, power tool, power transmission, and farm equipment fields are shown in Figs. 1, 2, and 3.

Spur gears (Fig. 1) are the most common type of P/M gear in use today. The P/M process is capable of producing spur gears of many tooth forms and modifications. If an EDM electrode can be ground or cut, the mating form can be duplicated in P/M dies.

Helical gears (Fig. 2) are also commonly made by the P/M process, though the helix angle of P/M gears is limited to approximately 35 deg. As with spur gears, many tooth forms and modifications are possible.

Bevel and face gears (Fig. 3) are readily made by the P/M process, but differ somewhat from spur and helical gears in mechanical property flexibility. Due to the nature of the compaction process, sections such as face gear teeth cannot be made to high densities

WALTER E. SMITH is Vice President—Engineering, Merriman Inc., a division of Litton Industries, Hingham, Mass. This article appeared in "Gear Manufacture and Performance," the Proceedings volume for the Metalworking Forum on Gear Manufacture and Performance, held at Troy, Mich., on October 29 to 31, 1973, and sponsored by the ASM Mechanical Working and Forming Division.

without resorting to copper infiltration or repressing. Most bevel and face gear forms can be made providing they are not undercut but maximum mechanical properties attainable will be somewhat less than those that can be achieved for spur and helical gears.

To design and specify P/M gears, one must understand what tolerances can be held, what mechanical properties can be obtained, and what to establish as inspection criteria. In order to cover these areas, this presentation will be divided into the following three sections: P/M Gear Tolerances, Load Ratings of P/M Gears, and Quality Control of P/M Gears.

P/M GEAR TOLERANCES

P/M gear tolerances are determined by the compaction tooling, compaction process, sintering techniques, and alloy. Tolerances can be divided into functional tolerances that directly affect gear function, and non-functional elements which pertain to portions of the gear that are more properly termed general machine-design considerations. Some of the non-functional configurations that are possible by the P/M process are shown in Fig. 4. The possible non-functional configurations follow the design rules found in "Powder Metallurgy Design Guide Book".[1]

Functional gear tolerances are first controlled by the compaction tooling, and then by the processing methods and gear material. Since most P/M gears are pressed in carbide dies, tooth form and dimensions remain relatively stable and constant throughout the process. Concentricity of the gear outside diam and

Fig. 1—P/M spur gears.

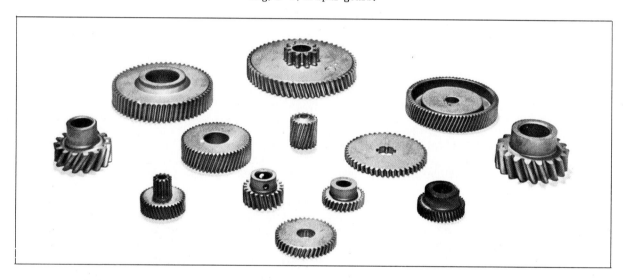

Fig. 2—P/M helical gears.

pitch diam to the bore are also controlled by the tooling. If the pressing core rod or punches are eccentric to the die, the parts will be eccentric. Once the part is pressed, the material characteristics and sintering process will determine the dimensional tolerance capability of the outside diam, pitch diam, test radius, and bore diam.

Because of the number of variables within the P/M process, it is not possible to state absolute dimensional tolerance capabilities. However minimum tolerances that one should expect for ferrous-based P/M gears without secondary operations are listed in Table I (spur and helical gears) and Table II (bevel gears).

LOAD RATINGS OF P/M GEARS

Choice of a P/M alloy is governed by the magnitude and nature of the transmitted load, the speed, the life

Table I. Minimum Tolerance Capabilities for Spur and Helical P/M Gears

Tooth-to-Tooth Composite Tolerance	AGMA Class 6
Total Composite Tolerance	AGMA Class 6
Test Radius	(0.002 in.) + (0.002 in.) × (diam)
Over Pin Measurement	(0.004 in.) + (0.002 in.) × (diam)
Lead Error	0.001 in./in.
Perpendicularity—Face to Bore	(0.002 in.) + (0.001 in.) × (diam)
Parallelism	(0.001 in.) + (0.001 in.) × (diam)
Outside Diameter	(0.004 in.) + (0.002 in.) × (diam)
Profile Tolerance	0.0003 in.

requirements, the environment, the type of lubrication, and the gear and assembly precision.

Most P/M gears are limited in performance by their impact or fatigue strength, as opposed to their compressive yield strength or surface durability. By fol-

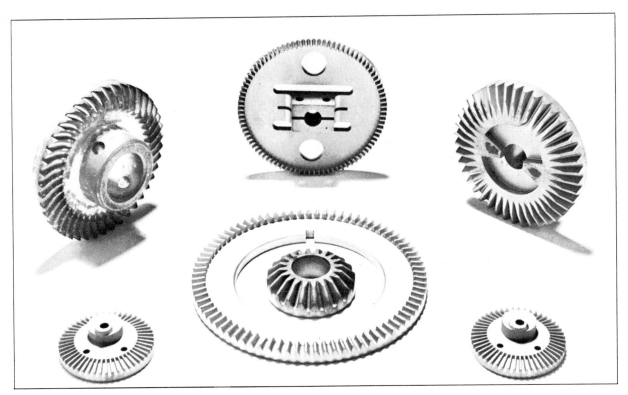

Fig. 3—P/M bevel and face gears.

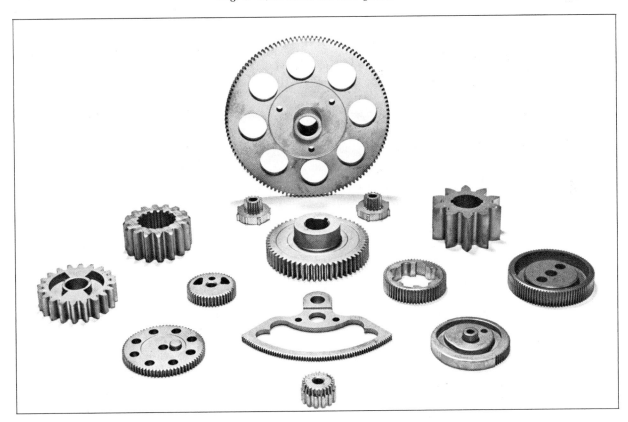

Fig. 4—Examples of P/M gear shapes.

lowing the procedure outlined below, the designer can narrow the choice of materials to select the most economic material for his application.

When selecting a material, the first step is to determine which P/M alloys have the necessary strength. In spite of the approximations involved, one can gener-

ally arrive at the correct material by using the methods outlined in the AGMA standards for rating the strength of spur, helical, and straight bevel gear teeth.[2,3,4]

From the equations and application data, one can predict the power which can be transmitted by a given gear set of particular materials. For example, the following

Table II. Minimum Tolerance Capabilities for Bevel P/M Gears

Tooth to Tooth Composite Tolerance	AGMA Class 6
Total Composite Tolerance	AGMA Class 6
Outside Diameter	(0.004 in.) + (0.002 in.) × (diam)
Perpendicularity–Face to Bore	(0.002 in.) + (0.001 in.) × (diam)

formula for allowable power is taken from AGMA 220.02 for spur gears.[2]

$$P_{at} = \frac{n_p d K_v}{126,000\,K_o} \times \frac{F}{K_m} \times \frac{J}{K_s P_d} \times \frac{K_L S_{at}}{K_R K_T}$$

where

P_{at} = allowable power of gear set, hp
n_p = gear speed, rpm
d = operating pitch diam of gear, in.
K_v = dynamic factor
K_o = overload factor
F = face width, in.
K_m = load distribution factor
J = geometry factor
K_s = size factor
P_d = diametral pitch
K_L = life factor
K_T = temperature factor
K_R = factor of safety
S_{at} = allowable bending stress for material, psi

Having the application and gear data, factors K_v, K_o, K_m, J, K_s, K_L, K_T, and K_R can be determined from sections of AGMA 220.02 and one only needs the allowable bending stress for the P/M material, S_{at}. In the absence of those data, it is a good approximation to use 0.3 of the tensile strength of ferrous based P/M materials ($S_{at} = 0.3 \times$ T.S.). The tensile strengths of various P/M materials are listed in MPIF Standard No. 35.[5] The only caution one should be aware of is that the data are typical, and the designer should consult the supplier for minimum property data. This is extremely important since minimum strengths determine mechanism functionality.

Assuming the material has sufficient strength to carry the bending loads, one next calculates the surface durability of the gear teeth. A good approximation of compressive load carrying ability can be found by using the methods outlined in the AGMA Standards for surface durability of spur, helical, and bevel gear teeth.[6,7,8] In addition to the application data, one must have the allowable contact stress (S_{ac}), the elastic modulus, and Poisson's ratio of the P/M material.

Tensile strength, elastic modulus, and Poisson's ratio of most ferrous-based P/M materials are also covered in MPIF Standard No. 35.[5] In the absence of these data, the allowable contact stress of ferrous-based P/M materials can be approximated by subtracting 10,000 psi from the tensile strength (S_{ac} = T.S. − 10,000 psi). In the absence of elastic modulus or Poisson's ratio data, values of ferrous-based P/M materials can be taken from Fig. 5. The elastic modulus and Poisson's ratio of ferrous-based P/M materials is proportional only to the density, not to the alloy or state of heat treatment for densities between 6.2 and 7.4 g/cc.[9]

In addition to rating a set of gears for strength and

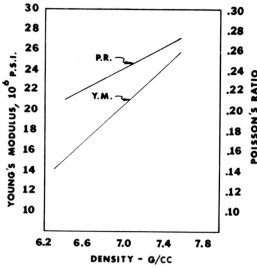

Fig. 5—Young's modulus (Y.M.) and Poisson's ratio (P.R.) rise with density of P/M parts.

durability, one must also be aware of potential scoring problems. Even though involute gears roll, enough inaccuracies exist so that the gear teeth surfaces slide. This effect can lead to scoring of gear teeth if the materials are incompatible.

In contrast to most metals, P/M materials have both a macroscopic and microscopic hardness as a result of porosity within the structure. Because the P/M structure consists of individual hard particles bonded together with interdispersed voids, the conventional Rockwell indention registers a composite hardness. Since sliding and surface contact involve microscopic contact (individual asperities), the particle or microhardness of the P/M material influences the score resistance of the P/M material more so than the macroscopic or conventional Rockwell hardness. As a generality, when a P/M gear is run against a wrought gear, the P/M particle hardness should be no harder than the hardness of the wrought gear to prevent wear of the mating gear.

QUALITY CONTROL

Inspection, one of the most important steps in the use of P/M gears, involves two stages: samples and production. In sample inspection, basic gear data such as tooth profile and size are checked because the pressing die is the main influence over these variables. The mechanical properties of the gear are also measured, and in many instances bench life and impact tests (or field tests) are made to check the gear functionality. Since one cannot measure the tensile strength of a gear directly, it is appropriate to establish a minimum tooth breakage strength that can be used as a production acceptance criteria from either sample gears made from a die or gears cut from a P/M slug.

One quick, accurate means of measuring the tooth strength of a spur gear is shown in Fig. 6. By mounting the gear between fixed supports and applying a vertical load, one can measure a tooth breakage strength. By testing a number of gears that perform successfully in the particular application, it is possible to calculate an X − 3σ lower limit for the tooth breakage strength that can be used as an acceptance test for fu-

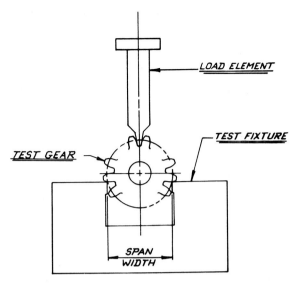

Fig. 6—Strength test for spur gear teeth.

ture material. Helical and bevel gears are usually evaluated in torque type tests, and the same type of statistical analysis is applied.

Use of statistics is particularly expedient in gear strength testing because one can use a sampling to project to the population, assuming the normal statistical assumptions hold. If data are normal, the use of an $\overline{X} - 3\sigma$ minimum acceptance criteria assures the customer that approximately 99.85 pct of his gears will have a tooth breakage above the selected minimum load. The use of a minimum tooth breakage strength acceptance has shown an excellent correlation to field performance.

For surface durability, it is proper to specify and measure a minimum Rockwell hardness because this macroscopic hardness relates to the material's compressive yield strength. The minimum Rockwell hardness depends upon the material, and can be established either from samples or from the supplier's information.

Where scoring may be a problem, in particular with heat-treated gears, it is appropriate to specify and measure a microscopic hardness. This is accomplished with a diamond indenter, and the microscopic or individual particle hardness of the material is measured. The particle hardness specification is generally determined from historical data for the producer's material.

In production, it is common practice to check the face width, outside diameter, over-pin measurement, total composite error, tooth-to-tooth error, perpendicularity of face-to-bore, parallelism, and inside diameter of the gears. Also, tooth strength and macroscopic hardness are usually checked. The sampling plan and frequency of measurement depend upon the criticality of the application.

SUMMARY

Spur, helical and bevel powder metallurgy gears are being used in all power transmission fields. Powder metallurgy, but nature a labor saving process that provides precision products with good structural properties.

Powder metallurgy technology has advanced considerably in the last 15 years, and today it is possible to predict P/M gear performance from available mechanical property data and a knowledge of the application. It is also possible to set quality control standards that will assure reliable, consistent performance.

REFERENCES

1. *Powder Metallurgy Design Guide Book,* Powder Metallurgy Parts Association, P.O. Box 2054, Princeton, New Jersey, 08540.
2. *AGMA Standard for Rating the Strength of Spur Gear Teeth,* AGMA 220.02, Aug., 1966, American Gear Manufacturers Association, 1330 Massachusetts Avenue, N.W., Washington, D.C. 20005.
3. *AGMA Standard for Rating the Strength of Helical and Herringbone Gear Teeth,* AGMA 221.02, July, 1965.
4. *AGMA Standard for Rating the Strength of Straight Bevel and Zerol Bevel Gear Teeth,* AGMA 222.02, Jan., 1964.
5. *P/M Materials Standards and Specifications,* MPIF Standard No. 35, Metal Powder Industries Federation, P.O. Box 2054, Princeton, New Jersey, 08540.
6. *AGMA Standard for Surface Durability of Spur Gear Teeth,* AGMA 210.02, Jan., 1965.
7. *AGMA Standard for Surface Durability of Helical and Herringbone Gear Teeth,* AGMA 211.02, Feb., 1969.
8. *AGMA Standard for Surface Durability Formulas for Straight Bevel and Zerol Bevel Gear Teeth,* AGMA 212.02, Jan., 1964.
9. W. E. Smith: *The Need for Non-Destructive Testing of Green P/M Parts,* presented at The Non-Destructive Testing of Green P/M Parts Seminar, May, 1970, Hartford, Connecticut.

The Integrity of P/M Forged Ni-Mo Steels Determined by Fracture Toughness

R. M. PILLIAR, D. R. HOLLINGBERY AND W. F. FOSSEN*

ABSTRACT

Plane strain fracture toughness tests of powder forged Ni-Mo steels in the quenched and tempered condition were used to determine the influence of process variables on properties. The effects of preform variables (powder type, size, density, sintering temperature), and forging variables (energy, temperature, rate) on K_{Ic} properties were studied and the results were used to suggest a modified powder forging route for improved properties.

Scanning and transmission electron microscopy and optical metallography were used to correlate variations in K_{Ic} to structural features within the forgings. The results indicated a strong dependence of K_{Ic} values on total oxygen level.

Introduction

Ni-Mo steel powders have been used extensively for the formation of heat treatable components by powder forging. The prealloyed powders used are readily hardenable after consolidation and, in general, the formation of oxides on powder surfaces prior to consolidation can be prevented. The standard practice in making parts from these powders is to mix elemental carbon to the carbon-free alloy powders, compact to the desired preform and then sinter and forge or directly hot forge the preform to full density. The commercial alloy powders used are low in carbon content and, therefore, highly compressible.

In earlier studies, the plane strain fracture toughness of Ni-Mo and Cr-Mn powder forged steel was determined[1,2,3]. The effect of surface oxides on powder particles in lowering the fracture toughness of the forged samples was reported. This effect was strongest with oxygen active, (Cr-Mn), steel powders. The success of the plane strain fracture toughness test in assessing the properties of the powder forged steels was demonstrated. The effect of microstructure on fracture toughness of the forged compacts was noted and related to powder characteristics and processing. In this study, a number of process variables and powder characteristics that could affect fracture toughness of Ni-Mo powder forged steels was investigated. The effect on K_{Ic}, (plane strain fracture toughness), of powder characteristics, (chemistry, shape, size), preform properties and processing, (density, sintering temperature), and forging parameters, (energy, temperature, rate), was studied. K_{Ic} values were related to fracture appearance and microstructure of the powder forged test specimens.

*Centre for Powder Metallurgy, Ontario Research Foundation, Mississauga, Ontario, Canada.

THE INTEGRITY OF P/M FORGED Ni-Mo
STEELS DETERMINED BY FRACTURE
TOUGHNESS
Pilliar, Hollingbery and Fossen

TABLE 1

	C	P	S	Si	Al	Ni	Mo	Mn	Fe
4600		.013	.011	.023	.01	1.68	.52	.20	bal.
4600-1	.51	.042	.012	.01	<.01	2.09	.61	.22	bal.
4600-2	.44	.040	.014	.03	<.01	.49	.60	.42	bal.

TABLE II

Particle Size Distribution Mesh Size	4600V	4600-1	4600-2 (−80 mesh)	4600-2 (−48 mesh)	4600-2 (−3 mesh)
+3					30.8
−3+8					24.8
−8+14					24.7
−14+28					12.0
−28+48					3.4
−48+65				46.2	1.3
−65+80	4.8	1.4	1.1	20.1	0.7
−80+100	5.5	6.3	12.5	11.8	1.0
−100+150	17.1	13.9	25.3	13.6	0.7
−150+200	22.5	19.8	25.2	6.2	0.1
−200+250	3.4	4.3	4.7	0.4	0.2
−250+325	20.8	20.2	16.8	1.4	0.2
−325	25.9	34.0	14.4	0.3	0.1
Apparent Density g cm^{-3}	3.18	3.06	3.69	4.44	3.86
Flow Rate, sec/50 g	22	24	18	N.F.	20

Experimental

Materials

The three powder compositions investigated are shown in Table I. The Hoeganaes 4600V powder is a commercially available, high compressibility (no carbon in solution) powder. Southwestern grade 1651 graphite was added to give a carbon level after forging of approximately 0.50 w/o carbon. The 4600-1 and 4600-2 powders are fully prealloyed, (carbon included), powders prepared in house by water atomization followed by a reduction anneal. The major difference between the 4600V, 4600-1 and 4600-2 powders is the higher Mn and lower Ni contents of the 4600-2 powder. The effect of higher Mn content on oxide formation and its subsequent effect on structural integrity of the powder forged parts was studied. Particle size distribution, shape, powder flow rate and apparent density for the minus 80 mesh size fraction were similar for the three powder types, (Table II). In addition to the minus 80 mesh size fraction, minus 48 and minus 3 mesh fractions of the 4600-2 powder were included in the program.

Processing Method

Two different methods of powder consolidation were used. The majority of preforms were compacted using a high energy rate forming machine, (Petro-Forge)[4]. By varying the charge pressures in the Petro-Forge, compacts of densities between 70 and 90% full density (approximate) were formed. Compacts were formed in a 0.05 m diameter cylindrical die and weighed approximately 0.450 kg. Zinc stearate die wall lubrication were employed for all compacts. The Petro-Forge provided a high input energy rate machine for powder forging. The total energy input, (and the rate), was varied by changing the combustion conditions. Energy inputs varying from 7450 joules to 15800 joules were achieved. Forging speeds up to 14.0 m sec^{-1} were attained.

The remaining samples were forged on a conventional hydraulic press. Forging speeds in this case were equal to approximately 2.8 m sec^{-1}. The coarser carbon-containing powders were compacted by conventional slow speed pressing with 0.3% Methofas binder addition because of their regular shape and higher hardness.

All preforms were sintered in high purity hydrogen atmosphere, (Dewpoint = 30 C), at 1010 C or 1200 C for 1/2 h, brought to the forging temperature in the same high purity H_2 atmosphere and then were transferred into a controlled atmosphere container to the die for hot forging. A 0.058 m cylindrical closed die was used for the hot forging operation. Transfer time between furnace and forging die was less than three seconds thereby minimizing decarburization or oxidation of the preform prior to forging.

Testing Methods

Compact tension test specimens for K_{Ic} determination in accordance with ASTM E-399 specifications were machined from the forgings. Charpy V-notch impact specimens were also made from some of the forgings. Both the K_{Ic} and Charpy samples were notched such that the resulting fracture plane was perpendicular to the forging plane and the crack propagated perpendicular to the forging direction. The machined samples were normalized at 875 C, austenitized at 830 C, oil quenched and tempered to the desired hardness level. The fracture toughness test specimens were fatigue pre-cracked and fracture toughness values were determined at room temperature. Charpy V-notch impact tests were also conducted at room temperature.

Samples for density and oxygen content determinations were cut from near the fatigue precrack of the tested K_{Ic} specimens. Oxygen determinations were made using a LECO analyzer. Each specimen was analysed three times to ensure reproducibility of results. Densities were determined by water displacement methods.

Fracture surfaces were examined by scanning electron microscopy and inclusions on the fracture surfaces were identified by SEM/X-ray methods and by selected area electron diffraction studies of carbon extraction replicas formed on the fracture surfaces.

Results

Charpy Impact vs K_{Ic} Test

The majority of K_{Ic} testing was done on samples tempered to a hardness level of 45 to 47 R_c. At this hardness level, all Charpy V-notch impact tests yielded impact resistances below 9 joules. At such low levels the significance of differences in impact resistance values is questionable.

To produce more meaningful Charpy impact tests results, samples were tempered to a hardness of 36 R_c. It was assumed that inclusions that might affect the 46 R_c tempered samples would also influence the lower hardness Charpy impact specimens although not necessarily to the same extent. An attempt to correlate K_{Ic} values to C_v values for the Ni-Mo steels tested is shown in Fig. 1.

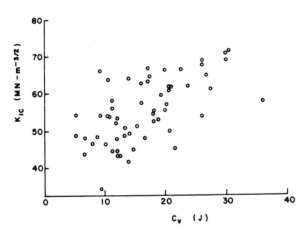

FIGURE 1 Graph indicating K_{Ic} vs Cv for compact-tension test specimen tempered to 46 R_c and Charpy V-notch specimen tempered to 36 R_c.

No strong correlation was observed. The K_{Ic} value represents an intrinsic material property and for this reason we concerned ourselves in subsequent studies with the effects of powder and process variations on this property only.

Material Variables - Composition and Powder Size

Metallographic examination of all heat treated samples sintered at 1200 C prior to forging indicated a tempered martensitic structure. Variations between samples because of inclusions within these tempered martensitic structures were observed, (Figs. 2 & 3).

FIGURE 2 Microstructure of 4600V forging after oil quenching and tempering treatment. Powder preform was sintered at 1200°C and forged at 1010°C. X500

The 4600V powder preforms sintered at 1200 C and forged at 1010 C resulted in a relatively clean tempered martensitic structure with very few inclusions. (Fig. 2). Prior particle boundaries were not evident. The 4600-1 samples appeared similar in microstructure.

In contrast the microstructure of forgings made from the 4600-2, minus 3 mesh powder is shown in Fig. 3. Large inclusions were observed at some of the prior particle boundaries and all boundaries were delineated by a darker etching zone. Higher magnification and electron microprobe studies of this darker etching boundary zone indicated a number of fine inclusions as well as a lower Mo and higher Mn content in this region. The 4600-2, minus 80 mesh powder preforms sintered and forged in the same way also showed some inclusions at prior particle boundaries but there was no evidence of a darker etching prior particle boundary zone.

The effect of higher Mn contents of the 4600-1 powder forgings on K_{Ic} is shown in Table III. The 4600-2, minus 80 mesh forged

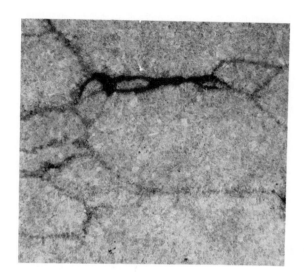

FIGURE 3 Microstructure of 4600-2, minus 3 mesh powder forging. The powder preform was sintered at 1200°C and forged at 1010°C. Oxide inclusions and structural inhomogeneities are seen at the prior particle boundaries.
X100

TABLE III Effect of Powder Composition on K_{Ic}

Material	O_2 ppm	K_{Ic} MN-m$^{-3/2}$	No. of Tests
4600-1	38	69±3	3
4600-2 (−80M)	279	57±1	2
(−48M)	370	54	1
(−3M)	646	64	1
4600V + 0.5C	83	59±4	3
Wrought	129	65	1

Sinter temperature = 1200°C
Forging energy = 13,150 joules
Preform density = 80% bulk density
Forging temperature = 1010°C

samples had lower K_{Ic} values, (57 MN-m$^{-3/2}$), than the 4600-1 samples, (69 MN-m$^{-3/2}$). Oxygen analysis indicated a large difference for the two types of powder forgings, (280 vs 38 ppm). Table III also indicates the higher K_{Ic} values for the fully prealloyed Ni-Mo powder forged steels compared to steels made

from high compressibility powders with elemental carbon added, (69 MN-m$^{-3/2}$ vs 59 MN-m$^{-3/2}$). The 4600-1 powder forgings also had lower oxygen content, (38 vs. 85 ppm).

The effect of particle size on K_{Ic} for the 4600-2 samples is shown in Table III. As indicated, powder size does not appear to significantly influence K_{Ic} for the higher Mn, 4600-2 forgings. However, oxygen content of the forgings was observed to increase with increasing powder particle size.

All samples in Table III were forged using the Petro-Forge at 13,150 joules from 80% dense preforms. Forged sample densities were 99.8+% of full density.

Preform Variables - Sintering Temperature and Preform Density

The 4600V powder preforms sintered and forged at 1010 C resulted in much lower fracture toughness properties compared to the 1200 C sintered samples, (Table IV). This influence of sintering temperature on fracture toughness was the strongest factor observed in these experiments. The lower sintering temperature resulted in K_{Ic} values less than two-thirds those of the 1200 C sintered

TABLE IV Effect of Sintering Temperature on K_{Ic}

Material	Sinter Temp. (°C)	O_2 (ppm)	MN-m$^{-3/2}$	No. of Tests
4600V + 0.5C	1010	602	36±3	3
4600V + 0.5C	1200	83	59±4	3

All samples forged at 13,150 joules using Petro-Forge
Forging temperature = 1010°C
Preform density = 80 bulk density

samples, (36 vs 59 MN-m$^{-3/2}$). The oxygen contents of these 1010 C sintered samples were very high, (>600 ppm). Optical micrography of these samples indicated the heavy oxide inclusion networks within these samples, (Fig. 4).

Tests on 4600V samples forged from 70, 80 and 90% full density preforms indicated no significant effect of preform density on K_{Ic} for the range of forging energies studied, (Fig. 5).

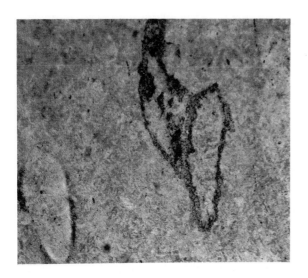

FIGURE 4 Microstructure of 4600V powder forged sample following heat treatment. The preform was sintered and forged at 1010°C. Particle boundary inclusions are seen in some regions. X500

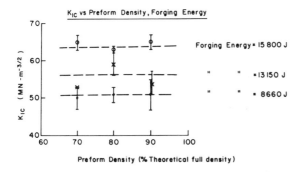

FIGURE 5 Graph showing the effect of preform density and forging energy on K_{Ic} of powder forged and heat treated 4600V samples.

Forging Variables - Rate, Energy and Temperature

In Table V, section i), K_{Ic} values for specimens formed using high energy rate methods, (Petro-Forge), and slower rate conventional hydraulic pressing methods are compared. For the powders studied there does not appear to be a significant effect of forging rate on K_{Ic}. As shown in Table V, section ii) and Fig. 5, increases in energy using the Petro-Forge resulted in higher K_{Ic}. The density of the lower energy forged samples was slightly less than that for the 13,150 and 15,800 joule forged samples, (99.2% vs 99.8%+ full density). Table V, section iii), and Fig. 6 show the effects of forging temperature on K_{Ic}. Increasing forging temperatures result in higher K_{Ic} values. The 815 C forged samples were found to be slightly lower in density, (equal to 98.4% full density). The samples forged at 875 and 1010 C were >99.8% full density.

In previous tests on Cr-Mn powder forged steels, a strong dependence of K_{Ic} on oxygen content was noted[3]. A similar correlation was observed in these studies on Ni-Mo steels. Fig. 7 represents a plot of K_{Ic} vs oxygen content for the Ni-Mo steels studied. The 4600-2, minus 3 mesh samples and the lower density samples do not conform to the general K_{Ic} vs oxygen correlation.

Fracture Appearance & Inclusion Identification

Scanning electron micrographs of the fracture surfaces for high and low oxygen containing samples are shown in Figs. 8a and 8b. In general, the lower oxygen samples (<300 ppm) were characterized by trans-particle fractures. The higher oxygen containing samples, (>300 ppm), showed regions of interparticle delamination mixed with the transparticle fracture regions. In

TABLE V Effect of Forging Variables On K_{Ic}

	Sinter Temp. (°C)	Preform Density (%)	Rate (m-sec^{-1})	Energy (Joules)	Forging Temp. (°C)	O_2 (ppm)	K_{Ic} (MN-m$^{-3/2}$)	No. of Tests
i) Forging Rate								
4600V	1010	80	14.0	14000	1010	602	36±3	3
4600V	1010	80	2.8	×	1010	622	35±2	2
4600-2 (−80M)	1200	80	14.0	13150	1010	279	57±1	2
4600-2 (−80M)	1200	80	2.8	×	1010	252	58±3	2
ii) Forging Energy								
4600V	1200	70	14.0	8660	1010	96	50±3	4
4600V	1200	70	14.0	13150	1010	221	53	2
4600V	1200	70	14.0	15800	1010	58	65±2	3
4600V	1200	80	14.0	8660	1010	71	51±2	3
4600V	1200	80	14.0	13150	1010	83	59±4	3
4600V	1200	80	14.0	15800	1010	72	63±1	3
4600V	1200	90	14.0	7450	1010	102	51±4	3
4600V	1200	90	14.0	13150	1010	306	54±3	2
4600V	1200	90	14.0	15800	1010	105	65±2	3
iii) Forging Temp.								
4600V	1200	80	14.0	13150	815	245	44±1	2
4600V	1200	80	14.0	13150	875	208	54±1	2
4600V	1200	80	14.0	13150	1010	83	59±4	3

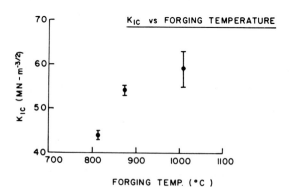

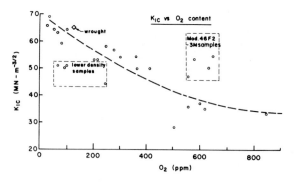

FIGURE 6 Graph showing the effect of forging temperature on K_{Ic} of 4600V powder forgings.

FIGURE 7 Dependence of K_{Ic} on oxygen content for Ni-Mo powder forged steels. All samples were heat treated to a hardness of 45 - 47 R_c. A wrought sample is included in the graph.

a

b

FIGURE 8 Scanning electron fractograph of powder forged Ni-Mo steels; a) high O_2 (649 ppm) 4600-2, minus 3 mesh sample b) low O_2 (38 ppm) 4600-2, minus 80 mesh sample. Note the zones of interparticle delamination in the high oxygen containing samples.
X40

these zones of interparticle failure, large non-metallic inclusions were evident on the fracture surfaces. Fig. 9 shows a scanning electron fractograph of a sample sintered at the lower temperature, (1010 C), and consequently having higher oxygen content. A zone of interparticle delamination is evident containing a number of flat, cleavage-like features, (A in Fig. 9). X-ray/SEM analysis of the fracture surfaces indicated oxygen containing inclusions in these regions.

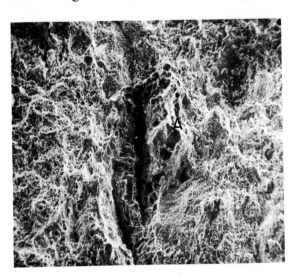

FIGURE 9 Scanning electron fractograph showing a region of interparticle delamination. Brittle, oxide inclusions can be seen at the delamination site, A, characterized by cleavage-type failure.
X420

Fracture of the larger oxide inclusions at the particle boundaries resulted in the cleavage-like features. Adjacent but still contained within the delamination zone, was a very fine dimpled structure with inclusion fragments at the base of the dimples. Fig. 10 shows such a zone for the 4600-2, minus 3 mesh sample: the transparticle fracture zones next to the interparticle delamination regions contained a coarser dimple network (Fig. 11). Inclusions were observed at the base of some of these coarser dimples. The source of these inclusions probably differed from the inclusions found in the interparticle delamination zones.

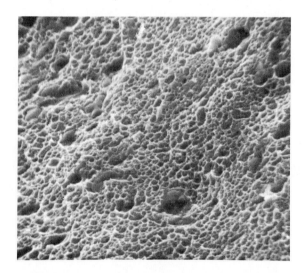

FIGURE 10 Zone of interparticle delamination failure showing the fine dimple size and fine inclusions associated with many of the dimples. Coarser inclusions are also seen. X3900

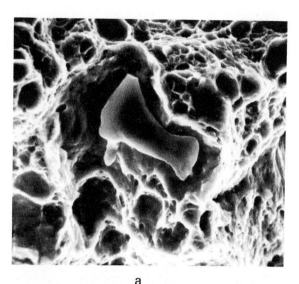

a

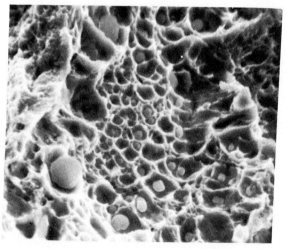

b

FIGURE 12 Scanning electron fractograph of high Mn containing, 4600-2 powder forged sample; a) containing a MnS inclusions and b) containing Mn_2SiO_4 inclusion fragments. X4100

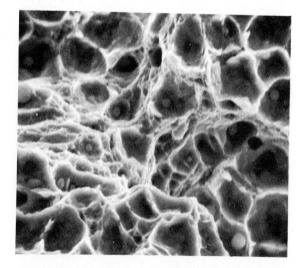

FIGURE 11 Zone of transparticle failure. The dimples size appears coarser than in Fig. 10. Some inclusions are observed. X3900

The frequency, size and distribution of the inclusions found in transparticle fracture regions could account for differences in fracture toughness of samples that failed primarily in this manner. These included samples with oxygen levels below 300 ppm. K_{Ic} values varies, however, between 50 and 69

$MN-m^{-3/2}$ for these samples. Fig. 12a and 12b show such inclusions in the 4600-2 powder forgings. SEM/X-ray analysis and selected area electron diffraction studies of extracted inclusions indicated that they were sulphides, silicates and oxide phases. As shown in Fig. 12b, fragmentation of inclusions could occur resulting in a finer dimpled structure.

The fracture surfaces for the lower Mn-containing 4600V and 4600-1 powder forgings had fewer apparent inclusions compared to the higher Mn 4600-2 forgings, (Fig. 13). This was expected because of the lower oxygen activity of the lower Mn containing powders.

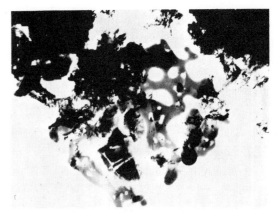

FIGURE 14 Transmission electron micrograph of a region of a carbon extraction replica formed from the fracture surface of a 1010° C sintered and forged P/M sample. Selected area electron diffraction of this inclusion indicated the central crystalline zone to be Mn_2SiO_4. Analysis using a wavelength dispersive unit indicated Ca and Si in the surrounding vitreous appearing zone. X2200

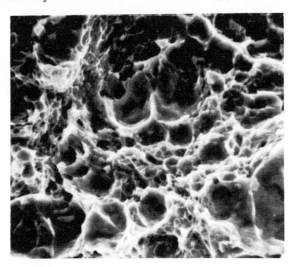

FIGURE 13 Fracture surface of a low Mn containing 4600-1 powder forging. Fewer inclusions are observed than in Fig. 12. X4200

The electron diffraction studies of some of the extracted inclusions yielded spot diffraction patterns that were identified as silicate and oxide phases, (Mn_2SiO_4, Fe_2SiO_4, Mn_3O_4). Some of the inclusions as seen in Fig. 14 took on a fused glassy appearance suggesting the presence of liquid phase inclusions during the sintering/forging process. X-ray analysis of this particular inclusion using a wavelength dispersive unit indicated the presence of Si and Ca.

The fracture appearance of a wrought Ni-Mo steel is shown in Fig. 15 for comparison. The dimpled fracture zone appears similar to that observed for most of the powder forged samples that failed in a transparticle mode. The inclusion appearance was similar to that for the lower oxygen containing 4600V powder forgings.

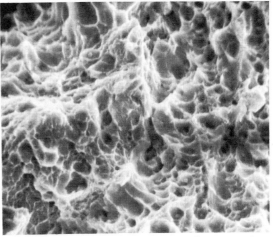

FIGURE 15 Scanning electron fractograph of wrought Ni-Mo steel specimen heat treated to a hardness of 45 - 47 R_c.

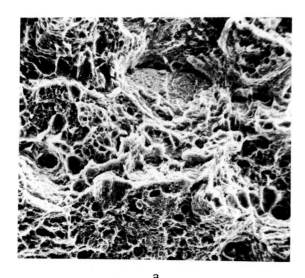

a

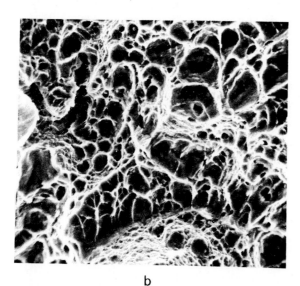

b

FIGURE 16 Scanning electron fractographs of forgings made from 4600V powders; a) forged at 15800 joules and b) forged at 8600 joules. X1750

As noted above, the lower forging energy, (7450 and 8660 joules), resulted in a lower K_{Ic} for the 4600V powder forged samples. Oxygen contents for the samples forged at the different energy levels were similar. Extensive fractographic examination of the samples forged at the different energy levels from the

80% density preforms did not indicate a significant difference in fracture appearance for the high (63 MN-m$^{-3/2}$), and low (51 MN-m$^{-3/2}$) K_{Ic} samples, (Fig. 16a and 16b).

Discussion

The results indicate that the total oxygen content, as demonstrated previously[3], strongly influences the fracture toughness of powder forged intermediate strength steels. The inclusions most influencing fracture toughness were located at particle boundaries. They probably were surface oxides on the prior powder particles, formed during or after atomization by powder oxidation. A second series of inclusions were found within the powder particles. These inclusions originated from the steel making practice independent of the atomization process. Complex oxides and manganese sulphides were identified within the powder particles. This group of inclusions had a less critical effect on K_{Ic} but did result in some differences for well consolidated powder forgings.

The various oxides phases identified at the interparticle failure zone represent weaker structures where easy crack initiation can occur resulting in lower K_{Ic} values. The aligned nature of these inclusions along the prior particle boundaries makes them very influential on crack initiation. These findings support those of earlier studies on both Cr-Mn and Ni-Mo powder forged steels[1,2,3].

Of particular interest in this study was the influence of powder composition and processing on total oxygen content and, consequently, K_{Ic}. The most significant parameter controlling K_{Ic} for the Ni-Mo steels studied was the sintering temperature used prior to forging.

Electron diffraction studies of replicas extracted from the fracture surfaces of the 4600V samples forged at 1010 C indicated the presence of silicate, (Fe_2SiO_4, Mn_2SiO_4), and oxide, (Mn_3O_4), inclusions. At the higher sintering temperature, (1200 C), total oxygen content was much lower, (Table IV), and inclusions were not detected easily on the extraction replicas. The depletion of these complex oxide phases after the higher temperature sintering probably was a result of the reduction of these oxides during the hydrogen atmosphere sintering. The presence of free carbon at the inclusion site would also contribute to this reduction. Because of the complex nature of the oxides, and the dynamic nature of the furnace atmosphere, (flowing H_2), it is difficult to speculate on the exact mechanisms of inclusion reduction but certainly the reactions would be favored by the higher sintering temperature.

The appearance of some glass-like inclusions, (Fig. 14), even after the 1010 C sintering, suggests that a second possible mechanism for oxide removal would be vapor phase reactions in the flowing H_2 atmosphere. Some silicate phases are known to have significant partial vapor pressures at 1200 C[5]. MoO_3, also exists in the vapor phase above 1100 C[6].

In a direct comparison of the lower Mn 4600-1 and the higher Mn 4600-2, minus 80 mesh samples sintered at 1200 C, the former were shown to have lower oxygen contents and higher K_{Ic}, (Table III). Manganese is relatively stable at 1200 C and it would appear best to limit the amount of this phase by controlling the Mn content of the alloy powders. It is recognized, however, that economic considerations might result in the use of the richer Mn and leaner Ni containing steel powders. With the 1200 C sintering treatment even the higher Mn steels have good fracture toughness, (i.e. 58 MN-m$^{-3/2}$).

Powders containing prealloyed carbon resulted in the lower oxygen contents after forging than did the samples with elemental

carbon additions. Although attention has been focused on powder surface oxide phases, the inclusions within the particles also will influence fracture toughness albeit to a lesser extent. The addition of carbon to the melt stock strengthens the reducing conditions during sintering thereby yielding a cleaner powder. K_{Ic} for the 4600-1 powder forgings in fact was comparable to the wrought samples studied. Therefore, with proper melting practice and powder handling during the forging operation, P/M forged steels can be as tough as wrought Ni-Mo steels.

The 4600-2 powder was used to study the effect of powder size on K_{Ic} of forged samples. The oxygen content of these samples increased with particle size. The higher oxygen content of the coarser particles probably results from the increased time for surface reaction of these larger particles during atomization. The slower cooling of the coarser particles also results in their more regular particle shape. The reduction anneal after atomization apparently was insufficient to cause the complete reduction of the heavier oxide layer on these large particles.

The higher Mn content detected at the particle boundaries of the forgings made from the minus 3 mesh fraction powders resulted from the high oxygen levels at the particle surface and the oxygen activity of Mn. Similarly Mo being a surface active ingredient, (high oxygen affinity), would diffuse preferentially to the particle surfaces and form MoO_3. This phase exists in the vapor form at temperatures above 1100 C[6]. The lower Mo content at the particle boundaries of the minus 3 mesh powder forgings result from the loss of MoO_3 by this vapor phase reaction during atomization, annealing and sintering.

Of particular interest in this study were the relatively high K_{Ic} values for the coarser size fraction 4600-2 powder forged samples.

Despite a very high oxygen content for the minus 3 mesh samples, (>600 ppm), the K_{Ic} values were equal to approximately 64 MN-m$^{-3/2}$. The fracture surface indicated a number of interparticle delamination zones normally associated with low K_{Ic} values. However, as seen in Fig. 3, the forging operation resulted in an elongation of particles perpendicular to the forging direction. The particle boundaries consequently became preferentially oriented in a direction perpendicular to the crack growth direction during K_{Ic} testing. Additionally, because of the large particle size, the crack initiation process as measured in the K_{Ic} test, might not be influenced by the particle boundaries in these samples if, for example, the fatigue precrack tip was situated within a particle. This would explain the high K_{Ic} values for these high oxygen forgings. It should be noted that crack extension in the direction parallel to the elongated particles, (i.e. parallel to the forging plane and perpendicular to the forging direction), should be much easier. The coarse size fraction forgings, although exhibiting high K_{Ic} values in the reported tests, would probably be highly anisotropic. This anisotropy of powder forged steels has been reported elsewhere[7].

The preform density within the range studied, (70 to 90%), did not affect K_{Ic}. This suggests that the use of 70% dense preforms would have economic advantages. It is important however that full, (virtual) density is achieved after the final forging. The effect of a small percentage of voids on lowering dynamic mechanical properties, (and fracture toughness), is well known[8,9].

This effect of a small amount of void on K_{Ic} might have accounted for the lower K_{Ic} values for the samples forged at 815 C. Both the 875 and 1010 C forged 4600V specimens were fully dense and K_{Ic} values were normal for the appropriate oxygen contents, (Figs. 6 and 7). The 815 C sample had an abnormally low K_{Ic} value, (44 MN-m$^{-3/2}$), even considering its higher oxygen content. The density of this sample forged below the recommended austenitizing temperature, (830 C for 4640 alloys), was equal to 7.73 g cm^{-3} or 98.4% full density. The samples forged at higher temperatures were about 99.8% full density. The increased resistance to deformation of samples forged at below the austenitizing temperatures resulted in the lower density and lower K_{Ic}.

In studying the effect of forging energy on K_{Ic}, the effects of a small amount of void again appeared of importance. As indicated in Table V, section ii), and Fig. 5, the 4600V samples forged at lower energies appeared lower in K_{Ic}. These samples also were lower in density, (equal to 7.80 g cm^{-3} or 99.2% full density), and the lower K_{Ic} could be attributed to the residual porosity within the forgings. Others have reported the strong effect of small void content on mechanical properties[10,11]. However, for the samples forged at 13,150 and 15,800 joules, densities were greater than 99.8% full density. The different K_{Ic} for the samples forged at these higher forging energies could not be explained by residual porosity effects. A relation has been proposed between K_{Ic} and spacing between dimples on fracture surface such that $K_{Ic} \sim \sqrt{2\sigma_y E\lambda_c}$ where σ_y, E and λ_c are the yield stress, Young's modulus and interdimple spacing respectively[12]. Accordingly, a higher K_{Ic} value should be associated with a finer dimpled fracture surface. Scanning electron microscopic examination did not indicate a significant difference in the fracture surfaces for the samples forged at different energy levels. We can only assume that the higher forging energy resulted in stronger interparticle bonding throughout the forging. This improvement in K_{Ic} has probably been gained at the expense of die life. In a commercial process, a minimum forging energy to give a required toughness should be included in the development program to give a cost efficient operation. The use of different forging rates did not influence fracture toughness of the powder forged samples, (Table V, section i)).

Conclusions

The significance of total oxygen content on the fracture toughness of powder forged Ni-Mo steels has been demonstrated. It was shown that toughness equal to wrought steels can be achieved for these Ni-Mo P/M ferrous alloys provided that proper processing conditions are maintained. Sufficiently high sintering temperatures to achieve low total oxygen either by oxide reduction or complex oxide vapor reactions have been shown to be important in producing powder forgings possessing high K_{Ic} values. The significance of a small amount of void on loss of toughness has also been demonstrated. Some improvements in K_{Ic} can be gained by increasing the forging energy but this has potential economic drawbacks. The importance of the melt chemistry during the steelmaking prior to atomization was indicated.

Finally, the usefulness of plane strain fracture toughness determinations in assessing the properties of powder forged materials has been demonstrated.

Acknowledgements

This work was supported in part by the Department of Industry, Trade and Commerce, Ottawa, and the Ministry of Industry and Tourism, Government of Ontario. The authors are grateful to Messrs. H. Paju and R. Blackwell for technical assistance in sample preparation and testing and to I. Murray and P. Richardson for assistance in scanning electron microscopy. The helpful discussion of Dr. W. J. Bratina is gratefully acknowledged.

References

1. R. M. Pilliar, W. J. Bratina, J. T. McGrath; Modern Developments in Powder Metallurgy, Vol. 7, p. 51, MPIF, Princeton, N.J., 1974
2. R. M. Pilliar, T. J. Ladanyi, G. A. Meyers, G. C. Weatherly; Proceedings of 4th Int. Conf. on Powder Metallurgy, Vysoke-Tatry, Czechoslovakia, Part I, p. 233, 1974
3. T. J. Ladanyi, G. A. Meyers, R. M. Pilliar, G. C. Weatherly; Metallurgical Trans. A, Vol 6A, p. 2037, 1975
4. R. Davies and R. H. T. Dixon; Powder Met., Vol. 14, p. 207, 1971
5. J. F. Elliott and M. Gleiser; Thermochemistry for Steelmaking, 1960, (Addison-Wesley Publishing Co.)
6. O. Kubaschewski, E. L. L. Evans, and C. B. Alcock; Metallurgical Thermochemistry, IVth Edition, 1967, (Pergamon Press)
7. W. J. Bratina, W. F. Fossen, D. R. Hollingbery, R. M. Pilliar; "Anisotropic Properties of Powder Forged Ferrous Systems", paper presented at 5th International Powder Metallurgy Conference, June, 1976, Chicago, U.S.A.
8. S. M. Kaufman and S. Mocarski; Forging of Powder Metallurgy Preforms, p. 131, (Metal Powder Industries Federation, Princeton, NJ 1973
9. J. D. Benedetto, F. T. Lally and I. J. Toth; Ibid. p. 104
10. B. L. Ferguson, Sang Kee Suh, A. Lawley; Int. J. of Powder Met. and Powder Technology, Vol. 11, No. 4, 1975, p. 263
11. B. L. Ferguson, A. Lawley; Modern Developments in Powder Metallurgy, Vol. 7, p. 485, MPIF, Princeton, N.J., 1974
12. G. T. Hahn and A. R. Rosenfeld; Met. Trans. A, 1975, Vol. 6A, p. 653

FATIGUE BEHAVIOR OF HOT FORMED
POWDER DIFFERENTIAL PINIONS

P. C. Eloff and L. E. Wilcox

Gleason Works
Rochester, New York, U. S. A.

INTRODUCTION

Much has been published during the last four years
(1-3)* concerning the production of automotive differential
bevel pinions by the hot forging of powder metal preforms.
Most of this information has concerned the economics or
process details of the new approach. There has been little
quantitative data, however, on the fatigue or impact
behavior of differential pinions made from powder preforms.

The purpose of this research was to establish process
guidelines for an integrated system to isostatically compact,
induction sinter, and hot forge powder preforms. Previous
research (4) had shown that powder compacts could be very
effectively deoxidized in relatively short times by using
induction heating. The program described here was intended
to produce actual components under varying induction sintering
conditions as a check on the earlier laboratory work and
to determine whether components produced under such conditions
would have acceptable physical properties.

The automotive differential pinion was chosen as the
test component for three reasons: 1) it is one of the
most severe impact applications in an automobile, 2) the
technology of bevel gears is well-understood at Gleason
Works, and 3) tests of other powder pinions had been made,
allowing a comparison of powder processes. The primary

*Numbers in parentheses refer to "References", listed at
 the end of the paper.

questions to be answered by this investigation were:
1) What is the effect of various time-temperature sintering
cycles on impact and fatigue behavior? and 2) Will an alloy
substantially leaner than the 4600 series provide adequate
impact and fatigue behavior?

Out of consideration for time and cost, the numbers
of samples were kept to a minimum and because of this,
only gross effects will be shown by the data levels.
The experiments were designed, however, to give sufficient
confidence that additional pinions produced by similar
processing could be submitted to the automotive industry
for exhaustive testing.

EXPERIMENTAL PROCEDURE

Processing of Powder Formed Pinions

Two prealloyed steel powders were chosen for this
study: Anchorsteel 4600V* and Anchorsteel 2000*, both
produced by Hoeganaes Corporation. The composition of
the 4600V powder was 0.005%C, 1.67% Ni, 0.46% Mo, and
0.18% Mn, with traces of S, P, and Si. The composition of
the 2000 powder was 0.018%C, 0.50% Ni, 0.48% Mo, and
0.23% Mn, again with trace elements similar to the 4600V.
The purpose of the 2000 experimental alloy is to obtain
hardenability equivalent to that of AISI 4600 series
steels, but at a lower cost owing to the reduction in
nickel content. An object of this investigation was to
learn whether the reduced nickel content would cause undue
penalties in toughness.

Because the prealloyed powders did not contain sufficient
carbon for the application, (0.15-0.20%), graphite had
to be admixed to the powders. The amount of graphite
to be used had to be determined in a separate set of
experiments. It was known from previous research (4)
that the amount of carbon consumed during short-cycle
induction sintering depends not only upon the oxygen content
of the powder, but also upon the sintering temperature
and time at temperature. Table I shows the amounts of
graphite added to various mixes depending on the intended
sintering cycle.

No compacting lubricants were added to the powders as
the preforms were produced by dry-bag isostatic compaction.
Figure 1 is a schematic of the method used for compaction
of the preforms and shows the basic geometry of the
preforms.

*Trademarks of Hoeganaes Corporation.

TABLE I

Graphite Contents of Powder Mixes Used in Study

Mix No.	Base Powder	Sintering Temp.	Time	Percent Graphite
6041	2000	2350°F	3 min.	0.28%
			6 min.	0.28%
6042	4600V	2100°F	3 min.	0.29%
			6 min.	0.29%
6043	4600V	2350°F	1 min.	0.31%
			3 min.	0.31%
6044	4600V	2350°F	6 min.	0.33%

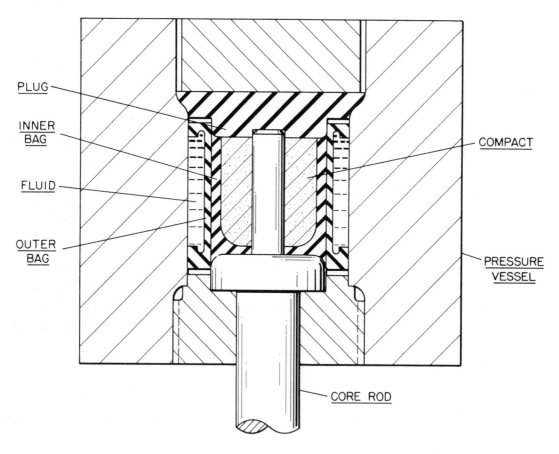

Figure 1. Schematic representation of dry-bag isostatic tooling.

As mentioned previously, several sintering cycles were used, ranging from 1 to 6 minutes at either 2100 or 2350°F. Heating was done using 3KHz induction power. The induction coil and preform fixture were enclosed in a chamber which was flushed with a protective atmosphere consisting of 90% N_2 - 10% H_2. The chamber was equipped to handle only one preform at a time. After the end of the sintering time was reached, the preforms were allowed to cool to approximately 1850°F and induction power was used to stabilize them at this temperature for one minute. The preforms were then removed from the sintering chamber and immediately transferred to the forging die. Because of the cylindrical preform shape, considerable metal flow was involved in forming the gear teeth; this was about 60%, based on the change in plan area during forming. Loads of about 80 tons per square inch, or about 240 tons total were required to completely form and densify the pinions. Formed pinions were allowed to cool in sand to prevent excessive oxidation or decarburization.

After deburring, all of the pinions were carburized in a single treatment at 1650°F to give an effective case depth of 0.040 - 0.050" (the depth at which the hardness is equivalent to 50 on the Rockwell "C" scale). The pinions were stress-relieved at 325°F after being quenched from the carburizing heat.

In order to eliminate gear mounting geometry as a factor in fatigue testing, the bore and spherical seat were precision ground to finish dimensions. The bores were ground using a pitch-line chuck to insure concentricity between the pitch cone and the bore. The pinions were then tested for contact pattern with a mating master gear using a Gleason T6A angular hypoid test machine. Readings on this machine indicated how much material had to be ground off the spherical portion of the pinions in order to have proper running contact with mating gears in the fatigue test.

For the purpose of comparison, conventionally machined production pinions and side gears were obtained from the Buick Division of G.M.C. The conventionally cut pinions and side gears used in the course of the investigation were cut from AISI 4615 steel using the Revacycle* process, and carburized to an effective case depth of 0.030" to 0.040". Pertinent dimensional data for the pinion and gear are listed in Table II.

*REVACYCLE is a trademark of Gleason Works denoting gears manufactured by a particular method of generating bevel gear teeth using a rotary broach type of cutter.

TABLE II

Differential Gear and Pinion Data

	PINION	GEAR
NUMBER OF TEETH	10	16
DIAMETRAL PITCH	4.708	4.708
FACE WIDTH	0.756	0.756
PRESSURE ANGLE	24°0'	24°0'
WHOLE DEPTH	0.430"	0.430"
ADDENDUM	0.227"	0.159"

Differential Fatigue Testing

Differential fatigue testing differs from impact testing in that it is primarily a running test and during each revolution of the pinion member, all possible loading positions are encountered. Consequently, by maintaining constant applied torque to the differential gear set, the maximum bending stress in the root fillet of a pinion tooth is always obtained and the test becomes insensitive to tooth load height position. Running tests have an additional advantage in that tooth surface wear resistance can be estimated by observing the amount of wear present during various stages of the differential test. One important disadvantage of the method is, however, that two pinion members and two side gear members are required to make up the differential assembly for one test, thus requiring twice the usual number of test specimens.

Figure 2 shows the test apparatus used for testing both powder and conventional cut pinion specimens. Pinion test specimens are run as pairs in a differential gear assembly with two side gears completing the assembly.

The differential assembly test box is shown in detail in Figure 3. The gear and pinion members are mounted on the ends of four large, cylindrical cartridges that fit inside the housing and are accurately aligned with the gear and pinion axes. Adjustments of the tooth contact pattern were obtained by shimming the four cartridges in or out along their axes.

After the test pinions and mating side gears were assembled in the differential test box, the differential

Figure 2. Differential pinion testing apparatus

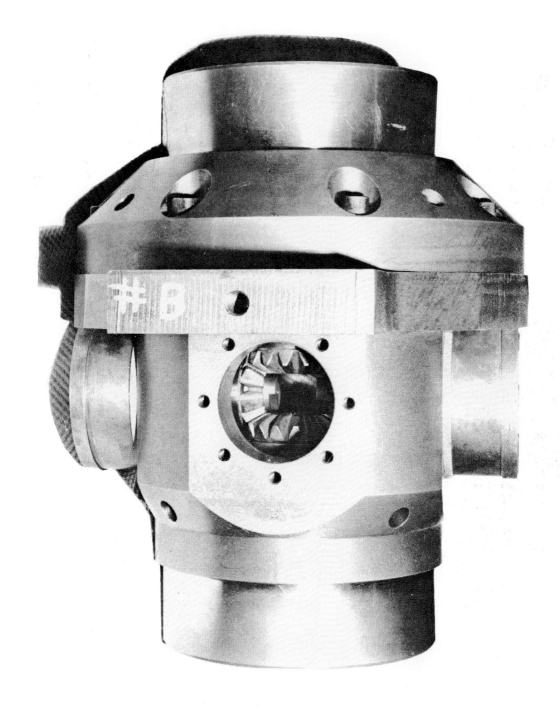

Figure 3.   Differential assembly test box

gear assembly was painted with marking compound and rotated by hand to establish the no-load contact pattern. The axial positions of the gear and pinion members were adjusted until a central-toe tooth contact pattern was obtained for no load conditions. The differential gear set was then cycled a few times under load to insure that the contact pattern remained essentially in a central-toe position for the duration of the test.

Figure 4 shows the arrangement of the differential gear assembly relative to the differential fatigue tester. Torque is applied to the differential set by rotating the differential housing about the axis passing through the side gear members of the differential assembly. Load cells are used to measure this force which can then be

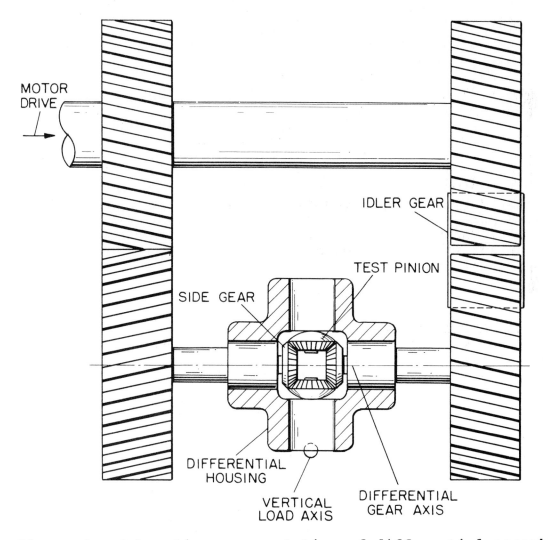

Figure 4. Schematic representation of differential assembly relative to differential fatigue tester.

used to calculate the torque applied to the pinion and gear teeth of the differential assembly. Applied torque was preset at 5,800 lb-in. and corresponded to a pinion bending stress of 92,400 psi. This stress level is known from previous testing to give a test life of approximately one hour.

All pinion test specimens were run under the following test conditions:

Applied Pinion Torque = 5,800 Lb-in.
Pinion Tooth Bending Stress = 92,400 psi
Pinion RPM = 41
Oil Temperature = 180°F

In order to minimize the occurrence of systematic errors, pinion test specimens were randomly selected from three test lots. Each test lot contained from six to twenty pinions with each test lot corresponding to one set of alloy and sintering time/temperature combinations. The first three test lots were exhausted and three new lots formed as experimental test pinions became available.

The termination of an individual test occurred when excessive vibration, caused by partial or complete fracture of one or more pinion teeth, was detected by an accelerometer attached to the differential housing. Substantial deformation of the pinion teeth and the associated vibration was usually found to preceed catastrophic tooth failure by less than a hundred cycles. Side gear wear was usually well advanced at the conclusion of a test as was pinion wear and therefore, new side gear members were installed for subsequent tests.

Impact Testing

It was initially desired to establish an impact fatigue type of test, using repeated blows, as described by DePaul (5). To perform this type of testing using actual bevel gears, a fixture to properly locate the gears is needed. After surveying the various methods of bevel gear impact testing used in industry, it was decided to make use of a fixture which held a bevel pinion only with no mating gear, and which incorporated a striker making line contact with one tooth of the pinion approximately along its pitch line. Testing showed that when using repeated blows of the same energy level, the striker tended to deform and/or fail before doing appreciable damage to the pinion. It was then decided to turn to a slightly different procedure, using the same type of fixture.

In this test, developed and used by a large automotive
manufacturer, a 20-pound drop hammer is dropped onto the
striker initially from a height of six inches. For each
successive blow, the drop height is increased by one inch
until failure occurs. While the data from this type of
test is somewhat more difficult to treat statistically,
the advantage of the test is that only a few blows (less
than 50) are required to fail the pinion, and two tests
can be run before the striker has to be re-ground to
correct the deformation.

Two powder-formed pinions of each alloy/sintering
cycle combination were impact tested. In three cases, there
was considerable disagreement between the two values, so
a third test was run and the outlying value discarded as
being due to experimental error. Three conventionally cut
pinions were tested for the purposes of establishing a
base-line; the values obtained in these tests agreed with
past history on similar pinions.

RESULTS

Differential Fatigue Testing

In order to establish the relative and absolute rankings
of pinion life data, all test data were analyzed using a
two-parameter Weibull cumulative distribution function.
Upper and lower bounds on confidence limits were set equal
to 95 percent and the B_{10} life, or the life corresponding
to ten percent cumulative failure, was selected as a basis
of comparison of life data.

Figures 5a and 5b show the fitted Weibull distribution
functions for each of the test lots, including the conventional
cut pinions. The Weibull slopes and the upper and lower
confidence bounds were estimated through the use of a General
Electric time sharing program, WEIBL$.

B_{10} lives for each pinion test lot were obtained by
reading the intersection point of the computed Weibull
distribution function for that test lot with the 10 percent
failure level. The Weibull slopes are seen to lie between
2.30 and 5.82 which is in good agreement with the slope value
of 3.79, corresponding to the conventional cut pinions. The
variability of the slopes is typical of the variability obtained
with lot sizes of four to nine pinions. Two of the Weibull
curves, corresponding to the 4600V alloy, six minutes sintering
time, and 2100 or 2350°F sintering temperature, contained
too few data points to be analyzed by computer and were
therefore curve fitted by eye. Consequently, no
confidence limits were obtained for these test lots.

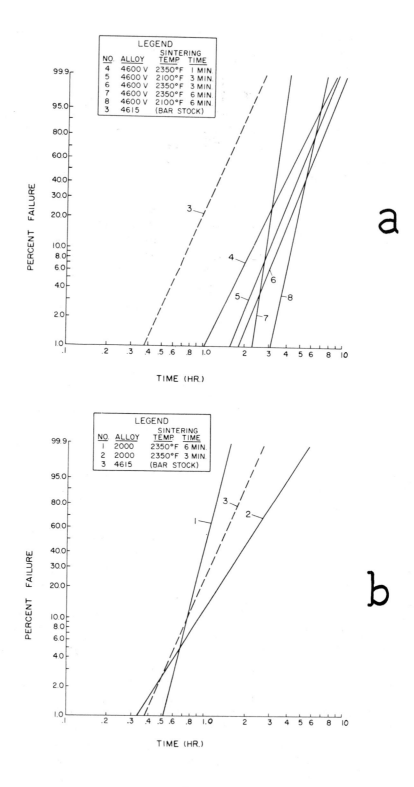

Figure 5. Weibull plots of fatigue test data, a) 4600V alloy pinions, b) 2000 alloy pinions.

Table III lists the B_{10} lives, along with the confidence limits for each of the test lots.

TABLE III

Estimates and Two-Sided 95% Confidence
Limits for Weibull Cumulative Distribution Function

Test Lot (Alloy Type)	Sintering Temp.	Time	B_{10} Life	Lower Limit	Upper Limit
4600V	2100°F	3 min.	2.92	1.56	5.47
4600V	2100°F	6 "	4.2*	----	----
4600V	2350°F	1 "	2.21	1.06	4.60
4600V	2350°F	3 "	3.42	2.08	5.64
4600V	2350°F	6 "	2.8*	----	----
2000	2350°F	3 "	0.92	0.39	2.16
2000	2350°F	6 "	0.74	0.49	1.11
4615	bar stock		0.76	0.49	1.79

*Confidence Limits Not Available

The results of the differential fatigue test show that the powder metal test pinions fall into two distinct groups. The first group is comprised of the 2000 alloy powder metal pinions with an average B_{10} life of 0.83 hours. The second group consists of the 4600V alloy powder metal pinions with an average B_{10} life of 3.2 hours. The average lives of the two groups compare to the conventional cut pinion B_{10} life of 0.76 hours. The greatest B_{10} life for a test group was obtained for the 4600V alloy, 2100°F sintering temperature and six minute sintering time, powder metal pinions where the B_{10} life of 4.2 hours is over 500% greater than the conventional cut pinion B_{10} life.

Impact Testing

Table IV lists the results of the impact fatigue tests. Impact fatigue values are given in terms of the cumulative blow energy absorbed by the sample prior to failure. As mentioned in the section on experimental procedure, two values of impact energy were measured for each sample type; where only one figure appears in Table IV, both samples had identical values. As was the case in differential

TABLE IV

Results of Impact Fatigue Tests

Alloy Type	Sintering Temp.	Time	Cumulative Impact Energy
4600V	2100°F	3 min.	20,400 in-lbs.
4600V	2100°F	6 "	18,620
4600V	2350°F	1 "	17,760-18,620
4600V	2350°F	3 "	16,920-20,400
4600V	2350°F	6 "	16,100-23,220
2000	2350°F	3 "	13,020-14,520
2000	2350°F	6 "	12,300
4615	bar stock		13,020-16,100

fatigue testing, the powder formed pinions made from 4600V alloy were superior to the conventionally machined pinions. The pinions made from the 2000 alloy were comparable to the conventional pinions from an impact fatigue standpoint.

Metallurgical Analysis

In order to more fully assess the results of this study, the pinions were metallurgically examined using standard techniques. Table V shows the results of carbon and oxygen determinations made on some of the tested pinions. The carbon contents of the powder pinions are seen to be slightly higher than that of the bar stock. The oxygen contents of the powder pinions can be seen to be a function of sintering time and temperature. The large spread in oxygen values in the 4600V samples sintered for one minute at 2350°F is not explained, but was probably caused by procedural error. It can be seen in the data that the Anchorsteel 2000V alloy did not respond as well to deoxidation as did the 4600V; this is probably related to the presence of more stable manganese oxides in the 2000 alloy. It will also be noted that the bar stock used for comparison was exceptionally low in oxygen. Metallographic examination of the bar stock pinions showed very small stringers, widely spaced, confirming the oxygen analysis.

The case depths and core hardnesses of some tested pinions were also checked for purposes of comparison.

TABLE V

Core Carbon and Oxygen Contents of Pinions

Alloy Type	Sintering Time	Temp.	Carbon Content	Oxygen Content
4600V	2100°F	3 min.	0.19-0.20%	200-285 ppm
4600V	2100°F	6 "	0.19-0.20	100-145
4600V	2350°F	1 "	0.20-0.21	26-305
4600V	2350°F	3 "	0.21-0.22	20-132
4600V	2350°F	6 "	0.17-0.19	7-13
2000	2350°F	3 "	0.19-0.21	179-196
2000	2350°F	6 "	0.19-0.20	153-157
4615	bar stock		0.14-0.17	16-23

Figure 6a-c shows representative case hardness profiles which represent the range of the samples tested. It will be noted that the 4600V alloy gave slightly deeper cases and somewhat higher core hardnesses than did the two other alloys. This point will be discussed later.

It was also decided to investigate the residual stress levels in the surfaces of some of the pinions since such stresses are known to have significant effects on fatigue behavior. The testing was done at Buick Motor Division, G.M.C., using an x-ray analyzer known as FASTRESS (R) developed by General Motors. This analysis showed that all pinions had residual compressive stresses on the order of 100,000 to 130,000 p.s.i. There was no correlation between fatigue life and residual stress over this small range.

DISCUSSION

Differential Fatigue Testing

The differential fatigue test results showed that all of the sample lots using 4600V powder produced fatigue lives superior to the conventional cut pinions. As shown in Table III and Figure 5a, all of the 4600V data fall in a narrow range. Curves 7 and 8 in Figure 5a show steeper Weibull slopes than the others and this may be due to the relatively long (6 minutes) sintering time used in both cases. It could be argued that the longer sintering times cause more

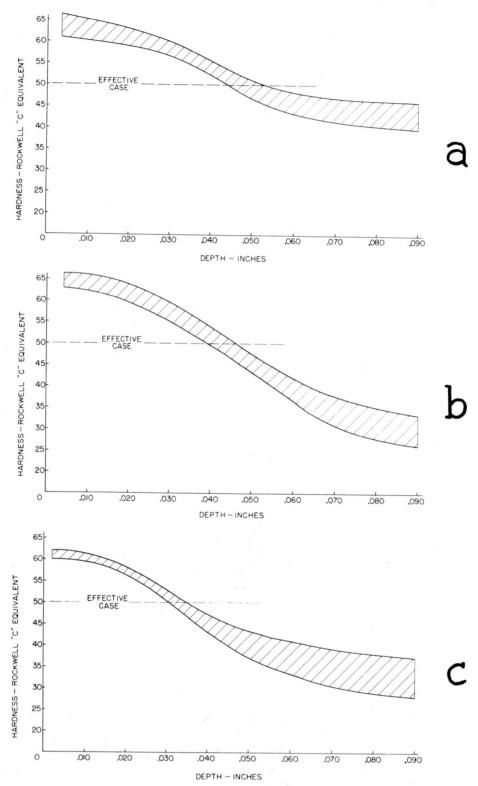

Figure 6. Case hardness profiles for differential pinions,
a) 4600V alloy, b) 2000 alloy, c) 4615 bar stock.

uniform deoxidation of preforms and therefore less scatter (steeper Weibull slope) in the fatigue data. In the case of the 4600V alloy, the sintering temperature should have little effect on deoxidation, since the major alloying constituents, nickel and molybdenum, are readily reducible by CO at temperatures even below 2100°F (6).

In the case of the 2000 alloy, the B_{10} lives corresponded closely to that of the conventional pinions, as shown in Table III and Figure 5b. Again, the longer sintering time used for the samples comprising curve 1 in Figure 5b has produced a higher Weibull slope than has the shorter sintering time. Lower sintering temperatures were not used with this alloy, since it contains a relatively high amount of manganese which is not easily reducible at these temperatures (6).

Since the data from the two powder alloys fell into two groups, it was decided to fit one Weibull curve to all of the data points from each alloy. This was done to obtain more data points for each curve. The results are shown in Figure 7 which also graphically indicates the 95 percent confidence limits on the B_{10} lives. It is plain that the 4600V pinions have superior fatigue life at the stress level of 92,400 psi, and the slope of the Weibull curve indicates

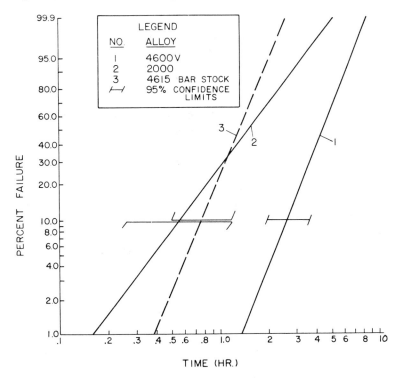

Figure 7. Replotted Weibull distributions for the three types of alloys tested.

a repeatability equivalent to that obtained with conventional pinions. Figure 7 also shows that the confidence limits on the B_{10} lives of conventional and 2000 alloy pinions overlap completely, indicating no significant difference in the two based on the limited data from this investigation. The slope of the 2000 alloy curve is lower than the other two, indicating greater scatter in the data. Closer analysis showed that this slope was due largely to one quite high value of test life, which is at least more reassuring than had it been due to one quite low value.

It is tempting to try to characterize the difference in fatigue behavior between the two powder alloys on the basis of their oxygen content. Indeed, as Table IV confirms, the 4600V alloy is much easier to deoxidize than is the 2000, probably because of the higher manganese content of the 2000 alloy. The oxygen argument begins to break down when we try to use it to explain the differences between the powder and the conventional pinions. Here we find the powder parts containing much more oxygen than the conventional ones, yet having similar or superior fatigue resistance. Thus, we are forced to look for another explanation for the comparatively good fatigue resistance of the powder pinions.

There are several possible reasons for a gear forged from powder to be superior in fatigue resistance to a gear cut from bar stock; these include:

1. finer grain size

2. random orientation and equiaxed morphology of inclusions

3. better surface finish

It proved difficult or impossible to attach any significance to the above factors. Many correlations of the data were tried, but most failed to show concrete trends probably because of the relatively small amount of data. The best explanation that could be found for the superior fatigue resistance of the 4600V alloy pinions was the higher core hardness and slightly deeper case depth than either the 2000 or conventional pinions (Figure 6). The powder alloy was really analogous to 4620 steel, whereas the conventional pinions were made from 4615 bar. The slightly higher carbon content of the powder alloy caused the higher core hardness. The 2000 alloy contained roughly the same amount of carbon as the 4600V, but because some of its manganese probably remained in the oxidized state, its hardenability was not sufficient to promote higher core hardness.

Further indication that core hardness and case depth were the major contributors to the differences between the 4600V and other types of pinions was gained from a close inspection of the failure modes of the pinions. With only two exceptions, the 4600V pinions failed in a normal bending manner even though the high stresses imposed in the tests tend to favor crushing of the case as a predominant failure mode. The conventional cut and 2000 alloy pinions all failed by the case crushing mode, indicating a core of insufficient strength to support the thinner case under the high applied loads.

Impact Testing

The results of impact-fatigue tests were basically similar to the rolling-fatigue tests, the 4600V alloy producing better impact resistance than the 2000 or 4615 bar stock. The 2000 alloy pinions did not quite equal the impact resistance of those made from bar stock, however, as was the case for rolling fatigue resistance.

It was interesting to note that of the 35 to 40 impact blows required to fail most of the pinions, initial cracks were observed after as few as 10 blows. Since the severity of the blows continually increased, this behavior suggested that the cores of the pinions acted as quite efficient crack arresters. Crack propagation rate data was not obtainable due to the nature of the test, however. If we assume that the cores were the most influential factor in differences in impact fatigue behavior, then the fact that the 4600V pinions had higher core hardnesses probably accounts for their superior performance. The combination of higher oxygen content and lower nickel content in the 2000 alloy pinions may account for the slightly lower impact resistance than exhibited by those of 4615 bar stock.

Differential Fatigue versus Impact Fatigue

It is useful at this point to discuss the relative results of the two types of mechanical tests used in this study. It was found by Pratusevich et al (7) that impact fatigue and rolling fatigue strengths of gears correlated well, but did not correlate as well to static strength or single-blow impact strength. Figure 8 shows the results of differential fatigue tests plotted versus impact fatigue results. There is definitely an upward trend in Figure 8, but too much scatter to define a good correlation. It should be pointed out that the differential fatigue results are in terms of fatigue life at constant stress, and not in terms of fatigue strength per se. The stresses employed in differential fatigue testing would fall in the steeply-sloped portion of the S-N curves for the three steels studied.

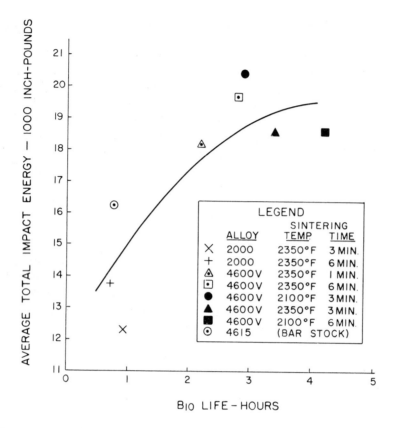

Figure 8. B_{10} lives versus impact energy for differential pinions tested.

Obviously, there is not necessarily a correlation between the slopes of the S-N curves and the endurance limits of the steel in question, which must be a factor contributing to the lack of correlation in Figure 8. The sparsity of impact-fatigue data points also contributes to the scatter in Figure 8.

Other Pertinent Data

Since the time during which this investigation was conducted, samples of differential pinions made by the same process have been submitted to various automotive manufacturers. Although the detailed data remains proprietary with the manufacturers, we can report that the 4600V alloy has continued to show superior results in all types of static, impact, and fatigue testing. Newer lots of Anchorsteel 2000 have been used for some samples with also superior results. The newer lots of the 2000 alloy contained more manganese and thus had more hardenability which promoted deeper effective cases and higher core hardnesses. The improved case and core properties therefore have brought the less costly 2000 alloy up to the level of 4600V in terms of mechanical behavior.

CONCLUSION

There is sufficient data in this investigation to conclude that differential pinions having fatigue and impact behavior equal or superior to conventionally produced pinions can be made from powder metal by a process using isostatic compaction; short-cycle, high-temperature sintering; and hot forging with sufficient metal flow to obtain full density. The relative effects of various time/temperature sintering cycles were somewhat overshadowed by minor variations in case depths and core hardnesses. The data did indicate, however, that longer sintering times, up to six minutes, promoted more uniform fatigue lives.

It can also be concluded that the Anchorsteel 2000 type of alloy shows definite promise for applications requiring high impact and fatigue resistance. This is especially important to the overall economics of the powder metal forming process, since this type of alloy contains lower amounts of nickel than the 4600 type and is therefore less costly.

ACKNOWLEDGEMENTS

Thanks are gratefully given to Messrs. T. J. Maiuri, H. E. Beckwith, J. B. Fortner, J. L. Reese, and M. R. Rowles for their excellent work in obtaining most of the data in this investigation. The helpful suggestions of Mr. W. Coleman and Dr. P. J. Guichelaar are also acknowledged.

The authors thank Mr. I. Beyerlein of Chevrolet Motor Division, G.M.C., for allowing them to purchase time on their impact-testing machine and for the useful suggestions on experimental technique.

The authors also express their gratitude to Messrs. D. Buswell and W. Dehaan of Buick Motor Division, G.M.C. for the residual stress measurements.

REFERENCES

1. Hense, V. C., "Forging Powder Metal Preforms for the Automotive Industry", presented at MPIF Full Powder Metallurgy Conference, Philadelphia, Pa., Oct., 1969.

2. Halter, R. F., "Pilot Production System for Hot Forging P/M Preforms", Modern Developments in Powder Metallurgy, V.4: Processes, Ed. H. H. Hausner, Plenum Press, New York, 1971.

3. Lusa, G., "Differential Gear by P/M Hot Forging", Modern Developments in Powder Metallurgy, Ibid.

4. Guichelaar, P. J., and Pehlke, R. D. "Gas-Metal Reactions During Induction Sintering", 1971 Fall Powder Metallurgy Conference Proceedings, Ed. S. Mocarski, MPIF, New York, 1972.

5. DePaul, R. A., "Impact Fatigue Resistance of Carburized Gear Steels - Development of a Testing Machine and Evaluation of Initial Test Results", Materials Research and Standards, Vol. 10, no. 3, 1970.

6. Eloff, P. C., and Kaufman, S. M., "Hardenability Considerations in the Sintering of Low Alloy Iron Powder Preforms", 1971 Fall Powder Metallurgy Conference Proceedings, op. cit.

7. Pratusevich, R. M., et al, "Impact-Fatigue Strength of Hardened Gears", Russian Engineering Journal, Vol. LI, no. 10, Oct., 1971.

HOT FORMED P/M - APPLICATIONS

T. W. PIETROCINI

GOULD INC., POWDER METAL PRODUCTS DIVISION

This paper will consist of applications for hot densified parts, principally an overview of the various parts we have reviewed during the past several years. The total number of parts involved exceed 300. They cover a wide spectrum of applications including gearing, fasteners, agricultural equipment, hydraulic components, magnetic parts and bearings. The product function consists of sliding, such as, in bearings, impact, fatigue, and mechanical properties, such as, in structural parts and gearing, sealing of hydraulic components, and the physical properties necessary for magnetic function.

METALLURGICAL FEATURES

Before getting into the specifics of each of the applications, let us consider some of the overbearing or metallurical features of this process as compared to the cast products and the wrought products which many of you are much more familiar with. First there is the question of residual porosity, primarily associated with the surface. Second there is the question of density variations within the structure. Third there are microstructural differences due to the manufacturing technique involved. Fourth the heat treatment of these parts does result in a somewhat different set of properties. Finally there is the question of machinability or providing a secondary operation to give adequate dimensional properties.

RESIDUAL POROSITY

The question of surface porosity is extremely important when we consider sliding applications such as, bearings or gearing wherein high Hertz stresses are developed which can cause surface fatigue or spalling. In 1971, we presented a paper on the heat treatment of what we term hot

densified powder metal alloys.[1] In this paper, a microstructure showing the porosity distribution at the surface was shown. A reprint of
this microstructure is shown in Figure 1.

This porosity was caused by the frictional behavior of the hot formed
material on the surface of the die. The porosity at the surface would
range as high as 4%. In order to avoid this surface porosity, one must
compensate his design such that sufficient flow takes place over the
surface. Generally speaking, 40% deformation would be required to
close this porosity. However, pore closure is very difficult to accomplish even with flow present due to the chilling effect of the die surface.

In castings, one has similar situation, in that there is shrinkage porosity present. In addition, there is a difference in grain size as
one moves away from the cast surface and goes further into the material.

Looking now at wrought materials we know that the forged surface quite
often is decarburized and has inclusions trapped in it. Generally for
best performance removal of that surface will be required prior to use
in the application.

Herein lies one of the difficulties and the major feature of hot forming of powder metal. We expect to use the as formed surface as the
operational surface therefore, this surface porosity is very critical
to us and we must take this into consideration during the design of the
preform, in the heating rate involved, the selection of lubrication, in
the selection of the die material and the temperature of its operation
to provide the best possible surface for the final product.

DENSITY VARIATIONS

Density variations can and do effect mechanical properties. Let us
look then at the tensile strength vs. density curve in Figure 2. Here
we can see very quickly that while there is an effect above 95% density
the tensile strength is over 200,000 psi and it is not going to be a
major problem in terms of overall structural performance. However, tensile strength alone does not tell the story. We further have to include
impact and fatigue strength.

There has been a lot of data published on the impact strength as a
function of density. The best known of this is data presented by
Harry Antes of Hoeganaes Corporation.[2]

This data shows that for atomized iron an eight fold improvement can be
made by going from 7.7 g/cc to 7.86 g/cc. Impact data on 4630 type
(1.8% Ni - 0.5% Mo) P/M steel alloy shows an improvement of three to
one for the same density range.

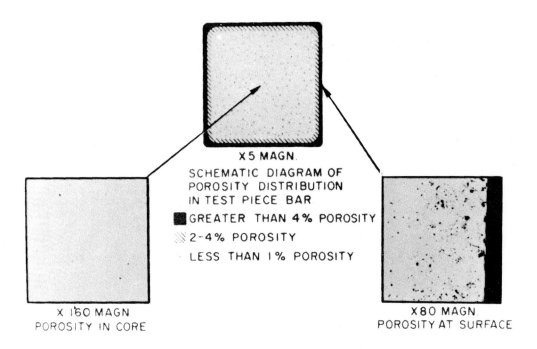

X 5 MAGN.

SCHEMATIC DIAGRAM OF
POROSITY DISTRIBUTION
IN TEST PIECE BAR

■ GREATER THAN 4% POROSITY

▨ 2-4% POROSITY

· LESS THAN 1% POROSITY

X 160 MAGN.
POROSITY IN CORE

X 80 MAGN.
POROSITY AT SURFACE

FIGURE 1 POROSITY DISTRIBUTION IN AS PRESSED TEST PIECE BAR

FIGURE 2

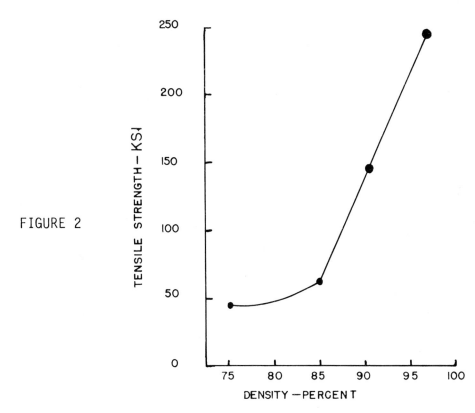

TENSILE STRENGTH VS. DENSITY

Source: *Modern Developments in Powder Metallurgy*, Vol. 7, MPIF, 1974

In our work on impact strength in 1970, we showed the identical values for 2% residual porosity with a 4630 P/M steel having a duplex microstructure.[3]

Thus, we've seen that while the tensile strength has not been grossly effected the impact strength certainly is and we have to take this into consideration. There are other reasons for low impact strength and these will be mentioned later.

Cast products have limited tensile properties. They suffer from density variations primarily due to shrinkage porosity as well as the inclusion present from casting of the alloy. While with wrought material, because the skin in general has been removed, density variations are not nearly as much a problem. However, there is a stringering effect of both carbides and inclusion which gives a difference in longitudinal versus transverse properties.

The effect of density variations on fatigue properties has not been sufficiently investigated to present any conclusive data. In general, these properties will be higher than those obtained from a P/M part unless sufficient working has taken place. By sufficient working, we are talking in the neighborhood of 60% deformation.

MICROSTRUCTURAL DIFFERENCES

Microstructure differences were also noted in our earlier paper wherein a master alloy was used to obtain certain mechanical properties and the resulting structure was duplex with areas of free ferrite as well as areas of high alloy due to the prealloyed master alloy.[1] Other types of structure are certainly possible. One can use mixed elementals which will result in high nickel or molybdenum rich areas. Another approach is to use fully prealloyed powders offered by several powder companies. The prealloy is preferable in the sense of having a homogeneous microstructure although, depending on processing characteristics and techniques used either type of structure has shown comparable properties.

Referring back to the duplex microstructure, the variation in alloy phases distributed throughout the part results in variations in the stress distribution particularly in fatigue type applications. The data we have to date indicates that the duplex structure provides improved fatigue resistance as compared to a homogeneous structure. We believe this is true because of the ability of the soft, free ferrite region to blunt fatigue cracks.

There are a lot of other idiosyncrasies that are not fully realized yet. These include the question of surface chemistry, particularly, the reaction of the iron carbon oxygen system and possible relationships to impact problems. There has been illusion made to getting around these problems with higher temperature forming as well as increasing the amount of deformation.

In castings, we again have a similar situation in which a duplex micro-structure can and quite often is present. This is related to the liquidus vs. solidus temperatures for the alloy system in question. This can cause a variation in the chemical make up. Wrought products are generally homogeneous due to the amount of working they have undergone to bring out the maximum mechanical properties.

HEAT TREATMENT EFFECTS

The heat treatment of hot formed P/M parts is superior to that of wrought alloys in that higher temperature can be used without grain coarsening. Whereas with conventional wrought materials, carburizing at elevated temperatures, such as 1750-1800⁰, grain coarsening will occur which will result in lower toughness. The advantage of high temperature heat treating is one of economics. The higher temperature aids carbon diffusion thus decreasing the time required to obtain the required case depth.

Carbon level in H/D:P/M materials can be anything you want to make it. Control of that carbon level is quite difficult. This process has not reached the control point of wrought in that the average spread is around ± .05% C as opposed to ± .02% C spread in wrought alloys. The fact does remain however, that since you are in general completely forming this part, one can work with higher carbon levels. Thus, induction hardening can be used. Parts can also be through hardened.

However, for many of our structural applications, such as gearing, the basic mechanisms found to work for wrought gears still hold true.

Case-core interaction is preferred to through hardening from the standpoint of surface fatigue. The primary reason for this is, that in carborizing, we provide a case which has residual compressive stresses.

These residual compressive stresses then help us resist the fatigue problems present in the application. In a one hour heat treat cycle, at 1700⁰F, one can develop .015 to .020 inch of case on hot formed P/M parts. This is shown in significant depth in Mr. Smith's paper mentioned earlier.

SECONDARY FINISHING EFFECTS

All of the normal metal removal techniques employed on wrought parts can be used. The most predominant of these is grinding. Grinding improves the dimensional characteristics as well as the surface finish. From the standpoint of machinability, we find that after hot forming an annealing operation or spheridizing operation must be performed to give maximum tool life. The material does turn freely and drilling and lathe work are no more difficult than for any of the comparable wrought materials.

APPLICATIONS

We have just talked about the general characteristics of these P/M parts. Let us now turn our attention to the types of applications which have been looked at and which hold potential promise **for** the future.

GEARING

In the area of gearing, we have dealt with differential pinions, agricultural bevels, chain saw sprockets, transmission sprockets, sliding clutches, planet gears, sector gears and Rezeppa joints. **Returning** then to the differential pinion, it can be noted that this part has been made by both hot densification and hot forging of P/M. **The two** methodologies differ in that in the one case we have an unshaped preform shown in Figure 3 and the other case a shaped preform shown in Figure 4. The final product has different mechanical properties dependent upon the preform. Basically with the reduced amount of metal flow, one does not achieve the impact strength necessary to meet the existing part, whereas the hot forged part does. However, the hot forged part suffers from some dimensional variation caused by weight control changes, these can be corrected by specific control of the preform weight. The hot densified part provides for the variation in weight by allowing some residual porosity. This porosity in turn accounts for the lower impact strength.

Here we have a paradox. We have a densified part that will meet the dimensional requirements and fit into the machine line set-up by the auto companies, but will not meet the impact requirement. On the other hand, we have a forged product that will meet the impact requirement, but because of the dimensional variations suffers from an economic limitation.

Rezeppa Joint

The pinion gear represents a type of product which has had an extensive amount of manufacturing cost extracted from it. The basic revacycle gear form was specifically developed to enable the use of a rotary broach type of cutting compared to a generated form cutter. This greatly reduced the cycle time. Thus, we have a very mature product wherein the necessary capital equipment already exists. The process is very well defined and is extremely difficult to compete against.

We know that there is a major movement to go toward smaller vehicles, much the same as Europe. One way of maximizing interior room while reducing the overall package, is to go to front wheel drive. The main drive component of such a system is a Rezeppa joint. This component has spherical ball tracks to allow turning and angular motion while maintaining power transmission. Figure 6 shows such a component.

DIFFERENTIAL PINION GEAR

FIGURE 3

PINION GEAR PREFORM - FORGING

FIGURE 4

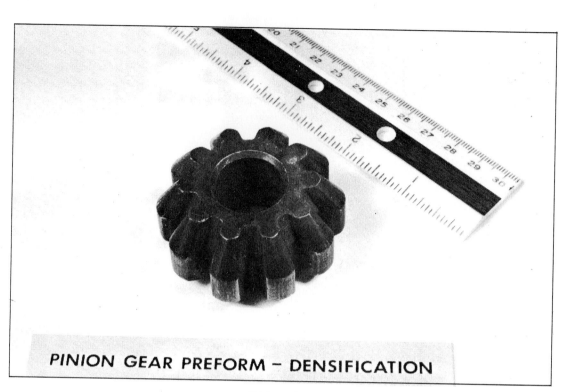

PINION GEAR PREFORM – DENSIFICATION

FIGURE 5

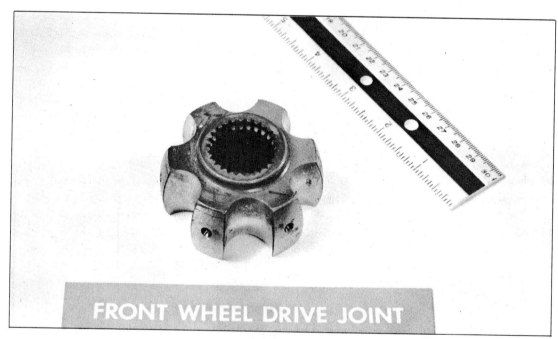

FRONT WHEEL DRIVE JOINT

FIGURE 6

From the standpoint of fabrication, a similar set of tooling motions are required by Rezeppa joints and pinion gears. Here, however; the economic picture is somewhat brighter since high volume production machining centers have not been purchased. Therefore, a capital expenditure is required. In view of this, any potential metal forming technique has an equal opportunity.

From an application standpoint, similar mechanical properties are required as in gearing -- resistance to high Hertz stresses and impact loading **are** required. Because of the contact stresses involved, a carburized alloy steel having a case depth of 0.060 inch minimum would be selected. Since the ball track groove requires very **tight** tolerance after heat treating, grinding would be needed. This would minimize any surface porosity problems and the P/M part could be heat treat**ed** at an elevated temperature to minimize the time required to form such a deep case. This component is currently undergoing evaluation tests.

Bevel Gear

Shown in Figure 7 is an agricultural bevel gear. This component operates in a dry environment where it is primarily subjected to abrasive wear. A density of 7.5 g/cc is adequate so long as severe **pounding** loads are not encountered. The material used for this application was 4620 carburized.

Drive Sprocket

Shown in Figure 8 is another wear application. This part is used as the drive sprocket for a chain saw. The material used was H/D:P/M 4650 through hardened to R_c55. This part had approximately one half the life of an investment cast Stellite part. The operating conditions consisted of extensive testing in sand. Failure occurred when sufficient gross wear had taken place such that the saw chain could no longer be driven.

Figure 9 shows another chain sprocket. This part is also 4650 through hardened to R_c55.

Sliding Clutch

Sliding clutch components, as well as planet gears have **also** been tested. These parts were made from 4620 material carburized to a case depth of 0.060 inch. The sliding clutch has passed preliminary laboratories tests and later this year will complete a 100,000 mile road test. The planet gear failed under bending stress due to residual porosity in the root area. These parts are currently made from carburized 8620 steel.

Sector Gear

Shown in Figure 10 is a sector gear which operates in an oxidizing

AGRICULTURAL BEVEL GEAR

FIGURE 7

CHAIN SAW SPROCKET

FIGURE 8

CHAIN DRIVE SPROCKET

FIGURE 9

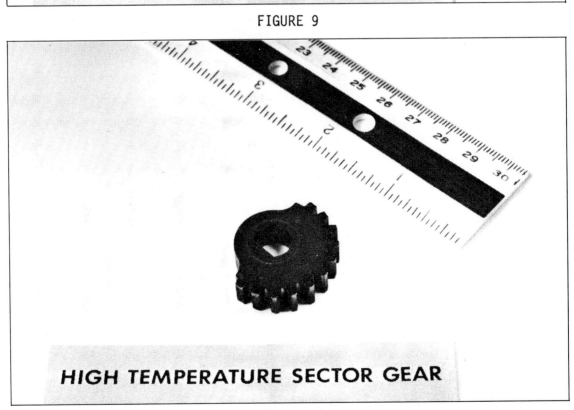

HIGH TEMPERATURE SECTOR GEAR

FIGURE 10

atmosphere at 1200°F. This material is Fe-15% Mo-15% Co which forms a coherent oxide film having self-lubricating characteristics. The operating range for this material is 800-1500°F. Below 800°F the oxide film is not self-generating while above 1500°F it begins to spall. This part replaced nitralloy which was failing due to fretting corrosion.

HYDRAULIC COMPONENTS

Piston

The piston shown in Figure 11 is used in a hydrostatic transmission. The material selected for this application is 4620 carburized to R_c60. Since a drilling operation is performed, the part is first annealed, then drilled, carburized, and ground. This component was previously cut from 1141 bar steel.

Gerotor Star

Gerotor pump components have been evaluated. These parts were made from 4620 carburized to a 0.060 inch case depth. There are two mechanical problems to contend with in this application. One is the strength of the I.D. spline and the other is spalling fatigue due to high Hertz stresses. While this type of pump produces up to 5000 psi, the major mechanical problem is spalling caused by localized loading due to dimensional variations. In general, hot formed powder metal will preform satisfactory if a grinding operation is performed following heat treatment. This grind insures dimensional accuracy as well as removal of surface porosity.

Hydraulic Coupling

Hot formed 1025 material has been used successfully as the starting stock for high L/D extrusions used as hydraulic couplings. The extrusion ratio of this material was found to be superior to that of AISI 1018. The primary reason for this appears to be the very low manganese content of the P/M material effectively reduces the work hardening coefficient. This application is not economically attractive.

Valve Lifter Body

Shown in Figure 12 is a valve lifter body. This part is currently in production. It is used in replacement sales. The OEM part consists of a cast iron body with a brazed on cap of chill cast white iron. This combination provides excellent machinability in the skirt while maintaining cap hardness necessary for adequate wear resistance.

The P/M component truly takes advantage of the hot forming process. The cap is 98% dense thus providing sufficient mechanical strength to avoid crushing under load. The skirt is 75% dense having interconnected porosity. Thus, this component provides lubrication to the side wall while in operation.

FIGURE 11

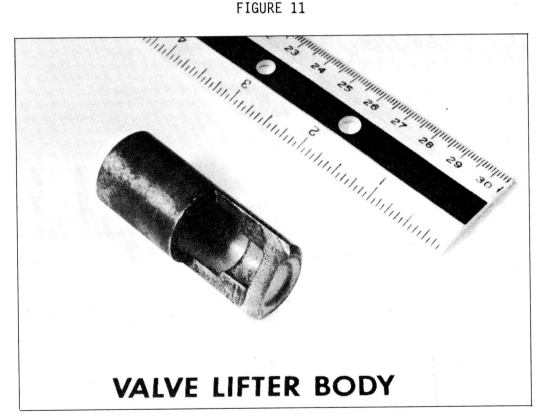

FIGURE 12

Source: *Modern Developments in Powder Metallurgy*, Vol. 7, MPIF, 1974

This part is manufactured from H/D:P/M 4670 using a 6.2 g/cc density sintered preform. Only the cap portion is heated and densified to 7.7 g/cc. The part is then annealed to allow machining of snap ring grooves, reaming of the cylinder bore, O.D. groove and cross drilled hole. The part is then carbonitrided to provide adequate wear resistance.

Cam Ring

Cam rings for power steering pumps have been tested. A modified 4650 alloy showed the best wear resistance for this application. This material had only 0.5% Ni and was through hardened to R_c55. The conventional material is gray cast iron.

MAGNETIC PARTS

While we have reviewed a variety of parts requiring good magnetic properties, our principle work has been on an alternator rotor pole. As is often the case in a manufacturing plant, our major effort was spent developing production methods of manufacture consistent with satisfactory performance as evaluated by our customer.

The magnetic properties obtained from hot formed P/M parts are intensely discussed in the literature.[4] This data clearly shows P/M to be superior to 1006 steel.

Shown in Figure 13 is a 70 amp alternator rotor pole. There are three requirements for this application. The first is to insure a clean internal structure after processing. The second is to avoid a metal flow condition that would create any folds in the area where the finger joins the flange. The third is to insure the density at the tooth tips. Once the basic manufacturing methods necessary to obtain these three conditions were obtained, satisfactory parts were made.

FASTENERS

We have worked on two fasteners. One is a threaded nut, the other is a valve spring retainer. The nut is made from H/D:P/M 1050 oil quenched and drawn to R_c50. While this part is adequate from a strength standpoint, it only finds economic use when the design dictates something other than a conventional hex nut.

The valve spring retainer is shown in Figure 14. This part replaces a 1040 steel forging and is made from H/D:P/M 4650 through hardened to R_c55. The major advantage in this application comes from the incorporation of the previously machined taper both on the O.D. and on the I.D. These tapers allow the O.D. of the part to seat into the spring while the I.D. taper is used to lock the keys in place.

FIGURE 13

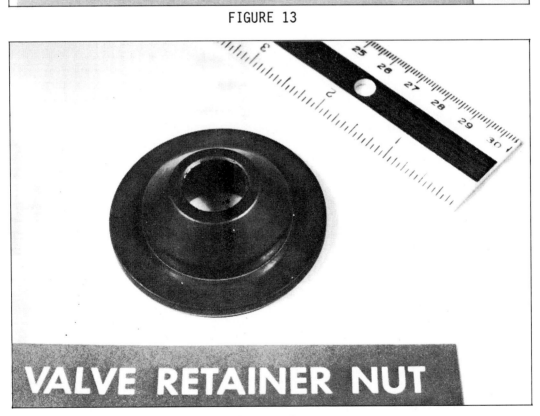

FIGURE 14

BEARINGS

Our experience to date has been with taper roller cones used for automotive front wheel bearings. All testing has been done by our customer. Limited life testing showed encouraging results. Spalling fatigue was noted however. The material tested was H/D:P/M 4620 carburized. It is felt that the present materials and manufacturing methods have not advanced to the point where this type of application has commercial value.

CONCLUSIONS

Resistance to wear can be provided by parts having a density of 7.5 g/cc through hardened to Rc55. A molybdenum steel powder is preferred since it provides good hardenability. A typical composition would be Fe - 0.5% Mo - 0.5% Ni - 0.5% C.

Impact resistance is attained in parts having a density of 7.8 g/cc. These are generally carburized to insure a good combination of a wear resistant surface with a tough core. A typical composition would be Fe - 0.5% Mo - 2.0% Ni - 0.25% C. However, density and chemistry above do not in themselves result in good impact. The processing parameters are equally important.

REFERENCES

1. K. Smith, "The Heat Treatment of Hot Densified Powder Metal Alloys," ASM Transaction, March 1972.

2. H. Antes, "Forming Metal Powders," SME, MF72-504.

3. T. W. Pietrocini, "Fatigue and Toughness of Hot Formed Cr-Ni-Mo and Ni-Mo Prealloyed Steel Powders," Modern Developments in Powder Metallurgy, Vol. 4, pg. 431.

4. S. Mocarski, et al, "Properties of Magnetically Soft Parts Made by Hot Forging of P/M Preforms," Modern Developments in Powder Metallurgy, Vol. 4, pg. 451.

Five Case Histories Reflect State of P/M Technology

*By SHERWOOD W. McGEE
and GUSTAV M. WALLER*

Here's a wealth of practical information on three conventional P/M parts and two made by the forging technique. A manifold blank, one of the conventional parts, has a shape that is difficult to machine or cast. It weighs 3.5 lb. A special nut, weighing 0.086 lb, is compacted to a preform density of 6.6 g per cu cm. Its over-all density after forging is 7.70 g per cu cm. Properties include a tensile strength of 150,000 psi and 4.0% elongation.

Current interest in commercial powder metallurgy centers on parts made with through-cavity tooling in presses with up to 1,000 tons of capacity. In sinterforging, the latest development, high-quality, precision parts are being made with largely the same press motions and identical raw powder materials.

To illustrate the capabilities of both conventional powder metallurgy and sinterforging (also referred to as forging preforms), we have selected five parts. Details are spelled out in each of the case histories.

● The manifold blank, Example No. 1, illustrates full use of "press-sinter-heat treat" operations in making a part with tensile properties about equal to those of a medium-strength gray cast iron. The shape is difficult to machine or cast; and machined tolerances are required. This part is notable for its degree of internal complexity, plus its dimensional precision.

● The transmission synchronizer hub, Example No. 2, is representative of the most advanced state of the art for a tooled combination of complex shape, dimensional precision, and mechanical properties competitive with those of machined bar stock. Because of its torque-carrying ability, this is a unique conventional P/M part.

● The spur involute gear, Example No. 3, shows the advance of

Mr. McGee is director of research and Mr. Waller is executive vice president-research and engineering, Burgess-Norton Mfg. Co., Geneva, Ill.

Example No. 1

Part: Manifold blank.
Weight: 3.5 lb.
Function: Distribution and control of high pressure oil in hydraulic motors.

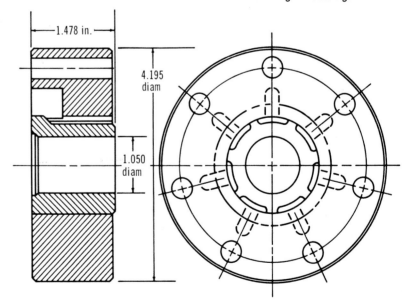

Dimensions and tolerances: OD, 4.00 in. ±0.010 in.
ID, 1.050 in. ±0.005 in.
Length, 1.478 in. ±0.005 in.
Slots to 0.250 in. ±0.010 in. and concentric with T-holes ±30'.

Production operations: Press two parts separately (housing and insert), green assemble, sinterbond and locally infiltrate, heat treat.

Alternative production methods: Precision investment cast and machine; machine and braze assemble, heat treat.
Advantage: Significant cost savings over machining or casting.

Material specifications: 5.0 Cu, 0.70 C, 0.3 Mn, bal Fe (nominal analysis); over-all density, 6.2 g per cu cm; PMPA FC0510 type; hardened for wear resistance; locally file hard in bore location. Typical tensile strength, 50,000 psi.
Type powder: Medium compressibility, reduced sponge iron, admixed with copper and graphite, locally copper infiltrated.

P/M engineering to large parts. Generally, high-density P/M gears such as this one are technically preferred to hobbed gears because of their more precise tooth geometry, tolerances, and surface finish. The operating torque in this case is 200 to 400 ft-lb. Peak torque is 800 ft-lb.

● The special nut, Example No. 4, is a relatively simple sinterforged part. It demonstrates that the process is competitive where it is justified in terms of quantity and the need for a combination of forged properties and machined precision.

● The track link end plate, Example No. 5, is a more complex sinterforging that has been tooled, produced in pilot quantities, and shown to be cost-competitive with conventional impression die forgings. All mechanical property requirements are met; and technical improvements have been obtained via better dimensional quality and the elimination of decarburization and draft.

Example No. 2

Part: Transmission synchronizer hub.
Weight: 1 lb.
Function: Pick up and carry full driveshaft torque input for medium-sized automotive transmission.

Alternative production method: Screw machine shape, broach inside diameter and outside diameter splines, heat treat, and grind.
Advantages: Cost savings, precision.

Dimensions and tolerances: Spline OD, 3.218 in.±0.005 in.
Bore spline major diameter, 1.8475 in. +0.001 in., −0.0000 in.
Bore spline minor diameter, 1.715 in. ±0.005 in.
Bore center point of diameter to OD, 0.003 in. TIR
Over-all length, 1.220 in. +0.000, −0.005 in.

Production operations: Press, presinter coin, resinter, heat treat, and tumble.

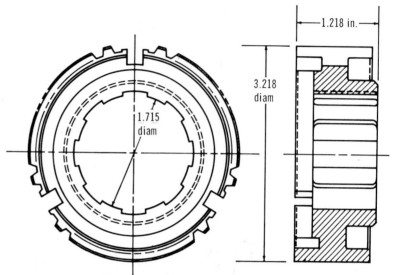

Material specifications: 2.0 Ni, 0.5 Mo, 0.45 C, bal Fe (nominal analysis); over-all density, 7.35 g per cu cm; PMPA FN0205-T (modified); hardened and tempered to Rc 30-35 on spline ends; typical tensile strength, 140,000 psi; 2.0% elongation.

Type powder: High compressibility, water-atomized iron, admixed and diffusionally alloyed with nickel, molybdenum, and carbon.

Example No. 3

Part: Spur involute gear.
Weight: 7.85 lb.
Function: Main drive gear in high-torque hydrostatic transmission.

Alternative production method: Part originally designed for powder metallurgy.
Advantages: Cost savings, precision.

Dimensions and tolerances: 85 tooth, 12 pitch, 20° angle involute spur gear.
Face width, 1.50 in.
ID to OD concentricity, 0.008 in. TIR
Maximum gear tooth lead error, 0.003 in.

Production operations: Press, presinter, coin, full sinter, size, heat treat, grind ID and one face.

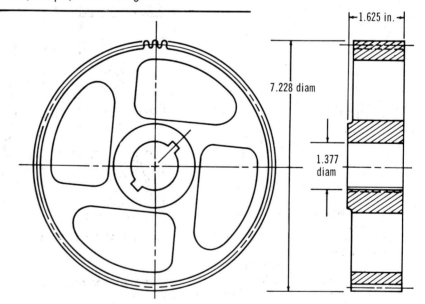

Material specifications: 2.0 Ni, 0.5 Mo, 0.45 C, bal Fe (nominal analysis); over-all density, 7.2 g per cu cm; PMPA FN0205-T (modified); hardened and tempered to Rc 37-50; typical tensile strength, 140,000 psi; 1.0% elongation.

Type powder: High compressibility, water-atomized iron, admixed and diffusionally alloyed with nickel, molybdenum, and carbon.

Example No. 4

Part: Special nut.
Weight: 0.086 lb.
Function: Retain piston pins in large diesel engines.

Alternative production method: Cut-off cold-drawn bar, mill slot, drill and countersink, tap, heat treat, and tumble.
Advantage: Cost savings.

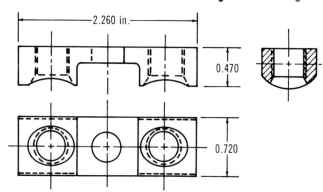

Dimensions and tolerances: Length, 2.260 in.±0.010 in.
Width, 0.720 in.±0.010 in.
Thickness, 0.450 in.±0.010 in.
Hole center true position, ±0.005 in.

Production operations: Press powder preform, presinter, forge, resinter, drill and countersink, tap, heat treat, tumble.

Material specifications: Sinterforged modified SAE 4640 analysis; minimum over-all density, 7.70 g per cu cm; hardened and tempered to Rc 28-33; typical tensile strength, 150,000 psi; 4.0% elongation.

Type powder: Prealloyed, water-atomized 4600 type, admixed with graphite and lubricant, pressed to a preform density of 6.6 g per cu cm.

Example No. 5

Part: End plate for tank track link.
Weight: 1.8 lb.
Function: Carry full load in medium-weight crawler track vehicle.

Alternative production method: Shear blank, forge, hot pierce and trim, cold coin, broach, assemble braze, heat treat, temper.
Advantages: Cost savings and possible design advantage in elimination of surface decarburization and forging draft allowance.

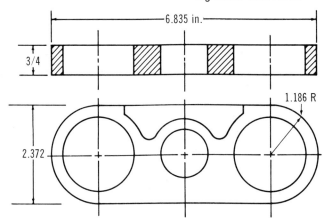

Dimensions and tolerances: Length, 6.835 in.±0.020 in.
Width, 2.372 in.±0.010 in.
Thickness, 0.750 in.±0.015 in.
Large holes, 1.827 in.±0.004 in.
Small holes, 1.233 in.±0.003 in.
True positions of holes to ±0.008 in.

Production operations: Press powder preform, presinter, forge, resinter, broach holes, assemble, braze, heat treat, temper.

Material specifications: Sinterforged modified SAE 4640 analysis; minimum over-all density, 7.80 g per cu cm; hardened and tempered to Rc 28-33; typical tensile strength, 165,000 psi; typical elongation, 12.0%.

Type powder: Prealloyed, water-atomized type 4600, admixed with graphite and lubricant, pressed to a preform density of 6.6 g per cu cm.

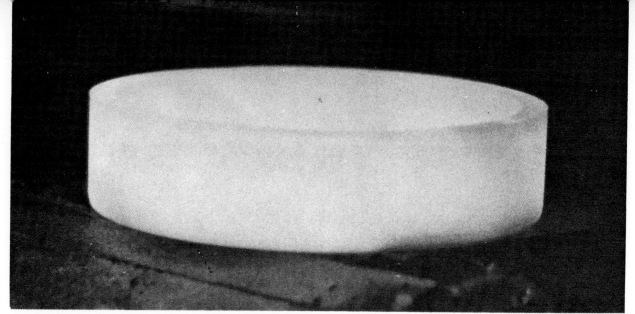

Sinta Forge bearing cup as it leaves the forging press.

Improved Bearings at Lower Cost via Powder Metallurgy

By John S. Adams and Douglas Glover

It HAS BEEN WELL KNOWN for many years that the strength of powder metal parts can be greatly improved by forging them to the density of wrought steel. The challenge was to develop a process for producing physically accurate parts by the thousands on a durable set of dies, day in and day out. After eight years of concentrated research and engineering development, Federal-Mogul succeeded.

In 1971, an efficient and reliable method was perfected for mass producing over-running clutch races for automatic transmissions by hot forging powder metal to the density of wrought steel. Since then, over 20 million of these races have been made by what was eventually named the Sinta Forge process. These parts have functioned with exceptional reliability in automatic transmissions in service worldwide.

Early in the development stages of the Sinta Forge process, Federal-Mogul's ball and roller bearing engineers began working with the powder metal engineers to develop durable powder metal bearing rings. The potential savings in raw materials, in processing costs, and in capital investment were strong incentives for continuing this research work without interruption.

The breakthrough in bearings came in 1972 when tapered roller bearing cups made by the Sinta Forge process, for the first time, in laboratory tests, matched, and actually ex-

ceeded, the theoretical fatigue life and durability objectives. The process for making bearings was rapidly converted from an experimental base with many laboratory improvisations to a full-scale manufacturing process with sophisticated production refinements and controls.

Manufacturing Control: Key to Quality

Common practice is to hot forge a sintered preform of 80 to 90% theoretical density (produced by conventional powder metallurgy practice) in a protective atmosphere. Correct forging practice and die design, along with good quality-control procedures, result in forgings of nearly full density (over 99.6% minimum theoretical) with engineering properties exceeding those of ingot-produced steel.

Federal-Mogul's Sinta Forge process includes the use of prealloyed powder, graphite, and a die lubricant; cold compaction; a high-temperature sinter; application of a protective coating; induction heating, and forging. There is an inspection procedure for each phase of the production process, to control density, oxide, and microstructural characteristics.

Bearing rings are made using water-atomized steel powder supplied to Federal-Mogul's specifications. Prealloyed nickel-molybdenum powders are blended with approximately 0.25% graphite, which results in a finished forging of a carburizing grade of steel similar to AISI 4625. Before production, a sample from each lot of blended

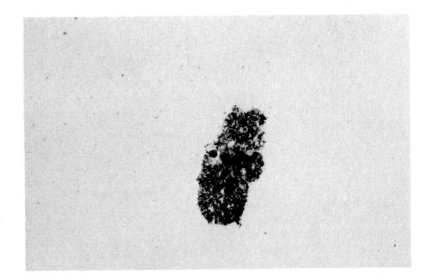

Fig. 1 — Microstructure of carburized case of a finished bearing ring. Large dark-etching constituent is bainite caused by plain iron. Fine dark constituent is normal bainite found in transition area between the case and core. (Lightly etched in 3% nital; 100X.)

powder is made into trial forgings and subjected to a battery of tests which includes checks for oxygen, carbon, plain iron, and nonmetallic inclusions.

Plain iron particles are present because commercial powder suppliers process both alloy and plain iron powders through the same equipment. With proper controls, the amount of plain iron is held to a small amount that does not interfere with the performance of the finished product. In the carburized case of a finished bearing ring, the iron may transform to bainite as illustrated in Fig. 1. The plain iron particles may show up in the lower carbon core structure as free ferrite as shown in Fig. 2.

Nonmetallic inclusions are of paramount interest to bearing manufacturers because of their influence on bearing fatigue life. Much of the nonmetallics in powder are in the form of atomized slag (Fig. 3). It has been proved by test that the atomized slag particles are much less influential on bearing fatigue life than the oxide inclusions found in ingot-produced bearing steel. For wrought bearing steels, the industrial standards are well established and defined by ASTM standards. However, for the Sinta Forge bearings it was necessary to develop new methods and standards for measuring and controlling nonmetallics.

Two methods of measuring nonmetallics in powder are employed by Federal-Mogul. One has a special magnetic device which removes nonmetallic particles from a 500 g sample. Nonmetallic particles are weighed with an analytical balance, and must be within established maximum limits. Powder suppliers have identical magnetic separators made by Federal-Mogul, which are used for source control of the powder.

In the second check for nonmetallic inclusions, a polished cross section of the trial forging, which is com-

pared with standards, is examined microscopically. Once a lot of powder is accepted, its identification is maintained throughout production and into the finished parts.

After acceptance, blended powder is cold compacted into hollow cylindrically shaped green briquettes. The weight and geometry of the briquettes are held closely to assure accurate die fill in the final step of the process.

Sintering is done at 2050 F (1121 C) in a continuous belt furnace, utilizing an endothermic atmosphere controlled to maintain a specific carbon potential. During sintering, the die lubricant is eliminated; a bond is established between the steel particles; and the graphite is diffused uniformly throughout the briquette. Preforms from each sintering furnace are periodically checked for carbon content.

Following sintering, preforms are coated with a graphite material to prevent oxidation during subsequent heating and to prevent excessive forging die wear. The coated preform is heated to forging temperature by an induction furnace which provides minimum time at temperature. Forging temperature control is maintained by an infrared pyrometer.

From the induction furnace the heated preform is automatically transferred to a mechanical forging press where a closed die produces the final shape and eliminates porosity. Control of the forging operation is maintained by a statistical sampling inspection plan. Samples are checked for minimum density by use of a rapid water displacement method capable of measuring ±0.005 g/cc. Metallurgical samples are prepared from the forgings and examined microscopically for structure and oxide penetration.

Each part in a forging lot is permanently identified, and is traceable to a specific powder lot. Subsequent manufacturing methods used to carburize, harden, and grind the bear-

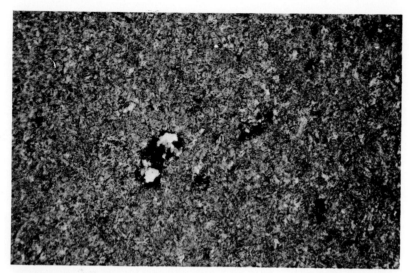

Fig. 2 — White-etching ferrite in core is due to plain iron. (Nital etch; 100X.)

Fig. 3 — As-polished cross section of Sinta Forge bearing race showing fine nonmetallic inclusions. (100X.)

ing rings are identical to those for conventional ingot-produced steel.

Tapered Cups Lead Development

Tapered roller bearing cups were chosen as the original bearing parts for Sinta Forge development. Cups are well suited for forging because of their wedge-shaped cross section. Furthermore, Federal-Mogul had acquired a great deal of technical experience in forging cups from wrought steel which was directly applicable to the Sinta Forge development. This technological background served to significantly shorten development time.

The main criterion for acceptance of bearings produced with a new material or by a new process is to match or exceed the durability of bearings already approved in accor-

dance with established standards. In the case of the development of Sinta Forge bearings, the original objective was to obtain an efficient manufacturing process and to increase the material choices; an increase in bearing fatigue life was a by-product.

It was somewhat surprising to find that the durability and fatigue-life test results of the first successful experimental lot of Sinta Forge cups were unusually good. Subsequently, extensive tests of cup specimens taken from production lots verified the consistently good fatigue life characteristics of the Sinta Forge product. To illustrate the life improvements, a typical Weibull plot is given in Fig. 4 for 595 series Sinta Forge production cup test results.

(Weibull plotting is the accepted method used by the bearing industry to statistically predict service life. Using a modified logarithmic scale, the Weibull plot shows the

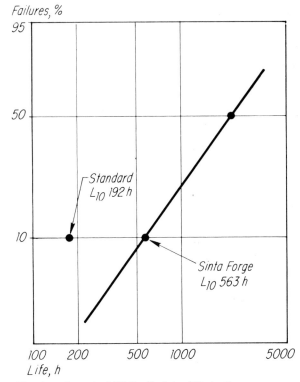

Fig. 4 — A typical Weibull plot of Sinta Forge cups showing L_{10} life of 563 h compared to standard L_{10} life of 192 h.

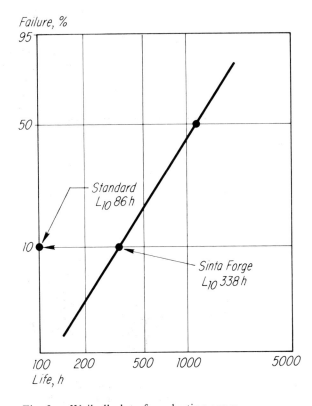

Fig. 5 — Weibull plot of production cones.

percentage of bearings in a population that are expected to fail at any particular time. It is common practice to use the 10% life (L_{10}) in comparing data. The standard rating shown in Fig. 4 is based on experience with previous tests of bearings made from ingot steel, and modified by the test conditions, such as, load, and shaft speed.)

Because of the differences in the degree of osculation between the rollers and the cup and cone roller tracks, the cone exerts a much greater influence on bearing life. Therefore, improvement in the fatigue life of cones is of much greater significance to improving bearing performance than an improvement in cup life.

Complete Sinta Forge Bearing Now Possible

As soon as the production of cups was well established, development work began on the production of tapered roller bearing cones (or inner rings) by the Sinta Forge process. By the end of 1975, the first experimental cones passed laboratory fatigue tests with an L_{10} life of 239 h, nearly three times standard ratings.

The first production cones were made late in 1976. Tests yielded a fatigue life exceeding three times standard ratings, verifying the excellent performance of the experimental cones.

The Weibull plot shown in Fig. 5 was made from production cone test data to illustrate the results.

The improvement in rolling contact fatigue life in both cups and cones is attributed to the fact that nonmetallic inclusions are uniformly dispersed throughout the material. In wrought steel, inclusions are often concentrated in specific zones. Consequently, the inclusions in wrought steel are much more likely to become the sites of high stress concentration where fatigue originates.

Laboratory tests were supplemented by field tests of Sinta Forge cups installed in over-the-highway truck tractor and trailer wheels in February 1974. The cups were removed from test after three years and almost 4 million miles of troublefree service. They were subsequently replaced with complete Sinta Forge cup and cone assemblies, and field tests were resumed.

Niche For Process: Sinta Forge joins the several proved methods of manufacturing bearing rings at Federal-Mogul, including cold and hot forging, hot ring rolling, and the conventional screw machine.

The selection of the manufacturing method for any specific ring size is based on maximum efficiency. Sinta Forge rings promise a significant increase in bearing reliability. ⊕

How Gearmaking Methods Compare

Alternatives to traditional machining of bar stock or rough forgings include cold forming, powder metal forming, and precision hot forging. Each has its niche.

Making gears by machining is expensive. Cutting times are long, particularly with difficult-to-machine alloys, and scrap losses run high.

In the performance area, cut gears have an unfavorable grain flow pattern with flow lines intersecting the gear tooth surface. Fatigue endurance and tensile ductility would be improved if the grain flow pattern ran parallel to the tooth surface.

These and other disadvantages of conventional, machined gears have led manufacturers to investigate, develop, and adopt alternative gearmaking processes.

Chief among these are the "chipless" processes of cold and warm forming, powder metallurgy, and precision hot forging.

All three boast high materials utilization factors. All three produce gears with fewer machining operations. Advocates of cold forming and precision hot forging claim significant performance plusses.

The discussion that follows is an examination of recent developments in these alternative processes. It's based on presentations made at "Gear Manufacture and Performance," a forum sponsored by ASM's Mechanical Working and Forming Division.

Ford's cold forming program, Merriman's P/M experiences, and two precision hot forging processes — Eaton's flow forging and the TRW process — are featured.

Cold Forming

The idea of cold forming gears is attractive to automakers because of the process' high-production potential. A well-engineered cold forming process should lead to savings in labor, material, and, often, to floor space compared with competitive manufacturing methods. Frequently, product quality is also improved.

Cold forming's disadvantages are seldom cited, state C. A. Stickels and S. K. Samanta, Scientific Research Staff, Ford Motor Co., Dearborn, Mich. These include:

- High development costs. Process and tooling design rely heavily on trial-and-error methods.
- Risk of high operating costs in a poorly engineered process.
- Inflexibility. A major effort is required to produce a new part.
- Need for high volume. In many instances, a volume of 100,000 parts per year would not be economical.
- The "crisis factor." A single machine accounts for a part's entire production. A major breakdown spells catastrophe.

In general, however, the economic plusses seem to outweigh the disadvantages and more and more parts are being looked at as candidates for cold forming.

Ten years ago, Ford did not specify cold forming in any gear manufacturing sequence. Today, cold forming processes are either used or being considered for use on nearly every engine and drive train gear.

Machines — Special, multistage presses with automatic parts transfer are required for high-volume applications. There are essentially two categories: cold heading machines and transfer presses (mechanical and hydraulic).

Headers, originally designed for forming nuts and bolts, come in a variety of sizes. High-speed, automatic operation makes them the most economical way to form small parts.

Three-stage models accept wire up to ¼ in. in diameter and can handle 2 in. shank lengths. Up to 300 parts per minute can be produced.

Giant headers have been built to feed wire up to 1⅞ in. in diameter. The high-speed production units can produce up to 35 parts per minute.

Seven-stage machines include a cut-off stage and six forming stages. Options are open to upset, extrude, coin, emboss, and trim, working at either end of the part.

Forward extrusion of well-lubricated wire in a header is limited only by the machine's relatively short stroke. Backward extrusion can be limited by lubrication problems, because the slug's cropped ends are lubricated by only the cooling oil sprayed in the machine.

Presses — Mechanical and hydraulic transfer presses are usually specified for parts weighing a pound or more. Production rates range from 40 to 140 parts per minute.

Pre-cropped and lubricated slugs are fed to the press. The additional steps add to costs, but lubrication on all slug surfaces means that more severe backward extrusions are possible.

Presses are more suitable than headers for parts requiring interstage anneals. Headers are ruled out if more than anneal is required.

A word of caution from Messrs. Stickels and Samanta: "This sophisticated equipment is of limited usefulness if similar sophistication and ingenuity are not exercised in planning the forming process and designing tools."

An idea of what's required is given in the examples that follow.

Starter Pinion — A part well suited for cold forming on a progressive header is the starter pinion gear made

Continued on p. 54

The automotive starter pinion gear is an excellent cold forming application. Ford makes 50 parts per minute from AISI 8620H on a National 1000 cold header. The forming sequence is shown here.

Table I — Minimum Tolerance Capabilities for Spur, Helical, and Bevel P/M Gears

Tooth to tooth composite tolerance	AGMA Class 6
Total composite tolerance	AGMA Class 6
Test radius*	(0.002 in.) + (0.002 in.) × (Diameter)
Over pin measurement*	(0.004 in.) + (0.002 in.) × (Diameter)
Lead error*	0.001 in./in.
Perpendicularity — face to bore	(0.002 in.) + (0.001 in.) × (Diameter)
Parallelism*	(0.001 in.) + (0.001 in.) × (Diameter)
Outside diameter	(0.004 in.) + (0.002 in.) × Diameter
Profile tolerance*	0.0003 in.

*Does not apply to P/M bevel gears.

Table II — Process Details for Precision Hot Forging of CH-47 Helicopter Spiral Bevel Gear Sets

Material	AMS 6265 CVM steel, hot-rolled finished bars
Billet size	
Gear	3½ in. in diam, 5½ in. long
Pinion	3½ in. in diam, 11⅞ in. long
Heating temperature and time	2,025 F ± 25 F, 45 min
Billet heating atmosphere	Endothermic, 7:1 ratio
Die temperature	400 F
Die lubricant	5% Aqua-Dag
Average transfer time	
Furnace to preform die	10 sec
Preform to coining die	12 sec
Observed coining temperature	1,900-1,950 F
Coining load	3.5 million lb

at Ford's Sandusky, Ohio, plant. Starting material is annealed and lime-coated, hot-rolled AISI 8620H wire. The steel is formed on a National 1000 cold header.

Forming sequence is:

- Draw to required dimensions.
- Shear slugs.
- Square.
- Back extrude cup.
- Pierce to remove web.
- Remove parts; anneal, phosphate, and soap; and feed to last header station.
- Forward extrude gear teeth.

One machine is used to make the part at a production rate of 50 per minute. Subsequent operations include machining, carburizing, and hardening.

Justification — Starter pinions don't require the same close dimensional tolerances or high surface finish demanded of transmission gears of similar size. Cold formed pinions more than meet requirements and, in fact, are better in these areas than pinions made by the methods formerly specified. Consequently, cold forming is justified in this application.

For an annual requirement of nearly 3 million parts, only 1,000 hr of operation at 100% efficiency are needed.

Gear Blanks — Large volume requirements and small size make automatic transmission pinion gears natural candidates for cold forming.

At Ford, the helical gears are manufactured by a combination of three processes: the blank is cold formed on a progressive header, and then the teeth are hobbed and finish rolled. Two of the principal shaping operations are by cold forming processes.

Less Material — After cropping, the slug is squared and indented, back extruded, pierced, and then the bore is sized. By back extruding the hole instead of machining it, material savings of 25% are realized.

The material is lime-coated, spheroidize annealed AISI 4027H.

Because the back extrusion is shallow, the conventional step of lubricating the slug ends can be omitted. Phosphate-plus-soap coating is also not needed because there is little movement of the part over the die.

Formed blanks are normalized before machining. Pearlite and low residual stresses promote good chip breaking characteristics and a good machined surface finish.

Similar blanks made by GM's Hydramatic Div. are warm formed on a progressive header. Messrs. Stickels and Samanta feel that a comparison is instructive.

Ductility — Hydramatic specifies vacuum degassed AISI 5140 wire. It's annealed to give a "blocky" pearlite structure that's good for machinability but bad for cold formability. However, heating the slugs to about 1,200 F gives even this microstructure adequate ductility for forming.

Initial forming steps — done cold — include drawing the wire, cropping and squaring slugs in the first two stations, and upsetting (by about 35%) to a pancake shape in the third station. Parts are then dropped out of the machine.

These steps are performed because it is difficult to crop slugs if their length is less than the wire diameter. On the other hand, slug diameter should approximate final part diameter. Consequently, wire diameter is selected so that $l/d > 1$ for cropping. Then the slug is upset to make $l/d < 1$ for backward extrusion.

Upset slugs are induction heated to 1,200 to 1,250 F, fed back into the header, backward extruded, and pierced. A subcritical anneal relieves residual stresses before machining.

Heat treating costs are higher for cold forming — spheroidize annealing and normalizing vs process annealing and subcritical stress relieving. Offsetting this is warm forming's greater complexity.

"Without a detailed cost analysis,

it would be difficult to choose between these two processes," claim Messrs. Stickels and Samanta.

Roll Forming — Considerable effort has been expanded attempting to develop a roll forming process for gears similar to the one used for splines.

Results of work to date indicate that "... if roll forming were to be used to make gear teeth, it would have to be followed by a shaving operation or some other metal removal operation to eliminate surface defects. This is not acceptable from a mass production standpoint."

Tooth finishing by roll forming has, however, been successful. After hobbing to establish tooth profile and helix angle, the Ford transmission pinions are finished by rolling through round dies.

The process not only gives an excellent surface finish, but is also used to crown the teeth. Dimensional changes are not significant and no surface flaws are generated. Costwise, the process is more economical than finishing by metal removal techniques. It also produces less part-to-part dimensional variations.

Trends — The cold forming experts see two important trends in gear-making:

- Attempts to cold form larger gear blanks.
- Attempts to cold form helical and hypoid gear teeth.

Many large gear blanks, currently hot forged or made on screw machines, can be cold or warm formed. These include transmission sun gears and differential side gears. The question is: Can they be made on headers?

Several factors appear to be working against the use of headers for large parts requiring large deformations. The Ford researchers point out that "... as larger parts are considered, it's likely that relubrication and annealing between forming stages will be necessary. Also, coiled wire is not available in diameters larger than 1½ in. This means that the headers would have to be fed with lubricated slugs."

The conclusion is that the prime advantage of headers — an entire fabrication sequence contained in one machine — is liable to be lost.

Gear Teeth — Distortion during heat treatment is likely to be more of a problem with cold formed gear teeth than with conventionally machined teeth. Grinding, for example, may be needed after hardening.

There are, however, some tooth forming processes under development that may prove feasible.

Although it has been shown that spline rolling can't be extended to gear rolling, it is possible that changes in the geometrical arrangement of forming die and workpiece can be devised to obtain satisfactory tooth quality.

Another possible cold tooth forming process is forward extrusion through a toothed die. Straight gears can be made this way. What's needed is a method for producing an accurate helical lead angle. There may also be problems preserving the lead angle near the slug's ends.

Hydrostatics — ASEA, a Swedish firm, has produced high quality cold-formed teeth on round, solid billets by hydrostatic extrusion. Tooth profiles and surface finish are very good. Extruded billets are then cut-to-length and each gear is bored and faced.

The problem here is preserving the quality of the finish on the teeth during handling and chucking. If teeth could be extruded on hollow billets (sizing the bore during extrusion), it might be possible to chuck on the inside diameter during cutoff and facing.

P/M Pressing and Sintering

The powder metallurgy process is a flexible method for gearmaking. It's capable of producing close-tolerance spur, helical, bevel, face, spur-helical, and helical-helical gears with strengths up to 160,000 psi.

"The process is particularly attractive when the gear contains depressions, through holes, levels, or projections," according to Walter E. Smith, Merriman Inc., Div. Litton Industries, Hingham, Mass.

Applications are found in the appliance, business machine, automotive, power tool, power transmission, and farm equipment industries.

Types — Spur gears are most common. If an EDM electrode can be ground or cut, the mating form can be duplicated in P/M dies.

A limitation on helical gears is that the helix angle can't exceed 35°.

Spur gears are the most common type of P/M gear made today. The only P/M spur gear design limitation is whether a suitable EDM electrode can be fabricated for die sinking.

Bevel and face gears have mechanical property limitations. Because of the nature of the compacting process, sections such as face gear teeth can't be made to high densities without copper infiltration or re-pressing, and undercuts are not permitted. In any event, maximum mechanical properties will be somewhat less than those possible with spur and helical gears.

Tolerances — Three basic areas needing investigation before specifying a P/M gear are outlined by Mr. Smith. They include tolerances, load ratings (mechanical properties), and quality control.

Gear tolerances are determined by compaction tooling, compaction process, sintering techniques, and the P/M alloy. There are also functional and nonfunctional tolerances.

Nonfunctional tolerances include hub design dimensions, keyways, set-screw holes, inside diameters for press fits, and other special body configurations. They're covered in the "Powder Metallurgy Design Guide Book," published by Powder Metallurgy Parts Assn., Princeton, N. J.

Functional tolerances are first controlled by the compaction tooling and then by processing methods and gear materials.

Tooth form and dimensions remain relatively stable throughout the process because most P/M gears are pressed in carbide dies. However, concentricity of the gear's outside diameter and pitch diameter to the bore are also controlled by the tooling. If core rods or punches are eccentric, parts will be eccentric.

After pressing, the material and sintering process determine tolerances

of the outside diameter, pitch diameter, test radius, and bore diameter.

Minimum — not absolute — tolerances for ferrous-based P/M gears (without secondary operations) are given in Table I.

Load Ratings — Selection of the proper P/M alloy depends on magnitude and nature of the transmitted load, speed, life requirements, environment, type of lubrication, and the gear and assembly precision.

Most P/M gears are limited in performance by their impact or fatigue strength instead of their compressive yield strength or surface durability. Consequently, the first step in P/M gear materials selection is determining which alloys have the necessary strength.

This is done by using the methods outlined in American Gear Manufacturers Assn., Washington, D. C., standards for rating the strength of spur, helical, and straight bevel gear teeth (AGMA 220.02, 221.02, and 222.02).

Equations that involve allowable bending stress can be solved using data from "P/M Materials Standards and Specifications" (MPIF Standard No. 35) published by the Metal Powder Industries Federation, New York.

Durability — Once the field has been narrowed to materials with sufficient strength to carry the bending loads, surface durability of the gear teeth is calculated. Methods outlined in AGMA 210.02, 211.02, 212.02 are used.

Equations calling for allowable contact stress, elastic modulus, and Poisson's ratio can be solved using data from MPIF Standard No. 35.

Gears made by precision flow forging (BLW process) have integrally formed teeth. Materials utilization is high — 80 to 95% — and tooth fatigue resistance exceeds that of cut gear teeth.

Microhardness — Potential scoring problems with incompatible materials must also be considered. P/M materials have both a macroscopic and microscopic hardness that results from having a porous structure.

Conventional Rockwell hardness readings actually reflect a composite hardness. Because sliding and surface contact involves microscopic contact between individual particles, the P/M material's microhardness is a more accurate indication of its scoring resistance.

A general rule is stated by Mr. Smith: "When a P/M gear is run against a wrought gear, the P/M particle hardness (microhardness) should not be higher than the hardness of the wrought gear to prevent wear of the mating gear."

Inspection — Quality control in P/M gearmaking is extremely important. Two stages — samples and production — are involved.

Sample inspections involve checking basic gear data such as tooth profile and size. These are essentially checks on pressing die condition. Gear mechanical properties are also measured and, in many instances, bench life and impact or field tests are run to check gear functionality.

A gear's tensile strength can't be directly measured. However, it is possible to establish a minimum tooth breakage strength and use this as a production acceptance criteria.

A simple method that works on spur gears is to mount the gear to be tested between fixed supports and apply a vertical load. Test a number of gears and apply statistical analysis (calculate an $\overline{X} - 3\sigma$) to arrive at a lower limit for tooth breakage strength. Helical and bevel gears are similarly evaluated, but in torque-type tests.

With normal data, use of this minimum acceptance criteria insures that 99.85% of the gears will have a tooth breakage strength above the selected minimum load. The criteria have also shown an excellent correlation to field performance.

Specify — A minimum Rockwell hardness is usually specified as a surface durability criteria because the value relates to the material's compressive yield strength.

In applications where scoring may be a problem, particularly with heat-treated gears, a particle microhardness specification is also included.

Production inspections include checking face width, outside diameter, over-pin measurement, total composite error, tooth-to-tooth error, perpendicularity of face-to-bore, parallelism, and inside diameter.

Sampling plan and frequency of measurement depends on how critical the application is.

Eaton's Flow Forging

The BLW Process was developed in Germany shortly after World War II by Bayerisches Leichtmetallwerk (BLW). Exclusive North American rights have been licensed to Eaton Corp. since 1963. The process has similarities with the precision forging processes used to make engine valves in that careful billet preparation is required and materials utilization runs from 80 to 95%. The high materials utilization factor can provide a decisive cost advantage in comparisons with conventional processes.

By precision forging gears with integral teeth, the process eliminates the machining of tooth forms and other intricate contours. Users get a random surface texture that avoids lines of stress concentration and, more importantly, a significantly increased gear tooth beam fatigue strength.

Any gear shape which can take a die impression without being picked up or pulled away by the receding die is a candidate for precision flow forging.

EDM — Die blanks are made of hot work tool steel. They're machined by conventional methods to basic contours and fully heat treated. Gear tooth contours are then added by electrical discharge machining (EDM).

The roughing die — providing about 90% of the hot work deformation — is usually a finished die that has worn beyond the ability to meet gear tooth tolerances.

Folds — Forging billets are machined to remove hot-rolled bark and sharp corners. The procedure eliminates subsequent folds or cold shuts, according to Thomas Russell, Mason Kelley, and Lou Danis, Valve Div., Eaton Corp., Battle Creek, Mich.

Billets are then induction heated in an endothermic atmosphere to the lowest possible temperature that will limit scale formation yet permit complete flow of metal into die recesses. The actual temperature rarely agrees with the handbook value.

Mechanical presses with capacities of up to 2,500 tons are used.

After forging, gears are discharged to a materials handling device that insures uniform cooling rates. The feature largely eliminates the need for process annealing.

Finish machining includes drilling, deburring, flash removal, and back facing. Machined gears have a variety of surface textures and a thin scale on tooth profiles. Surface textures are blended and the scale is removed by grit blasting.

Heat treating and additional finishing operations, if needed, follow.

Testing — Before Eaton decided to go precision flow forging, a number of comparisons were made between gears cut from bar stock and gears made by the BLW process.

All samples were made from a bar of 4⅝ in. in diameter hot-rolled AISI

8620H steel. Test configuration was a seven-pitch spur gear with 29 teeth, a 20° pressure angle, and a 4.1428 in. pitch diameter.

The spur gear – not a good shape for commercial BLW forging because of short die life – was selected because it's a simple shape to test.

All specimens (cut and forged) had the same case depth (0.050 in.), case hardness (Rc 61 to 62), core hardness (Rc 45), and general microstructure. Before fixture testing, the case was ground off each side of the gear (0.060 in. metal removal) to an 0.875 in. face width and the center bore was finished.

Other, roller-shaped specimens were prepared for assessing the effects of surface texture on forged and machined cylindrical surfaces. Specimens were rolled against each other in a radioisotope wear fixture at very high compressive stress levels until failure by surface pitting occurred.

Fatigue – A special test fixture was designed for testing gear tooth fatigue beam strength. The apparatus applied fatigue stresses to the roots of the gear teeth. High stress level was set at 120,000 psi, low stress level at 90,000 psi.

Single-blow impact tests were also specified because some gears must withstand sudden load applications in service.

The Eaton engineers report that test results show that "... in most respects, the forged gears were equal to or better than the conventionally machined gears."

Advantages – As expected, microstructure and hardness of both forged and cut gears were nearly identical.

Results of grain flow examinations were also predictable. Grain flow in cut gears was determined by that of the hot-rolled bar stock. Grain flow in forged gears followed the tooth contour.

A slight advantage for forged rollers was noted in the surface fatigue test results. However, Messrs. Russell, Kelley, and Danis make no claim to an inherent and consistent advantage.

Impact test results revealed a 15% advantage for the forged parts. Average energy absorbed was 154 ft-lb for forged and 132 ft-lb for cut gears. In this instance, it's not felt that the limited test results can be used to qualify impact strength as a process selection criteria.

Gear Teeth – The Eaton engineers do, however, have a high degree of confidence in the gear tooth beam fatigue test results (see graph below). They feel that the advantage shown by forged gears could be "... regarded as decisive ..." in a process selection decision.

Significant is the value of 78,000 psi for 50% failure of forged gears compared with a value of 66,000 psi for 50% failure of cut gears. They conclude that "... integrally forged gear teeth formed by the precision flow process are about 20% higher in fatigue strength than cut teeth."

TRW's Precision Hot Forging

Fabricating costs for high-performance spiral bevel gears are an important factor in the high over-all cost of aircraft power transmission trains.

Conventional machining methods appear to have reached their limits in terms of gear performance and cost effectiveness.

Precision hot forging could eliminate many of the time consuming and costly tooth cutting operations by directly preforming gear teeth. The process also produces a metallurgically superior gear.

What follows is a discussion of a precision hot forging process developed by TRW (not identical to the BLW process) and how it has been used to make main transmission spiral bevel gears and pinions for the CH-47 helicopter.

Comparing – The CH-47 gear set is made of AMS 6265 steel and is mounted in the helicopter transmission nose gear box. It transmits engine power at high rpm to the speed reducing gear train to drive the main motor.

The input pinion operates at 14,720 rpm and transmits 3,750 hp, developing a pitch line torque of 16,056 in.-lb. It's expected to operate 1,100 hr between overhauls. Calculated maximum tooth bending stress is 29,000 psi with a tooth contact stress of 205,000 psi. Design load per tooth is 4,523 lb.

Design, materials, tolerances, finishes, and other technical requirements for the precision forged gear set were identical to those of the conventionally machined gear set currently in production. This made an accurate comparison of processing costs and gear performance possible.

Process – The TRW precision forging process is based on modified, single-action crank presses that have been used for production precision forging for more than 30 years. The press selected for this program had a 2,000 ton capacity.

Parts are typically produced with one or more preform blows followed by a coining blow, although several operations can be performed simultaneously during each press cycle.

"The multiblow method is usually selected because it lets us use small-diameter stock," state J. L. Lazar and R. R. Skrocki, Materials Technology, TRW Inc., Cleveland.

Results of beam fatigue tests indicate that precision flow forged gear teeth are about 20% higher in fatigue strength than cut teeth. Spur gears, forged and cut from AISI 8620H, were tested.

Small-diameter material can be more accurately and economically cut into billet lengths which are then upset to produce larger-diameter preformed forgings for final coining.

Both gear and pinion were forged with a hot preform blow followed by a hot coining blow that forms tooth details.

Preforming — In bevel gear forging, the starting billet is first converted from a 3½ in. in diameter, 5½ in. long cylinder into a flat, conical disk measuring approximately 9 in. in diameter and 1¾ in. long.

The preform is then transferred to the hot coining station. The lower die block is of completely open design, permitting rapid placement and centering of the preformed part and insuring that sufficient residual heat is maintained.

This important feature minimizes metallurgical deterioration of the surface by allowing both preforming and coining operations to take place during the same heating cycle.

Die Life — TRW engineers estimated that a billet temperature of 2,000+ F would be necessary to provide sufficient plasticity for filling of the tooth portion of the gear forging. However, at these temperatures, prolonged contact between billet and die would have a deleterious effect on die life.

To minimize contact time, the tooth form was designed into the punch portion of the die set. Consequently, any delay occurring after the preform was placed in the hot coining station would not result in heat being transferred to the tooth portion of the cavity.

Pinion — The approach to forge tooling for the spiral bevel pinion was similar to that for the spiral bevel gear. Differences were caused by the pinion's integral shaft or stem — a more complex configuration.

The effect was to increase the number of forging operations from two to three by adding another upsetting stage. The additional step was eliminated in the development program by machining the first preform shape from the starting bar material, reducing tooling costs and forging development time. The extra upset forge tooling couldn't be economically justified in light of the small number of parts to be made in the program.

Pinion preforming converts the portion of the billet protruding above the lower die from a 3½ in. in diameter, 5½ in. long cylinder into a flat conical disk roughly 7 in. in diameter and 2 in. thick.

Upper dies at both preform and

Another process for precision hot forging gears with integral teeth has been developed by TRW. The as-forged spiral bevel gear set shown is made of AMS 6265 steel. Application: the CH-47 helicopter's main transmission.

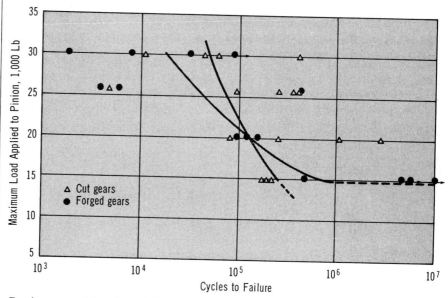

Teeth on precision forged AMS 6265 helicopter pinions have a higher fatigue limit than cut teeth. Loads shown are applied actuator loads. Tooth loads are approximately 33% greater.

coining stations contact the hot part for very short periods, again limiting heat transfer to the critical part of the tooling containing the gear tooth configuration.

Parameters — A total of four developments for the gear and three for the pinion were needed to arrive at the final process parameters shown in Table II on p. 54.

The success of the forging program in obtaining desired tooth configuration was borne out by data on the amount of material that had to be removed in finish machining operations on precision forged and ma-chined semifinished gears and pinions. The amount of stock removal needed was comparable in all cases.

Evaluation — Tests were performed on forged and cut gears of the same material and heat treatment. This made the main variable the method of obtaining tooth configuration — forging or machining.

Mechanical tests consisted of comparative single-tooth fatigue evaluations under unidirectional loading conditions. All fatigue tests were conducted at R = 0 (minimum load = 0). Cyclic frequency depended on the maximum applied load and varied between 10 and 20 Hz. Results are shown in the graph (p. 59). Note that loads on the graph are the applied actuator loads. Tooth loads are approximately 33% greater.

Failure – The data indicate to Messrs. Lazar and Skrocki that the endurance limit of forged gears is significantly higher than that of cut gears when the results at 15,000 lb load are examined.

Three of the four forged gears tested at this level did not fail above 4.6 million cycles, while three cut gears failed below 300,000 cycles. Results at high load levels did not indicate a significant difference.

However, it is the results at the low load levels – near the design point – that are of importance, particularly when design torque is considered. The engineers point out that "... design torque per tooth is approximately 16,000 in.-lb at maximum horsepower. The torque applied at the lowest test level load is 60,000 in.-lb per tooth."

Future tests will be aimed at determining a precise endurance limit and will be concentrated at lower load levels, decreasing from the 15,000 to 20,000 lb range.

Structure – Another advantage of precision forging – favorable flow line patterns – was also observed in structural investigations. Core structure and carbide distribution were identical for forged and cut gears. However, carburized case depth was consistently lower for forged gears. The observation is unexplainable in light of the similar heat treatments.

Cost Data – An economic analysis was made comparing forged and machined gears. Time to make a machined gear was 2.272 standard hours; to make a forged gear, 1.514 hours. Cost per machined gear came out to $29.88; per forged gear, $12.00.

The data indicate that precision forging is practical and adaptable to production applications. However, adoption by industry, and the aircraft industry in particular, isn't likely to occur until the FAA and other regulatory bodies such as AGMA qualify the process.

Precision forging's potential is great. The relatively low forging forces required suggest that gears measuring up to 30 in. in diameter could be produced on existing presses if the required die sinking electrodes and EDM equipment were available. ✦

The properties and structure of some powdered metal gun parts

When parts are complex or small or both it is often necessary to develop special testing procedures for inspection and quality control.

R. J. DEANGELIS *
T. R. ROBE **

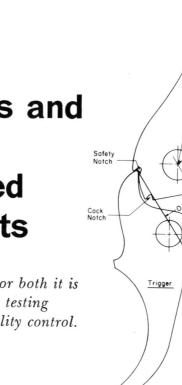

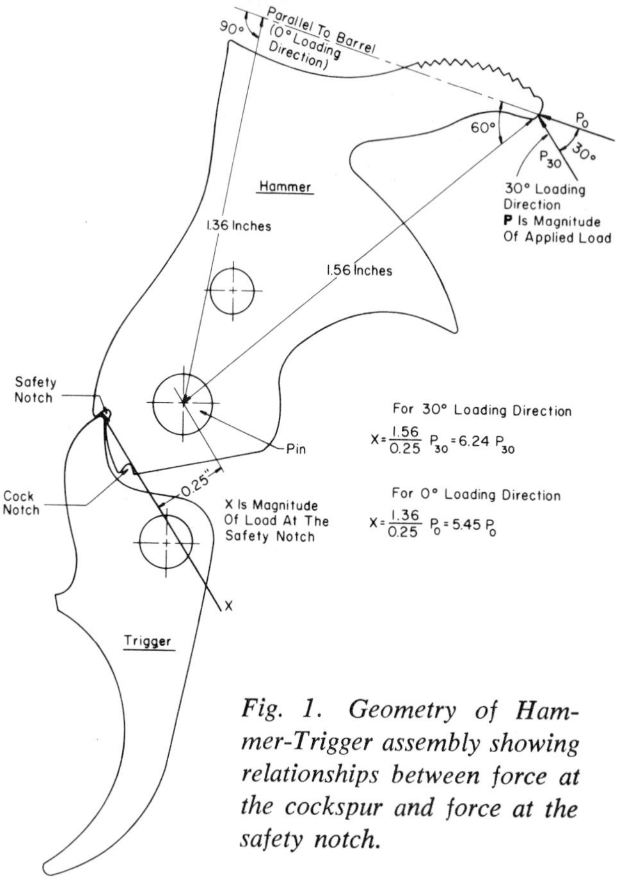

Fig. 1. *Geometry of Hammer-Trigger assembly showing relationships between force at the cockspur and force at the safety notch.*

In a recent article, Weiner (1)† discussed the increasing use of powdered metal parts in the manufacture of guns. It is quite evident that, as powder metallurgy technology continues to advance, gun manufacturers will increase the use of P/M parts.

Some of these parts are very critical to safety and gun reliability. They must perform at design expectations. The parts vendor must supply parts to meet the gun manufacturer's minimum specifications on tensile strength, elongation, and hardness, and within chemical composition range. The only real assurance that these specifications are being met and that a high quality of material is being supplied is to perform tests on actual parts.

In practice, either the geometry is so complex or the piece size prohibits preparation from the part, of specimens for mechanical testing. Test results are often determined from specimens prepared from the same initial material which has been processed with the powder metal parts. Another approach is to make hardness determinations on actual parts and to correlate them with tensile properties of the material by using a predetermined calibration curve. These procedures, although they are accepted as standard practice, introduce a degree of uncertainty. Material specifications alone may not ensure a consistent quality part. Especially in the case of critical parts, such as hammers and triggers, the manufacturer's specifications give little indication of the expected performance of the part in the gun assembly under loading conditions which may be actually encountered in the field.

The work described here reports methods and results of mechanical tests performed on powder metal produced hammers and triggers and the behavior of these parts in the actual gun assembly. In addition to the mechanical test results, observations on the metal-

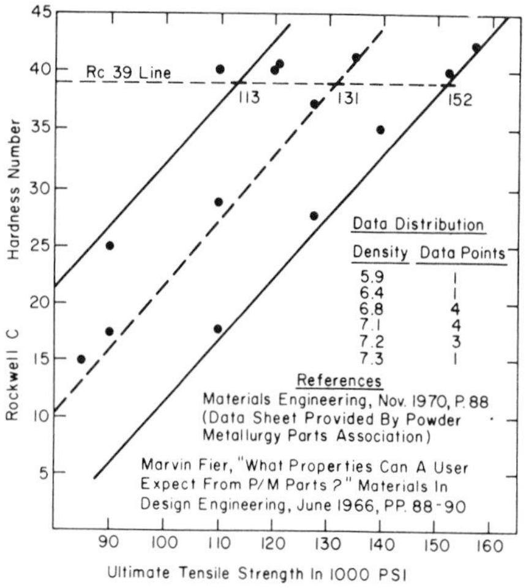

*Department of Metallurgical Engineering and Materials Science, University of Kentucky.

**Department of Engineering Mechanics, University of Kentucky.

†References are at the end of the report.

Fig. 2. *Relationship between macro-hardness and ultimate tensile strength of low alloy powdered metal steels.*

lurgical structure of the parts used in the testing are reported.

It is not our function to judge, for or against, the use of powder metal produced hammer-trigger assemblies. The purpose is to show critical, meaningful tests and the results obtained from these tests. The results are likely to be interpreted somewhat differently by the part producer, the gun manufacturer, and the sportsman.

The major objectives of this investigation were to determine the static mechanical properties of powder metal parts presently used in a typical hammer trigger assembly, the dynamic mechanical response of the entire assembly, and the correlations between the metallurgical structure and the mechanical properties of the powder metal parts.

The specific hammer-trigger assembly chosen for this investigation is shown in figure 1. The mechanics of the assembly are also shown. The most critical region of the assembly is at the safety notch, therefore this area was studied. The hammers and triggers tested are the standard replacement parts for the Park Chandler Inc., (formerly Harrington and Richardson Inc.) Topper single shot shotgun series. The hammers were made of an iron base alloy having a nominal composition of 2.0% Ni, 0.18% Mn, 0.01% C, 0.011% Si, 0.018% S, 0.004% P. The parts are produced by pressing at 40 tons per sq. in. followed by a 2050°F sinter for one hour. The parts have a minimum sintered density of 6.8 gm/cc. Heat-treatment is oil quench from 1550°F and tempered at 250°F for 1.5 hours.

TEST PROCEDURES AND RESULTS

Hardness determinations. The uncertainties associated with measuring the hardness of powder metallurgy parts are discussed by Feir (2). Furthermore, the difficulties of correlating hardness readings with tensile strength are also pointed out (2,3). However, there exist sufficient data to indicate there is a correlation between hardness and tensile strength of powder metal steel parts (4,5) even though the correlation is not as well defined as in the case of wrought steels. Figure 2 shows this correlation for low alloy steels. Hardness determinations on the gun parts are given in Table I. Taking the Rockwell hardness of C-39 (from Table I), figure 2 gives an estimated ten-

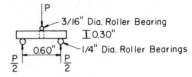

(a) Simple Beam Test Geometry

Fracture Load P = 4275 lbs.

(b) Typical Fractured Part

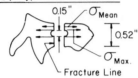

Fig. 3. Simple beam best on hammer.

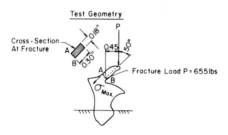

Fig. 4. Cantilever beam test on hammer.

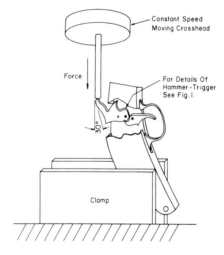

Fig. 5. Mechanics of Dynamical test assembly.

TABLE I

SUMMARY OF HARDNESS TEST RESULTS ON POWDER METAL HAMMER		
Ident Location	Vickers Hardness Number	Rockwell Hardness Numbers
Interior Material	316	39R$_C$/ 69R$_A$/ 96R$_G$
In region between notches	Not determined	39R$_C$/ 69R$_A$/ 96R$_G$
At outer surface	291	Not possible to determine
At surface of center hole	255	Not possible to determine

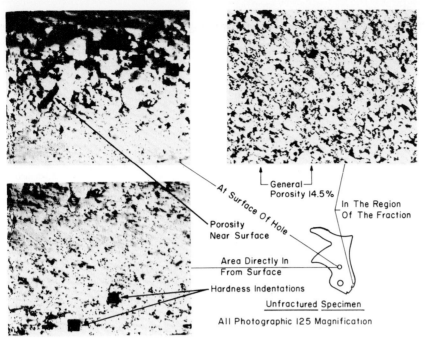

General
Porosity 14.5%

At Surface Of Hole

In The Region
Of The Fraction

Porosity
Near Surface

Area Directly In
From Surface

Hardness Indentations

Unfractured Specimen

All Photographic 125 Magnification

Fig. 6. Photomicrographs of powdered metal hammer.

load values remain constant even though the loading rates change by one hundred fold.

The load carrying capacity of the safety notch assembly prior to trigger-tip failure is between 560 and 948 lbs. This corresponds to direct loading at the cockspur of 90 to 152 lbs. These values also are independent of loading rate.

The total work done on the hammer-trigger assembly prior to failure is between 5.4 and 11.1 inch-lbs.

METALLURGICAL STRUCTURES

Two hammer specimens, one in virgin conditions and one in which the safety notch was fractured in dynamic testing, were prepared for metallographic examination. Polishing removed more than 0.005 inches from the original surface. The structures observed on the unfractured specimen are shown in Figure 6 and it is noted that there exists a greater amount of porosity near the outer surface of the specimen. Also visible in these micrographs are the Vickers hardness indentations. These structures show that the reduced hardness near the surface of the part (see Table I) is due to the increase in the porosity near the surface. Also the porosity in the region where the hammer fractures occur in dynamic testing is about 14.5%. This is a greater amount of porosity than observed in the interior of the specimen.

The several relatively simple test procedures described show that meaningful data on the mechanical behavior of small complex powdered metal gun parts both in and out of the gun assembly can be developed. These could lead to useful quality control measures.

sile strength of 113,000 psi to 152,000 psi, with an estimated median value of 131,000 psi.

Static mechanical tests. Simple beam tests were employed to determine directly the static mechanical properties of the hammer material. The geometry of the test is shown in figure 3a. The specimen was loaded to fracture at 4275 lbs. The loading-displacement curve is linear up to the fracture point. A sketch of the fractured specimen is shown in figure 3b. Taking into account the stress concentration factor of 2.38 at the hole (6) and the calculated mean maximum stress of 117,500 psi, one may conclude that the ultimate strength of the material is well above 117,500 psi.

In addition a cantilever beam test was performed on the hammer. In this test the cockspur of the hammer is loaded (as shown in figure 4) to a fracture load of 655 lbs. As in the simple beam test the load-displacement curve remains linear up to fracture. The maximum tensile stress at fracture was calculated to be 174,000 psi.

When compression tests were performed on sections of the hammer, no signs of failure, either visible or from the load-displacement relationship, were observed while loading to a stress of 200,000 psi.

Dynamic mechanical tests. Tests to determine the dynamic mechanical properties of the powder metal hammer-trigger assembly were made using a testing machine operating in compression. Figure 5 shows the essential features of the testing arrangement. The tests were performed by loading hammers and triggers in a cut away gun assembly. For each test the loading direction, the loading rate, the failure load and the part (hammer or trigger) in which the failure occurred was recorded. From these data, the following quantities were calculated:

(a) the distance the hammer cockspur moved before failure
(b) the total impulse of failure
(c) total work done on assembly through cockspur prior to failure
(d) the magnitude of the force at the safety notch at failure.

The approximate fracture areas on the failed parts were measured and used to calculate an average shear stress at fracture. This calculation gives only a rough approximation of the actual stress at the time of fracture.

From the data secured, the load carrying capacity of the safety notch at fracture of the hammer is between 940 and 1620 lbs. and correspond to loads at the cockspur of between 165 and 265 lbs. Note that the fracture

REFERENCES

(1) Alfred N. Weiner, "Powder Metallurgy Makes Good Gun Parts, Precision Metal, Vo. 29, No. 4, p. 35 (1971).

(2) Marvin Feir, "Pick The Right Method to Test Hardness of Powder Metal Parts," Materials Engineering, March 1969, p. 60-61.

(3) "Introduction to Powder Metallurgy," Joel S. Hirschhorn, American Powder Metallurgy Institute, New York, 1969.

(4) Symposium on Powder Metallurgy, Iron and Steel Institute, London, Special Report No. 38, 1947.

(5) Symposium on Powder Metallurgy, Iron and Steel Institute, London, Special Report No. 58, 1954.

(6) "Design and Machine Elements," V.M. Faires, 4th Ed., Macmillan Company, New York, 1965. pm

Heavy-Duty Sintered Parts of Complicated Shape

F. J. ESPER, J. KREISSIG and H. MICHELMANN*

Abstract

Over the years sintered structural parts have attained a firm foothold in many branches of industry. Today it is possible to produce highly stressed complicated powder metallurgy parts more economically than conventional parts due to improved iron powder quality and progress in powder

Hochbeanspruchte Sinterteile mit komplizierter Form

Maschinenbauteile aus Sinterwerkstoffen haben mittlerweile einen festen Platz in vielen Industriezweigen erobert. Durch neue verbesserte Eisenpulverqualitäten und durch Weiterentwicklung der pulvermetallurgischen Technologie können heute auch hochbeanspruchte komplizierte Formteile pulvermetallurgisch wirtschaftlicher hergestellt werden als nach konven-

Pièces frittées de forme compliquée à tolérance élevée

L'emploi de pièces frittées pour la construction de machines a acquis au cours des années une place reconnue dans beaucoup de secteurs industriels. Grâce à une amélioration récente des poudres ferreuses et aussi grâce au développement ultérieur de la technologie des poudres métalliques, on peut aussi réaliser aujourd'hui des formes compliquées de haute

Introduction

The necessary tool machinery becomes steadily more sophisticated and offers constantly new possibilities. Even complicated shapes can be successfully produced by automation today. But capital expenditure necessarily increases with the convenience of the equipment. Material savings – if possible at all – are usually neglected, and the shorter working periods are offset in many cases by longer preparatory periods or down time.

In many cases powder metallurgy offers the possibility of producing economically complicated shapes that often are highly stressed. During the last few years great progress has been made in this direction. Furthermore, improvements in mechanical properties were developed concurrently, and consequently a considerable selection of sintered materials with sufficient strength, toughness and hardness, or wear resistance, respectively, is being offered today.

It should be pointed out that often a sintered material has been rejected unjustifiedly because it was believed that it could not fulfill the respective specifications, and therefore the economic advantages were unnecessarily disregarded. By adaption of the design to the specific properties of the material, this difficulty can often be circumvented without trade-off of any other kind of disadvantage. In some cases sintered materials could even be used in existing designs. Stresses acting on the part were simply overestimated. It will be shown on hand of a number of examples how complicated or highly stressed parts can be shaped.

Auxiliary Air Valve Housing of an Automobile Engine

In order to compensate for the higher friction losses in a cold engine it is necessary to increase idling speed. For this purpose an auxiliary air valve is attached to the fuel injection pump in parallel to the throttle. In the cold engine the auxiliary valve is open and with

metallurgy technology.

The parts described in this paper have been selected from various fields of application. Most are built into subassemblies that are used in automotive vehicles. Even so-called safety parts may be found among them.

tionellen Herstellverfahren.

Die hier beschriebenen Formteile sind aus unterschiedlichen Anwendungsgebieten ausgewählt worden. Die Mehrzahl wird in Aggregaten eingebaut, die in Kraftfahrzeugen verwendet werden. Es befinden sich sogar Sicherheitsteile darunter.

tolérance de manière plus économique qu'à l'aide de procédés de fabrication conventionnels.

Les pièces ici décrites sont tirées de plusieurs différents domaines d'application. La plupart de ces pièces sont introduites dans des ensembles qui sont utilisés dans des véhicules automobiles. On trouve même des pièces dites de sécurité à la base.

increasing engine temperature it closes gradually. The auxiliary valve may also be constructed as a rotary gate valve that is activated by means of a bimetallic coil spring. In an air cooled engine the bimetallic spring is in contact with the motor oil in the crankcase while in a water-cooled engine ventilation is controlled by the cooling water temperature.

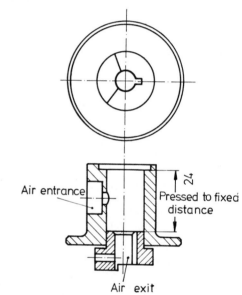

Fig. 1 b
Design with
sintered materials.

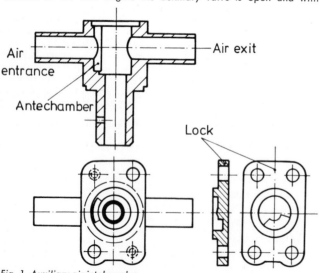

Fig. 1 Auxiliary air intake valve.
Fig. 1 a Conventional design with solid materials.

* Department of Materials Research, Robert Bosch GmbH, Stuttgart, Germany

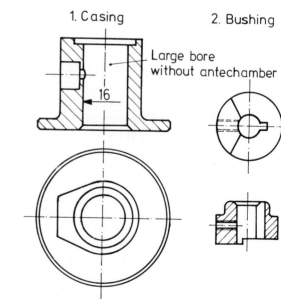

Fig. 1 c Individual sintered parts.

Figure 1a shows the design of an air intake valve housing with connecting lugs. The chamber for the rotary valve lies between them. A screw-on cover contains the stops for the angle of rotation. This design is unsuitable as a sintered part because of molding difficulties.

After some intermediate designs, a final design permitting powder metallurgy production of this part was successfully developed (Fig. 1b). Air is introduced laterally through a cylinder penetrating the exit wall. The rotary valve register is attached to the air exit opening at one side of the hollow cylinder. The antechamber which had a very complicated shape in the original design now has a simple cylindrical shape. In the powder metallurgy version the cover is not screwed on, but flanged. Therefore it can be considerably simplified.

In order to avoid escape of air, antechamber and rotary valve must fit together tightly, i. e. close tolerances are necessary for both parts. For production reasons it is advantageous to produce the housing in two parts. Production as a single part would not permit sufficient accuracy of the different diameters of the inner bore and the length of the large bore.

The two-part production also offers the following advantages:
1. The full length bore of a diameter of 16 mm and a tolerance of G 7, that takes in the valve, can be sized without difficulty.
2. The housing is press-fit into this bore. It is not fastened by other means.
3. The exact length dimension 24 is maintained by press-fitting the bushing to a stop.
   This is the only critical length dimension. Since it is adjusted during assembly, the tolerances of all length dimensions of the individual parts correspond to those resulting from the production process without need for special precautions.

Centrifugal Weight Support for an Injection Pump Governor

The centrifugal weight support is a structural part for the governor of a Diesel fuel injection pump (Fig. 2). Four centrifugal weights are connected to the centrifugal weight support with bolts that also serve as bearings for the centrifugal weights. They form a unit of the governor that regulates the idling of the engine and protects the engine from overloads during full load.

The shape of the centrifugal weight support is already designed to the specific peculiarities of the sintering process. The complicated shape with the recesses, "a" which are essential for the free pendulum movement of the centrifugal weight, the eight bolt bearings "b" of a tolerance quality D_9 and the four penetration bores "C" are all produced in one pressing operation. There exists no processing method that can produce these complicated shapes more economically to the required accuracy without chipforming machining. The centrifugal weight support is stressed by a constant torsional

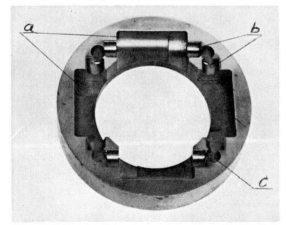

Fig. 2 Centrifugal weight support (because of its complicated shape a typical sintered part).

oscillation of a total axial load of approximately 1000 N. A sintered material alloyed with phosphorus and a density of $\varrho \geq 6.9$ g/cm³ was selected that has a particularly good combination of strength and elongation and performed well during fatigue tests.

This part shows very clearly the economc superiority of the sintering process as compared to other methods of production.

Adapter for a Fuel Injection Pump Governor

The conventional idea that parts stamped from sheet cannot be beaten in price finds its limitation either by an excess of rejects or

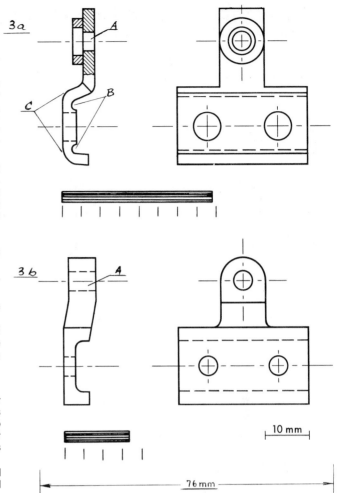

Fig. 3 Guide for fuel injection governor.
Fig. 3a Conventional design with welded-on bearing.
Fig. 3b Sintered design.
The bars under the respective descriptions facilitate a cost comparison.

by the need for further operations in final production. Bore A of the adapter in Fig. 3 constitutes a bearing surface. It has to be reinforced by a ring which has been produced by stamping in a preliminary separate process. The ring is welded onto the adapter.

The shape of the lower half of the adapter part has to be so complicated in order to guarantee a well adhering surface for the grooves. But during stamping cracks often occurred in the places designated by C which ran transversely across the material.

Figure 3 shows the construction of a sintered part. The shape has been adapted to sintering technology. By increasing the cross-section of the part with the bearing surface A the subsequent reinforcement by the welded-on ring is made superfluous. The grooves B can also be eliminated because a larger cross-section can also be selected for the lower part. The pressed plain surface fulfills the demands, and the entire part corresponds to all test specifications. Further minor dimensional changes adapt the part completely to powder technology.

A cost comparison shows that the adaption saved 60% of the cost.

Serrated Disk for Percussion Drill Machine and Grinding Disk for Coffee Mill

Figure 4 shows serrated disks for a percussion drill machine in conventional form and made from powered metals. The complicated shape of the disk that is highly stressed can be manufactured relatively economically by cold forging of rod sections. But the machining of other surfaces is very expensive. It must be centrally picked up in the profile of the teeth and the bore and external finish must undergo chipforming machining. The extent of the chipforming machining operation becomes more obvious when one compares the weights of the raw material and the finish part. The ratio is 60 g to 20 g.

As opposed to this, powder metallurgy permits production without scrap or subsequent chipforming machining. Therefore a sintered serrated disk should soon replace the bar stock material. For a long time this could not be risked because no equivalent sintered steel was available that could compare in tensile strength to the C 15 K material.

This example shows that a comparison of tensile strength values should not necessarily be the decisive factor in determining the usefulness of sintered steel. Steel C 15 K was apparently selected originally for this purpose because it fulfill the technical demands and could be machined easily, and also because it was economical. Subsequently the capability of this material became standard for the particular application. In this connection it was overlooked that for

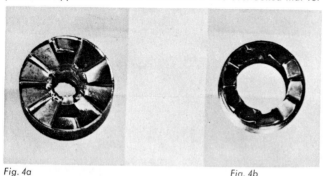

Fig. 4a   Fig. 4b

Fig. 4 Serrated disk for percussion drilling machine.
Fig. 4a Cold forged from casting (a disk weighing 48 g was made from 60 g cast stock. After chipforming machining the finished part weighs only 20 g).
Fig. 4b Part made from sintered metal (ready for assembly).

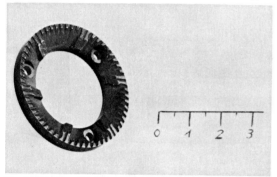

Fig. 5 Grinding disk for coffee mill.

Fig. 6 Compressor piston and connecting rod made from cast iron and sintered material in various manufacturing stages. Upper row, left to right: Cast piston – Blank sintered piston – Blank Sintered piston – machined
Piston with connecting rod (ready for assembly)
Lower row, left to right: Cast connecting rod – Blank Sintered rod – Blank.

all practical purposes the limits were set too high. Therefore, for a long time nobody dared to conduct any experiments with sintered metals which have lower characteristics. Consequently a proper rationalization was not applied.

But a "daring" experiment with a chip-formed machined sintered disk, which was designated as nonsentical by experts, showed that the tensile values specified were too high.

Since that time a double pressed and sintered serrated disk of a density of $\varrho \geq 7.2$ g/cm³ that is subsequently case hardened, is used in percussion drill machines (Fig. 4b). As shown by this example, each case should be examined individually to determine whether the material strength requirements have not been set too high.

Based upon this experience with the serrated disk for percussion drill machines, and the wear characteristics of sintered materials, grinding disks for coffee mills were manufactured from sintered materials. Such a grinding disk is pressed powder metallurgically with many sharp-edged teeth and 3 fastening holes (Fig. 5). A sintered steel alloyed with C and Cu with a density of $\varrho \geq 6.6$ g/cm³ is used. The necessary wear resistance is achieved by hardening of the grinding disk after sintering. The costs cannot be beaten by any other method of production.

Piston and Connecting Rods for Refrigeration Equipment

Todate the pistons and connecting rods for refrigeration compressors have been manufactured from cast iron. The as cast raw parts demand major cost-intensive machining.

A conversion of these two elements to sintered metal parts called for design changes of rod and piston. In addition to better economics the following conditions have to be met:
1. Sufficient machinability for both parts
2. Absolute tightness of the bottom of the piston
3. Uniform phosphate layer on the finished parts.

The final shape was approached as far as possible in the sintered piston and connecting rod (Fig. 6). In the sintered piston only the eye in which the connecting rod is fastened must be bored perpendicular to the direction of compacting, and the piston surface must be ground because the tolerance between the piston and cylinder wall is but 5 μm. In contrast to the sintered piston, the cast piston must also be machined internally.

In the connecting rod a bore must be cut in the longitudinal axis to provide lubricants for the bearings. The bearing surfaces themselves are finish machined so that they are plane parallel. Subsequent machining was difficult with the standard Fe-Cu alloy because the tools dulled rapidly. Any addition of sulfur to improve machining of the sintered steel was ruled out. It was feared that the iron sulfide formed during sintering may react with the refrigerant "Frigen". If these difficulties with subsequent machining of the sintered parts could not be overcome in some other way, economic production of piston and connecting rods from sintered steel was questionable. The solution to the problem was the use of sintered Fe-Cu-Sn alloys. These alloys have sufficient mechanical strength, machinability and improved wear resistance. The latter is due to hard Cu-Sn phases in the grain boundaries and in the angles between the Fe-grains. The required gas tightness is achieved by a steam treatment.

In spite of the necessary post-sintering processing steps the production costs of sintered pistons and connecting rods are substantially lower than those of their cast counterparts.

Magneto Cam in Automotive Engines

The ignition spark in Otto engines is produced on the secondary winding of the ignition coil in such a way that the primary winding is interrupted by means of a contact. The timing of this ignition spark must be very accurate. This is achieved by a rotating cam which breaks the contact.

Conventional production of this cam was very troublesome because the external configuration of the cam is not in the shape of a simple geometry. The following work procedures that had to be conducted in part on special tool machines were necessary to produce the cams from the as received unalloyed free-cutting steel bar stock (Fig. 7):

1. Turning and cutting
2. Milling of the grooves
3. Deburring
4. Striking of the directional arrows
5. Case hardening
6. Grinding of the curve shape
7. Polishing of the curve shape

The cam is subjected to high wear because of the high sliding velocity and pressure of 4 to 5 N. This is the reason for case hardening of the cam.

The sintered steel cam shown in Fig. 7 is pressed into finished shape including the directional arrows. It is sintered and sized. The sintered part has the cam shape over its entire length because the chucking extension of the part can be eliminated. The required dimensional accuracy and low roughness depth of 5 μm can be achieved by hard metal inserts in the press- and sizing tools.

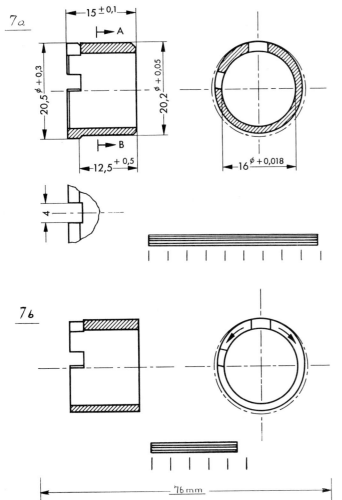

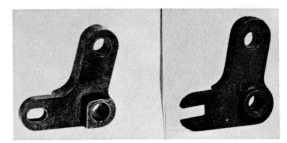

Fig. 8a   Fig. 8b
Fig. 8 Angle lever for fuel injection pump governor.
Fig. 8a Sintered angle lever with closed slot.
Fig. 8b Open slotted angle lever made from low carbon steel.

Fig. 7 Cam for oscillating magnetic ignition.
Fig. 7a Conventional design from solid material.
Fig. 7b Design for sintered material.

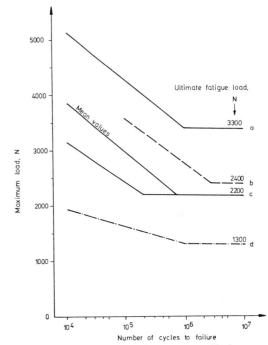

Fig. 9 Wöhler curves of various angle levers for governors of fuel
injection pumps.
a) Wöhler curve of double-pressed and double-sintered angle
levers,
b) Wöhler curve of single-pressed and single-sintered angle levers
of lower density,
c) Wöhler curve for angle levers as produced formerly from drawn
rolled steel (All levers are case hardened),
d) Wöhler curves of a lever of group b) with slit slot.

The sintered steel cams need not be case hardened. Lead additions serve to improve the sliding properties of the sintered cam to the extent that wear stays within permissible tolerances for the cam as well as for the fiber-reinforced Duroplastic part of the contact-breaker linkage.

Therefore the advantage of the sintered part is as follows: the entire complicated shape of the cam can be produced in one operation, and the require surface finish and tolerances can be achieved by sizing. Suitable selection of the material, i. e. lead additions, eliminates case hardening which in turn eliminates subsequent machining otherwise required because of distortion during hardening.

This simplified production of the sintered cam results in considerable cost savings, as can be seen from the Figures.

Angle Lever for Centrifugal Governor of Fuel Injection Pumps

The amount of the fuel in Diesel fuel injection pumps is regulated be centrifugal weights as a function of engine rpm. An angle lever serves to transfer the centrifugal forces of the weights to the governing elements. These must essentially fulfill two conditions:

1. The diameters of the two bores must be very accurate and their axes must run truly parallel.

2. The levers must have sufficient vibratory fratigue strength. Fracture of the lever causes excessive rpm which endangers operational safety.

As shown if Fig. 8, the lever has a complicated shape. If made from low carbon steel high production costs of the part ensue. The angle levers that were made initially from low carbon steel had an open slot at one lever arm in order to reduce manufacturing costs. Fatigue strength of the material used was 2200 N (Fig. 9). The first levers made from sintered steel fractured at the slot. Subsequently the slot was closed thus increasing strength. Generally speaking, the angle lever withstood the required stress. In order to be absolutely

sure, a fatigue load of at least 3300 N was demanded years ago when the sintered steel toggle lever was first introduced (Fig. 9a). This requirement could be satisfied only by a double-pressed and double-sintered material, that was also still case hardened like the low carbon steel.

In order to achieve further cost savings, it was attempted to produce the angle levers by means of a simple pressing and sintering operation. Since meanwhile an improved uniformity in strength of the levers was facilitated by improved technology, any danger of severe damage by tearing was eliminated. But the confidence in sintered parts was still so low that the minimum value of 2400 N (Fig. 9b) for the fatigue strength was unacceptable, even though this value was still 200 N above that for angle levers made from low carbon steel. Only after the fatigue stress could be accurately determined and was found not to exceed 100 N, were low cost sintering methods found acceptable.

This angle lever example shows distinctly 3 points:

1. Exaggerated safety requirements and non-confidence in sintered steel lead to exaggerated strength requirements.

2. Actual stresses acting on the structural parts should be determined as accurately as possible so that the correct material may be selected.

3. Correct utilization of the powder metallurgy process by using material-cognizant designs also permit high stressing of sintered structural parts.

Source: Powder Metallurgy International, Aug 1974

P/M PARTS IN THE FARM EQUIPMENT INDUSTRY‡

T. L. BURKLAND*

ABSTRACT

A few of the parts used currently in each of several types of production farm equipment will be presented. The farm machinery chosen as examples are the Synchronous Crop Thinner, and one model each, of a planter, combine, grain drill and tractor. The parts and machines discussed may be considered as typical of those used throughout the industry. The discussion will include some description of the function of the farm equipment selected in order to relate the P/M parts to that function. The P/M parts will be covered in as much detail as possible including grade, density, mechanical property requirements and any pre or post sintering operations involved.

Some thoughts on the specifying of a P/M part and the development of new parts are expressed.

INTRODUCTION

The title of this article is "P/M in the Farm Equipment Industry"; however, I would like to emphasize that I will discuss only that portion of the industry as represented by Deere and Company. The rest of the industry is, I am sure, using P/M parts in similar or in other applications peculiar to their own design. I have elected to take several of our production machines, show some of the more interesting structural applications and discuss the reasons for conversion to, or original design in P/M.

Deere and Company has ten factories on the North American continent, including one each in Canada and Mexico. We have expanded outside North America and now have production facilities in South America, Europe and Africa. My comments will be confined to P/M parts used in farm equipment produced in the United States.

Our John Deere Des Moines Works manufactures Corn Pickers, Cotton Pickers, Cultivators, Chisel Plows, Sprayers, Crop Thinners and Grain Drills. I will discuss several parts in current use in the Crop Thinner and the Grain Drills.

Our John Deere Plow-Planter Works manufactures Plows, Bedders, Tillers, Harrows and Planters. My remarks here will be confined to P/M parts used in one of the planters. The John Deere East Moline Works manufactures Combines and the John Deere Waterloo Tractor Works manufactures Tractors; parts from each machine will be discussed.

There are two predominant factors that have been primarily responsible for our initial and increasing use of structural sintered steel powder parts. The first factor is economic; we simply were and are seeking a lower cost means of obtaining any given part. The second factor has been the gradual change in our gray iron foundries, both in the state-of-the-art and in economics. In many of our earlier machines and this is particularly true of our grain drills and planters much of the operating mechanisms such as spur and bevel gears and sprockets to name a few were made of gray iron and used in the as-cast condition or with a very little machining or other finishing. If there was a wear problem involved, the wear portion would be cast in a chill resulting in both a smooth and a very hard surface. This was possible because practically all of our factories operated their own foundries and the sand molds in the large part were individually made, largely by hand and with considerable care. Due to changing technology and economics many of these foundries

‡ *Presented at 4th International P/M Conference, Toronto, Canada.*
* *Materials Engineering Department Deere and Company, Moline, Illinois 61265*

have been shut down. The remaining foundries are highly automated and the small, precise castings made in the old type foundries are either no longer available, or have become too costly. Even during this transition period design and service requirements were demanding closer tolerance parts. Under these conditions the P/M process offered an attractive alternative to the more costly machined castings or wrought steel parts. We are now in what might be called a third stage in the use of P/M parts in that the design engineers are now thinking of P/M in the initial stages of design as to what it can contribute not only in lower cost but also in improved function or reliability.

A discussion of some of the P/M parts in use on the farm equipment mentioned follows; included is some comment on the function of the machine itself and the function of the P/M part as a part of that machine, also some details on the part and the material and process specifications.

Planter   A typical modern day planter does more than just plant seeds. The planter shown in Fig. 1 is equipped with tanks for liquid fertilizer (they can be equipped with hoppers for solid fertilizer in place of the tanks) with the seed hoppers and planting mechanism directly behind followed by two rows of hoppers, one for a herbicide and the other for an insecticide. Planters are used for the planting of the larger seeds such as corn, soybeans, cotton, maize, etc. A considerable number of P/M structural parts are used in the manufacture of the planters; however, we will discuss only two of them.

The floating cam, which is shown in Fig. 2, is composed of three P/M parts consisting of the two pins and the cam plate. These are assembled in the green condition and sinter bonded together. Specifications are as follows:

	Cam Plate	Pin
Density, Gm/cc	6.6	6.6
Carbon, %	0.6–1.0	0.6–1.0
Copper, %	1.9–2.0	2.75–3.0
Nickel, %	1.65–1.9	1.0–1.2
Moly, %	0.5–0.65	—
Weight, Gms	228–232	17–18
Total Assembly Weight, Gms	263–267	—
Hardness	Rc 60 (file)	—

This part was originally designed in P/M as a half circle and it was necessary to bolt it into assembly, in addition being only a half circle a slip clutch assembly was necessary in the power train to prevent damage if the planter was backed up. With the new floating cam design, the cam is held in place by the two pins eliminating the need for bolting and with the cam surface being a full circle the clutch mechanism became unnecessary and was eliminated. The alloy mixture used for the cam plate develops alloy rich areas during sintering that are martensitic after sintering and therefore provides the wear characteristics necessary to withstand the sliding wear involved.

The cutaway view shows the bottom or flat side of the floating cam as it is assembled in the finger pickup mechanism. The fingers operate in a clockwise

FIGURE 1. Corn Planter equipped for application of liquid fertilizer, herbicide and insecticide as well as the planting of corn.

FIGURE 2. Floating Cam Used in Plateless Planter

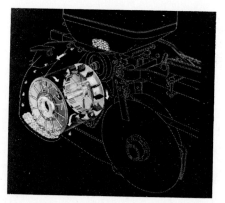

FIGURE 3. Cutaway view of finger-pick up mechanism, Plateless Planter

direction and are opened and closed as they travel over the cam face. The fingers pick up, hold and release the corn kernels or other seeds. (Fig. 3).

Tooth Driven Sprocket   This is an assembly consisting of a sheet metal sprocket plate with a powder metal hub. The hub originally was of wrought steel and machined to final shape. It was changed to P/M not only for a cost reduction but also to eliminate a difficult machining operation. The hub is inserted in the hex shaped hole in the sprocket plate and is then swaged to hold it in place. (Fig. 4) Specifications are as follows:

Density	6.9–7.1 Gm/cc
Carbon	0.24% max.
Control Weight	91–93 gms.

The low carbon, iron-carbon grade was chosen because of the relatively good ductility, at the density level specified, for the swaging operation.

Grain Drill   A typical modern day grain drill is shown in Fig. 5. This drill is equipped with two large hoppers—one for fertilizer and one for the principal seed crop—and with two smaller hoppers—one for seeding a grass cover crop and one for brome-grass. Grain drills are used to plant seeds of the grains such as

FIGURE 4.   36 Tooth Driven Sprocket

FIGURE 5.   Grain Drill equipped to apply fertilizer, plant grain and a grass cover crop.

wheat, oats, rice, etc., and seeds of the grasses. There are many P/M structural parts in use in the drills today; however, we will confine our discussion to only a few.

An assembly consisting of the feed roll and the shut-off in a seed cup is shown in Fig. 6. The feed roll and shut-off are in P/M and the seed cup is gray cast iron. The feed roll rotates during the planting operation and meters out the seed; the shut-off does not rotate and therefore seed does not flow through where the shut-off extends into the seed cup. The rate of seed flow is controlled by the distance the shut-off extends into the seed cup.

For many years the feed roll and shut-off parts of the fluted force feed mechanism were made of gray cast iron and used in the as-cast condition with very little finishing necessary to accomplish assembly and operation. This was possible because the factory producing grain drills had its own foundry and took great pride in producing castings of sufficient accuracy and surface finish for the intended applications. This foundry was closed down in the sixties and from that time on it became increasingly difficult to obtain castings of the same quality at an acceptable cost. Within a short time we found it necessary to turn to P/M for parts with the desired accuracy and surface finish and at an acceptable cost. Fig. 7 illustrates the feed roll and shut-off parts used in the seeding of grains, a smaller variety is used for the grasses.

FIGURE 6.   Grain Drill Fluted Force Feed Assembly

FIGURE 7. Feed Roll and Shut-off

Specifications are as follows:

| | Shut-off | Feed Roll | |
		Flute	Stem
Density, Gms/cc	6.1–6.5	5.7–6.1	5.8–6.2
Carbon, %	0.60–1.00	0.25–0.60	0.30 max.
Copper, %	1.00–2.00		2.00–6.00
Control Weight, gms	286–288	454–458	

The Feed Roll is made in two sections assembled in the green and sinter bonded. The P/M assembly has functioned very well in service, however excessive wear has been encountered in the planting of rice because of the abrasiveness of the rice hulls. This wear problem has been solved by heat treating from a carbonitriding atmosphere resulting in a file hard wear surface. The parts are then steam treated for improved corrosion resistance.

Grain Drill Planetary Gear Assembly

The planetary gear assembly shown in Fig. 8 is an original design in P/M and contributed to a change in the mechanism of metering and distributing fertilizer during the planting operation. Prior to the development of the P/M planetary gear assembly, the application rate was regulated by the opening or closing of a gate valve which controlled the gravity flow of the fertilizer. This method was subject to the many problems that affect gravity flow i.e., build up of fertilizer on the gate, change of ground slope which would affect the gravity flow, humidity, etc. The planetary gear assembly is part of a change to a system that provides positive feed of the fertilizer using driven feed wheels; application rate is adjusted by changing the speed of the feed shaft on which the feed wheels are mounted. Specifications are:

	Sun Gear	Planet Gear	Internal Gear	Carrier
Density, Gm/cc	6.1–6.5	6.1–6.5	6.1–6.5	6.1–6.5
Carbon, %	.60–.90	.60–.90	.60–.90	.60–.90
Copper, %	1.75–3.00	1.75–3.00	1.75–3.00	1.75–3.00
Control Weight, Gm.	67–68	97–98	532–536	211–213

FIGURE 8.   Grain Drill Planetary Gear Assembly

FIGURE 9.   Synchronous Crop Thinner Attached to Tractor

Synchronous Crop Thinner

This gear assembly has performed satisfactorily in service; a problem was encountered early in production of the carrier due to cracking at the base of the pins during ejection from the die; however, this was soon remedied.

The synchronous crop thinner is a relatively new development in farm equipment; a side view of the thinner as mounted on a tractor is shown in Fig. 9 which depicts the knife in position to cut out excess plants leaving the desired and most efficient growing population. The thinner is used for the thinning of sugar beet, cotton and vegetable crops. The thinner operates through an electronic sensing probe which when it contacts the plant left standing actuates a knife which cuts out the undesired plants ahead. The heart of the thinner is the actuator which is a hydraulic motor; the actuator drives the knife when so instructed by the electronic probe to cut out the excess plants. There are two P/M parts used in the actuator, the piston and the connecting rod. These two parts are shown in Fig. 10 and 11. Specifications are as follows:

	Con Rod	Piston
Density, Gms/cc, min., before Cu infiltration	6.2	6.1
Density, Gms/cc, min., after Cu infiltration	7.2	7.1
Carbon, %	60–1.0	.60–1.0
Weight, Gms, before infiltration	614	378
Weight, Gms, after infiltration	744	437

The connecting rod is made as a double and is cut in half and machined; the final weight of a single connecting rod is 284 grams.

FIGURE 10. Actuator Piston

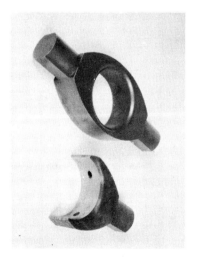

FIGURE 11. Actuator Connecting Rod

Both parts are heat treated in a carbo-nitriding atmosphere to a minimum apparent hardness of Rockwell C 35. The use of P/M in these two parts has contributed markedly to the successful operation of the actuator. A good bearing material in the curved portion of the Con Rod which acts to drive the crankshaft is a necessity. The rod operates in an oil bath and in the initial development work considerable difficulty was encountered in finding a suitable bearing material; a steel bearing surface was found to gall, standard bronze bearing materials extruded under service loads, aluminum bronze did not extrude but wore out, and oil grooves in the bearing material were not effective. The copper infiltrated powder metal surface polishes to a high finish when operating in oil and will scuff somewhat if the oil supply is low; however, it will repolish when the oil supply is renewed. The success of the power metal material is attributed to the surface porosity which is still present despite the copper infiltration and which retains and supplies oil over the entire bearing surface.

Verification of the surface and internal porosity can be seen in the Scanning Electron Microscope photograph in Fig. 12.

Tractor
P/M structural parts and bearings have been used in our tractors for many years. The bearings and oil pump gears came first, the use of P/M structural parts started slowly but the momentum is picking up sharply and we expect our use to show a marked increase in the years ahead. I will discuss only a few of the many P/M parts that are currently part of the very complex mechanism that is a tractor. Except for an oil pump gear, the P/M parts we will look at are in the hydraulic system.

Rockshaft Servo Cam and Cam Follower
The Rockshaft is a hydraulically operated crank at the rear of the tractor that is used to raise, lower and control a wide variety of integral implements and 3-point hitch tools. The rear end of a tractor showing the rockshaft and 3-point hitch elements is shown in Fig. 13.

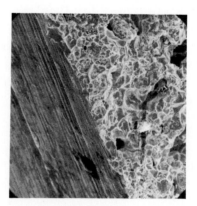

FIGURE 12. Connecting Rod at 500×
Scanning Electron Microscope Photomicrograph. Bottom portion—ground bearing surface showing porosity extending to the surface. Top portion—fractured cross section showing internal porosity

FIGURE 13.   Rockshaft and Hitch Elements Mounted on Rear of Tractor

The degree of lift of the Rockshaft is controlled through a linkage system by the Rockshaft Servo Cam and Cam Follower both of which are in P/M.

The Rockshaft Servo Cam and Follower are shown in Fig. 14 and 15; specifications are as follows:

	Servo Cam	Follower
Density, Gms/cc	6.35 min.	6.1–6.5
Carbon, %	.60–1.0	.60–1.0
Copper, %	2.0–6.0	2.0–6.0
Part Weight, Gms	445	98
Heat Treatment		

The cam is induction hardened on the cam surface, the surface is to be hard to a Rc 52 file (a file tempered to a Rc 52 hardness).

The follower is through hardened to the same hardness requirement as the cam.

These two parts have been in P/M since their adoption in 1965; they would be considerably more expensive if machined from a cast or wrought material.

FIGURE 14. Rockshaft Servo Cam

FIGURE 15. Rockshaft Cam Follower

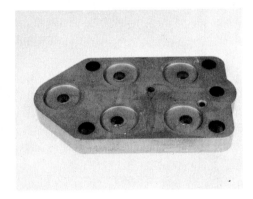

FIGURE 16. Selective Control Valve Housing Cap

Selective Control Valve Housing Cap The cap shown in Fig. 16 was originally machined from gray cast iron; changing to P/M eliminated most machining but not all, it did result in a 34% reduction in cost. This cap is part of the hydraulic system that supplies hydraulic power through remote cylinders to the farm implement attached to or being pulled by the tractor. Specifications are:

Density, Gms/cc	6.1–6.5
Carbon, %	.60–1.0
Copper, %	2.0–6.0
Part Weight, Gms	718

The cap is plastic impregnated to withstand 2,500 psi in assembly.

Oil Pump Gear One of the oil pump gears in use in our tractors is shown in Fig. 17.

Density, Gms/cc	6.20 min.	(6.1–6.5)
Carbon, %	0.60–1.00	
Copper, %	2.00–6.00	
Part Weight, Gms	1350	
Heat Treatment	Particle Hardness Rc 55 min.	
	Apparent Hardness Rc 25 min. (at pitch diam. on end of tooth)	

P/M was chosen for oil pump gears for cost reduction reasons. Transmission oil pump gears have always been in P/M; engine oil pump gears were originally in steel and were changed to P/M in 1965. Engine gears operate normally at 60 psi transmission gears at 200 psi.

Combine Grains, Maize, Corn and Soybeans are the principal crops that are "combined." As can be seen in Fig. 18, a combine is a compact piece of materials handling equipment. The grain is first cut by the mower bar at the front of the platform, augered to the center of the platform, conveyed through the feeder house to the cylinder and concave where the grain is threshed from the stalk, the grain falls through the straw walkers and is augered to the cleaning shoe, the clean grain is augered to the clean grain elevator, raised to the grain tank level and augered into the tank; the grain is unloaded from this tank into a wagon or truck by augers. Corn and beans go

FIGURE 17. Oil Pump Gear

FIGURE 18.   Cutaway View of Combine with Corn Head

through the same process. One of the principal uses found for P/M in the combine has been in bevel gears. We currently have three bevel gears which are almost identical; there are 21 of these bevel gears in use in a combine and they are used to drive the straw spreader at the rear of the combine and the various augers in use in the combine.

One of these bevel gears is shown in Fig. 19; specifications are as follows:

	Auger Gear
Density, Gms/cc,	
before infiltration	6.0–6.4
after infiltration	7.1–7.6
Carbon, %	.60–1.0
Copper, %	15–25
Part Weight, Gms	277–279
Tooth Weight, Gms	74
Tooth Strength (Push-off test) lbs.	
before heat treatment	3500
after heat treatment	4000
Heat Treatment	
Quench from Carbo-nitriding atmosphere	
Temper 400°F	
Hardness	
Teeth—Hard to a Rc 60 file	
Hub Rb 95 min.	

FIGURE 19.   Auger Bevel Gear

There are many other P/M structural parts in use in our combines; we have a very active value analysis and value engineering program in action right now seeking cost reduction or a better functioning part through change to our new design in P/M. Our design engineers see many good applications that involve the arc welding of P/M components together or to wrought steel. We have recently performed some development work on the arc welding of P/M and believe it is entirely feasible to rely on it as a production process.

In the past, our design engineers have relied to a great extent on the powder metal parts producer's sales and engineering personnel to assist in the conversion of parts to P/M or the development of a new design in P/M. This assistance covered all facets such as design, composition, density levels, coining, finishing, coating, etc. This assistance has been excellent in the majority of cases leading to satisfactory parts that perform well in service. On a few occasions however the advice and recommendations have led to less than satisfactory results. Much of this is to be credited to a lack of knowledge on our part or to a less than complete understanding of the end requirements. A major problem sometimes arises when the density and consequent mechanical properties of a part from production tooling varies from the density and properties of a part machined from a P/M slug. This has happened to us in bevel gear applications where the density achieved at the root of the gear tooth in the production gear was less than in the gear machined from a P/M slug. This resulted in a weaker tooth that would not stand up in service whereas the original prototype gears machined from slugs did perform satisfactorily.

This particular problem was overcome by copper infiltration. However, a better understanding of the P/M process and design requirements on our part and a clearer understanding of our needs by the P/M parts producers may have produced a satisfactory part without the need for a "fix."

CONCLUSION   Our design engineers are gradually becoming more competent in the design and specification of powder metal parts; however, we will undoubtedly always depend on the expertise of the P/M sales and engineering personnel.

We are finding that it is necessary to be more specific in our requirements for each P/M part. With the help of P/M sales and engineering personnel, we expect to establish the important criteria for each part and enter them into the part specifications. Some of these will be:

1. Weight

    A. A quality control weight range for a given P/M part will be established and agreed upon and noted on the part drawing. This may be used by receiving inspection as a basis for acceptance or rejection. The weight range basis may be:—as sintered, after copper infiltration, after impregnation with oil or plastic, etc.

    B. Where considered necessary the weight range for a portion of a P/M part (such as the teeth of a gear) may be established and the method of determining agreed upon.

2. Composition

    The composition ranges as carried in the various society standards and also in our own company standards are wide and we consider them too wide to represent good quality control. Where considered necessary for the satisfactory performance or reliability of a part, a narrower range will be established subject to agreement between supplier and purchaser and entered on the part drawing.

3. Testing

    When considered necessary, destructive tests simulating field loading will be established again upon agreement of the parties involved and noted on the part drawing. These will be used as a basis for acceptance or rejection.

    Standard requirements such as apparent hardness, particle hardness, impact strength are specified when those properties are important.

Properties of Stainless Steel P/M Filters

G. HOFFMAN* AND D. KAPOOR**

ABSTRACT

Investigation in the problems of highly porous stainless steel filter materials. Correlation between the type of metal powder and the characteristics of the filter. Advantages of isostatic compaction. A new coining operation is proposed with no effect on the characteristics of filters. The necessity for a standard test method for determination of filter properties is emphasized.

1. Introduction

Corrosion resistant sintered metallic filters manufactured from AISI 316 are increasingly in demand for the purification of aggressive mediums (1, 2, 3). Major applications are in petrochemicals, nuclear power, food industry and air pollution control. Considering the economic aspects, the required characteristics for optimum sintered filters are as follows (4):

1. Uniform distribution of porosity over the active filter area and high filtrating rate (filtration efficiency).
2. Highest permeability for a given pore size (energy efficiency).
3. High mechanical strength and good

regenerability (exploitation of raw materials).

The uniform density distribution can be obtained with the help of isostatic compaction, which allows the manufacture of seamless filter elements with larger dimensions. The first requirement is therefore dependent on the processing technique, while the other two are combined with material properties.

This paper deals with the systematic investigation of various filter materials. Their characteristics were determined by using different powders and the effect of particle size and shape were studied. Furthermore, the influence of pore forming material (PFM) on filter properties was investigated. It was aimed at developing filter elements with optimum properties and high regenerability.

2. Experimental Procedure

2.1 Powders

The investigation was carried out with three types of atomized stainless steel powders supplied by La Floridienne, Brussels, Belgium. The characteristics of the powders are given in Table 1 and the different particle shapes of the powders are shown in Fig. 1.

316 L-F is a commonly used powder with irregular particle shape. It has good compressibility and excellent green strength. 316 B-F consists of a mixture of

* Formerly R & D Dept., Sintermetallwerk Krebsöge, now with Laboratorium für Betriebsfestigkeit, Darmstadt, W. Germany.
** R & D Dept., Sintermetallwerk Krebsöge, W. Germany.

TABLE 1   Characteristics of the Investigated Powders

Powder Type		316 L-F				316 L-B				316 L-P		
Chem Analysis [wt-%]		C Cr Mn Mo Ni Si Fe				C Cr Mn Mo Ni Si Fe				C Cr Mn Mo Ni Si Fe		
75–100 µm	Apparent Density [g/cm³]	2,5				2,45				3,7		
	Sieve Analysis [wt-%]	+100 2,8	+88 56,3	+74 33,0	-74 7,0	+100 4,3	+88 61,8	+74 27,6	-74 6,3	+100 6,0	+88 35,6	+74 50,4 -74 8,0
100–200 µm	Apparent Density [g/cm³]	2,35				2,37				3,5		
	Sieve Analysis [wt-%]	+240 0,5 +177 8,6 +150 24,7 +125 30,4 +100 32,3 -100 6,5				+240 8,0 +177 28,5 +150 37,0 +100 13,0 -100 3,5				+200 4,6 +177 17,7 +125 52,0 +100 17,7 -100 7,0		
200–300 µm	Apparent Density [g/cm³]	1,95				2,22				3,2		
	Sieve Analysis [wt-%]	+300 2,6 +250 44,0 +200 46,7 -200 9,7				+300 2,2 +250 44,2 +200 49,2 -200 4,4				+300 7,0 +250 46,8 +200 38,6 -200 7,6		
300–500 µm	Apparent Density [g/cm³]	1,85				2,05				3,0		
	Sieve Analysis [wt-%]	+500 2,6 +420 20,4 +350 37,5 +300 31,5 -300 8,0				+500 2,9 +420 28,8 +350 29,7 +300 34,6 -300 8,0				+500 7,6 +420 15,9 +350 46,1 +300 24,4 -300 6,0		

h). After sintering, the discs were coined between two parallel plates. In order to obtain the same coining grade, independent of different shrinking behaviour and

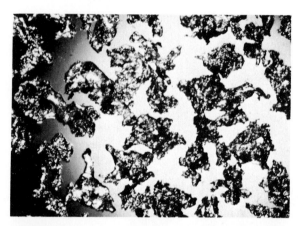

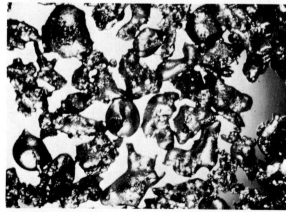

FIGURE 1   Particle shape of the investigated powders a) 316 L-F b) 316 B-F c) 316 L-P.

10% spherical particles and 90% of the particles have a high L/D ratio. This powder still has good compressibility and green strength. Under certain atomizing conditions, it is possible to manufacture a powder type 316 L-P, which has a smooth particle surface, nearly spherical shape (potato-shape) and a high apparent density. Due to the regular shape of particles, the compaction can only be carried out with the help of an organic binder which increases the adhesion between the particles. The resulting green strength allows an adequate ejection from the die. In order to further increase the porosity, 316 B-F and 316 L-P powders were mixed with PFM-material. Ammonium bicarbonate was selected because of its low dissociating temperature and was used in various sieve fractions.

2.2 Determination of Properties

2.2.1 Test Specimens

The evaluation of the filter properties was carried out with 56 mm dia, 2.6 mm thick discs. The samples were compacted and sintered under production conditions in two different furnaces (Elino walking beam furnace, 1280 C, NH₃, approx. 50 min; Degussa vacuum furnace, 1240 C, 2.5

distortion, the movement of the upper punch was exactly limited by a 2.5 mm hardened ring. The final volume of the sample therefore, is constant and the weight of the compacted powder is directly proportional to the required density. Hence, the time consuming final density measurement is eliminated, whereby the flow rate and pore size is not.

2.2.2 Property Measurements

The schematic diagramm of the measuring apparatus for determining flow rate characteristic is presented in Fig. 3. The permeability constants (ϕ = viscous permeability coefficient, Φ = inertia permeability coefficient) were calculated according to the equation (5, 6):

$$\frac{A}{s \cdot \eta} \cdot \frac{\Delta p \left(p_2 + \dfrac{\Delta p}{2} \right)}{Q_2 \cdot p_2} = \frac{1}{\phi} + \frac{1}{\Phi} \quad \frac{Q_2 \cdot \rho_2}{A \cdot \eta} \quad (1)$$

where

A = effective filter area [cm²]

s = thickness of the filter [mm]

η = viscocity of air at atmospheric pressure and room temperature [N-sec/m²]

ρ_2 = density of air at atmospheric pressure and room temperature [g/cm³]

p_2 = atmospheric pressure [bar]

Δp = measured pressure drop [mbar]

Q_2 = measured flow rate at atomospheric pressure [m³/h]

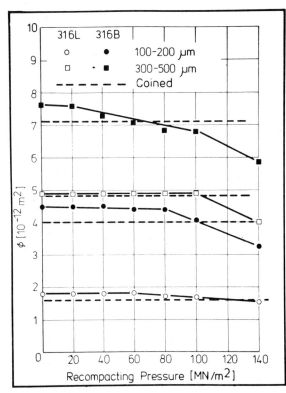

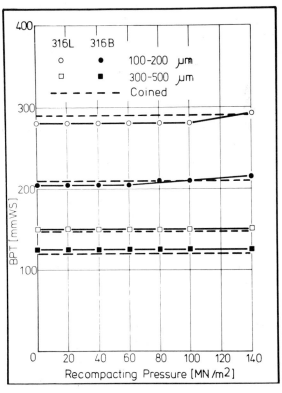

FIGURE 2 Characteristics of different filter materials as a function of coining pressure a) Viscous permeability coefficient b) Bubble-Point.

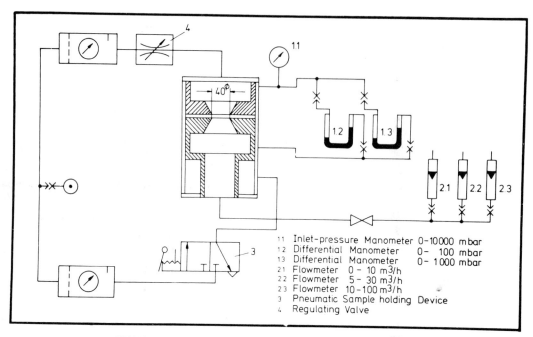

FIGURE 3 Schematic diagramm of the Flow Test apparatus.

The filter grade (d_{GBT}) was measured by glass-bead-transmission test (7) and correlated with the maximum pore size (d_{BPT}) as determined by the bubble-point test (8). The resulting factor κ is a criterion for the pore geometry and the probability of a continiuous pore capillary:

$$\kappa = \frac{d_{GBT}}{d_{BPT}} = \frac{d_{GBT}}{4 \cdot \gamma} \cdot \Delta p_{eff} \qquad (2)$$

where

d_{GBT} = filter grade according to glass-bead test [μm]

d_{BPT} = maximum pore size according to bubble-point test [μm]

γ = surface tension of the test liquid [N/m]

Δp_{eff} = effective minimum differential pressure at which constant bubbling first occurs [mbar]

The shear strength test was selected as a standard for determining the mechanical strength of the filter materials. According to (4) the clearance between the punch and the die should be greater than twice the diameter of the largest particle. In this investigation the value of 0.3 mm was used as an ideal compromise for all the tested sieve fractions. Preliminary tests were carried out to establish a correlation between shear and tensile strength. Fig. 4 shows that coarser particle size results in a higher shear strength for the given tensile strength. Comparing shear and tensile strength as a function of density (Fig. 5), it indicates, however, that the coarser particle size reduces the tensile strength, while the shear strength is unaffected. This could be possibly due to the influence of notch stresses on tensile strength. For this reason, shear strength is preferred as a characteristic for the material behaviour. Further, prior to the shear strength test, the same specimen is used for determining flow rate and filter grade.

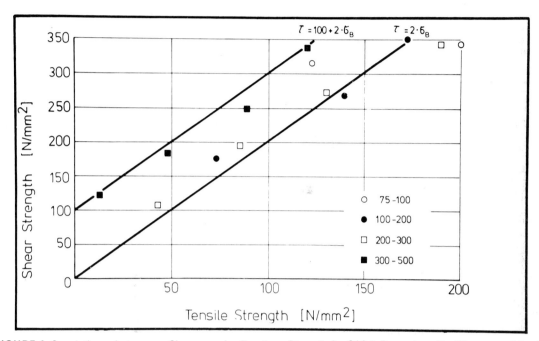

FIGURE 4 Correlation between Shear and Tensile Strength for 316 L-F powder with different particle size.

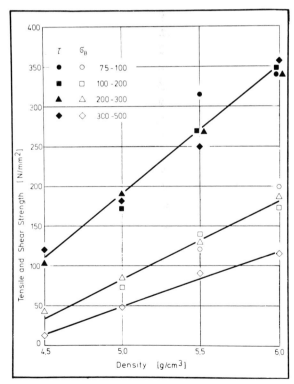

FIGURE 5 Tensile and Shear Strength of 316 L-P powder as a function of density.

3. Results

3.1 Permeability and Filter Grade

3.1.1 Powder Type 316 L-F

Properties of filter materials based on individual powders depend on the porosity and particle size. The permeability is effected to a greater extent with varying porosity than by varying the particle size. Fig. 6 presents the viscous permeability coefficient and filter grade as a function of porosity for the investigated sieve fractions. It is shown that due to the irregular particle size, only a very small variation in filter properties occur.

3.1.2 Powder Type 316 B-F

Mixtures of elongated particles containing 10 wt.% of spherical particles result in a remarkable improvement of the filter properties with a very small effect on compressibility and green strength. With

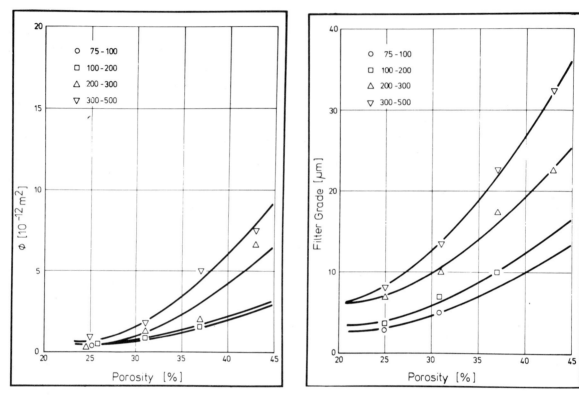

FIGURE 6 Filter characteristics of 316 L-F powder as a function of porosity a) Viscous permeability coefficient b) Filter grade by Glass-Bead-Test.

this powder type, the number of pores per given area increases simultaneously with decreasing pore size. Therefore for a particular filter grade, the permeability compared with that of 316 L-F is higher; the result is a higher energy efficiency. The permeability and filter grade of 316 B-F is plotted as a function of porosity in Fig. 7. Compared with Fig. 6, it is shown that with similar powder fraction and porosity, the permeability coefficient is nearly doubled, whereby the filter grade remains nearly unaffected. Surprisingly, the finest sieve fraction (75–100 μm) shows the same behaviour for both the powders. This indicates that particles with a high L/D ratio improve the filter properties only above a particular particle size.

Mixtures of PFM and metal powder offer

a possibility to increase the porosity and therefore expands the range of filter characteristics. Fig. 8 shows that the porosity of approximately 50% could be obtained and correspondingly an increase in filter grade up to 60 μg. The expected increment of pore size by using coarser sieve fractions of the PFM is only observed in connection with the coarser particle size. The smaller the particle size of the metal powder, crushing of the PFM particles during compaction increases. For these powders the behaviour of PFM is nearly independent of the selected sieve fraction.

3.1.3 Powder Type 316 L-P

The nearly spherical particle shape of the potato-shaped powder is the main reason for a high apparent density. Therefore, a variation of porosity during the compaction is limited. The application of this type of powder is only significant by mixing it with PFM. Fig. 9 shows the effect of the amount of PFM with different particle size on porosity and filter grade. Further, the compacting pressure was varied between 200 and 400 MN/m². As expected, the porosity depends only on the amount of PFM and the compacting pressure. Because of the smaller compacting ratio as compared with that of 316 B-F, crushing of the PFM is eliminated. Filter grade, and pore size distribution respectively can therefore be influenced by varying the sieve fraction of the PFM. This additional process parameter allows optimization of the required filter properties. Viscous per-

meability coefficients vs. porosity are plotted in Fig. 10. It is shown that for a given porosity, there is no effect of various PFM sieve fraction on viscous permeability coefficient. It is therefore possible to manufacture filter materials with a constant permeability and varying filter grade.

Filter materials manufactured from potato-shape powder offer a considerable widening of the feasible filter properties, better possibilities for optimizing the filter characteristics and a considerable improvement in regenerability due to the smoothness of pore capillaries.

The development of the processing technique for the potato-shaped powder leads to a new generation of sintered filter materials and opens up a new field for industrial application.

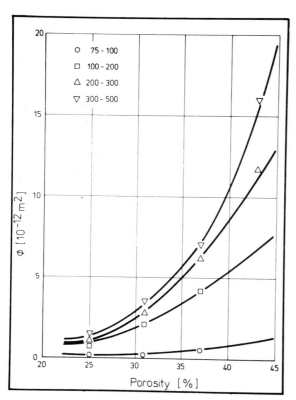
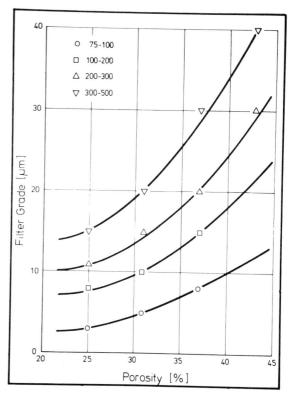

FIGURE 7 Filter characteristics of 316 B-F powder as a function of porosity a) Viscous permeability coefficient b) Filter grade by Glass-Bead-Test.

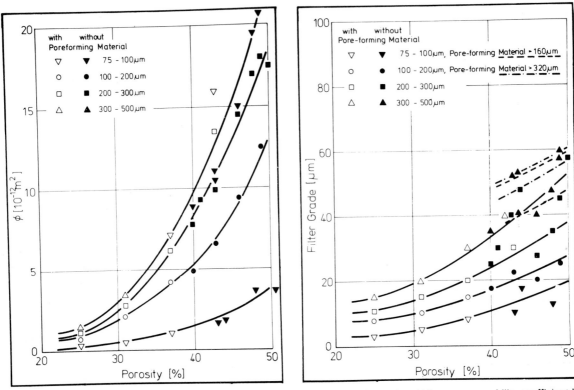

FIGURE 8 Filter characteristics of 316 L-P powder + PFM as a function of porosity a) Viscous permeability coefficient b) Filter grade by Glass-Bead-Test.

3.2 Shear Strength

The selected shear strength test allows us to differentiate the behaviour of the filter materials manufactured from various powders. The results as a function of porosity are given in Fig. 11. It is shown that particularly at higher grades of porosity, 316 L-P powder leads to an increment in shear strength. This is possibly due to either an improved sintered contact between the individual particles or additive elements, which influence the surface tension of the melt and are required for the atomization process of the potato-shape powder.

3.3 Optimum Filter Materials

This investigation permits optimization of filter materials based on different types of powders. The resulting optimum materials, arranged according to their filter grades are summarized in Tables 2–4. These tables include the sieve fraction and porosity, where the permeability and shear strength are maximum for a given filter grade. This investigation indicates that altering the characteristics of the powders, has a strong effect on the properties of the filter material. Not only the energy efficiency can be improved, but also a wider range of filter characteristics are obtained. The smoother the surface of the powder particles, the more requirements of the filter materials are fulfilled. Particularly, the use of potato-shaped powder particles results in an additional improvement of regenerability, consequently leading to new types of filter materials. Further, there is an increment in filter grade, which is important and required for filtration of high viscous liquids or sound damping.

4. Discussion

Many mathematical models for describing the behaviour of the sintered filter elements are presented in the literature. Two of these correlations were considered for this investigation: the correlation of H. Geominne et al. (9, 10) and the correlation of J. Bukowiecki (11).

They describe the correlation between permeability, filter grade, porosity and the characteristics of powders. The intention is to predict the behaviour of newly developed materials, recognize the effect of the different parameters and facilitate the testing of a production run.

4.1 Correlation of H. Geominne et al.

Considering the validity of the Darcy's equation and assuming capillaries instead of regular pores, H. Geominne gives an empirical relationship between the viscous permeability coefficient and the product of filter grade (FF) times porosity (P).

$$\phi = k \cdot (P \cdot FF)^n \qquad (3)$$

The results of this present investigation were evaluated according to H. Geominne's relation. Log-log plots of the mean values of all the investigated materials are shown in Fig. 12. These results confirm the above mentioned correlation not only for the pure powders, but also for compacted chopped fibers, powder-fiber mixtures as well as fiber felts. This figure indicates that the

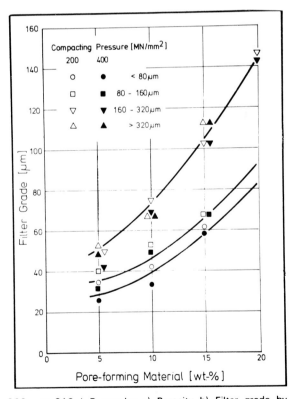

FIGURE 9 Influence of the amount of PFM on 200 - 300 μm 316 L-P powder a) Porosity b) Filter grade by Glass-Bead-Test.

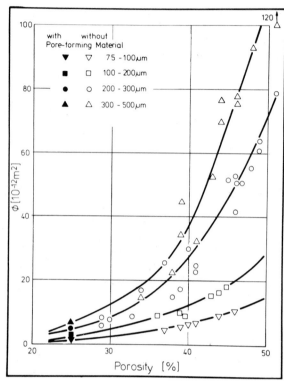

FIGURE 10 Viscous permeability coefficient of 316 L-P powder + PFM as a function of porosity.

use of potato-shape powder results in a maximum permeability at a given $(P.FF)$-value for all materials based on powders. A higher permeability can only be obtained by using fibers. The determination of the constants k and n in equation (3) from the experimental data permits calculation of the filter grade from porosity and permeability. Both these values are simply determined from experimental procedures for small and complicated shapes as well as for very large isostatic compacted filter elements. As shown in Fig. 12 the experimental results allow us to differentiate between the individual powders. However, there is no influence of the sieve fractions on the function of Geominne. For practical purposes it would be helpful if particle size was to be considered.

4.2 Correlation of J. Bukowiecki

Buckowiecki correlates the maximum pore size, determined by the bubble-point test (d_{BPT}), the porosity (P) and maximum and minimum particle size of the used

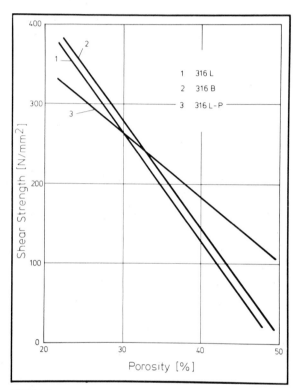

FIGURE 11 Shear strength of the investigated materials as a function of porosity.

TABLE 2 Optimum Filter Materials, Based on 316 L-F

Filter-Grade [µm]	Powder-Fraction [µm]	Porosity [%]	Spec. Permeability φ[10⁻¹²m²]	φ[10⁻⁷m²]	Bubble Point [mmWS]	τ_{NH_3} [N/mm²]	$\tau_{Vak.}$ [N/mm²]
3	75 - 100	23-27	0.4	0.06	580-590	345	370
5	75 - 100	29-33	0.7	0.6	470-480	315	245
8	100 - 200	29-33	0.7	2	400-410	270	250
10	100-200	35-39	1.5	11	300-310	180	185
	200 - 300	29-33	1.5	3	270-280	270	225
15	200 - 300	35-39	2	14	205-210	190	160
20	300 - 500	35-39	5	7	150-155	185	170
25	200 - 300	41-45	7	30	150-160	105	100
30	300 - 500	41-45	8	25	115	120	90

powder (D_{max}, D_{min}) with the viscous permeability coefficient:

$$\phi = \frac{d_{BPT} \cdot D_{min}}{96 \cdot D_{max}} \cdot P$$

Fig. 13 compares the permeability coefficients determined experimentally with the calculated values from equation (4). It is shown that up to a maximum particle size of 200 μm, there is a good agreement with the correlation. For coarser particle size, the calculated value is twice that of the experimental determined permeability. This is in accordance with the work of Bukowiecki.

With the help of this correlation, an estimation of the permeability is possible and the influence of different particle size and porosity can be easily determined. For production purposes, the particle size of the powder and porosity is known, whereby the determined bubble-point allows the calculation of permeability.

The use of PFM, however, particularly with the potato-shaped powder, leads to discrepancies between the experimental and calculated values, which limits the application of the above equation (Fig. 14). The different sieve fractions of the PFM affect the pore size distribution, and therefore either an emperical correlation factor is necessary or precise pore size distribution must be determined.

4.3 Determination of the Pore Size Distribution by the Sigma-Flow Method (12)

This method combines the flow rate test and mathematical models for describing flow and pore size distribution. The apparatus as shown in Fig. 3 was used for determining permeability of the filter material in a dry and a completely wet state. Iso-propanol was used as a wetting agent. The details of the test method and the data reduction for mathematical models are presented in (12).

The Sigma-Flow Method was applied for materials used in this investigation and the results were compared with experimental data by the glass-bead test. All three types of powders were used with filter grades, between 6–10 μm and approximately 20 μm.

The results as shown in Fig. 15 fit a normal distribution and therefore, the linear ($\Delta p - \Delta p_{BPT}$) – probability Q_{WET}/Q_{dry} plot was selected. This is in contradic-

TABLE 4 Optimum Filter Materials, Based on 316 L-P

Filter-Grade [μm]	Powder-Fraktion [μm]	Porosity [%]	Spec Permeability φ[10⁻¹²m²]	φ[10⁻⁷m²]	Bubble Point [mmWS]	τ_vak [N/mm²]
8	75 - 100	23 - 27	1	7	395 - 400	300
10	45 - 100	23 - 27	2	9	270	315
	100 - 200	37 - 41	4	38	175	165
15	75 - 100	37 - 41	6	30	165 - 170	180
20	100 - 200	37 - 41	10	50	130	170
30	100 - 200	41 - 43	16	170	120	160
40	200 - 300	34 - 38	18	30	90 - 100	270
50	200 - 300	35 - 39	20	50	75 - 80	250
70	200 - 300	38 - 42	70	110	60	190
100	300 - 500	42 - 46	80	250	40 - 45	130
120	500 - 700	43 - 47	100	400	35 - 40	150
150	200 - 300	52 - 56	170	1000	30	65
	500 - 700	49 - 53	140	5000	25 - 30	100

TABLE 3 Optimum Filter Materials, Based on 316 B-F

Filter-Grade [μm]	Powder-Fraktion [μm]	Porosity [%]	Spec Permeability φ[10⁻¹²m²]	φ[10⁻⁷m²]	Bubble Point [mmWS]	τ NH₃ [N/mm²]	τ vak [N/mm²]
3	75 - 100	23 - 27	0,2	0,3	605	380	360
5	75 - 100	29 - 33	0,5	2	510	300	270
8	100 - 200	23 - 27	1	1	280	390	350
10	100 - 200	29 - 33	2	6	240	325	250
15	100 - 200	35 - 39	4	30	205 - 210	210	170
20	200 - 300	35 - 39	6	22	150 - 160	210	140
30	200 - 300	41 - 45	14	70	105	130	90
40	300 - 500	41 - 45	16	93	80	150	130
50	300 - 500*	46 - 50	20	100	75	-	100
60	300 - 500*	48 - 52	26	110	60 - 65	-	50

* with Pore-forming Material

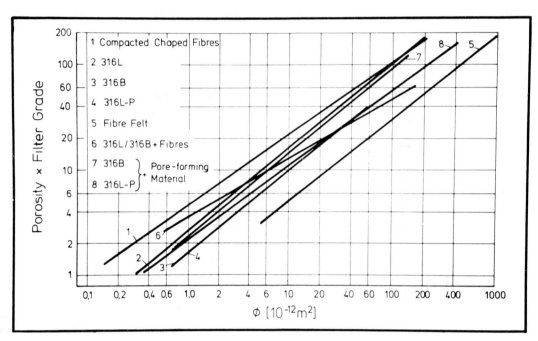

FIGURE 12   Log ϕ - log (P $\times$ FF) plot according to equation (3).

tion with the results of Cole (12), where log-normal distribution was considered which should correspond with the pore size distribution.

By using equation (5), the pore size distribution can be evaluated.

$$d = \frac{4 \cdot \gamma}{(\Delta p - \Delta p_{BPT}) + \Delta p_{BPT}} \qquad (5)$$

Sigma-Flow implies that the pore size corresponds with the probability of 15.87 % (standard deviation), which is in accordance with the filter grade. This value as shown in Table 5, leads to discrepancies with the filter grade determined by the glass-bead test. Better consistency was obtained with the $\pm$ 6 (84.13%) value. Further experiments are necessary to define an additional factor which will describe the geometrical shape of the pore capillary. Thus, allowing to obtain a better agreement between the various experimental results.

As shown in Fig. 15, different particle sizes and shapes result in various pore size distributions. It must be specified that the behaviour of the filter material cannot be described with only the Sigma-Flow value. The effect of the filter element thickness on the results of the "Sigma-Flow method" should also be investigated.

In spite of these uncertainties, this method should have a good opportunity in future as a standard method for describing the filter characteristics for engineering application.

5. Industrial Applications

Highly porous sintered stainless steel elements are used for filtering of solid particles from gaseous and liquid mediums. Further, they are employed for throttling and stabilising flow mediums, as well as for sound damping and air pollution control. The following examples demonstrate the opportunities of sintered metallic filters and show that advantages of isostatic compaction technique in connection with the use of potato-shape powder.

Fig. 16 shows flame arresters as a safety device for explosive gaseous mixtures. They were manufactured by axial compaction and sintering.

Powders with an irregular particle shape were used (316 L-F and 316 B-F) for

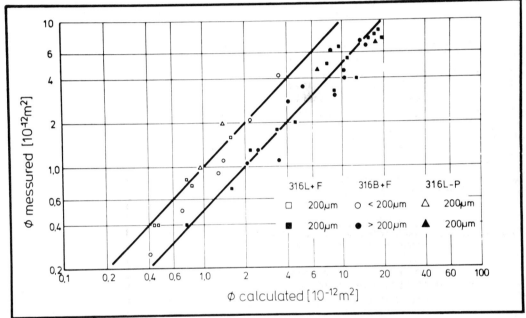

FIGURE 13   Relation between measured and calculated viscous permeability coefficient according to equation (4).

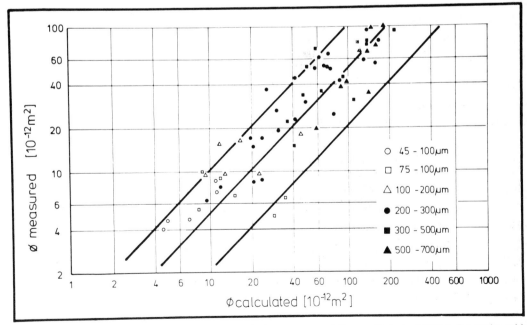

FIGURE 14 Relation between measured and calculated viscous permeability coefficient of 316 L-P powder with PFM.

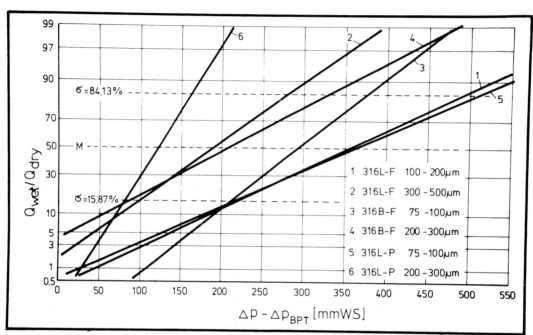

FIGURE 15 Linear $(\Delta p - \Delta pbpt)$-probability Q_{wet}/Q_{dry} plot for the Sigma-Flow method (12).

TABLE 5 Comparison of Different Pore Sizes, Calculated from "Sigma-Flow Method" with the Glass-Bead-Test

Powder Type	Sieve Fraction [μm]	Filter Grade (GBT) [μm]	d_{BPT} [μm]	$d_{15.87}$ [μm]	d_M [μm]	$d_{84.13}$ [μm]
316L-F	100 - 200	10	27	16	13	11
	300 - 500	20 - 25	45	30	23	18
316B-F	75 - 100	8	21	14	12	11
	200 - 300	20	55	35	24	18
316L-P	75 - 100	6 - 7	23	15	12	10
	200 - 300	20	59	38	32	28

obtaining maximum green strength for larger dimensions and irregular pore capillaries and therefore a minimum extinguishing gap. In order to optimize the safety of the flame arresters for a particular application, density, density distribution, particle size and shape must be adjusted.

Fig. 17 shows isostatic compacted seamless filter elements and demonstrates the possibility of manufacturing filters with larger dimensions. These elements were installed in fruit presses. Besides, there are other applications in the food and petrochemical industries. The uniform density distribution obtained by isostatic compaction technique lead to uniform permeability and filtration rate: the efficiency therefore is increased.

Fig. 18 demonstrates distinctly the advantages of the isostatic compaction technique for manufacturing complicated shapes. It shows a filter element, where the active surface is significantly increased. Because of the application for the purification of high viscous polymers, the use of potato-shaped powder (316 L-P) is absolutely necessary. The regenerability of the potato shaped material remarkably increases durability, reduces maintenance and raw material costs. Further, the maximum permeability for the filter grade minimizes the required energy loss.

FIGURE 16 Flame arresters.

6. Summary

The necessity of a standard test method for determining filter characteristics has been discussed and an appropriate method has been suggested. A significant simplification was obtained by introducing a coining operation, which does not influence the test results. With the help of this operation, the required effort for systematic investig tions is substantially reduced. This standard test permits also an easier supervision of the production run.

Three different types of stainless steel powders with various particle shape (316 L-F, 316 L-B and 316 L-P) were investigated. The effect of porosity and particle size on the filter characteristics (permeability, filter grade and shear strength) were determined. The results indicated that particularly with nearly spherical shape (potato-shaped powder) filter materials could be manufactured with the following characteristics:

a) a remarkable widening of the filter characteristics,

b) possibility of an optimum adjustment between filter material and the requirements, whereby energy efficiency must be considered,

c) regenerability of the filter elements,

which permits optimum exploitation of the material.

Extensive experimental data was used to examine certain theoretical correlations between the individual filter characteristics. It was shown, that in certain ranges a

FIGURE 17 Seamless isostatic compacted filter elements.

FIGURE 18 Filter elements with a complicated shape, manufactured by isostatic compaction of 316 L-P powder.

consistency between measured and calculated values exists. This opens a possibility for testing some production runs with minimum experimental effort. The Sigma-Flow Method for determining the pore size distribution was investigated and discussed. It is believed that this method requires some modifications which may offer a good opportunity as a standard test for determining filter characteristics.

The advantages of the isostatis compaction technique above all in connection with the potato-shaped powder is demonstrated.

7. Acknowledgement

This paper is a part of industrial research work carried out on behalf of Forschungsgemeinschaft Pulvermetallurgie, Schwelm, and was financially sponsored by Arbeitsgemeinschaft Industrielle Forschungsgemeinschaften, Cologne. The authors express their sincere thanks to both the organizations for their support. The help of Mr. K. Dalal in preparing this manuscript is gratefully acknowledged.

References

1. C. Agte, K. Ocetek, Metallfilter, Akademie-Verlag, Berlin, 1957.
2. R. Dietz, J. Niessen, G. Zapf, Jahrbuch 1969 der Forschungsberichte des Landesamtes für Forschung des Landes Nordrhein-Westfalen, 1969, Westdeutscher Verlag, Opladen, W. Germany.
3. W. R. Johnson, Machine Design, 1974, July, 97–101.
4. V. T. Morgan, Symposium on Powder Metallurgy, London 1954, 81–89.
5. V. T. Morgan, Symposium sur le Métallurgie des Poudres, Paris, 1964, June, 419–430.
6. G. Dörr, A. Kirste, DEW - Technische Berichte, 1973, 13, 2, 143–151.
7. MIL - F - 8815 B, 1967, August 10.
8. ISO-Normentwurf, Doc. Nr. 146 E, 1973, June.
9. H. Geominne, R. de Bruyne, J. Ross, E. Aernoudt, Filtration and Separation, 1974, July/August, 1–5.
10. H. Geominne, R. de Bryne, J. Ross, Preprint, 4th European Symposium on Powder Metallurgy Grenoble, 1975, May 13–15.
11. J. Bukowiecki, Prace Instytutu Metall Niezelaznych, 1974, 3, 2, 81–87.
12. F. W. Cole, Filtration and Separation 1975, Jan./Feb., 17–22.

The Preparation of High-Conductivity Compacts from Copper Powder

P. W. Taubenblat and G. Goller

INTERNATIONAL CONFERENCE

1970 AMSTERDAM

Production of powder-metallurgy (PM) parts for conductive applications is discussed using conventional sintering, vacuum sintering, and processing of preforms by forging, rolling, or drawing. Conductivities of 100% IACS and mechanical properties similar to those of wrought copper were achieved by further processing of preforms. Simple pressing and sintering resulted in a part with an electrical conductivity of ~ 90% IACS. For a given high-purity material (electrolytic copper powder), the variable most affecting conductivity is the density of the final PM part.

The use of copper powder for the production of intricate parts for high-conductivity electrical applications is relatively new and growing. Several PM producers in the United States are now making on a large scale, complex parts such as armature bearing blocks, contacts for circuit breakers, shading coils for contactors, and heavy-duty contacts weighing as much as 1·5 lb (0·68 kg) for circuit-breakers, from copper powder. Typical parts are shown in Fig. 1. Other manufacturers are actively engaged in pilot-plant production of parts for both the electrical and electronic industries. This investigation is intended to demonstrate that proper processing of high-purity copper powder will yield copper parts of high conductivity and with mechanical properties similar to parts produced from wrought copper. It further shows that fabrication of copper powder preforms to a density close to that of wrought copper can be achieved on conventional processing equipment. This processing includes:

(1) Conventional pressing and sintering.
(2) Vacuum sintering.
(3) Hot forging of preforms.
(4) Hot rolling and drawing of preforms.

Experimental

The specimen used was in the form of a small bar ~ 0·4 × 0·5 × 3 in (10 × 13 × 76 mm) long. "LO"-type high-purity electrolytic copper powder, produced by the United States Metals Refining Co., was used throughout. This powder has a minimum copper content of 99·5% and contains the typical impurities shown in Table I.

The physical properties of the loose powder and the green properties of pressed compacts are noted in Table II.

Test-specimens were prepared by mixing the copper powder with 0·75 wt.-% lithium stearate lubricant and compacting at 20 tonf/in² (276 MN/m²). The compacted bars were then sintered at 1000° C (1275 K) for either 5 or 30 min in an atmosphere of dissociated ammonia or for 30 min at 950–1000° C (1225–1275 K) in a vacuum of 120–140 μm. Control specimens of cast OFHC-brand copper were processed side

Manuscript received 2 March 1970. Contribution to an International Conference on "Copper and Its Alloys" to be held in Amsterdam on 21–25 September 1970. P. W. Taubenblat, B.Sc., M.S., is Manager of New Product Development and Physical Metallurgy, AMAX Base-Metals Technical Department, Carteret, N.J., U.S.A. G. Goller is Senior Technician in the same organization.

Table I

Chemical Analysis (%) of "LO" Copper Powder

Antimony	0·0015	Nickel	0·00038
Arsenic	0·0003	Phosphorus	0·001
Bismuth	< 0·0001	Sulphur	0·005
Cadmium	< 0·001	Tellurium	< 0·001
Carbon	0·008	Tin	0·0002
Insoluble	0·002	Wt. Loss in	
Iron	0·006	Hydrogen	0·08
Lead	0·05	Zinc	< 0·001
Manganese	0·0001	Copper	Bal.

Fig. 1 Typical PM parts for electrical applications.

by side with the PM preforms in final fabrication comprising one or another of the following treatments:

(1) Hot forging to ~ 0·2 in- (5 mm) thick with a reduction of 50–70% followed by annealing in steam at 500° C (775 K). The density of the preforms was determined after the forging step, and conductivity after annealing.

(2) Hot rolling to 0·25 in -(6·3 mm) dia. rod with a reduction of 73%, followed by annealing at 980° C (1255 K) for 30 min under charcoal, and finally cold drawing to 0·08 in- (2 mm) dia. wire with a reduction of 90–95%. Tensile properties were determined on the hard wire. Additionally, one set of

TABLE II

Physical and Green Properties of " LO " Copper Powder

Screen Analysis, %	
(Tyler Mesh No.)	
+ 100	0·1
− 100 + 150	4·3
− 150 + 200	14·1
− 200 + 325	29·4
− 325	52·1
Apparent Density	2·6 g/cm³
Flow	29·2 g/50 s
Green Strength at 20 tonf/in²*	2000 lbf/in²
(276 MN/m²)	(13·8 MN/m²)
Green Density at 20 tonf/in²	7·2 g/cm³
(276 MN/m²)	

* The tonf values quoted in this paper are based on the American "short ton" (=2000 lb).

cold-worked wire samples was annealed at 850° C (1125 K) for 30 min in hydrogen for the bend-test determination, while a second set of wires was annealed at 500° C (775 K) in steam for tensile tests and conductivity measurements.

(3) Hot rolling to 0·05 in- (1·3 mm) strip with a reduction of 85–90% and annealing at 980° C (1255 K) for 30 min under charcoal. Tensile properties and electrical conductivity were determined on these samples.

The effects of sintering and working on the density and conductivity of the copper preforms are shown in Table III. Sintering in vacuum produced a part with a density and conductivity similar to conventional sintering. Density attained was 8·0 g/cm³ and conductivity was 87%. However, regardless of the sintering conditions, working produced comparably dense parts with substantially the same conductivity. The electrical conductivity of the PM materials after hot forging or drawing to wire was ∼ 100% IACS, i.e. only slightly below that of wrought copper.

Fig. 2 shows how conductivity varies with density. The graph also includes published data previously obtained in this laboratory[1] with sintered, coined, and resintered specimens produced from " LO " copper powder. Our latest data permitted an extension of the original curve to > 100% conductivity.

As shown in the comparison in Table IV, the tensile properties of the materials produced from preforms were

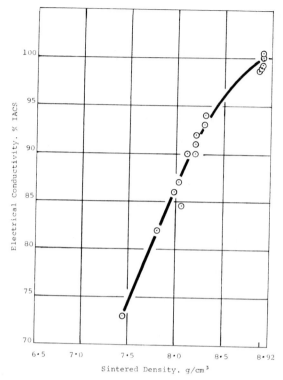

Fig. 2 *Relationship of electrical conductivity and sintered density of compact produced from copper powder.*

comparable to those of wrought material made from cast copper.

Table IV also shows bend-test results after annealing wire specimens in hydrogen in accordance with the ASTM embrittlement test.[2] This test is designed to indicate the presence or absence of oxygen in copper. If oxygen is present, the copper fails the bend test by its inability to withstand more than a few reverse bends.

The microstructures of the preforms, as sintered, and after hot forging, are shown in Figs. 3 and 4. It can be seen that the PM material, as sintered, contained considerable porosity but that most of the pores were then closed during hot working so that the density approached that of the wrought copper.

TABLE III

Comparison of the Density and Conductivity of PM Materials with Wrought Copper

	PM Material				Wrought Copper	
	Sintered 5 min		Sintered 30 min			
Condition	Density, g/cm³	Conductivity, % IACS	Density, g/cm³	Conductivity, % IACS	Density, g/cm³	Conductivity, % IACS
As sintered:						
Vacuum	—	—	8·03	87·2	—	—
Dissociated ammonia	7·86–7·95	82·6–84·5	8·06	86·8	—	—
Hot forged	8·86	98·8*	8·89	99·1*	8·95	101·5*
Hot-rolled and annealed strip	8·8	99·3	8·8	98·7	8·9	101·5
Hot-rolled rod, cold drawn to wire, and annealed	8·8	100·4	8·8	100·4	8·9	101·4

* Annealed in steam at 500° C (775 K) after forging.

TABLE IV

Comparison of Tensile and Bend Properties of PM Materials with Wrought Copper

Material and Condition	Tensile Strength, lbf/in²	MN/m²	Yield Strength, (0·01% offset) lbf/in² (MN/m²)	Elongation, % in 2 in (50 mm)	No. of Bends
Strip, hot rolled, annealed at 980° C (1225 K) under charcoal:					
PM material, sintered 5 min	39 000	269	4 800 (33)	44·8	—
,, ,, sintered 30 min	37 800	261	5 700 (39)	48·4	—
Wrought copper	35 300	244	4 400 (30)	41·0	—
Wire, cold drawn 90–95%:					
PM material, sintered 5 min.	60 400	416	53 700 (370)	3·4	—
,, ,, sintered 30 min.	60 800	418	50 100 (345)	4·1	—
Wrought copper	59 000	406	51 500 (355)	4·0	—
Wire, annealed at 500° C (775 K) in steam:					
PM material, sintered 5 min.	36 600	252	9 500 (65)	39·8	—
,, ,, sintered 30 min.	36 100	249	9 400 (65)	42·5	—
Wrought copper	35 100	242	8 700 (60)	41·1	—
Wire, annealed at 850° C (1125 K) in hydrogen:					
PM material, sintered 5 min.	—	—	—	—	7/9
,, ,, sintered 30 min.	—	—	—	—	9/10
Wrought copper	—	—	—	—	10/14

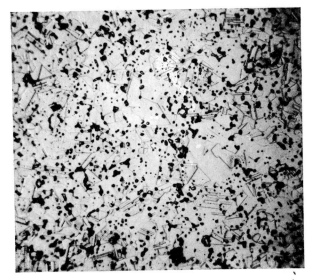

Fig. 3 Structure of preformed copper-powder compact after sintering for 30 min. at 1000° C (1275 K). CrO₃ etch. × 120.

Fig. 4 Structure of hot-forged copper-powder preform. CrO₃ etch. × 120.

The major differences in appearance between the wrought PM material and the wrought cast copper is in grain size. After each of the working stages, the grain size of the PM material became markedly finer than that of the wrought copper (typically, 0·035 as against 0·4–0·6 mm, respectively). This effect is reflected in the tensile-test results. In all the tests, the tensile strength of the wrought copper was less than that of the PM material, which would be explained by the difference in grain size.

Conclusion

PM parts with excellent electrical conductivity can be made from high-purity copper powder by vacuum- or conventional sintering techniques. Coupled with forging of preforms and further processing of the preforms to sheet or wire, conductivities and mechanical properties similar to those of wrought copper can be achieved. In addition, powder-metallurgy techniques for fabrication of copper parts can often obviate all need for machining. Cost savings effected by eliminating the loss of raw material and machining operations can offset the usually higher cost of metal powders.

References

1. P. W. Taubenblat, *Internat. J. Powder Met.*, 1969, **5**, 89.
2. "Standard Specification for Oxygen-Free Electrolytic Copper Wire Bars, Billets, and Cakes", *ASTM Designation* (**B170–67**), ASTM Standards (1969), Part 5.

The Properties of P/M Electrical Contact Materials

H. SCHREINER* AND W. HAUFE*

ABSTRACT

This paper deals with the development and production of electrical contact materials. AgNi 10 represents the metal-metal system having components which are insoluble in each other; AgCdO 12 represents the system of metal-metaloxide, the oxide being finely and uniformly dispersed in the base metal; AgC 3 stands for the metal-non metal system. These contact materials are applied as electrical contacts in low voltage contactors in the field of power engineering. By amount of additives, by special combinations of the base components, and by changing the powder metallurgical processing steps, the properties of contact materials may be varied within a wide range. The conditions of manufacture are specified. The measurement of weld strength, erosion rates and contact resistance were carried out under conditions close to practice. An appropriate test equipment is described. Special emphasis is given on accuracy of results, good reproducibility of the measurements and economy of test procedure. This test apparatus helps the development of new contact materials, and the pre-selection and quality control of the specimens. The properties of the above mentioned contact materials are given in comparison with those of pure silver.

Introduction

With regard to heterogeneous materials the importance of powder metallurgy techniques lies in the development of special material structures and in the many possibilities of combining the components widening the field of application of P/M materials. Pores as structure elements are of special interest. Materials having a penetrating pore system may be used as filters or as contact disks in semi-conductor-rectifiers. P/M materials with pores filled with a lubricant, are used as self-lubricating bearings. P/M parts with high accuracy of shape are used for structural elements due to their economical mass production. The porosity of the structural parts is reduced if high strength is required.

Higher density compacts can be obtained by hot forming or forging. The pore system of the sintered specimens can be infiltrated by liquid metals. These infiltration materials are practically pore free. Metals having a high melting temperature, hard metals and magnetic materials may be produced by powder metallurgy methods. Of particular interest are composite materials consisting of metals that are insoluble in each other in the liquid state, combinations of a metal and a metal compound as well as a metal and a non-metal (e.g. graphite). Composite materials in the production of electrical contacts have a great importance because of the numerous variations of material structure and composition, resulting in a wide spread property range (1–5). Ratio of density and

---

* Siemens AG, Nürnberg, West Germany

porosity are important values from the viewpoint of electrical contact applications: the erosion rate (loss of material due to arcing) is diminished, resulting in a prolongation of the service life of the contacts in the switching apparatus. For this reason, a method of producing electrical contacts by P/M techniques is used to insure that the density is at its maximum with minimal porosity. The special characteristics which are desired for these electrical contacts are easier to achieve with heterogeneous sintered materials than with materials produced by melting.

Composite materials i.e. those from the systems of Ag-metal oxide and Ag-non metal, have favorable engineering properties. They show, however, an insufficient wettability versus liquid metals. By hard soldering or welding these materials can be joined with the support metal with difficulty. Double-layer-contacts with a well solderable or weldable second layer are obtained by means of powder metallurgy methods (6). These economically produced specimens show a high strength of the interface.

The production steps of some P/M contact materials used for low voltage contactor applications are outlined. AgNi 10, AgCdO 12 and AgC 3 stand for the systems of metal-metal, metal-metaloxide and metal-non metal, respectively. Determination of engineering properties requires special testing procedures.

Base Materials and Production Conditions

Properties of the Metal Powders

The properties of the powders (7) used for the production of AgNi 10, AgCdO 12 and AgC 3 are summarized in Table 1. The AgNi 10 and AgC 3 base powders were

TABLE 1

Base powders	Method of production particle shape	Particle size in μm	Flow rate t_F in s/100g	Bulk density ρ_B in g/cm³	Tap density ρ_T in g/cm³
Ag	Electrolysis dendritic	<100 $\hat{x} = 5$	∞	1, 9	2, 9
Ni	Decomposition of Ni-carbonyl spherical	<7 $\hat{x} = 5$	∞	2, 3	3, 4
Graphite	Electrographitic irregular	<8 $\hat{x} = 5$	∞	0, 16	0, 23
AgNi 10 Granulated powder		<300 $\hat{x} = 100$	$t_{F6} = 13, 4$	1, 9	2, 5
AgCdO 12 Atomized powder, internal oxidation		<200 $\hat{x} = 100$	$t_{F4} = 16, 4$	4, 2	5, 1
AgC 3 Granulated powder		<300 $\hat{x} = 100$	$t_{F6} = 15, 0$	2, 0	2, 6

$\hat{x}$ = mode.

t_F-Index = diameter of nozzle in mm.

subjected to a heat granulation in order to achieve good flow behaviour. AgCdO 12 is composed of an internally oxidized atomized powder (8).

Production Steps

When manufacturing AgNi 10 and AgC 3 contacts from thermally granulated powders by pressing, sintering and re-pressing the production steps are almost the same (3, 9). AgNi 10 compacts of high edge strength are obtained at a pressure of 100 MN/m², AgC 3 compacts at a pressure of 200 MN/m². The vacuum sintering of AgNi 10 specimens takes place at 850 C, hydrogen sintering of AgC 3 at 820 C. The applied pressure for repressing is 800 MN/ m² in both cases. The manufacture of AgCdo 12 is as follows: internally oxidized atomized composite powder is compressed at 400 MN/m², the compacts are then sintered in air at a temperature of 850 C. The hot repressing at a pressure of 800 MN/m² results in a low porosity of the samples (10). To achieve a low gas content, the compacts are heat treated at 700 C in nitrogen. The final shape of the contacts is obtained by cold repressing at 800 MN/m². The AgCdO 12 and AgC 3 contacts show an insufficient wettability versus liquid hard solder due to the relatively high amounts of CdO and C, respectively. In consequence they are provided with a well solderable Ag-layer (0, 3 mm) by double layer pressing technique.

Structure and Physical Properties

The AgNi 10 and AgC 3 contact materials produced from heat granulated powders have a structure with uniform distribution of Ni- and graphite dispersions in

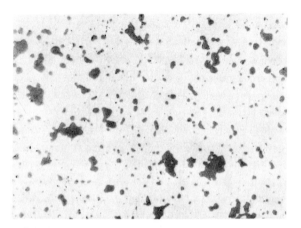

FIGURE 1 Contact material AgNi 10 Base powder: heat granulated AgNi 10-powder.

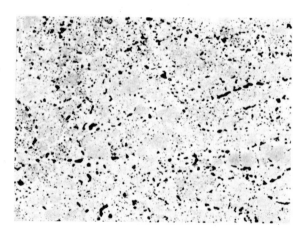

FIGURE 2 Contact material AgCdO 12 Base powder: atomized, internally oxidized composite alloy.

the silver base metal (Figs. 1 and 3). The mean Ni particle size is about 5 μm, the largest particles are of about 20 μm diameter. The diameters of the graphite dispersions are in the range of 3 to 10 μm.

Pressing and sintering the internally oxidized AgCdO 12 composite powder results in very fine, uniformly dispersed CdO segregations with an average particle size < 3 μm in the silver matrix (Fig. 2). The uniform distribution of the CdO particles is achieved even when the dimensions of the contacts are rather large. Single CdO segregations with diameters from 3 to 10 μm are visible at the grain boundaries. Due to hot re-pressing of the sintered compacts, the photomicrograph shows no pores in the structure.

The values of density, density ratio, hardness and electrical conductivity (11, 12) alone are not sufficient data for characterizing a contact material (see Table 2). The knowledge of weld strength, erosion rate and contact resistance is necessary. The next part of this article deals with a test fixture permitting the determination of the engineering properties. The test results are discussed by comparison with

the corresponding properties of pure Ag, produced by melting.

Measurement of Properties

The described P/M contact materials are mainly applied to power engineering switches and frequently also in the field of communication techniques.

For power engineering applications, the significant properties are (a) welding (13–24), (b) erosion behaviour (25–31), (c) contact resistance (32–38). The requirements referring to these properties may be quite different for each type of power switch. Testing contact materials in a real contactor means a significant expenditure of time and cost. An economical way of testing is achieved by a testing apparatus working under closely controlled conditions. The comparison of test results of different contact materials requires previously fixed test conditions.

Design and Operation of Test Apparatus

A convenient testing device for measuring the above mentioned contact properties is shown in Figs. 4 and 5 (39). Construction includes four main elements: (1) The arrangement of the contacts, (2) the switching units for contact making and (3) con-

FIGURE 3 Contact material AgC 3 Base powder: heat granulated AgC 3-powder.

TABLE 2

Contact material	Density ρ in g/cm³	Density ratio R in %	Brinell-hardness HB2,5/31,25/10	Electrical conductivity σ in m/Ωmm²
Ag1000	10,50	100,0	104	60
AgNi10	10,21	99,1	90	49
AgCdO12	10,11	99,6	91	43
AgC3	9,33	99,0	60	37

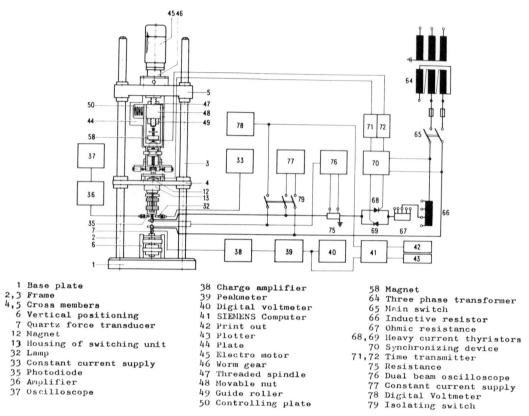

1 Base plate	38 Charge amplifier	58 Magnet
2,3 Frame	39 Peakmeter	64 Three phase transformer
4,5 Cross members	40 Digital voltmeter	65 Main switch
6 Vertical positioning	41 SIEMENS Computer	66 Inductive resistor
7 Quartz force transducer	42 Print out	67 Ohmic resistance
12 Magnet	43 Plotter	68,69 Heavy current thyristors
13 Housing of switching unit	44 Plate	70 Synchronizing device
32 Lamp	45 Electro motor	71,72 Time transmitter
33 Constant current supply	46 Worm gear	75 Resistance
35 Photodiode	47 Threaded spindle	76 Dual beam oscilloscope
36 Amplifier	48 Movable nut	77 Constant current supply
37 Oscilloscope	49 Guide roller	78 Digital Voltmeter
	50 Controlling plate	79 Isolating switch

FIGURE 4 Scheme of test device.

tact breaking, (4) the motor driving system. The apparatus permits a choice of testing conditions within a wide range and is easily installed.

Contact making is initiated by de-energizing an electromagnet. The movable contact is then accelerated by a contact spring until point of closure. By changing this spring and the accelerated mass, both closing velocity and duration of bouncing can be adjusted: The fixed contact contains a quartz force transducer by which the weld strength is measured. A second system of spring-mass-electromagnet effects the contact breaking process. The positioning of the movable contact is performed by the motor driving system. All the switching and measuring functions are controlled automatically.

Recording of the movement of the movable contact when making and/or breaking is done by a photodiode. Current-time function and arc voltage are stored by a dual beam oscilloscope. Contact resistance is measured in accordance with the current-voltage drop method. The calculation of the cumulative frequency of both weld strength and contact resistance (40) is performed by a SIEMENS-computer-system.

Test Conditions

Closing a contact is often accompanied by some bouncing that may lead to contact

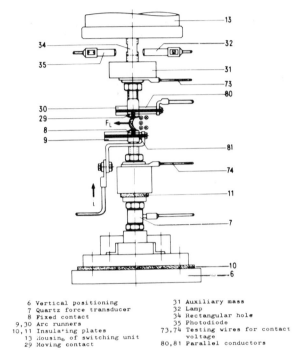

6	Vertical positioning
7	Quartz force transducer
8	Fixed contact
9,30	Arc runners
10,11	Insulating plates
13	Housing of switching unit
29	Moving contact

31	Auxiliary mass
32	Lamp
34	Rectangular hole
35	Photodiode
73,74	Testing wires for contact voltage
80,81	Parallel conductors

FIGURE 5 Contact assembly with conductors.

welding due to arcing. "Weld strength" therefore refers to the force required for separating the welded contacts. The material loss by evaporation during arcing is called "erosion" (41). Its rate depends on several material constants, current level and duration of arcing, and is commonly given as the loss of volume per number of operations. By means of a magnetic blow the arc may be driven out of the contact gap on the arc runners. The interaction of contact surfaces and arcing at contact breaking affects weld strength (42).

Heating of the current carrying contacts is closely related to the contact resistance.

The test conditions as applied to the measurement of the properties of contact materials are summarized in Table 3. Operation with the test apparatus involves closing of the contacts under load, separating the usually welded contacts, reclosing

and circuit breaking. Both, contact closing and breaking are performed synchronously to a predetermined phase of the voltage. With respect to alternating current, the instant of closure is at $\alpha_{m1} = 0°$ (less than 0, 1 ms after current zero), breaking occurs at $\alpha_{b1} = 80°$ ($\pm$ 0, 1 ms). When testing contact materials for use in AC contactors, the polarity of test current is reversed after each operation. In case of DC applications the polarity stays constant. Contact resistance is measured after contact closing (R_1) and after contact breaking (R_2).

Test results

Studies on weld strength, erosion and contact resistance were carried out with the test apparatus on Ag 1000, AgCdO 12 and AgC 3 contact materials. Fig. 6 shows cumulative frequency plots of the weld strength. For a comparative assessment of contact materials the maximum value and the slope of the curves represent relevant data. The reproducibility of the single plots fall within approximately 5%. Statistical evaluation reveals the following order of materials. AgNi 10 and Ag 1000 show elevated values of weld strength, AgCdO 12 has medium, AgC 3 low weld strength. The histograms in Fig. 7 illustrate the $F_{99,8}$ values of weld strength, the volume losses ΔV and the maximum values of R_1 and R_2, respectively. Data is based on 500 switching operations. The bright areas of the histograms indicate the variations of the maximum values, while arithmetic means are marked by cross-lines. The diagram indicates the superiority of both AgC 3 and AgCdO 12 contact materials over Ag 1000 and AgNi 10 with respect to weld strength. These contact materials are therefore applied whenever a high making

TABLE 3   Test Conditions when Measuring Properties of Contact Materials for Contactor Applications

Contact Dimensions

Area of cross section	square shaped	
Edge length	a = 10 mm	
Area of stationary contact	convex, r = 80 mm	
Area of movable contact	convex, r = 80 mm	

Mechanical Test Conditions

Closing velocity	v_{c1}	= 1 m s^{-1}
Contact force	F_C	= 60 N
Duration of bouncing (3 Rebounces)	t_b	= 5 ms
Velocity at contact make	v_m	= 7,4 $\cdot$ 10^{-3} m s^{-1}
Velocity at contact break	v_b	= 0,8 m s^{-1}
Contact gap	a	= 10 mm

Electrical Test Conditions

Voltage	U	= 220 V
Frequency	f	= 50 Hz
Current at contact make	$\hat{I}_m$	= 1000 A
Current at contact break	$\hat{I}_b$	= 1000 A
Power factor	cos φ	= 0,4

Operation 1

Instant of contact make	α_{m1}	= 0° (0,1 ms after current zero)
Instant of contact break	α_{b1}	= 80° ($\pm$ 0,1 ms)

Operation 2

Instant of contact make	α_{m2}	= 180° (0,1 ms after current zero)
Instant of contact break	α_{b2}	= 260° ($\pm$ 0,1 ms)

Operation 3 = Operation 1

Magnetic flow per 100 A	$\dfrac{B}{I}$	$= \dfrac{50 \cdot 10^{-4}\ T}{100\ A}$
Current-time function	i	$= \hat{I} \cdot \sin\omega t$
Number of operations	n	= 500

Mechanical and Electrical Test Conditions when Measuring Contact Resistance

Closing Velocity	v_{c_2}	= 7,4 $\cdot$ 10^{-3} m s^{-1}
Voltage	U	= 5 V
Current	I	= 10 A

capacity is required. AgCdO 12, in comparison, shows the lowest erosion rates resulting in a considerably longer service life of the contacts (43).

Considering contact resistance the R_1 values are in every case lower than the corresponding values of R_2. The ratio R_2/R_1, can be considered a measure of layer-formation on the contact surfaces due to evaporation of material at contact opening.

This ratio is 66 for Ag 1000, 84 for AgNi 10; AgCdO 12, 34; AgC 3, 2.

Quality of Solder of Contact and Support Metal

Some properties of contact materials, such as weld strength and erosion are influenced by occurring soldering imperfections of the contact and support metal (44). The erosion rate is increased by the factor of three due to imperfect soldering. Contacts having a soldered area of less than 70% were unsoldered totally or partly during operation.

Unsoldering rate of contacts should not exceed the value of 10% at sufficient quality of solder.

Assessment of Contact Properties

For a reasonable assessment of two or three different contact properties a standardized $(F + \Delta V)$ or a standardized $(F + \Delta V + R)$-value are recommended. A contact material is suitable for a special application only if the single property values do not exceed previously fixed upper limits, in which case the sum of the standardized values should be minimum.

Depending on the particular application the erosion rate is required to be as low as possible in this way improving the service life of the contacts, i.e. number of operations. On the other hand priority may be given to a low weld strength when a high making capacity of the contactor is considered.

By the pre-selection of contact materials by means of the described testing apparatus both testing time and costs are consid-

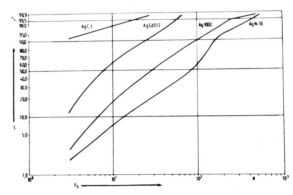

FIGURE 6 Cumulative frequency of weld strength of contact materials.

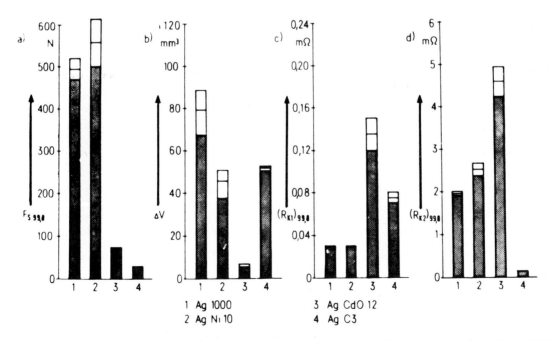

1 Ag 1000
2 Ag Ni 10
3 Ag CdO 12
4 Ag C3

FIGURE 7 $F_{99,8}$-values of weld strength, volume losses ΔV; $R_{1(99,8)}$ and $R_{2(99,8)}$-values of contact resistance of contact materials.

erably reduced. The final testing of the contact material is performed in a production type contactor.

References

1. Holm, R., Electric Contacts Handbook, Springer-Verlag, Berlin 1958.
2. Keil, A., Werkstoffe für elektrische Kontakte, Springer-Verlag Berlin 1960.
3. Schreiner, H., Pulvermetallurgie elektrischer Kontakte, Springer-Verlag Berlin 1964.
4. Shibata, A., Silver-Metal Oxide Contact Materials by Internal Oxidation Process, Proc. 7th Intern. Conf. Electr. Contact Phen., Paris (1974) pp. 214–220.
5. Slade, P. G., C. A. Andersson and R. Kossowsky, The Use of Ag-W-CdO an Ag-Si$_3$N$_4$ as Contact Materials, Proc. Holm Sem. Electric Contacts (1975) pp. 99–104.
6. Schreiner, H., Two Layer Compact-Sinter-Infiltration Technique for Producing Contact Materials for Power Engineering Purposes, Powd. Met. Int. 7 (1975) pp. 21–24.
7. Hausner, H. H., "Modern Developments in Powder Metallurgy," Vol. 4 Plenum Press, New York/London (1971) pp. 23–28.
8. DBP 1.029.571 v. 13. 4. 1961.
9. Stolarz, S., Materialy na styki elektryczne, Wydawnicmwa Naukowo-Techniczne, Warszawa 1968.
10. DBP-Anmeldung 2.446.698 v. 30. 9. 1974.
11. Kosco, J. C., The Effects of Electrical Conductivity and Oxidation Resistance on Temperature Rise of Circuit-Breaker Contact Materials, IEEE Trans. Parts, Materials and Packaging Vol. PMP-5, No. 2, June 1969, pp. 99–103.
12. Pedder, D. J. and F. S. Brugner, The Effect of Density and Oxide Particle Size on the Electrical Conductivity of Silver-Cadmium Oxide Electrical Contact Material, Proc. Holm Sem. Electric. Contacts (1975) pp. 47–52.
13. Merl, W., Weld Tests on Silver-Contacts with the ASTM Weld Test Fixture, ASTM-Committee B.4 Meeting, New York (1965).
14. Hyzer, W. G. and K. Le Baron, Weld Signatures Reveal New Information on how Electrical Contacts Weld, 5. Intern. Conf. on Electric Contacts, München (1970), pp. 158–161.
15. Geldner, E., W. Haufe, W. Reichel und H. Schreiner, Prüfschalter zur Messung der Schweißkraft von Kontaktwerkstoffen für die Starkstromtechnik, ETZ-A, Bd. 92 (1971) S. 637–642.
16. Geldner, E., W. Haufe, W. Reichel und H.
Schreiner, Ursachen der Schweißbrückenbildung und Einflüsse auf die Schweißkraft elektrischer Kontaktstücke in der Energietechnik. ETZ-A, Bd. 93 (1972) S. 305–306.
17. Hueber, B. F., Joule Heating and Development of Temperature in a Symmetric Metallic Current-Constriction, Proc. 6th Int. Conf. Electr. Cont. Chicago 1972, S. 31–39.
18. Geldner, E., W. Haufe, W. Reichel und H. Schreiner, Schweißkraft von Reinsilber, Reinkupfer und verschiedenen Kontaktwerkstoffen auf Silberbasis, ETZ-A, Bd. 93 (1972) S. 216–220.
19. Haufe, W., W. Reichel, H. Schreiner und R. Tusche, Einfluß der Schaltzahl and Polarität des Prüfstromes auf die Statistik der Schweißkraftwerte von Reinsilber bei synchronem und asynchronem Schliepen der Kontaktst., Bull. SEV 63 (1972) 461–467.
20. Haufe, W., W. Reichel, H. Schreiner und R. Tusche, Modelluntersuchungen zur statistischen Verteilung der Schweißkraftwerte von Reinsilber, Bull. SEV 63 (1972) S. 1033–1036.
21. Haufe, W., W. Reichel, H. Schreiner und R. Tusche, Einfluß der Prelldauer und der Kontaktkraft auf die Statistik der Schweißkraftwerte von Ag1000 und AgCdO10, Bull ASE 64 (1973) S. 500–504.
22. Turner, C. and H. W. Tuner, Minimum Size of Silver Based Contacts Proneto Dynamic Welding, Proc. 7th Intern. Conf. Electric. Contact Phen., Paris (1974) pp. 163–167.
23. Geldner, E., W. Haufe, W. Reichel und H. Schreiner, Schweißkraft verschiedener Kontaktwerkstoffe beim dynamischen und statischen Offnen der Kontaktstücke, Bull ASE/UCS 65 (1974) S. 236–240.
24. Schröder, K. H. und E. D. Schulz, Einschaltverschweißen von Kontakten bei verschiedenen Prelldauern, ETZ-B 27 (1975) S. 213–215.
25. Melaschenko, I. P., Kontakte in der Elektrotechnik; Verschleißfeste Unterbrecherkontakte für Niederspannungsschaltgeräte. Akademie-Verlag, Berlin, 1965, S. 103–110.
26. Erk, A. und K.-H. Schröder, Über den Materialverlust homogener und heterogener Kontaktwerkstoffe für Schaltgeräte mit magnetischer Lichtbogenbelastung. ETZ-A, Bd. 89 (1968) S. 373–377.
27. Belkin, G. S., Methode zur Berechnung der Erosion von Starkstromkontakten bei der Lichtbogeneinwirkung Elektrichestvo (1972) 1, S. 61–65.
28. Haufe, W., W. Reichel und H. Schreiner, Abbrand verschiedener WCu-Sinter-Tränkwerkstoffe unter Öl bei hohen Strömen Z. f. Metallkunde 62 (1971) S. 592–595.
29. Haufe, W., W. Reichel und H. Schreiner, Ab-

brand verschiedener WCu-Sinter-Tränkwerk-stoffe an Luft bei hohen Strömen, Z. f. Metall-kunde 63 (1972) S. 651–654.

30. Schröder, K. H. und E. D. Schulz, Der Einfluß des Trennaugenblickes auf den Lichtbogenab-brand öffnender Kontaktstücke beim Wechsel-stromschalter, Metall 28 (1974) S. 463–468.

31. Schröder, K. H. und E. D. Schulz, Über den Einfluß des Herstellverfahrens auf das Schalt-verhalten von Kontaktwerkstoffen der Energie-technik, 7. Int. Tagg. über elektrische Kontakte, Paris 17.-21.6.1974, S. 38–45.

32. Merl, W. und W. Siegmar, Der Widerstand me-tallisch reiner und fremdschichtbehafteter Kon-takte und seine statistische Beschreibung, 4. Int. Tagg. über elektrische Kontakte, Swansea, 15.-18.7.1968, S. 50–53.

33. Dietrich, B., Zum Verhalten geschlossener Kon-taktstücke mit Fremdschichten bei Stromfluß, 5. Int. Tagg. über elektrische Kontakte Mün-chen, 4.-9.5.1970, Bd. 1, S. 19–22.

34. Merl, W. und M. Mittmann, Eine Apparatur zur wiederholten Messung und Registrierung des Kontaktwiderstandes für statistische Untersu-chungen, 5. Int. Tagg. über elektrische Kontakte München, 4.-9.5.1970, Bd. 1, S. 320–323.

35. Mano, K. und T. Oguma, The frequency charac-teristics of sliding precious metal contact noise, 5. Int. Tagg. über elektrische Kontakte, München, 4.-9.5.1970, Bd. 1, S. 297–300.

36. Tittes, E., Über die Anwendung statistischer Methoden auf die Auswertung von Versuchen mit elektrischen Kontakten, 7. Int. Tagg. über elek-trische Kontakte, Paris, 17.-21.6.1974, S. 326–332.

37. Neumeyer, V., Kontaktwiderstandsänderungen in isolierstoffgekapselten Schaltkammern, 7. Int. Tagg. über elektrische Kontakte, Paris, 17.-21.6.74, S. 404–409.

38. Bär, G., Ermittlung und Aussagekraft statis-tischer Kontaktwiderstandswerte, Elektrie 28 (1974) S. 375–377.

39. Schreiner, H. und W. Haufe, Messung der Schweißkraft, des Abbrandes und des Kontakt-widerstandes mit einem Prüfschalter zur Beur-teilung von Kontaktwerkstoffen für die Ener-gietechnik, Z. f. Werkstofftechnik, to be pub-lished.

40. Koepke, B. G. und R. I. George, A Study of Welding of Medium Energy Electrical Contacts, 6. Intern. Conf. on Electr. Contact Phenomena (1972) pp. 15–24.

41. Farral, G. A., Arcing Phenomena at Electric Contacts, Proc. Holm. Sem. Electr. Cont. Phe-nom. (1969) pp. 119–144.

42. Kossowsky, R. and P. G. Slade, Effect of Arcing on the Micro Structure and Morphology of Ag-CdO-Contacts, 6. Intern. Conf. on Electr. Contact Phenomena (1972) pp. 117–127.

43. Shen, Y. S. and R. H. Crock, A Study of the Erosion Modes of Ag-CdO10 Contact Material, Proc. 7th Intern. Conf. Electr. Contact Phen., Paris (1974) pp. 31–37.

44. Schreiner, H., Güte der Lötung bzw. Schwei-ßung von Kontaktstücken auf dem Trägerme-tall-Prüfung und Beurteilung nach Beschalten im Prüfschalter, Vortrag gehalten vor der Ar-beitsgemeinschaft "Kontaktverhalten und Schalten" des VDE, Karlsruhe, 1.10.75.

Magnetic Behavior of High Density P/M Bodies

S. ISSEROW* AND H. P. HATCH*

ABSTRACT

Magnetic permeability of high density P/M bodies was investigated for possible application to nondestructive characterization of residual porosity. The same iron base bodies were used for measurements of electrical resistivity and ultrasonic velocity. The a-c permeabilities plotted against density showed considerable scatter whereas similar plots of resistivity and ultrasonic velocities showed consistent linear relationship, as previously reported.

Introduction

The current trend in powder metallurgy is away from the conventional pressed-and-sintered parts with density about 85% of theoretical to parts having densities above 90% or 95%.[1] Densities close to theoretical are essential if P/M parts are to be used in critical dynamic components. The dynamic mechanical properties such as toughness and fatigue are highly sensitive to density.[2-4] Residual porosity severely impairs these properties, which fall off sharply as the porosity increases. Hence, interest is strong in test methods to qualify P/M parts for critical applications by measuring the extent of porosity and establishing that it does not exceed the porosity tolerable for the application.

The physical properties usually providing the basis for nondestructive testing show a relationship with density (or porosity) that approaches linear, especially as the porosity disappears. Among such properties are electrical resistivity and ultrasonic velocity. The static mechanical properties (yield strength, tensile strength, modulus) similarly approach a linear relationship with density. In contrast, the dynamic properties depart strongly from linearity and simulate an exponential relationship. It is obvious that the application of high density P/M parts would be advanced by identification of a physical property having similar nonlinear sensitivity to density and lending itself to evaluation of a part, preferably by a rapid method adaptable to production rates.

A clue to a suitable property was suggested by the doctoral dissertation of Youssef.[5] His work on iron and nickel compacts included magnetic measurements which showed that at low magnetic fields the magnetization and permeability increase steeply with density. This sensitivity was related by Youssef to the effect of pores on reversible displacement of domain boundaries.

A program was therefore undertaken to define the magnetic behavior of a set of ferrous materials, representing a range of densities. Magnetic measurements were preformed with alternating rather than direct

*Army Materials and Mechanics Research Center
Watertown, Massachusetts

current (see Youssef's measurements) to obtain data for a method applicable with reasonable speed in a production line.

Procedures and Results

Our technical report[6] provides details regarding the following: preparation of samples, method of magnetic measurements to obtain permeability data, processing of data to obtain true permeability as a function of density for fixed magnetic fields, and the data plots obtained by these means. The principal features are summarized here.

Square bars, 0.30 or 0.39 inch on a side, were prepared by the various means needed to achieve densities ranging from 80 to 100% of theoretical. All the bars were prepared from high purity atomized iron powder (A. O. Smith-Inland grade 300M) without any carbon additions. Full density was achieved by hot forging. The lower densities resulted from various combinations of pressing and sintering conditions. A comparison set of full density samples was also prepared from wrought stock (Armco iron).

The electromagnetic measurements were made at a frequency of 25Hz. The basic electrical quantities measured were the current in the primary coil, from which the applied field H_O is deduced, and the voltage induced in the secondary coil, related to the induced magnetism B. The actual field H is obtained by correcting H_O for the demagnetizing field created by the sample. The apparent permeability μ' for these a-c measurements is B/H and is used to calculate the true or corrected permeability μ or μ_{rel} taking into account the effects of frequency and the electrical resistivity of the sample. For the latter, the resistivity of each sample was measured and the linear relationship to density or porosity was confirmed.

In the sequence outlined above, the electromagnetic measurments first give a family of plots of apparent permeability μ' versus actual field H with density as the parameter. Correction of the permeabilities leads to similar plots of μ_{rel} versus H. This family of plots is now used to obtain a plot of μ_{rel} versus density at fixed field H, set here at 2.5 and 12.5 Oersted. The final plots of μ_{rel} versus density represented the objective of this program. These plots showed a trend for permeability to increase with density but the scatter was too great for use of permeability to characterize density or porosity. The scatter even prevents a statement regarding linearity of the relationship between permeability and density. Similar scatter was found in plots (at 12.5 Oersted) of the apparent permeability μ', the quantity that would show up directly (without a-c corrections) in a practical test of a component.

A check of the samples by the ultrasonic techniques established by Brockelman[7] verified the reliability and consistency of the test samples. The ultrasonic velocities in directions parallel and perpendicular to the direction of pressing of the compact showed excellent linear correlation with the density as reported previously. Ultrasonic measurements were thus shown to be superior to the permeability measurements for detection of density differences. Conceivably permeability is more sensitive to other factors which can be viewed either as undesirable interferences or as features whose characterization may be sought. Such factors may include grain size, pore size, pore morphology, and oxide(s) either at the surface or inside a compact.

References

1. *The Trend: Denser, Larger Parts.* Metal Progress, v. 105, no. 4, April 1974, p. 92.
2. Squire, A. *Density as a Criterion of the Mechanical Properties of Iron Powder Compacts.* Watertown Arsenal Laboratory, Experimental Report No. WAL 671/16,

October 31, 1944; also, *Density Relationship of Iron Powder Compacts.* Trans. AIME, v. 171, 1947, p. 485-505. Squire's plots have been reproduced in various standard treatments of powder metallurgy including Goetzel, W. D. Jones, and Hirschhorn.

3. Jenkins, I. *Some Aspects of Residual Porosity in Powder Metallurgy.* Powder Metallurgy, v. 7, no. 13, 1964, p. 68-93.

4. Kaufman, S. M., and Mocarski, S. *The Effect of Small Amounts of Residual Porosity on the Mechanical Properties of P/M Forgings.* International Journal of Powder Metallurgy, v. 7, no. 3, 1971, p. 19-30.

5. Youssef, H. *Etude des proprietes magnetiques des metaux ferromagnetiques frittes et contribution a l etude de leurs proprietes mecaniques et electriques,* Metaux. v. 45, 1970, p. 99-121 and p. 140-153; see Chemical Abstracts, v. 73, 1970, 112137p.

6. Isserow, S. and Hatch, H. P., *Magnetic Behavior of High-Density Powder Metallurgy Bodies. AMMRC TR 76-10, March 1976.*

7. Brockelman, R. H., *Dynamic Elastic Determination of the Properties of Sintered Powder Metals in Advanced Experimental Techniques in Powder Metallurgy,* edited by J. S. Hirschhorn and K. H. Roll, Plenum Press, N. Y., 1970.

SECTION XII
Metallographic Preparation and Microstructure

SELECTION AND PREPARATION OF METALLOGRAPHIC SPECIMENS

by James Marsden

INTRODUCTION

Metallography is the science concerned with the analysis of the constitution and structure of metals and alloys as revealed by the microscope. Metallographic analysis should be attempted only with carefully selected and properly prepared specimens.

This text will cover in detail the selection and preparation of specimens for microexamination. Although there may be some phases in the procedures which could be related to both wrought and powder metallurgy, most of the techniques that will be discussed are unique to metal powders and parts produced from these powders.

The major difference between parts made from metal powders and those made from wrought metal is the voids which exist in the metal powder parts. Since parts made from metal powders are not usually compacted to full density, they contain voids which are commonly called pores. If moisture is entrapped in these pores during preparation of the specimen, it could bleed out during etching and cause a staining reaction, masking the true microstructure of the metal. The techniques described in this report are unique to the metallography of metal powders. For standard metallographic procedures, refer to Kehl's, "The Principles of Metallurgical Laboratory Practice."

PROCEDURES

SPECIMEN SELECTION

A specimen for metallographic examination should be characteristic of the material or problem in question. If the specimen is not representative, then neither will the microstructure be. For example, if a sample of powder is to be examined, the sample should be mixed thoroughly before mounting to guarantee a better and more equal distribution of particle size and shape. In the examination of parts that have failed in operation, the section adjacent to or including the region of failure should be selected if possible. When a part is subjected to furnace atmospheres during a sintering or heat treating cycle, an entire cross section should be selected to determine what effect, if any, the atmosphere had on each surface of the part. If a part has blistered during the heating cycle, the region containing the blister should be prepared, etc.

Source: Technical Bulletin D178, Hoeganaes Corp., Oct 1972

403

Figure 1

Macro-section of specimen on which a coolant was applied during sectioning 12X

Figure 2

Macro-section of specimen sectioned without using a coolant   12X

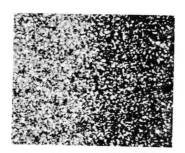

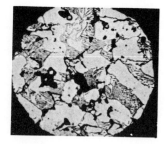
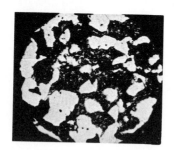

Figure 3

Top: Low magnification micrograph (100X) showing a normal and burned cross section which occurred during sectioning. Left: Normal microstructure of pearlite and ferrite. Right: Burned region shows martensite and ferrite

Figure 4

Condensing unit used to wash porous specimens

CUTTING SPECIMENS

After the specimen has been selected, it will most likely have to be sectioned to expose the surface to be examined, such as a cross section, or to reduce it to a workable size or both.

Specimens are cut with a hacksaw, band saw or abrasive wheel usually referred to as a cut-off wheel. If the material is relatively soft, such as pure iron, bronze or brass, a hacksaw should be sufficient. However if the material is somewhat harder, a band saw or cut-off wheel may be required.

When power tools are used, it is advisable to use a liquid coolant while cutting to help avoid overheating the specimen. Since the microstructure of a material is often a function of the temperature it has experienced, overheating may introduce microstructural changes and thereby introduce errors in the metallographic analysis. Figures #1 and #2 show the difference between a specimen sectioned using a coolant and a specimen which has been overheated or burned. Figure #3 shows a microstructural transformation due to overheating during cutting.

WASHING SPECIMENS

The specimens should be washed after sectioning to remove any foreign material which may have infiltrated the pores. The two units which are successful in removing contaminants from the pore regions are the extraction condenser and the ultrasonic cleaner.

The Extraction Condenser

The extraction condenser is more efficient and the least expensive of the two. This unit consists of a flask, extraction cup siphon and a condensing unit which fits on top of the flask (see Figure #4). A solvent, such as acetone, is placed in the flask and the parts to be washed are placed in the siphon cup. A cold water line is connected to the condensing coil and the flask is heated. The solvent is evaporated and when the gas comes in contact with the cold condensing coil, the solvent condenses and drips into the siphon cup. When the siphon cup becomes filled to a predetermined level, it will empty. This recycling process allows a continual flow of clear solvent over the specimen while the oil or foreign material removed from the pores remains at the bottom of the flask.

The Ultrasonic Cleaner

The ultrasonic cleaner consists of a power supply and a small tank which holds the solvent bath. The power source sets up high frequency waves in the bath. These waves force the solvent into the pores of the specimen, thus removing any foreign substance from these regions. However, since the specimen is placed directly into the bath, most of the washing operation is taking place in a contaminated solution.

MOUNTING SPECIMENS

Specimens are usually mounted if they are too small to be safely held by hand during preparation and if the edges must remain flat for micro-examination. There are several types of mounting media on the market. The application and material to be mounted play a major part in determining the medium to be used. Some of the more common media application and problems that may be encountered are:

Figure 5

Shows overmixed epoxide with air bubbles with the resulting cured micromount.

Figure 6

Shows epoxide mounting media properly mixed and the resulting clear mount.

Figure 7

Mounting press by Buehler Ltd.

Figure 8

Photograph showing bakelite mounts. (a) Mounted using proper heat but pressure was not maintained. (b) Mounted using the proper pressure but low temperature (170°F) and (c) Mounted using proper temperature and pressure.

Epoxide

Epoxide comes as two solutions: epoxide resin and the catalyst which accelerates the hardening reaction. The two liquids should be mixed according to the specifications supplied by the manufacturer. However, when mixing together, special care must be taken so that the solution is not overmixed. Overmixing can introduce an excessive amount of air into the solution and cause bubbles to form (see Figure #5).

Epoxide resin is the most desirable medium for mounting samples of metal powder particles. Since it is liquid before the curing cycle, each individual particle will be coated with resin assuring a better bonding action between particle and resin.

Epoxide is also excellent for infiltrating the pore regions when preparing low density specimens. Using a glass plate and mounting mold (a section of aluminum or copper tube is used as a mold), first coat the glass plate and mold with a release agent so that the mount can be ejected easily from the mold after curing. Then place a thin layer of resin in the mold, add the specimen and place the unit in a vacuum for approximately 10 minutes. Remove the specimen from the vacuum and add enough resin to the mold to produce a mount that will be comfortable to handle during the preparation stages. After the mount has cured for a period of 6 to 8 hours at room temperature (the curing time can be reduced by applying heat as specified by the manufacturer), the mount can be ejected from the mold and is ready for processing. If the directions are followed properly, the mount should be hard, translucent, and bubble free as shown in Figure #6.

Bakelite

Bakelite usually comes in powder form or as premolds. It is much more convenient and not as time consuming to use as epoxide. However, there must be equipment available to heat and to apply pressure during the mounting operation. There are mounting presses and molds available specifically for this application (see Figure #7).

The specimen is placed in the mold and powder poured around the specimen. The mold is then placed in the press and heat and pressure are applied to it. It is extremely important that the molds be heated to the temperature specified by the manufacturer (usually 270 to 300°F) and that the required pressure be maintained during the heat cycle. If the pressure is not maintained, the gases given off by the bakelite at elevated temperatures will remain entrapped in the mold. Figure #8 shows the results of underheating, failure to maintain pressure during the heating cyle, and a mount produced using the heat and pressure as directed.

When mounting powder particles in bakelite, it is advisable to first crush amber bakelite to -100 mesh. Mix the metal powder particles with the crushed bakelite and place the mixture in a mold. Add a softer bakelite for a backing. The amber bakelite is rather hard and brittle and when used alone has a tendency to form cracks.

Since the bakelite is molded using a press and since the mold used is tightly sealed, it is not practical to apply a vacuum. Therefore, the only way bakelite can be infiltrated into the pores is by using the pressure applied during mounting. Unfortunately, the bakelite does not penetrate very deeply and is rather ineffective in closing the pore regions.

Source: Technical Bulletin D178, Hoeganaes Corp., Oct 1972

ROUGH GRINDING

The specimen surface to be examined should be made planar (flat) and free from any mount media that might cover areas of the specimen surface. This can be accomplished by using either a file or a motor driven emery belt, the latter being the fastest and simplest method. When using a motor driven emery belt, overheating the sample can be prevented by using a coolant to reduce the build up of heat created by friction between the emery belt and the specimen.

The grit size of the emery belt to be used will depend upon the amount of material to be removed from the specimen surface. If a specimen requires the removal of a considerable amount of material, it is advisable to use a coarse (80 grit) belt. On the other hand, if only a few thousandths of an inch of surface material is to be removed, a fine (240 grit) belt would be more desirable. The grit size to be used depends largely on the skill and judgement of the operator.

When using the belt grinder, special attention should be given to the amount of play in the belt. If the belt is taut, the specimen surface will be relatively flat. If the belt is loose, it will tend to have a flapping motion thus producing an unevenness of the specimen surface.

Since the motor-driven belt produces an unevenness in the specimen surface a 240 grit belt is usually the finest motor driven belt used. For grinding below 240, hand grinding or use of a lead lap wheel is advisable.

FINE GRINDING

Fine grinding is normally a hand grinding operation. The emery papers used must be of the highest quality, particularly with respect to uniformity of the size of the grit particles.

For hand grinding, emery papers of four grit sizes are used starting at 240 grit and progressing in fineness through 320, 400 and 600 grit. The sample should be held with both hands and approximately equal pressure applied to the overall area of the mount. While applying pressure move the mount across the emergy paper using single motions away from the body. Continue this procedure until the mount is flat and scratches from the previous grit size are eliminated. The sample should be turned 90° when progressing from one grit to another; this will clearly show any scratches left by the previous grit size.

The lead lap can also be used for fine grinding. This method consists of grinding the specimen on a series of lead laps, each lap is impregnated with a suitable emery abrasive of decreasing particle size.

The laps are usually made of a 50-50 lead-tin alloy that is grooved in the form of a spiral (see Figure #9). The spiral is formed opposite to the direction of rotation so that the grooves of the spiral will retain any solutions applied while the wheel is in motion.

There are several advantages to lead lap grinding: the ability to retain a very flat surface from edge to edge during grinding; the amount of disturbed metal at the specimen surface is held to a minimum; and it is good for retaining non-metallic inclusions in the specimen.

Figure 9

Load Lap Wheel
50-50 Lead-Tin Alloy

ROUGH POLISHING

Rough polishing is an intermediate step between fine grinding and fine polishing. Rough polishing is required since the scratches left by fine grinding are too coarse to be removed by the fine polishing step without leaving a layer of smeared or disturbed metal on the polished surface.

There are several media available, such as diamond paste, magnesium oxide, aluminum oxide, etc. that can be used in the rough polishing operation. However, the grit size of the compound, regardless of which one may be selected, usually varies between 1.0 to .3 microns and is used with a nap-free cloth such as, nylon, usually referred to as a "hard" cloth.

There are at least three advantages at this stage to using a nap-free cloth rather than a cloth containing a nap: it will aid the abrasive solution to produce a better cutting reaction, thus removing the metal rather than smearing it; it results in better retention of non-metallic inclusions; and it produces a flatter surface.

Even though use of a nap-free cloth will accomplish these three things, it will not produce a scratch free, mirror finish and therefore is not recommended for use in final polishing, except in special cases where a scratch free surface has to be sacrificed to retain the inclusion material.

One method that has proven very successful uses a wheel covered with a nylon cloth and an aqueous solution containing 1 micron alumina polishing compound (1 part alumina to 4 parts distilled H_2O by volume). The wheel should be rotated at approximately 400 to 500 rpm and rather light pressure applied to the sample during polishing. The solution is applied while the wheel is in motion. This can be accomplished very nicely by using a plastic squirt bottle. This method, if done properly, will remove the scratches left from the fine grinding (600 grit) in approximately 1 to 2 minutes, with a minimum of smearing (disturbed metal) on the specimen surface.

Diamond paste is a good polishing medium and will accomplish the same results as described for alumina but is considerably more expensive.

Source: Technical Bulletin D178, Hoeganaes Corp., Oct 1972

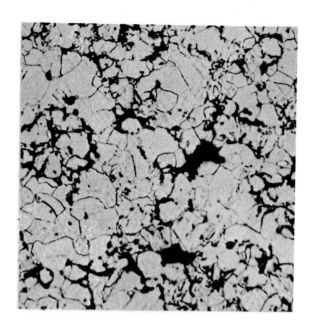

Figure 10

Micrograph shows distortion of microstructure due to presence of smeared
metal. 2% Nital etch. 300X

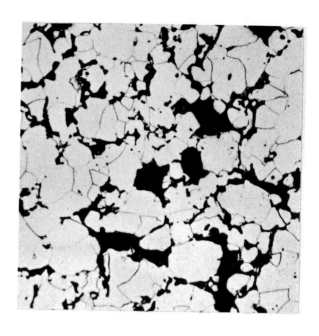

Figure 11

Micrograph shows microstructure after
etching and repolishing several times.
2% Nital etch. 300X

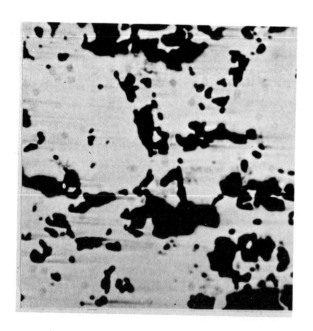

Figure 12

Shows "Comet Tails" produced during polishing operation. Harder constituents do not
polish at the same rate as the base metal.
The above condition results when specimen
is not rotated during polishing.

Figure 13

Same specimen as Figure 12
rotated during polishing operation. 300X

FINE POLISHING

Fine polishing is the final step in specimen preparation and should produce a scratch free surface. There are several types of cloths and polishing compounds that can be used. They must be chosen to accomplish three major objectives: the production of a scratch free surface; the retention of non-metallic inclusions; and the production of a minimum of disturbed metal.

The first objective is to produce a scratch free surface. This would suggest that a soft cloth with a very fine grit media would be the most desirable. The second objective is to retain the non-metallic inclusion material. This would suggest a short napped cloth because the shorter the nap of the cloth, the less chance of the nap pulling out the inclusions from the specimen surface. The third objective is to relieve the specimen surface of any smeared or disturbed metal. This would require that the combination of cloth and media produce a cutting action of the specimen surface. A method used to remove the disturbed metal is repeated etching and repolishing on the final wheel. The etching reagent will attack the metal and in turn will create less resistance to the cutting action of the abrasive. However, all the metal that has been attacked by the reagent should be removed in the repolishing operation. If this is not accomplished, the areas where all the metal was not removed will show signs of a heavier attack by the reagent on final etching and thus a non-uniform microstructure.

The amount of disturbed metal is usually determined by the degree of hardness of the material. For instance, a specimen of hard iron-carbon will tend to have less smearing of the surface during polishing than a sample of soft pure iron. Figures #10 and #11 show a comparison of microstructures with and without disturbed metal present.

A combination of a very soft, short nap cloth, such as Astromet, and an aqueous solution of 0.5 micron alumina (1 alumina to 4 distilled H_2O by volume) is recommended for fine polishing. The specimen should be rotated in a direction counter to the rotation of the polishing wheel. This will produce a more evenly polished surface. If the specimen is held stationary, polishing defects known as "comet tails" will form around the "harder" phases within the microstructure. This is due to the difference in the rate of metal removal between the two phases. Figures #12 and #13 show the difference between a sample polished in one direction (showing comet tails) and a sample which has been rotated during polishing.

The final polishing operation should take approximately 3 to 5 minutes of actual polishing time (overpolishing will tend to round the surface of the pores thus destroying the pore structure of the specimen) and produce a scratch free, mirror like finish to the specimen surface. Figure #14 shows the results of overpolishing.

POWDER METAL PARTICLES

Powder metal particles are undoubtedly among the most difficult specimens to prepare. They must be flat to retain any surface condition that may exist and also reveal the clear, true microstructure of the particles. Both of these conditions can be accomplished in a single operation by using the following procedures.

After mounting the powder (as explained on Page 407), omit all rough or fine grinding of the specimen surface because this will cause an undesirable relief of the particles; round the edges of the mount so that the sharp edge does not tear

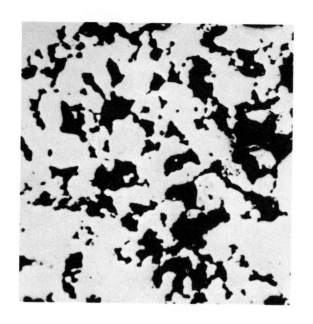

Figure 14

Shows rounding of the pore structure
which results from over-polishing.
300X

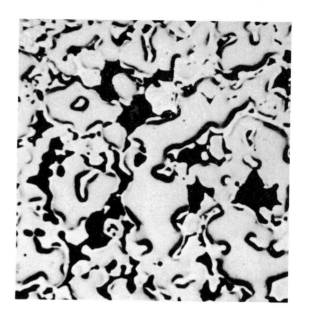

Figure 15

Shows results of over-polishing in
an automatic polisher. 300X

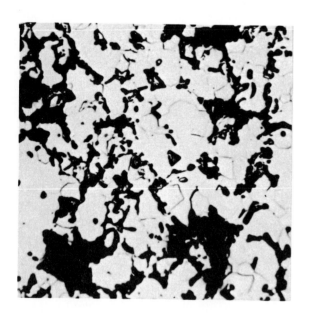

Figure 16

Shows results of polishing in a mild
acidic solution (PH 5-6). 300X

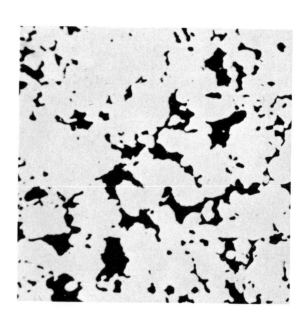

Figure 17

Shows a surface which has been properly
polished in a neutral polishing solution
(PH 7).
300X

the cloth. Go directly to the rough polishing step. The specimens should be polished for approximately two to three minutes (time will vary with the type of material) on a wheel covered with a hard nap-free cloth (such as the Texmet cloth by Buehler) and a 1.0 micro alumina aqueous solution. Over polishing will cause relief. Next, wash the sample thoroughly and continue polishing, this time using a nylon cloth and the 1.0 micron alumina solution. Apply a very light pressure for a very short period of time (approximately 30 seconds). Extended time periods will cause rounding of the particles. The final polishing step requires a wheel covered with an "Astromet" cloth and a .05 micron alumina aqueous solution (1 part alumina to 4 parts H_2O), again apply a very light pressure. Inspect the specimen by microscope periodically to determine when the specimen is suitable for examination.

AUTOMATIC POLISHING

The automatic polisher is intended for the rough and fine polishing steps of specimen preparation. The automatic polisher works by rotating the sample across a polishing cloth in an abrasive solution by a vibrating motion. This vibrating action must propel the samples in a smooth rotating motion to produce a flat, evenly polished surface. If the vibrating action causes bouncing of the specimen, it will produce an uneven polishing of the specimen.

Aqueous abrasive solutions are usually used in this method of polishing, therefore particular care should be taken with materials that corrode easily. Special care should be taken to control the pH of the solution (pH should be approximately 7 for neutral solution) so that an etching reaction does not take place during the polishing operation. Figures #15 through #17 show microstructures which are (15) overpolished, (16) etched during polishing and (17) polished properly.

There is undoubtedly a vast combination of abrasives and cloths that can be used in the automatic polishers. However, Page 417 of the text contains several designated materials and the combination of cloths and abrasives which have proven successful in their preparation.

Figure #18 shows the Vibromet polisher manufactured by A. B. Buehler, Inc., one of several units available for this type of polishing.

Figure 18

Vibromet Polishing Unit
by Buehler Ltd.

Source: Technical Bulletin D178, Hoeganaes Corp., Oct 1972

ETCHING

The etching of the specimen surface after the final polishing operation reveals the structure of the metal for observation by microscope by attacking certain constituents preferentially within the metal structure.

The composition of the reagent will vary depending on the composition of the metal or alloy and the treatment which the specimen has experienced.

The etching reagents used for metallographic specimens are usually composed of organic or inorganic acids or alkalies. The general behavior of these etchants is related to one of the three following characteristics:

1. Hydrogen ion concentration (acidic solution)
2. Hydroxyl ion concentration (basic solution)
3. The ability to preferentially stain one or more of the structural constituents

For instance, the hydrogen ion solution should be used for etching iron base materials whereas, the hydroxyl ion solution would be used to etch copper, etc. Etching not only requires a knowledge of what reagent should be used for a specific metal, but also a knowledge of what constituents will most likely be present in the microstructure of a metal or alloy.

For example, if a microstructure is overetched, certain characteristics of the structure may be attacked too vigorously and thus removed from the structure. However, if the sample is underetched, there may be constituents in the structure that are not attacked and thus are not visible for observation.

Pages 415 through 417 contain photomicrographs of several different materials, such as iron, iron-carbon, iron-copper-carbon, etc. in the following conditions:

1. Underetched
2. Normal Etch
3. Overetched

Also listed is the type of reagent used for each specimen.

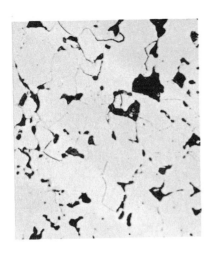
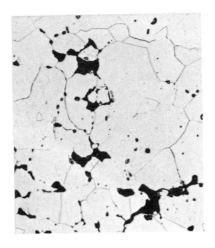

Underetched

Very poor definition of
grain boundary regions.

Good Etch

Easy to distinguish grain
boundary regions, etc.

Overetched

Shows very heavy attack in
grain boundary regions and
also pitting within the grains.

Iron: Sintered 30 minutes at 2050°F in an atmosphere of dissociated
ammonia. Microstructure is ferrite. 300X Magnifications. 2% Nital etch.

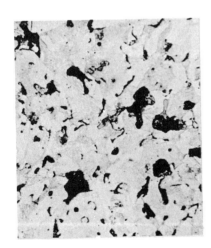
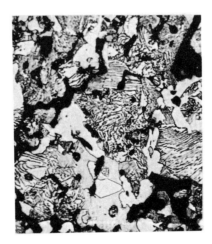

Underetched

Very difficult to dis-
tinguish between the
ferritic and pearlitic
regions.

Good Etch

Shows very good detail in
the pearlitic regions and
also good definition be-
tween the phases present.

Overetched

Heavy attack on the pearlitic
and boundary regions, causes
distortion of the true microstructure.

Iron-Carbon: Sintered 30 minutes at 2050°F in an atmosphere of dissociated
ammonia. Microstructure shows a pearlitic matrix with some ferrite grains
present. Combined carbon level is approximately .60%. 300X Magnifications.
2% Nital etch.

Source: Technical Bulletin D178, Hoeganaes Corp., Oct 1972

 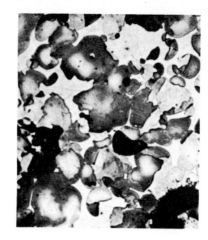 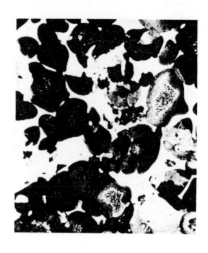

Underetched

Very poor detail show-
ing depth of Cu diffu-
sion and also poor de-
finition between Fe
and (liquidus) Cu
boundaries

Good Etch

Shows depth of Cu diffu-
sion into the Fe and also
rounding of the Fe parti-
cles indicating some of
the Fe is being dissolved
in the liquidus Cu

Overetched

Again heavy attack on the
iron-copper regions caused
distortion of the true microstructure

Iron-Copper: Iron + 20% copper sintered 30 minutes at 2050°F in an
atmosphere of dissociated ammonia. Microstructure shows two conditions
exist - (1) copper has diffused into the iron and (2) some of the iron
has been dissolved in the liquidus copper. The molten copper can dissolve approximately 3% iron at sintering temperature. 300X Magnifications.

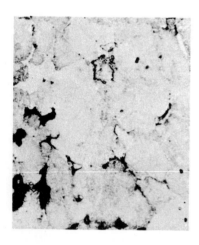  

Underetched

Difficult to distinguish
various constituents
within the structure,
lacks detail.

Good Etch

Very good detail, easy to
distinguish different
phases present.

Overetched

Heavy attack to certain constituents within the structure, creates a very distorted
image.

Iron-Copper-Carbon: Fe + 5% Cu + 1% C, sintered 30 minutes at 2050°F
in dissociated ammonia atmosphere. Microstructure shows a pearlitic
matrix with some copper diffused into the pearlite. Also note the
free copper in the boundary regions. 300X Magnifications.

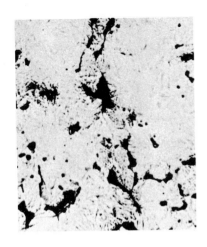 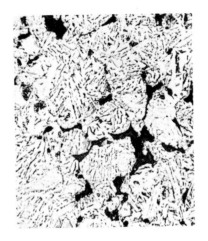 

<u>Underetched</u>

Very hard to distinguish between different constituents present in the microstructure.

<u>Good Etch</u>

Shows good detail and definite differences between the phases present.

<u>Overetched</u>

Heavy attack of certain phases causes distortion within the microstructure.

Low Alloy: Fe, Ni, Mo, Mn, C sintered 30 minutes at 2050°F in an atmosphere of dissociated ammonia. Microstructure shows a matrix of ferrite and very coarse pearlite. Approximately .40% combined carbon present. 300X Magnifications.

CLOTHS & ABRASIVES - "AUTOMATIC POLISHER"

Rough Polishing

Step #1 - Use a Texmet cloth with 1.0 micron alumina aqueous solution. This operation normally will take approximately 15 to 20 minutes. Specimens should be checked by microscope periodically to prevent over polishing and/or possible etching.

Step #2 - Use a Texmet cloth with .3 micron alumina aqueous solution, follow same procedures as listed in Step #1.

Final Polishing

Use a micro cloth with 0.05 micron alumina aqueous solution. Considering Step #1 and #2 of the rough grinding procedures were carried out properly, this operation will normally take approximately 15 minutes. The specimen should be checked periodically to prevent over polishing and/or etching of the specimen.

Source: Technical Bulletin D178, Hoeganaes Corp., Oct 1972

The following is a list of materials which have proven successful for the previously described applications.

Mounting Materials

Bakelite (Red)	A. B. Buehler, Ltd.	No. 20-3200
Bakelite (Amber)	A. B. Buehler, Ltd.	No. 20-3500
Epoxide Resin	A. B. Buehler, Ltd.	No. 20-8130
Epoxide Hardener	A. B. Buehler, Ltd.	No. 20-8132
Release Agent	A. B. Buehler, Ltd.	No. 20-8185

Rough Grinding (belt grinder)

Silicon Carbide Abrasive Belts	A. B. Buehler, Ltd.	No. 16-5100

Fine Grinding

Emery Papers	A. B. Buehler, Ltd.	No. 30-5160	240 grit
Emery Papers	A. B. Buehler, Ltd.	No. 30-5160	320 grit
Emery Papers	A. B. Buehler, Ltd.	No. 30-5160	400 grit
Emery Papers	A. B. Buehler, Ltd.	No. 30-5160	600 grit

Rough Polishing

Abrasives

Alumina (alpha)	A. B. Buehler, Ltd.	No. 40-6310	1.0 micron
Alumina (alpha)	A. B. Buehler, Ltd.	No. 40-6305	0.3 micron

Cloths

Texmet	A. B. Buehler, Ltd.	No. 40-7618 8" polishing wheel
Texmet	A. B. Buehler, Ltd.	No. 40-7628 8" automatic polisher
Nylon	A. B. Buehler, Ltd.	No. 40-7058 8" polishing wheel

Final Polishing

Abrasives

Alumina (gamma)	A. B. Buehler, Ltd.	No. 40-6301	0.05 micron

Cloths

"Astromet"	Precision Scientific	No. 87944 8" polishing wheel
Micro-Cloth	A. B. Buehler	No. 40-7228 automatic polisher

Microstructure of Ferrous Powder Metallurgy Alloys

By Athan Stosuy*

IRON AND STEEL powder metallurgy parts have been commercially produced for many years with the use of the mixing, pressing and sintering techniques and equipment described on pages 449 to 464 in Volume 4 of this Handbook. Metallography is an important aid to the investigation and control of these techniques. In addition to providing the same kinds of information as for other metals, metallography of powder metallurgy specimens also reveals particle configuration, uniformity of mixing, interparticle porosity and its distribution, degree of particle bonding, and degree of diffusion alloying.

Ferrous powder metallurgy parts are made from iron powders, alloy steel powders, or mixtures of: (a) iron and graphite powders; (b) iron (or steel) and copper powders; or (c) iron, copper and graphite powders (with or without powders of other metals).

Iron Powder. Sintering of plain iron powder sequentially involves the establishment and growth of bonds between the particles of powder at their areas of contact, grain growth and migration of the grain boundaries formed at the bonds, spheroidization of the pores between the particles, and the elimination of small pores (and, possibly, the growth of large pores). The formation of bonds is opposed by residual material from the lubricant, by impurities and surface oxides, and by poor contact.

The density and size of the green compact change during sintering. As the sintering temperature increases, porosity decreases (see micrographs 1648 to 1651) and shrinkage increases. High density and an acceptable distribution of porosity can be obtained by pressing, sintering, re-pressing and re-sintering (see micrograph 1652).

Mixtures of iron and graphite powders are sintered by three mechanisms: (a) establishment and growth of iron-to-iron bonds; (b) diffusion of carbon in, and combination of carbon with, iron; and (c) spheroidization of pores.

The oxide content of the iron is an important factor in the reactivity of the iron with graphite. During sintering, the major portion of the oxide must be reduced by the graphite and the sintering atmosphere before the iron and graphite can combine. To ensure high reactivity with the iron for rapid reduction of oxide and fast carburization, fine graphite powder, free of silicon

carbide and having low ash content, is normally used. Although iron powder of low oxide content is not required, it is important to know the oxide content, because the graphite addition will have to be large enough to allow for reaction with the oxide and loss to the atmosphere and still leave enough to produce the desired iron-carbon alloy.

Apart from bonding, the main factor affecting the properties of a sintered steel part is the amount of combined carbon formed in the steel. Over-all expansion during sintering is directly proportional to the amount of combined carbon. As with rolled steel, the strength of sintered steel increases rapidly with increasing combined carbon content, but a maximum is reached near the eutectoid content of 0.8%; between 0.9 and 1.0%, the strength drops markedly (see micrographs 1653 to 1658). Thus, most sintered steel has a combined carbon content of 0.8 to 0.9%, and a graphite addition greater than 0.9%. Carbon loss is controlled by maintaining the carbon potential of the sintering atmosphere at 0.7 to 0.9%.

As the sintering temperature or time is increased, spheroidization of the pores causes the strength of a compact to continue to increase after carburization of the iron to an all-pearlite structure is complete, and the resulting densification of the compact causes a reduction in the over-all growth of the part (see micrographs 1660 to 1667).

Mixtures of Iron and Copper Powders. The sintering of iron-copper mixtures involves (a) solid bonding of iron to iron, (b) solid bonding of copper to iron, (c) melting of copper, (d) solution and diffusion of copper in solid iron, and (e) solution and precipitation of iron in liquid copper.

At the usual sintering temperature of 2000 to 2050 F (1093 to 1121 C), 7.5 to 9.0% copper is soluble in iron. However, with this copper content and the usual sintering conditions, some of the molten copper will remain undissolved, or free (see micrographs 1671 to 1674), and can dissolve about 3% iron.

The solubility of copper in iron decreases with decreasing temperature and is less than 0.1% at room temperature. Therefore, copper dissolved in iron at the sintering temperature must precipitate as the compact is cooled, thus hardening it. Faster cooling lowers the temperature at which the precipitate forms, which makes the precipitate finer and increases its hardening effect (see micrographs 1681 and 1682).

The solution of copper in iron causes growth of the compact; the solution of iron in the free copper causes shrinkage. These processes go on simultaneously, with solution of copper predominating only in the early stages of sintering. For example, the addition of 7.5 to 10% copper causes significant amounts of growth, but a 20% addition increases the amount of free copper to the extent that growth is no greater than for a 7.5% copper addition (see micrographs 1671 to 1674).

Mixtures of Iron, Copper and Graphite Powders. The sintering of iron-copper-graphite mixtures involves the mechanisms associated with both iron-graphite and iron-copper mixtures. Diffusion of carbon is usually complete before the melting point of copper is reached. Carbon has little effect on the solubility of copper in solid iron, but it decreases the rate of solution. Thus there is usually more free copper remaining when carbon is present, which acts to reduce the amount of growth.

Sintered alloy steels are made from three types of powders: admixed, semi-alloyed, and fully alloyed.

Sintering of a common admixed iron-copper-nickel-graphite alloy (see micrograph 1694) involves (a) diffusion of carbon; (b) melting, solution and diffusion of copper; and (c) solution of nickel in liquid copper, and solution and diffusion of nickel in solid iron. Diffusion of nickel is comparatively rapid along the surface and along grain boundaries, but sluggish within grains. Nickel often diffuses incompletely, leaving areas of austenite that are visibly nickel-rich. The periphery of the nickel-rich areas sometimes dissolves enough carbon to form martensite during normal cooling from sintering.

"Semialloyed" sintered steel usually contains nickel, molybdenum and copper powders that have been partly alloyed during co-reduction of oxides (see micrograph 1695). The diffusion of copper and molybdenum is relatively fast, whereas the diffusion of nickel is slow. This leaves numerous nickel-rich areas of austenite, sometimes surrounded by high-carbon martensitic areas.

Fully alloyed sintered steel usually is made of atomized alloy steel powder (see micrograph 1696). The steel powder commonly contains nickel, molybdenum and manganese. Copper powder is sometimes mixed with the steel powder to limit shrinkage of the compact.

Hardness values were measured, and are reported here, in Rockwell B.

*Manager of New Product Development, Hoeganaes Corp., a subsidiary of Interlake, Inc.

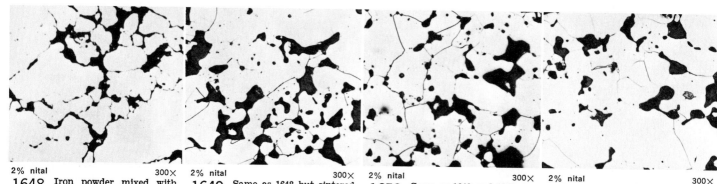

2% nital   300×

1648 Iron powder mixed with ¾% of lubricant and pressed to a density of 6.5 g per cu cm; as-pressed condition. The multigranular particles are completely separated by voids (black). See also micrographs 1649 to 1652.

2% nital   300×

1649 Same as 1648 but sintered in dissociated ammonia for 30 min at 1950 F (1066 C). Many bonds have formed between particles during sintering, but numerous particle boundaries remain. Pores (voids) are angular.

2% nital   300×

1650 Same as 1648 and 1649 except compact was sintered at 2050 F (1121 C). The bonds between the particles are more numerous and extensive than in 1649; some grain growth and spheroidization of pores have occurred.

2% nital   300×

1651 Same as 1648 to 1650 except compact was sintered at 2150 F (1177 C). Only a few particle boundaries remain, and considerable grain growth, grain-boundary migration, and spheroidization of pores have taken place.

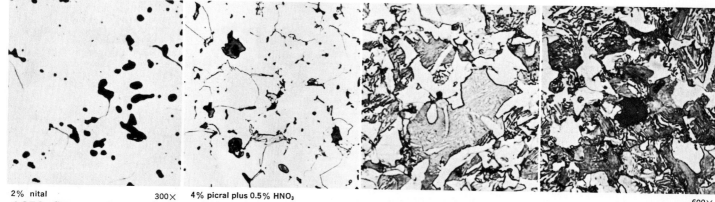

2% nital   300×

1652 Same as 1650 except the compact was re-pressed at 50 tsi (689 MPa) and resintered at 2050 F (1121 C). No particle boundaries remain; some grains have grown beyond the original particle size. The pores have become distinctly spheroidal.

4% picral plus 0.5% HNO₃   600×

1653, 1654, 1655 Iron powder mixed with increasing amounts of graphite (see also micrographs 1656 to 1659) to produce various contents of combined carbon after the mixtures were pressed to a density of 6.3 g per cu cm and sintered in dissociated ammonia for 30 min at 2050 F (1121 C). **Micrograph 1653 (left):** Trace of combined carbon; structure consists of essentially all-ferrite grains; transverse-rupture strength is 59,000 psi (407 MPa). **Micrograph 1654 (center):** 0.4% combined carbon; structure consists of equal amounts of pearlite (mottled) and ferrite (light); transverse-rupture strength has increased to 70,000 psi (483 MPa). **Micrograph 1655 (right):** 0.6% combined carbon; amount of pearlite (mottled) has further increased, raising transverse-rupture strength to 78,000 psi (538 MPa).

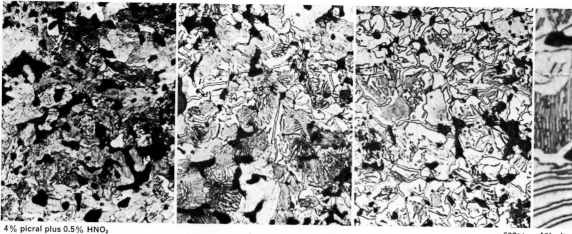

4% picral plus 0.5% HNO₃   600×

1656, 1657, 1658 Continuation of the series that originates, and is described, in 1653 to 1655. **Micrograph 1656 (left):** 0.8% combined carbon; the structure consists entirely of fine pearlite (black areas are pores); transverse-rupture strength has increased to 87,000 psi (600 MPa). **Micrograph 1657 (center):** 1.0% combined carbon, resulting in the formation of free cementite (white) at grain boundaries, which reduced the transverse-rupture strength to 73,000 psi (503 MPa). **Micrograph 1658 (right):** 1.2% combined carbon, with the result that the network of free cementite has become continuous, reducing transverse-rupture strength to 58,000 psi (400 MPa); see micrograph 1659 for a higher-magnification view.

4% picral plus 0.5% HNO₃   2000×

1659 Same specimen as in 1658 but shown at higher magnification, which reveals details of the lamellae of fine pearlite, the continuous network of free cementite at the grain boundaries, and the rounded corners of the pores.

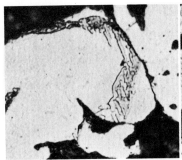

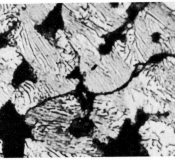

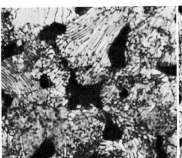

4% picral plus 0.5% HNO₃   800×

1660 Iron powder mixed with 1.25% graphite, pressed to 6.1 g per cu cm, sintered 30 min at 1850 F (1010 C). Slight pearlite formation (0.10% combined carbon); remaining carbon is graphite in pores (dark). Transverse-rupture strength is 20,000 psi (138 MPa).

4% picral plus 0.5% HNO₃   800×

1661 Same as 1660 except the sintering temperature was increased to 1900 F (1038 C), which resulted in combined carbon content of 0.75%, an essentially all-pearlite structure, and an increase in transverse-rupture strength to 51,000 psi (352 MPa).

4% picral plus 0.5% HNO₃   800×

1662 Same as 1660 and 1661 except sintering temperature was increased to 2050 F (1121 C), which resulted in increases in particle bonding and pore spheroidization, and a transverse-rupture strength of 80,000 psi (552 MPa). Combined carbon content, 0.75%.

4% picral plus 0.5% HNO₃   800×

1663 Same as 1660 to 1662 except sintering temperature was increased to 2150 F (1177 C). This further increased particle bonding, pore spheroidization; transverse-rupture strength increased to about 95,000 psi (655 MPa). Combined carbon, 0.75%.

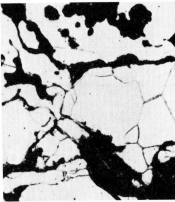

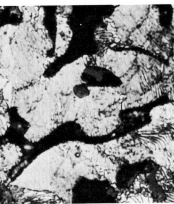

4% picral plus 0.5% HNO₃   800×

1664 Iron powder mixed with graphite, and pressed to a density of 6.1 g per cu cm; as-pressed. No bonding of the multi-granular particles occurred, and no pearlite formed. Combined carbon content is zero; graphite remains in pores (dark). See 1665 to 1667.

4% picral plus 0.5% HNO₃   800×

1665 Same as 1664 except sintered for 5 min at 2050 F (1121 C), which resulted in combined carbon content of 0.70%, an almost all-pearlite structure, some bonding, and transverse-rupture strength of about 61,000 psi (421 MPa). See 1666 and 1667.

4% picral plus 0.5% HNO₃   800×

1666 Same as 1664 and 1665 except sintered for 30 min, which resulted in some spheroidization of the pores and further increases in bonding and transverse-rupture strength (to about 80,000 psi, or 552 MPa). Combined carbon remained at 0.70%. See 1667.

4% picral plus 0.5% HNO₃   800×

1667 Same as 1664 to 1666 but sintered 120 min, resulting in almost complete bonding, increased pore spheroidization, and greater transverse-rupture strength (91,000 psi, or 627 MPa). Strength equals that obtained by sintering 10 min at 2200 F (1204 C).

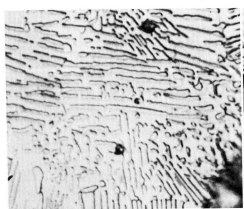

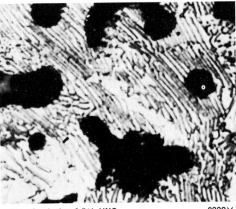

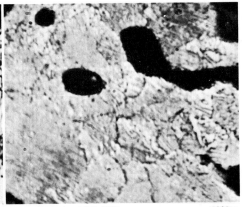

4% picral plus 0.5% HNO₃   2000×

1668 Iron powder mixed with graphite, pressed to a density of 6.1 g per cu cm, sintered 30 min at 2050 F (1121 C) and cooled through the range of 1350 to 1000 F (732 to 538 C) at 3.5 F (1.9 C) per minute. Result is an almost all-pearlite microstructure with coarser spacing than that in 1669 and 1670.

4% picral plus 0.5% HNO₃   2000×

1669 Same as 1668 except cooled from 1350 to 1000 F (732 to 538 C) at 115 F (64 C) per minute. The faster cooling resulted in an almost all-pearlite structure with medium spacing, and in greater transverse-rupture strength (82,000 psi, or 565 MPa, vs 67,000 psi, or 462 MPa, for 1668). Pores are black. See 1670.

4% picral plus 0.5% HNO₃   2000×

1670 Same as 1668 and 1669 except cooled from 1350 to 1000 F (732 to 538 C) at 225 F (125 C) per minute. The further increase in cooling rate resulted in an almost all-pearlite structure with spacing too fine to be resolved, and in an increase in the transverse-rupture strength to 87,000 psi (600 MPa).

Iron-Copper: Effects of Copper Content, Sintering Temperature, Sintering Time

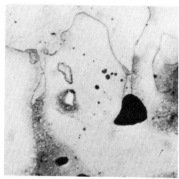

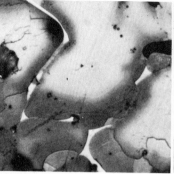

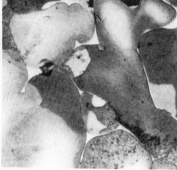

2% nital   700×

1671 Mixture of 98% iron, 2% copper, powders, pressed, and sintered for 30 min at 2050 F (1121 C). Some copper dissolved in the iron (gray areas); little free copper (white) is present. Tensile strength, 29,000 psi (200 MPa); hardness, R$_B$ 4. Pores are black.

2% nital   700×

1672 Same as 1671 except mixture contained 7.5% Cu. Much more copper was dissolved in the iron, resulting in tensile strength of 31,000 psi (214 MPa) and raising hardness to R$_B$ 20. Small areas of free copper are present. See also 1673 and 1674.

2% nital   700×

1673 Same as 1671 and 1672 except mixture contained 10% Cu. Even more copper was dissolved in the iron, increasing tensile strength to 31,500 psi (217 MPa); more free copper is present, decreasing the hardness slightly (to R$_B$ 19). See micrograph 1674.

2% nital   700×

1674 Same as 1671 to 1673 except mixture contained 20% Cu. Still more copper was dissolved in the iron, further increasing the tensile strength (to 34,000 psi, or 234 MPa); much more free copper is present; greatly reducing the hardness (to R$_B$ 12).

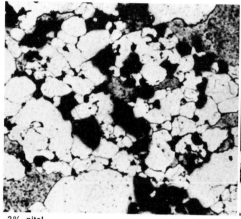

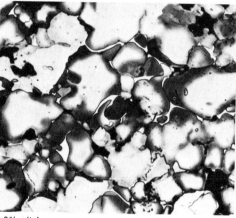

2% nital   300×

1675 Mixture of 92.5% iron and 7.5% copper multigranular powders, pressed, and sintered 30 min at 1950 F (1066 C). At this low temperature, the copper was not melted and the compact had low tensile strength (13,000 psi, or 90 MPa) and hardness (R$_B$ 2.5). See 1676, 1677.

2% nital   300×

1676 Same as 1675 except sintering was done at 2000 F (1093 C), which melted the copper. Some copper was dissolved in the iron (gray areas), which increased the tensile strength of the compact to 32,000 psi (221 MPa) and the hardness to R$_B$ 25. Pores are black. See 1677.

2% nital   300×

1677 Same as 1675 and 1676 except sintering was done at 2050 F (1121 C). More copper was dissolved in the iron, and little free copper remains. The hardness of the compact was increased to R$_B$ 29.5, but tensile strength was about the same as for the compact in 1676.

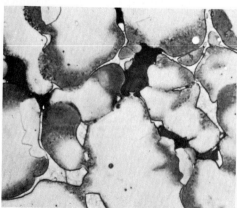

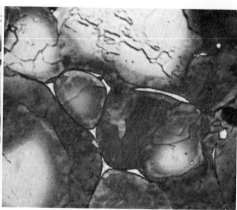

2% nital   700×

1678 Mixture of 92.5% iron and 7.5% copper multigranular powders, pressed, and sintered 5 min at 2050 F (1121 C). Some copper dissolved in the iron (gray areas); much free copper (white) remains. Tensile strength, 15,000 psi (103 MPa); hardness, R$_B$ −6. See 1679, 1680.

2% nital   700×

1679 Same as 1678 except sintered for 30 min. Considerably more copper was dissolved in the iron, which increased the tensile strength of the sintered compact to 31,000 psi (214 MPa), and less free copper remained, which increased the hardness to R$_B$ 20. See 1680.

2% nital   700×

1680 Same as 1678 and 1679 except that the compact was sintered for 60 min, dissolving almost all of the copper. This resulted in further increases in the tensile strength of the sintered compact (to 35,000 psi, or 241 MPa) and the hardness (to R$_B$ 31).

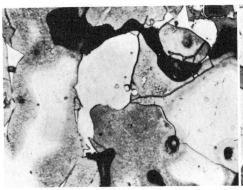

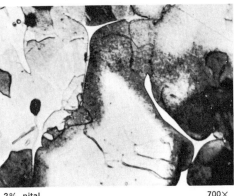

2% nital   700×

1681 Mixture of 92.5% iron and 7.5% copper powders, pressed, sintered 30 min at 2050 F (1121 C), and cooled at a rate normal for a small part. With this relatively fast cool, the copper dissolved in the iron precipitated as a fine dispersion (gray areas). Tensile strength, 31,000 psi (214 MPa); hardness, R_B 23. See 1682.

2% nital   700×

1682 Same as 1681 except compact was cooled from sintering temperature at a rate normal for a large part. This relatively slow cool produced coarser copper precipitate (as indicated by darker shade of gray than in 1681), slightly lower tensile strength (30,000 psi, or 207 MPa), and much lower hardness (R_B 14).

2% nital   700×

1683 Mixture of 97% iron, 2% copper and 1% graphite powders, pressed, and sintered 30 min at 2050 F (1121 C). Enough carbon diffused into the iron to produce an all-pearlite structure; presence of carbon retarded diffusion of copper. Tensile strength, 70,000 psi (483 MPa); hardness, R_B 79. See 1684-1686.

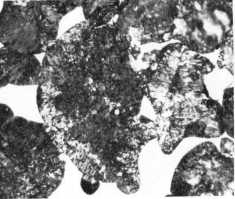

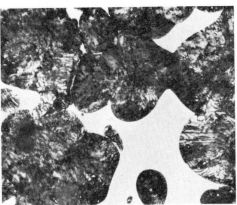

2% nital   700×

1684 Same as 1683 except the mixture contained 5% Cu. Some copper was dissolved in the iron (darker areas of the pearlite matrix), and some free copper (white) remains. Tensile strength and hardness are the same as for the 2% Cu compact in 1683. Pores are black.

2% nital   700×

1685 Same as 1683 and 1684 except the copper content was 10%. More copper was dissolved in the iron; larger areas of free copper (white) are present, reducing tensile strength of the sintered compact to 63,000 psi (434 MPa) and the hardness to R_B 78. See 1686.

2% nital   700×

1686 Same as 1683 to 1685 except the copper content of the mixture was 20%. The areas of free copper (white) have markedly increased in size; this has reduced the tensile strength of the sintered compact to 60,000 psi (414 MPa) and the hardness to R_B 70.

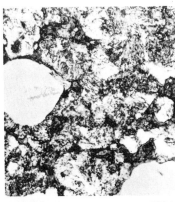

2% nital   300×

1687 Mixture of 94% iron, 5% copper, 1% graphite powders, pressed, sintered 30 min at 1950 F (1066 C). Carbon diffusion produced an all-pearlite structure; the copper did not melt. Tensile strength, 42,500 psi (293 MPa); hardness, R_B 65. See 1688 to 1690.

2% nital   300×

1688 Same as 1687 but sintered at 1985 F (1085 C). Copper (white) began to melt and dissolve in the iron (dark gray areas), increasing tensile strength of the sintered compact to 50,000 psi (345 MPa) and hardness to R_B 69. Pores are black. See 1689 and 1690.

2% nital   300×

1689 Same as 1687 and 1688 but sintered at 2000 F (1093 C). Melting of the copper was complete, and little free copper remains. The large increase in dissolved copper increased tensile strength to 65,000 psi (448 MPa), hardness to R_B 75. See also 1690.

2% nital   300×

1690 Same as 1687 to 1689 except the compact was sintered at 2050 F (1121 C). The copper was almost completely dissolved in the iron, with the result that the tensile strength was increased to 67,000 psi (462 MPa) and hardness was increased to R_B 78.

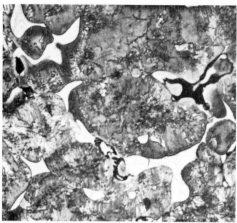

2% nital   700×

1691 Mixture of 94% iron, 5% copper, 1% graphite powders, pressed, sintered 5 min at 2050 F (1121 C). Diffusion of carbon produced an all-pearlite structure; some copper dissolved in the iron. Tensile strength, 59,000 psi (407 MPa); hardness, R_B 71. See 1692, 1693.

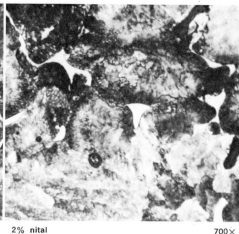

2% nital   700×

1692 Same as 1691 except sintered for 30 min. More copper dissolved in the iron (darker areas of the pearlite matrix); this reduced the amount of free copper (white) and increased the tensile strength to 68,000 psi (467 MPa) and the hardness to R_B 77.5. See 1693.

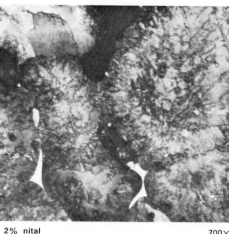

2% nital   700×

1693 Same as 1691 and 1692 except sintered 60 min. The copper almost completely dissolved in the iron, which increased the tensile strength of the sintered compact to 73,000 psi (503 MPa) and the hardness to R_B 82.5. Only a small amount of free copper remains.

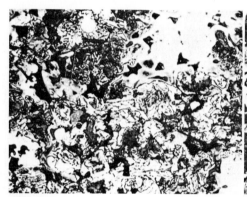

2% nital   300×

1694 Alloy steel made by pressing and sintering mixed powders of unalloyed iron, copper, nickel and graphite (94 Fe, 1 Cu, 4 Ni, 1 C). Structure: pearlite matrix, with austenite in nickel-rich areas at grain boundaries; some martensite at periphery of these areas.

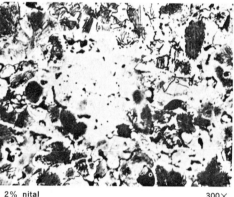

2% nital   300×

1695 "Semialloyed" steel made by pressing and sintering iron, copper, nickel and molybdenum powders partly alloyed during coreduction of oxides, and graphite powder. Structure is same as that of compact in 1694, which was produced from completely unalloyed powders.

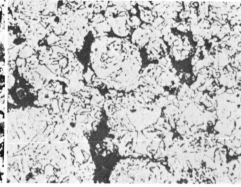

2% nital   300×

1696 Alloy steel made by pressing and sintering fully alloyed powder of Ni-Mo steel, with some copper powder added for size control. Structure consists of fine pearlite, and free ferrite; with this powder, there are no nickel-rich areas as in 1694 and 1695.

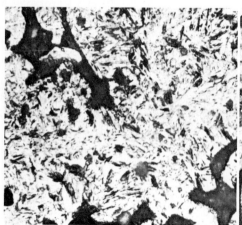

2% nital   700×

1697 Sintered alloy steel compact fully hardened by austenitizing at 1450 to 1600 F (788 to 871 C) and oil quenching. Structure consists mainly of martensite. Compare 1698, 1699.

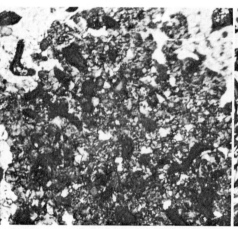

2% nital   700×

1698 Same as 1697 except more slowly cooled from the austenitizing temperature. This resulted in the formation of very fine pearlite, and consequently in lower hardness.

2% nital   700×

1699 Same as 1697 except retained austenite is present. Possible causes: high combined carbon content, alloying with nickel or manganese, use of carbonitriding atmosphere.

INDEX

429